Handbook of Ecological Models Used in Ecosystem and Environmental Management

T0179234

Applied Ecology
and Environmental Management

A SERIES

Series Editor
Sven E. Jørgensen
Copenhagen University, Denmark

Handbook of Ecological Indicators for Assessment of
Ecosystem Health, Second Edition
Sven E. Jørgensen, Fu-Liu Xu, and Robert Costanza

Surface Modeling: High Accuracy and High Speed Methods
Tian-Xiang Yue

Handbook of Ecological Models Used in Ecosystem and
Environmental Management
Sven E. Jørgensen

ADDITIONAL VOLUMES IN PREPARATION

Handbook of
Ecological Models Used
in Ecosystem and
Environmental Management

Edited by

Sven Erik Jørgensen

CRC Press
Taylor & Francis Group
Boca Raton London New York

CRC Press is an imprint of the
Taylor & Francis Group, an informa business

CRC Press
Taylor & Francis Group
6000 Broken Sound Parkway NW, Suite 300
Boca Raton, FL 33487-2742

First issued in paperback 2019

© 2011 by Taylor and Francis Group, LLC
CRC Press is an imprint of Taylor & Francis Group, an Informa business

ISBN-13: 978-1-4398-1812-1 (hbk)
ISBN-13: 978-0-367-86583-2 (pbk)

Visit the Taylor & Francis Web site at
http://www.taylorandfrancis.com

and the CRC Press Web site at
http://www.crcpress.com

Contents

Preface...vii
Acknowledgment...xi
Editor...xiii
Contributors..xv

Section I Models of Ecosystems

1 Introduction...3
   Sven Erik Jørgensen

2 Lake Models...15
   Sven Erik Jørgensen

3 Estuary Ecosystem Models...63
   Irene Martins and Joao C. Marques

4 Models of Rivers and Streams..137
   Sven Erik Jørgensen

5 Models of Coral Reefs 153
   Sven Erik Jørgensen

6 Models of Forest Ecosystems..161
   Sven Erik Jørgensen

7 Grassland Simulation Models:
   A Synthesis of Current Models and Future Challenges.....................175
   Debra P. C. Peters

8 Agricultural Systems...203
   Søren Hansen, Per Abrahamsen, and Merete Styczen

9 Aquaculture Systems...241
   Bruno Díaz López

10 Models of Wetlands 257
   Sven Erik Jørgensen

11 **Wastewater Systems**..277
    Krist V. Gernaey, Ingmar Nopens, Gürkan Sin, and Ulf Jeppsson

Section II Models of Environmental Problems

12 **Eutrophication Models**..325
    Sven Erik Jørgensen

13 **Modeling Oxygen Depletion in a Lagunar System:**
    The Ria de Aveiro..347
    José Fortes Lopes

14 **Models of Pollution by Heavy Metals**..393
    Sven Erik Jørgensen

15 **Models of Pharmaceuticals in the Environment**.................................415
    Sven Erik Jørgensen and Bent Halling Sørensen

16 **Models of Global Warming and the Impacts of Climate**
    Changes on Ecosystems 439
    Sven Erik Jørgensen

17 **Landscape-Scale Resource Management:**
    Environmental Modeling and Land Use Optimization for
    Sustaining Ecosystem Services..457
    Ralf Seppelt

18 **Simulation of Fire Spread Using Physics-Inspired and**
    Chemistry-Based Mathematical Analogs..477
    Alexandre Muzy, Dominique Cancellieri, and David R. C. Hill

19 **Air Pollution** 499
    Kostadin Ganev

20 **Acidification**..557
    Tomáš Navratil

Index 601

Preface

This book is the second edition of a handbook that was published as first edition in 1996, and gives an overview of models applied in ecosystem and environmental management. The first edition, which was called *Handbook of Environmental and Ecological Modelling*, was edited by Sven Erik Jørgensen, B. Halling Sørensen, and Søren Nors Nielsen. A statistical survey of the ecological and environmental models that have been published up to 1994 showed that approximately 2000 models have been presented in the literature during the period 1975–1994, either in books or in relevant journals such as *Ecological Modelling & Environmental Software*. A questionnaire was sent to the authors of about 1000 models, and the handbook was written based on the most representative models and the answers from 440 modelers.

It would be an almost impossible task to produce a book with detailed information about a reasonable and representative fraction of the models developed up to year 2011, because many more models have been developed in the past 16 years. A statistical survey of models published during a period of 6 months in 2008 alone shows that 42% of all models are published in the scientific journal *Ecological Modelling*; and as *Ecological Modelling* has published about 300–400 models every year from 1994 to 2010, it is estimated that about 1000 new ecological and environmental models are added to the scientific literature every year. This means that at present probably about 20,000 models have been reported in the international peer-reviewed literature, covering the period 1970–2010. In 1995, reasonably detailed information about one model required about 1.5 pages. This means that about 30,000 pages would be needed to provide information on the 20,000 models that one could possibly find in the scientific literature in the past 40 years.

It has therefore been decided to produce a handbook of models used in ecosystem and environmental management that is very different from the 1996 handbook—that is, a handbook that attempts to draw the state-of-the-art development of models for the ecosystem and environmental management. It is possible to distinguish between ecosystem models and models focusing on environmental management problems. These models focus on entire ecosystems or on specific environmental problems, respectively. The title of the handbook underlines the selected core topic: *Handbook of Models Used in Ecosystem and Environmental Management*. The most modeled ecosystems and environmental problems are presented in the handbook in two parts, covering models of ecosystems and environmental problems.

Section I covers the models of ecosystems, but the most rarely modeled ecosystems, such as savannas, mountain ecosystems, polar ecosystems, open sea ecosystems, and deep sea ecosystems, are not included. Coastal ecosystems, in general, are covered by the comprehensive chapter on models of estuaries, Chapter 3. Chapter 2, which concentrates on lake models, presents the general considerations that are valid for all freshwater ecosystems. Chapter 3 gives details about model development that are important for all models of marine ecosystems. Similarly, Chapter 7, whose main focus is on grassland, touches on the general features of models of terrestrial ecosystems, and Chapters 8 and 9, which delve on deal with models of agricultural and aquaculture systems, respectively, illustrate how to develop models for man-managed ecosystems. Chapter 11 focuses on models of wastewater treatment systems, and describes the factors that should be considered for completely controlled systems. The models of two other aquatic ecosystems—freshwater ecosystems (rivers and streams) and marine ecosystems (coral reefs)—are covered in Chapters 4 and 5, respectively. Chapter 6 gives an additional example of terrestrial ecosystems by presenting forest models. Chapter 10 covers wetland models. Wetlands can be considered an ecosystem with the characteristic properties of both aquatic and terrestrial ecosystems.

Section II is devoted to models of environmental problems. Chapters 12, 13, and 20 cover "out of balance" problems—eutrophication models, models of oxygen depletion, and acidification models. The environmental management problems highlighted in these chapters are discharge of nutrients, organic matter, and acidic components, respectively. Basically, these three chapters are all focusing on water pollution problems. Chapters 14 and 15 cover pollution by toxic substances, namely, heavy metals and organic toxic components. Global warming models are reviewed in Chapters 16. Chapter 17 focuses on models used in landscape planning. Chapter 18 presents models of fire and spreading of fire—what we could call a natural pollution problem. Air pollution models are the topic of Chapter 19. Air pollution problems are very different from water pollution problems, because of the aerodynamics considerations. Chapter 20 looks into models of acidification—an important environmental problem for aquatic ecosystems.

With the selection of ecosystems and environmental problems that are presented in detail in these 20 chapters, it is my hope that this handbook will provide a good overview of models that are applicable today for management of ecosystems and for the solution of the most pressing environmental problems. All chapters have a very comprehensive list of references, which should facilitate the possibility of finding the basic literature for the development of almost all possible ecological and environmental models.

Sven Erik Jørgensen
Copenhagen

MATLAB® is a registered trademark of The MathWorks, Inc. For product information, please contact:

The MathWorks, Inc.
3 Apple Hill Drive
Natick, MA 01760-2098 USA
Tel: 508-647-7000
Fax: 508-647-7001
E-mail: info@mathworks.com
Web: www.mathworks.com

Acknowledgment

Mette Vejlgaard Jørgensen has provided the photos used for the cover.

Editor

Sven Erik Jørgensen is professor emeritus in environmental chemistry at the University of Copenhagen. He holds a doctoral degree in environmental engineering (Karlsruhe University) and a doctorate of science in ecological modelling (Copenhagen University). He is an honorable doctor of science at Coimbra University (Portugal) and at Dar es Salaam University (Tanzania), and was also named as an Einstein Professor of the Chinese Academy of Science. He was the editor-in-chief of *Ecological Modelling* from 1975 to 2009. From 1978 to 1996, he served as the secretary general of the International Society of Ecological Modelling (ISEM). Since 2006, he has been the president of ISEM. He has received several awards, including the Prigogine Prize, the Pascal Medal, and the prestigious Stockholm Water Prize. He has published more than 350 papers, of which 260 were published in peer-reviewed international journals; he has also edited or authored 66 books. He was editor-in-chief of the *Encyclopedia of Ecology*, which was published in 2008. He has taught courses in ecological modelling in 32 different countries.

Contributors

Per Abrahamsen
Basic Science and Environment
    Faculty of Life Sciences
University of Copenhagen
Copenhagen, Denmark

Dominique Cancellieri
Research Laboratory SPE UMR 6134
Università di Corsica –
    Pasquale Paoli
Grossetti, Corti, France

Kostadin Ganev
Institute of Geophysics
Bulgarian Academy of Sciences
Sofia, Bulgaria

Krist V. Gernaey
Department of Chemical and
    Biochemical Engineering
Technical University of Denmark
Lyngby, Denmark

Søren Hansen
Basic Science and Environment
    Faculty of Life Sciences
University of Copenhagen
Copenhagen, Denmark

David R. C. Hill
Research Laboratory LIMOS
    UMR CNRS 6158
Blaise Pascal University
Aubiere, France

Ulf Jeppsson
IEA
Lund University
Lund, Sweden

José Fortes Lopes
Departamento de Física
Universidade de Aveiro
Aveiro, Portugal

Bruno Díaz López
The Bottlenose Dolphin Research
    Institute BDRI
Golfo Aranci, Italy

Joao C. Marques
Institute of Marine Research
Coimbra University
Coimbra, Portugal

Irene Martins
Institute of Marine Research
Coimbra University
Coimbra, Portugal

Alexandre Muzy
Research Laboratory LISA UMR
    CNRS
Università di Corsica – Pasquale
    Paoli, Campus Mariani
Corti, France

Tomáš Navrátil
Czech Geological Survey
Prague, Czech Republic

Ingmar Nopens
BIOMATH
Ghent University
Ghent, Belgium

Debra P. C. Peters
USDA ARS
Jornada Experimental Range
    and Jornada Basin Long Term
    Ecological Research NMSU
Las Cruces, New Mexico

Ralf Seppelt
Department Computational
    Landscape Ecology
UFZ-Helmholtz Centre for
    Environmental Research
Leipzig, Germany

Gürkan Sin
BIOMATH
Ghent University
Ghent, Belgium

Bent Halling Sørensen
Pharmaceutical Faculty
Copenhagen University
Copenhagen, Denmark

Merete Styczen
Basic Science and Environment
    Faculty of Life Sciences
University of Copenhagen
Copenhagen, Denmark

Section I

Models of Ecosystems

1

Introduction

Sven Erik Jørgensen

CONTENTS

1.1 Progress in Ecological and Environmental Modeling 3
1.2 Outline of the Handbook..9
    1.2.1 Models of Different Ecosystems 9
    1.2.2 Models of Environmental Problems 11
References...13

1.1 Progress in Ecological and Environmental Modeling

This book is the second edition of a handbook that was published in 1996 as the first edition, and gives an overview of models applied in ecosystem and environmental management. The first edition, which had the title *Handbook of Environmental and Ecological Modeling*, was edited by Sven Erik Jørgensen, B. Halling Sørensen, and Søren Nors Nielsen. A statistical survey of the ecological and environmental models that have been published up to 1994 showed that approximately 2000 models have been presented in the literature during the period 1975–1994, either in books or in relevant journals such as *Ecological Modelling & Environmental Software*. A questionnaire was sent to the authors of about 1000 models, and the handbook was written based on the most representative models and the answers from 440 modelers. The answers to the following questions were presented in the handbook for the 440 models:

What is the model title?

What type of model is developed? (biogeochemical model, population dynamic model, ecotoxicological model, ANN, IBM, SDM, etc.)

Which type of ecosystem was modeled?

What is the purpose of the model?

Which are the state variables and the forcing functions?

How has the model been developed?

Which software was applied for the model development?

Is the model available? How?

Which are the most relevant publications?

It would be an almost impossible task to produce a book with detailed information about a reasonable and representative fraction of the models developed up to year 2011, because many more models have been developed in the past 16 years. A statistical survey of models published during a period of 6 months in 2008 shows that 42% of all models are published in the scientific journal *Ecological Modelling*; and as *Ecological Modelling* has published about 300–400 models every year from 1994 to 2010, it is estimated that about 1000 new ecological and environmental models are added to the scientific literature every year. This means that at present probably close to 20,000 models have been reported in the international peer-reviewed literature, covering the period 1970–2010. In 1995, reasonably detailed information about one model required about 1.5 pages. This means that about 30,000 pages would be needed to provide information on the 20,000 models that one could possibly find in the scientific literature in the past 40 years.

It has therefore been decided to produce a handbook of models used in ecosystem and environmental management that is very different from the 1996 handbook—specifically, a handbook that attempts to draw the state-of-the-art development of models for the ecosystem and environmental management. It is possible to distinguish between ecosystem models and models focusing on environmental management problems. These models focus on entire ecosystems or on specific environmental problems, respectively. The title of the handbook underlines the core topic: *Handbook of Models Applied in Ecosystem and Environmental Management*.

In the modelling literature, it is feasible to find models of the following ecosystems:

1. Lakes
2. Estuaries
3. Rivers
4. Coastal zones
5. Coastal lagoons
6. Open sea ecosystems
7. Grasslands
8. Savannas
9. Forests
10. Polar ecosystems
11. Mountain ecosystems
12. Coral reefs

13. Wetlands (various types of wetlands, wet meadows, bogs, swamps, forested wetlands, marshes, and flood plains)

14. Deserts

15. Agricultural systems

16. Aquacultures

17. Wastewater systems

It is furthermore possible to find models focusing on the following environmental problems:

1. Oxygen depletion

2. Eutrophication

3. Acidification

4. Pollution by toxic organic compounds, including pharmaceuticals and endocrine disruptors

5. Pollution by heavy metals

6. Control of fishery

7. Pollution of groundwater

8. Planning of landscapes

9. Global warming and climate changes

10. Decomposition of the ozone layer and its effects

11. Loss of biodiversity

12. Spreading of fire

13. Air pollution

It is possible in the modeling literature to find papers dealing with almost all combinations of these ($13 \times 17 = 221$) types of models, although it will be difficult to find combinations of pollution problems and polar ecosystems or mountain ecosystems, at least for mountain ecosystems above the timberline. Furthermore, for models of the decomposition of the ozone layer, mainly the chain of chemical processes in the ozone layer has been modeled. It implies that if we classify models in accordance with the ecosystem and the pollution problem they attempt to solve, we will have approximately 200 classes of models—slightly less than the above-mentioned 221 possibilities.

We can divide the ecosystem models, 1–17, into five groups according to how many different models one can possibly find in the literature and not counting minor modifications as a change of model:

I. Ecosystems that have been modeled heavily and where it is possible to find several hundreds of different models in the model literature: rivers, lakes, forests, and agricultural systems

II. Ecosystems for which it is possible to find in the order of hundreds of different models: estuaries, wetlands, and grasslands

III. Ecosystems that have been modeled many times, but of which less than 100 different models have been developed: wastewater systems and aquaculture systems

IV. Ecosystems that have been modeled more than 10 times but less than 25–30 times: coral reefs

V. Ecosystems that have only modeled a couple of times up to a handful of times: polar ecosystems, savanna, and mountain ecosystems above the timberline

Similarly, we can divide the models of pollution problems into five groups:

I. Heavily modeled pollution problems: oxygen depletion, eutrophication, organic toxic substances including pharmaceuticals in the environment, air pollution problems, and global warming including impacts of climate changes

II. Environmental problems that have been modeled approximately a hundred times: heavy metal problems, acidification, and groundwater pollution

III. Environmental problems that have been modeled many times, but of which less than 100 different models have been developed: fire spreading and overfishing

IV. Environmental problems that have been modeled more than 10 times but less than 25–30 times: endocrine disruptors

V. Environmental problems for which it is not possible to find more than at the most a couple of models or a handful of models in the model literature: application of gene modified organisms (GMO)

A number of new model types have been developed mainly in the past two decades to answer a number of relevant modeling problems or questions that arose as a result of the increasing use of ecological models in the 1970s. Seven relevant modeling questions already formulated about 1980 as a result of the model experience gained in the 1970s were attempted to be answered with the new model types, which were mainly developed from 1990 to 2010:

1. How can we describe the spatial distribution that is often crucial to understand ecosystem reactions and to select the best environmental strategy?

2. Ecosystems are middle number systems (Jørgensen, 2002) in the sense that the number of components is magnitudes smaller than the number of atoms in a system. All the components are different

and that is often important for a proper description of the ecosystem reactions in order to consider the differences in properties among individuals.

3. The species are adaptable and may change their properties to meet the changes in the prevailing conditions—it means the forcing functions. Furthermore, the species may—by more drastic changes—be replaced by other species better fitted to the combinations of forcing functions. How to account for these changes? Even the networks may be changed if more biological components with very different properties are replaced by other species. How to account for these structural changes?

4. Can we at all model a system that has a poor database—very few data of only low quality?

5. The forcing functions and several ecological processes are in reality stochastic. How to account for the stochasticity?

6. Can we at all develop a model, when our knowledge is mainly based on a number of rules/properties/propositions?

7. Can we develop a model based on data from a wide spectrum of different ecosystems—it means that we have only a very heterogeneous database.

These problems could not be solved by the "old" model types: biogeochemical models, population dynamic models, and steady-state models.

Spatial models often based on the use of Geographical Information System (GIS) have been developed to formulate an answer to question 1.

Individual-based models (IBMs) are able to give an answer to question 2. Software that can be used to develop IBMs is even available to facilitate IBM development. The software can also be utilized to cover spatial distribution (see question 1).

Structurally dynamic models (SDMs) have been developed to solve the problem expressed in question 3.

Fuzzy models can be used to make models based on a poor and maybe only a semiquantitative database.

Stochastic models were applied in the 1970s in very few cases, but they have been further developed today, although the application is still not widespread—probably because there is no frequent and urgent need to include stochastic processes in ecological models.

IBMs can, in many cases, meet the demands expressed in question 6.

Artificial neural networks (ANNs) provide a good solution to the problem formulated as question 7.

Sometimes, ecotoxicological models are considered a special type of models, although they are developed similar to other biogeochemical models. They have been widely used, particularly in the past 10–15 years, because the need for these models is very obvious: for environmental risk assessment of chemicals. The two chapters on heavy metal pollution and medicine (pharmaceuticals) in the environment represent this type of models. All the other types of models can be found in the various chapters as examples presented in more detail or mentioned in the model overview given in most chapters.

Table 1.1 gives a summary of model statistics, based on the number of publications in the journal *Ecological Modelling*. The percentage application of the most general model types are compared for the period 1975–1980 with the period 2001–2009 (see also Jørgensen, 2009). Ecotoxicological models are included as a model type even though they are constructed similar to biogeochemical models; they are—as indicated above—often considered as a separate model type, because they are an important subset of biogeochemical models and they have some characteristic properties. In the table, we have distinguished between nine model types:

1. Dynamic biogeochemical models
2. Steady-state biogeochemical models
3. Population dynamic models
4. Spatial models
5. SDMs
6. IBMs
7. Ecotoxicological models
8. Fuzzy models
9. ANN models

TABLE 1.1

Application of the Most General Model Types for the Period 1975–1980 with the Period 2001–2006

	% Application 1975–1980	% Application 2001–2010
Dynamic biogeochemical models	62.5	30.8
Steady state biogeochemical models	0	1.8
Population dynamic models	31.0	24.2
Spatial models	0	19.9
Structurally dynamic models	1.5	8.2
Individual based models and cellular automata	0	5.9
ANN and use of artificial intelligence	0	5.4
Fuzzy models	0.5	1.8
Ecotoxicological models	0	2.0

The number of papers published during the period 2001–2010 is about nine times the number of papers published per year during the period 1975–1980. This means that the number of papers on dynamic biogeochemical models published recently is more than 4.5 times the number published during the late 1970s, and that the number of papers on structurally dynamic modeling has increased by a factor of almost 50 during the past 40 years. It is of course not surprising as the new types of models have been developed because there was an urgent need to answer the seven modeling problems listed above.

With the present spectrum of model types, it is possible to solve all the major modeling problems that had been raised in the 1970s. This development has, of course, increased the application of ecological models in general and, in particular, the use of new model types. However, it is also clear that all these problems cannot be resolved to everyone's complete satisfaction. There are still a few problems that may not be possible to solve by use of a single model type. A very complicated problem may require the use of a hybrid model, that is, by a combination of the model types presented here.

New model types will inevitably be developed during the coming decade to solve the most complicated ecological and environmental problems that we still have not solved properly. There is, however, a general agreement among ecological modelers that at present we have sufficient model types to solve most of the ecological modeling problems that we are currently facing. A few modeling case studies are still needed, however, to reflect proper model developments in real situations.

1.2 Outline of the Handbook

This handbook is divided into two parts: "Models of Different Ecosystems" and "Models of Environmental Problems."

1.2.1 Models of Different Ecosystems

Ten different ecosystems are covered in this part of the book. The selected ecosystems represent the entire spectrum: from completely natural to completely man-controlled ecosystems. Models of marine ecosystems, freshwater ecosystems, and terrestrial ecosystems are included to illustrate the three main types of ecosystems. Very frequently modeled ecosystems include lakes, rivers, forests, agricultural systems, and wetlands are presented, and rarely modeled ecosystems such as coral reefs are included to illustrate the modeling problems when modeling experience is limited. The problem of model development for different ecosystems is very well represented and illustrated by the 10 selected ecosystems.

The chapter on lake models (Chapter 2) gives an overview of the spectrum of lake models, describes the fundamental equations applied in most lake models, and illustrates lake models by giving a more detailed example and an example of using SDMs in lake modeling. This example also shows how SDMs, in general, can be applied to ecological modeling.

The chapter on estuaries (Chapter 3) thoroughly covers all the processes that are characteristic for marine ecosystems, including coastal lagoons and coastal zones. The chapter gives a comprehensive overview of the applicable submodels to describe the primary production in marine ecosystems, which is very much the focus in ecosystem and environmental management.

Chapter 4 covers river models. It provides an overview of river models together with the most applied equations in river models and a detailed example of a very characteristic, not-too-complex river model developed via the application of STELLA, a widely applied modeling tool.

Marine ecosystems are also represented in Chapter 5, specifically coral reefs. It is an ecosystem that has not been frequently modeled, but it has several characteristic processes that are different from most other ecosystems and was therefore included in this handbook. A few recent models are mentioned to illustrate the spectrum of coral reef models, and an example is presented in more detail to show the characteristic features of coral reef models.

Forest ecosystems are modeled heavily, and a good overview of the spectrum of forest models is presented in Chapter 6 together with a characteristic example and a less characteristic, nonorthodox example based on thermodynamics. This example is included to demonstrate that by using a new creative method, it is sometimes possible to develop good and simple models.

Chapter 7 gives an overview of grassland models. The chapter also discusses the grassland processes in detail, which will facilitate the development of other grassland models in general.

Chapter 8 covers agricultural systems; the models are, of course, different from natural ecosystems, because management issues can more easily be included. The focus of these models is often on the use of fertilizers and how to achieve a high yield with the right amount of fertilizers applied and how to reduce the loss of fertilizers to the environment.

Aquaculture systems (Chapter 9) are similar to agricultural systems—they are straightforward to manage, and this, of course, is reflected in the models developed and in processes included in the illustrative examples given in this chapter. Many of the equations covering fish growth are also applicable in fishery models.

Wetland models are presented in Chapter 10. A good overview of the spectrum of models and the types of wetlands modeled is presented. Two examples are given in more detail: an example of the selection of natural wetlands to treat agricultural drainage water and construction of an artificial (constructed) wetland. This chapter illustrates the typical applications of wetland models in environmental management.

Chapter 11 focuses on wastewater systems, a completely man-controlled system. The models emphasize how a model development is carried out when the forcing functions are completely manageable in contrast to Chapters 2–7, where the natural forcing functions play a more significant role than the man-controlled ones. The chapter gives a nice overview on how to model different important wastewater systems. As the processes in wastewater systems are controlled, it is possible to apply many of the wastewater treatment models as good approximations for the development of models of natural ecosystems. For instance, a model for the anaerobic digestion of sludge is presented in this chapter, and the models of the core processes can also be used in wetland models, because the same anaerobic conditions are present in many wetlands. Similarly, the biological decomposition processes that are characteristic for the activated sludge plant can be found in a number of natural ecosystems.

Models of all freshwater systems, marine systems, terrestrial ecosystems, man-managed ecosystems, and man-controlled systems are based on some basic and important considerations. These are covered in Chapters 2, 3, 7, 8–9, and 10, respectively.

1.2.2 Models of Environmental Problems

Nine environmental problems are covered in this part of the book. They represent water pollution problems, air pollution problems, solid waste pollution problems, toxic substance pollution problems (both heavy metals and organic toxic substances, represented by pharmaceuticals), landscape planning problems, and the two important regional and global problems: spreading of fire and global warming. All problems involved in developing models for various environmental and pollution problems are covered in these nine chapters.

Chapter 12 covers the eutrophication problem, which requires a good description of plant growth, including the growth of phytoplanktons and submerged vegetation, as a function of the nutrient concentrations. The possible use of equations to cover these processes is discussed. An overview of various applicable eutrophication models is presented together with concrete examples. The examples encompass the application of SDMs.

Models of the environmentally important process—oxygen depletion—which is very important for aquatic ecosystems due to the low solubility of oxygen in water at normal temperature, is presented in Chapter 13. The mathematical formulation of the processes determining the oxygen depletion is discussed in detail.

Models of heavy metal pollution is discussed in Chapter 14, by giving an overview of the possibilities and by presenting in more detail examples of heavy metal pollution in an aquatic ecosystem and in a terrestrial ecosystem. The chapter also touches on the pollution of groundwater by heavy metals. An example of the use of SDM for heavy metal pollution of an aquatic ecosystem is also included. Copper pollution is changing the size of zooplanktons in aquatic ecosystems, which is possible to model using a SDM.

Chapter 15 illustrates models of pollution by toxic organic substances by focusing on pharmaceuticals in the environment. Other organic pollutants without a medical effect as pharmaceuticals are modeled similarly, but it was preferred to focus on pharmaceuticals in this chapter to cover the extra dimension of these chemicals: because they have an important useful application in the society, their elimination would have a negative effect. Pharmaceuticals are, furthermore, biologically very active compounds, which clearly illustrates the associated pollution problems. The chapter discusses the characteristic properties of ecotoxicological models. An overview of pharmaceutical models is presented and generally illustrates the most applied types of ecotoxicological models. A detailed example is given to illustrate the considerations that are needed to develop models of toxic organic matter.

Chapter 16 focuses on global warming and the impacts of climate changes on ecosystems. A good overview is given for the spectrum of models applied to predict the temperature change as a result of emission of greenhouse gases. One of the comprehensive models is presented in detail to illustrate this type of model, which are often characterized by having a high number of different processes because of the complexity of the problem. Lately, many models of the impacts of climate changes on ecosystems have been developed, and in this chapter it is discussed which characteristics such models should include. A few recent models of the impact of climate change on nature are cited to illustrate this very important model application.

Recently, it has been acknowledged that landscape planning is important for the functioning of various ecosystems in the landscape. Therefore, it has, of course, also been discussed how models can be applied to perform a proper planning of the entire landscape, considering the interactions among the ecosystems that make up the landscape. This is the core topic of Chapter 17.

Spreading of fire has recently become a serious environmental problem. Fire has destroyed millions of hectares in Spain, Portugal, Greece, California (USA), and Australia. It was therefore found of importance to develop models that could predict the spreading of fire. Chapter 18 illustrates with a characteristic and detailed example how to solve this modeling problem.

Control of air pollution requires the application of models that are illustrated in Chapter 19. A good overview of the possibilities of applying models to follow and reduce air pollution is presented in this chapter, together with an overview of the spectrum of air pollution models.

Acidification of aquatic ecosystems is most often due to air pollution, the so-called acid precipitation. It is possible, by using models, to find a solution to this problem, as illustrated in Chapter 20, which also discusses the equations that can be used in different acidification models.

References

Jørgensen, S. E., B. Halling Sørensen, and S. N. Nielsen. 1996. *Handbook of Environmental and Ecological Modelling*, 672 pp. Boca Raton, FL: CRC

Jørgensen, S. E. 2002. *Integration of Ecosystem Theories: A Pattern*, 3rd ed., 428 pp. Dordrecht: Kluwer Academic Publishers.

Jørgensen, S. E. 2009. *Ecological Modelling: An Introduction*, 190 pp. Southampton: WIT.

2

Lake Models

Sven Erik Jørgensen

CONTENTS

2.1 Introduction 15
2.2 Selection of Models as Management Tools 17
2.3 Integration of Hydrodynamics and Ecology 26
2.4 Ecotoxicological Lake Models 29
2.5 Lake Acidification Models 32
2.6 Lake Fisheries Models 33
2.7 Structurally Dynamic Models Applied to Reveal the Ecosystem
   Properties of Lakes 34
2.8 Latest Development in Lake and Reservoir Modeling 50
References 55

2.1 Introduction

Lakes and reservoirs are important ecosystems, because they provide a valuable storage of water—in most cases, of fresh water. They are crucial for the other ecosystems in the landscape and for the human society.

Lakes and reservoirs offer numerous services of great importance to humans. They serve as resources for drinking water and important recreational sites, provide food by fishery, facilitate transportation, ensure a higher biodiversity in the landscape, purify water and air, and enhance the natural recycling of important elements—to mention the most important direct and indirect ecological services. A good function of lakes and reservoirs is a prerequisite for a provision of these ecological services in full scale. Therefore, lakes and reservoirs have been among the main focuses of environmental management.

The following environmental management problems of lakes and reservoirs have been identified:

1. Eutrophication
2. Acidification
3. Toxic substance pollution

4. Water level changes, most often volume shrinkage (and often accompanied by increased salinity)

5. Siltation

6. Introduction of exotic species

7. Overfishery

The problems are rooted in changes—most often manmade changes—in the drainage area. Models have been increasingly used in lake management, primarily because they are the only tools that are able to quantitatively relate the impacts on an ecosystem with the consequences for the state of the ecosystem. Many ecosystem models have been developed during the past few decades, and it is not surprising that they have found a wide application in lake and reservoir management efforts.

This chapter gives an overview of the applicable lake models. Chapter 12 will present in more detail the application of eutrophication models to cope with problem number 1 in the list of environmental problems presented above. The eutrophication models presented in Chapter 12 can of course be used in other aquatic ecosystems than lakes and reservoirs, when eutrophication is the problem, for instance, estuaries, bays, and lagoons, but eutrophication models have found their widest application in lake and reservoir management. Similarly, Chapter 20 presents acidification models, which deal with problem number 2 on the list. These models can, of course, also be used generally, but have frequently found application in environmental management of lakes, reservoirs, and rivers. Pollution by toxic substances is also covered in two more chapters in Section II of this book. Several of the models presented in these two chapters can also be used for management of lakes and reservoirs.

Models mainly focusing on the other four problems in the list and ecological models used to reveal the characteristic features of lakes and reservoirs as ecosystems are presented in this chapter together with a more general overview of modeling problems relevant to ecological and environmental management of lakes and reservoirs.

Based on these considerations, this chapter will uncover the following issues:

1. In the next section, the problem of selecting a good lake management model will be discussed.

2. Lakes are aquatic ecosystems and this implies that development of a good model often means integration of hydrology with ecology. Section 2.3 presents this important step in lake modeling.

3. A short overview of ecotoxicological lake models is given in Section 2.4. The short overview should be viewed as a supplement to Chapters 14 and 15.

4. Section 2.5 describes the use of acidification models in addition to what is presented in Chapter 20. The emphasis here is on the use of acidification models for lake management.

5. Section 2.6 deals with lake fishery models.

6. Section 2.7 describes the structurally dynamic models (SDMs) that can be used to reveal ecosystem properties, which are of importance for a better understanding of ecosystem reactions and for better lake management. In addition, SDMs are applied for environmental management of eutrophication, and other SDM examples are also mentioned in Chapter 12.

7. The last section gives an overview of the latest development in lake modeling, which includes a wider application of two- (2D) or three-dimensional (3D) lake models. The future directions of lake modeling will be briefly discussed in context.

It is the purpose of this chapter, together with the overview of eutrophication models in Chapter 19, to cover the state of the art of lake modeling today.

2.2 Selection of Models as Management Tools

A management problem to be solved can often be formulated as follows: If certain forcing functions (management actions) are varied, what will be the corresponding changes on the state of an ecosystem? The model is used to answer this question or, in other words, to predict what will change in a system when the control functions that are managed by humans are varied over time and space.

Typical *control functions* are the consumption of fossil fuel, regulation of water level in a river by a dam, discharge of wastewater, agriculture in the drainage area, or fisheries policy.

In the case of eutrophication, the management problem often is: Which method or combination of methods should be selected among the existing possibilities to reduce the discharge of nutrients? This question can be answered by comparison of the corresponding scenarios obtained by the model runs.

It is important that the manager should take part in the entire development of a management model, since he will ultimately define the modeling objectives and select the modeling scenarios. The success of the application of a management model, to a large degree, is dependent on an open dialog between the modeler and the manager.

A further complexity is the construction of ecological–economic models. As we gain more experience in constructing ecological and economic models, more and more of them will be developed. It often is feasible to find a relation between a control function and economic parameters. If a lake is a major water resource, an improvement of its water quality will inevitably result in a reduction in the treatment costs of drinking water if the same water quality is to be provided. It is also possible to sometimes relate the value of a recreational area to the number of visitors and to how much money they spend on average in the area. In many cases, however, it is difficult to assess a relationship between the economy and the state of an ecosystem. For example, how can we assess the economic advantages of an increased transparency in a water body? Ecological–economic models are useful in some cases, but should be used with caution, and the relations between the economy and environmental conditions critically evaluated, before the model results are applied.

Data collection is the most expensive component of model construction. For many lake models, it has been found that needed data collection comprises 80–90% of the total model costs. Because complex models require much more data than simple ones, the selection of the complexity of environmental management models should be closely related to the costs involved in the environmental problem to be solved.

Thus, it is not surprising that development of the most complex environmental management models have generally been limited to large ecosystems, where the economic involvement is great.

The predictive capability of environmental models can always be improved in a specific case by expansion of the data collection program, and by a correspondingly increased model complexity, provided that the modelers are sufficiently skilled to know in which direction further expansion of the entire program must develop in order to improve the model's predictive capabilities.

The relation between the economy of the project and the accuracy of the model is presented in the form shown in Figure 2.1. The reduction in the discrepancy between model predictions and reality is lower for the next dollar invested in the project. But it is also clear from the shape of the curve that the associated errors cannot be completely eliminated. All model predictions have a standard deviation associated with them. This fact is not surprising to scientists, but it is often not appreciated by decision-makers, to whom the modeler typically presents his or her results.

Engineers use safety factors to assure that a building or a bridge will last for a certain period, with a very low probability of breakdown, even under extreme conditions. No reputable engineer would propose using a smaller, or no, safety factor to save some concrete and reduce the costs. The reason is obvious: Nobody would want to take the responsibility for even the smallest probability of a building or bridge collapse.

When decision-makers are going to make decisions on environmental issues, the situation is strangely different. Decision-makers in this situation

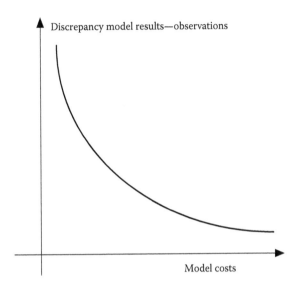

FIGURE 2.1
The model costs increased when a lower discrepancy between model results and observations is required.

want to use the standard deviation to save money, rather than assuring a high environmental quality under all circumstances. It is the modeler's duty, therefore, to carefully explain to the decision-maker all the consequences of the various decision possibilities. A standard deviation of a prognosis for an environmental management model, however, cannot always be translated into a probability, because we do not know the probability distribution. It might be none of the common distribution functions, but it is possible to use the standard deviation qualitatively or semiquantitatively, translating by the meaning of the model results by the use of words. Civil engineers are more or less in the same situation, and have been successful in the past in convincing decision-makers of the appropriate steps to be taken in various situations. There is no reason that environmental modelers cannot do the same.

It often is advantageous to tackle an environmental problem in the first place with the use of simple models. They require much fewer data, and can give the modeler and decision-maker some preliminary results. If the modeling project is stopped at this stage for one reason or another, a simple model is still better than no model at all, because it will at least give a survey of the problem.

Simple lake models, therefore, are good starting points for the construction of more complex models. In many cases, the construction of a model is carried out as an iterative process, and a stepwise development of a complex model. The first step is development of a conceptual model. It is used to get a survey of the processes and state variables in the ecosystem of concern. The next step is development of a simple calibrated and validated model. It is used to establish a data collection program for a more comprehensive effort closer to the final selected version. The third model will often reveal specific model weaknesses, the elimination of which is the goal of the fourth version of the model. At first glance, this seems to be a very cumbersome procedure. However, because data collection is the most expensive part of modeling, constructing a preliminary model for optimization of the data collection program will ultimately require fewer financial resources.

A first, simple mass-balance scheme is recommended for all biogeochemical models, particularly for lake models, where it is important to identify the limiting factors as function of time. The mass balance will indicate the possibilities for reducing or increasing the concentration of a chemical or pollutant, which is a crucial issue for environmental management.

Point sources of pollution are usually easier to control anthropogenic nonpoint sources which, in turn, are more easily controlled than natural pollutant sources, as shown in Table 2.1. Distinction can be made among local, regional, and global pollutant sources. Because the mass balance indicates the relative quantities from each source, it is possible to identify which sources should receive the initial attention (e.g., if a nonpoint regional source of pollutants is dominant, it would be pointless to concentrate first on eliminating small, local point sources, unless the latter might also have some political influence on regional decisions).

It has already been recognized that the modeler and the decision-maker should communicate with each other. It is recommended, in fact, that the decision-maker be invited to follow the model construction process from its very first phases, in order to become acquainted with the model strength and limitations. It also is important that the modeler and the decision-maker together formulate the model objectives and interpret the model results. Holling (1978) has demonstrated how such teamwork can be developed and used phase by phase. It is denoted mediated modeling or institutionalized modeling. For further details about this approach and its advantages, see Jørgensen, Chon, and Recknagel (2009). The conclusions are clear: The modeler and decision-makers and all parties interested in the lake management problem should work together in all phases of the modeling exercise. Having

TABLE 2.1

Examples of Sources of Water Pollutants

Source	Examples
Point sources	Wastewater (nitrogen, phosphorus, biochemical oxygen demand); sulfur dioxide from fossil fuels; discharges of toxic substances from industries
Nonpoint manmade sources	Agricultural use of fertilizers; deposition of lead from vehicles; contaminants in rain water
Nonpoint natural sources	Runoff from natural forests

the modeler first build a model, and then transfer it to a decision-maker accompanied by a small report on the model, is not recommended.

Communication between the decision-maker and the modeler can be facilitated in many ways, and it is often the primary responsibility of the modeler to do so. If a model is built as a menu system, as presented by Mejer (1983), it might be possible to teach the decision-maker how to use the model during a few hours, thereby also increasing his or her understanding of the model and its results. The effect of this approach is increased by the use of various graphic methods to illustrate the best possible decision in regard to what happens with the use of various management strategies. Under all circumstances, it is recommended that time be invested in developing a good graphic presentation of the model results to a decision-maker. Even if he or she has been currently informed about a model project through all its phases, the decision-maker will not necessarily understand the background and assumptions of all the model components. Thus, it is important that the model results, including the main assumptions, shortcomings, and standard deviations underlying them, are carefully presented with the use of an illustrative method.

A good environmental model can be a powerful tool in the decision-making process for management actions. A wide range of environmental problems has been modeled in the past 10–15 years. They have generally been an important aid to decision-makers. With the continuing rapid growth in the use of environmental models, the situation will only improve in the future. However, we have not achieved the same level of experience for all environmental problems.

The use of models in environmental management is definitely growing. They have been widely used in several European countries, North America, and Japan. Furthermore, environmental agencies in more and more countries are making use of model applications. Through such journals as *Ecological Modelling* and the *International Society for Ecological Modelling* (ISEM), it is possible to follow the progress in the field. This "infrastructure" of the modeling field facilitates communication and accelerates the exchange of experiences, thereby enhancing the growth of the entire field of ecological modeling. It appears there will soon be a need for a model "bank" or database, where users can obtain information about existing models, their uses, characteristics, etc. At the same time, it is not possible to readily transfer a model from one case study to another, as is reiterated in this book. It also is difficult to transfer a model from one computer to another unless it is exactly the same type of computer. However, it is often very helpful to learn from experiences of others involved in similar modeling situations in other parts of the world. Journals such as those mentioned above facilitate such exchanges.

The issue of the generality of models also deserves discussion. Few models have been used on multiple case studies, in order to gain wide experience on this important matter. The eutrophication model presented by Jørgensen (1994) has been used on a wide range of lakes in the temperate and the tropical zone, as well as for shallow and deep lakes, and even fjords. The experience gained in these case studies is illustrative, but does not necessarily represent the general properties of models. A further discussion on the generality of models is presented in Chapter 12 for eutrophication models.

Management models are used in water quality management of lakes and reservoirs for the purposes outlined below, with reference to the type of management models:

General

- To estimate pollution sources in the watershed, using simple calculation models

For lakes and existing reservoirs

- To predict possible future water quality conditions when control functions in the watershed are altered by human activities
- To provide estimates for decisions between different water quality options for use in long-term planning
- To support short-term operational management decisions regarding water quantity and quality
- To optimize sampling schedule investigations and control of water quality
- To assess the environmental risks associated with discharge of micropollutants

Before reservoir construction

- To estimate budgets of major water quality components of rivers that will enter the reservoir, as well as in the reservoir and the reservoir outflows
- To provide reasonable estimates between alternative construction sites, dam heights, and outflow and outlet structures so that decisions are supported
- To predict conditions in future reservoirs, and the consequences of different management options on water quality within the watershed

The first step in selecting a model is specification of the goals for which the model will be used. The selected model must be one intended to answer the questions of interest, as defined by the lake manager. Another consideration in selecting a model is the specific data available for use with the model. Using most advanced models is impossible, for example, if the quantity and quality of the water inflows have not been measured. In some cases, it may be possible to obtain data required for the model, but this inevitably involves expenditures of money and time, since most lake inflows vary considerably over time. The availability of personnel skilled in the use of models also can be a significant limitation. Learning the modeling procedure and understanding the applied basis for model development can be both difficult and time consuming. Thus, appropriate model selection requires balancing the importance of the problem, money and time invested in the model development, the available personnel, and the existence of appropriate models.

As emphasized in the previous sections, models only provide a gross simplification of reality. Thus, caution is always necessary when considering model results. Moreover, three levels of uncertainties influence the use of even the best model (Hilborn, 1987), including:

- Noise—the natural variability that occurs sufficiently frequently to be routine (various sampling schemes and statistical analyses are required to accommodate this uncertainty)
- The state of nature that is not well known
- Surprises—unanticipated events (flexible management strategies cope with unanticipated events more effectively than rigid, dogmatic strategies)

These uncertainties are inherent in any complex system, and may occur in any specific case. By proper use of validation techniques, it is possible in most cases to quantify the uncertainty, which is very valuable for the interpretations of the model predictions.

Models useful for lake quality management may be classified as follows:

- **Simple static calculation models** consisting of algebraic equations or graphs. These models are based predominantly on the statistical elaboration of data sets, sometimes large ones. Thus, they are limited by the extent of the materials covered in the analysis. For example, many models based on empirical relationships between phosphorus and chlorophyll presented in the literature. Thornton et al. (1999) provide details on models for estimating nonpoint source pollution loads.
- **Complex dynamic models** providing analysis of the timing of water quality conditions. Almost of all of the models presented in this volume belong to this class, and represent a wide spectrum of complexity.
- **Geographical information systems (GIS)** used for problems requiring spatial resolution. This is typical for estimating pollutant loads, particularly for nonpoint source pollution. The basis of GIS is computerized maps and procedures for entry and treatment of spatial data. For example, a specific watershed can be included in the database and corresponding pollution sources indicated. Through the use of models (usually more complex dynamic models), the expected pollution input can be calculated. Several attempts to apply GIS in

watershed management (DePinto and Rodgers 1994) or combined watershed models with reservoir models have emerged in recent years.

- **Prescriptive models** can be used to calculate lake water quality conditions without indicating appropriate management options for a given situation. By means of scenario analysis, it is possible to test management alternatives and to predict the potential consequences of water quality. This can be useful in selecting the most appropriate management possibilities. These models are based either on simple static calculations or dynamic simulations, depending on whether a time-independent or time-dependent solution is required.

- **Optimization models** incorporate selection procedures to choose the most suitable option, based on a set of criteria. Such models, which are often based on complex dynamic models, can allow simultaneous analysis of several management alternatives or goals. The major component of an optimization model is called a goal function, which is a function the user seeks to minimize or maximize critical water quality variables (e.g., oxygen concentration in a lake), or the money spent on attaining a specified level of water quality improvement. Optimization with constraints means that some, or all, of the management parameters are limited (i.e., they are forced to remain within specified limits due to natural conditions, management limitations, etc.). The following examples illustrate these kinds of constraints:

  1. Inflow phosphorus levels cannot be reduced more than the capabilities of treatment plants or other available reduction methods.
  2. It is impossible to mix a water body beyond its greatest depth; thus, there is a natural depth of mixing.
  3. There are no feasible reasons to reduce the chlorophyll concentrations, or increase the oxygen concentrations, above a certain limit.

The *standard optimization* procedure is used when a management decision involves only one variable (e.g., reservoir outflow), and is determined by one characteristic of this variable (e.g., outflow rate by turbines). The goal in this case, for example, can be to keep the reservoir water level within certain limits. *Multiparameter* optimization also concerns one variable, although the optimal performance is searched for in a multidimensional parameter space. It can be asked, for example, to what extent should the bottom outflow gate of the reservoir be opened, and a turbine run, to keep the water level in the reservoir within certain limits. The constraints are given by the maximum capacity of the bottom outlets and the turbines, and by the minimum and maximum flow rates prescribed for the river downstream of the reservoir. The model formulation with respect to water quality, for example, may be retaining some water quality variable below a certain limit, while also modifying several parameters characterizing the use of various management options. As an example, the model GIRL OLGA is intended for making dynamic, time-dependent estimates of the best combination of the time sequence of using five different management options, each characterized by one parameter that can be manipulated within certain limits. *Multivariable* optimization simultaneously takes into account several variables. Examples would include keeping the oxygen concentration, quantity of algae, and the content of organic matter in a lake within certain limits. *Multigoal* optimization is the most complex, with several goals having to be simultaneously achieved. In this latter case, a compromise set is sought, rather than a unique optimal solution, with the model user driven to make a proper selection. A combination of multiparameter, multivariable, and multigoal formulation also is possible, although such a solution has not yet been attempted for water quality modeling. Up to the present time, more complex formulations have generally been used to address water quantity problems (e.g., optimal operation of a reservoir cascade). Nevertheless, progress in developing optimization techniques is moving forward rapidly, and useful water quality formulations will doubtless be available in the future. Table 2.2 provides a listing of optimization lake quality management models.

It is emphasized that optimization procedures only allow to select among the possibilities included in the model, and is limited by the validity of the model, including its assumptions and formulations and its imposed constraints.

Thus, the model conclusions should be used with caution, with the user considering the model limitations, possible inadequacies, and possible insufficiency of input data. The model can be run before any decision is taken, and several alternatives are investigated. Alternatively, the model can be

TABLE 2.2

Optimization: Water Quality Management Models

- Dynamic optimization of eutrophication by phosphorus removal. Used for a Japanese lake (Matsumura and Yoshiuki 1981).
- Optimal control by selective withdrawal (Fontaine 1981).
- Optimizing reservoir operation for downstream aquatic resources. Applied on Lake Shelbyville, IL, USA (Sale et al. 1982).
- Optimizing reservoir operation for downstream aquatic resources. Applied on Lake Shelbyville, IL, USA (Sale et al. 1982).
- GIRL OLGA for cost minimization of eutrophication abatement, using time-dependent selection from five management options (Kalceva et al. 1982; Schindler and Straskraba 1982). Applied on several reservoirs in the Czech Republic.
- Stochastic optimization of water quality (Ellis 1987).
- COMMAS for prediction of environmental multi-agent system (Bouron 1991).
- DELWAG-BLOOM-SWITCH for management of eutrophication control of shallow lakes (van der Molen et al. 1994).
- GFMOLP, a fuzzy multi-objective program for the optimal planning of reservoir watersheds (Chang et al. 1996).

connected to automatic devices that activate water quality management options. As examples, chlorophyll concentrations and meteorological variables can be automatically recorded and put into a computer model. Short-term predictions by the model can be used to switch water mixing devices on or off, or to specify the intensity of phosphorus purification.

2.3 Integration of Hydrodynamics and Ecology

Some of the commonly applied models for practical water quality management of lakes and reservoirs, particularly deep, stratified ones, belong to the category of coupled hydrodynamic–ecological models (Orlob 1983, 1984; Varis 1994). The reason for coupling detailed hydrodynamics and basic chemical and biological processes is that hydrodynamic conditions vary widely, depending on changes of inflows and meteorology. The variability is particularly great in reservoirs under various operational regimes. The depth and time distribution of the chemical compounds in them is highly affected by currents and other movements of water masses, and the biological processes react on physical conditions both directly and indirectly via chemistry.

The coupled hydrodynamic–ecological models are used to simulate time-dependent water quality conditions under different meteorological conditions, eutrophication or its abatement, increasing or decreasing pollution, in order to evaluate the consequences of different management strategies. Furthermore, the conditions existing shortly after a reservoir is filled (the aging process) can be simulated, and also the conditions in reservoirs subject to different operation regimes, undergoing dam reconstruction and/or turbine repair. With modern personal computers, the simulation of 1-year cycles can be done in a few minutes, and the *scenario analysis* (comparative analysis of a number of model runs) can be used as a background for making decisions. The manager can see, with the graphical model outputs, the probable consequences of different variants of management actions under different hydrological and meteorological situations.

The input conditions needed to run the models are lake morphometry, annual cycles of hydrology and meteorology, and chemical and biological conditions of the inflows. Particularly important is the knowledge (or reasonable assumptions) about the inflow water density, as determined by temperature and salinity. The morphometry is characterized particularly by the area and volume curves. In the case of reservoirs, it is also characterized by the slope of the bottom and sides of the reservoir. The characteristics of the outflow structure and operation regimes are, of course, important as well.

The *hydrodynamic part* of the model provides information on the physical structure of the water masses under different external conditions. Its final output is usually the water temperature stratification. However, for the chemical and biological components, the intensity of mixing and exchange among different strata is the decisive output used. The hydrodynamic models are categorized according to the space dimensions being considered, as 1D, 2D, or 3D.

One-dimensional models provide the depth distribution of the physical properties in the deepest part of the water body and the water movements. In addition, *2D models* provide the longitudinal profile, which is important for reservoirs. The most complicated are the *3D models*, which treat the full volume of a water body in both the horizontal and vertical directions. The 1D models are adequate for most management purposes, but 2D models may be necessary if the knowledge of the longitudinal distribution in a reservoir is important for a manager. The use of a 3D model may be necessary for large lakes, as well as lakes and reservoirs with complicated shapes.

The *ecological part* of the model provides information on the chemical and biological components of interest to the manager. The biological part need not be directly used by the manager, but must be considered because of the close interactions between water chemistry and biology. Only the conservative chemical components (e.g., salinity) do not strongly interact with the biological components, although most chemical variables are either directly used and/or released by organisms, or are affected by other chemical changes produced during photosynthesis, respiration, and decomposition reactions. As an example, changes in pH and oxygen conditions in a lake or reservoir affect most chemical reactions. There is also a feedback between lake thermics and phytoplankton, in that higher radiation adsorption caused by phytoplankton leads to more heat accumulation in the upper water layer, surface temperatures increase, and the thermocline is located at a shallower depth. Phytoplankton growth is enhanced by the higher temperature (Straskraba 1998).

Few of the existing models have found broad application (Table 2.3). In Russia and Ukraine, similar models have been developed by Vladimir Lavrik (Kiev) mainly from the mathematical perspective (e.g., Lavrik et al. 1990). In addition to the models listed in Table 2.3, some of the more complex ecological models without elaborate hydrodynamics have been used for water quality management decisions, particularly in the more simple, shallow water bodies. A few of these models should be mentioned:

- The model of Jørgensen et al. (1978), which was first applied to Glumsø Lake, has been applied in 25 cases as a management tool (see Chapter 12).
- AQUAMOD, the AQUAtic MODel by Straskraba (1976), is a simple eutrophication-oriented model. In the extended version, the model GIRL OLGA (mentioned below) was used for Czechoslovak reservoirs. The model also exists in a two- and three-layer version, including sediments (Straskraba and Gnauck 1985). Gnauck et al. (1990)

TABLE 2.3

Coupled Hydynamic–Ecological Models

- LAKE ONTARIO model (Scavia and Chapra 1977) is a multipurpose ecological model developed specifically for the Great Lakes, in particular Lake Ontario.
- SALMO (Benndorf and Recknagel 1982) includes schematic distinction into epilimnion and hypolimnion, interactions between phytoplankton biomass and changing external nutrient load, temperature, light, mixing, and zooplankton.
- SALMOSED (Recknagel and Benndorf 1995) includes exchange of phosphorus between water and sediments.
- RESQUAL by Stefan and Cardoni (1982) was applied for Lake Chicot and was the basis for further models designed by the group around Heinz Stefan at the University of Minnesota (see MINLAKE below).
- CE-QUAL-RIV1 is a derivative of the model CE-QUAL for water quality in larger rivers (Bedford et al. 1983). The model was originally destined "for crude planning analysis of reservoir eutrophication" (Wlosinski and Collins 1985) but was continuously extended into 2D form. The model is distributed free by the USEPA and is broadly used, particularly in Americas (e.g., Diogo and Rodrigues 1997).
- FINNISH THREE DIMENSIONAL WATER QUALITY-TRANSPORT MODEL by Virtanen et al. (1986) concentrates first of all on longitudinal flows in complex shaped reservoirs (Koponen et al. 1998).
- WASP4 (Ambrose et al. 1988): includes hydrodynamics, conservative mass transport, eutrophication, oxygen, and toxic chemical–sediment dynamics. The model is also distributed free by USEPA, including documentation. The model is used in United States for a number of predictions, a recent list being mentioned (e.g., in Tufford et al. 1998).
- MINLAKE (Riley and Stefan 1988) was recently extensively modified to allow estimation of the effect of climate changes on U.S. lakes (Stefan and Fang 1994a) and dissolved oxygen conditions for regional lake analysis (Stefan and Fang 1994b).
- MIKE is originally a model with a simplified two-layer representation. A new version MIKE11 was taken over by the Danish Hydraulic Institute, Agern Alle 5, Horsholm, and represents a 2D vertical hydrodynamic flow model coupled with an eutrophication model covering 12 state variables belonging to inorganic nutrients, phytoplankton, zooplankton, detritus, and dissolved oxygen. A number of additional models for estimating the loads from watershed (CATCHMENT, LOAD, WETLAND), and other water quality processes are also available. Both 2D and 3D models available from DHI as MIKE 22 and MIKE 33, and they can also be coupled with ecological components in a very flexible manner by use of the software denoted ECOLAB.
- ASTER and MELODIA (Salençon and Thèbauld 1994a, 1994b, 1995) were applied in France to model stratification and biological conditions in Pareloup Reservoir. The hydrodynamic part represents a separate model, EOLE (Salençon 1997).
- DYRESM—WQ, the DYnamic REServoir Model—Water Quality (Schladow et al. 1994; Hamilton and Schladow, 1997; Schladow and Hamilton 1995, 1997; Hamilton and De Stasio 1998; Hamilton 1999) is based on the hydrodynamic model DYRESM. The water quality model simulates stratification and flow conditions, phytoplankton, nutrients (different forms of phosphorus and nitrogen), oxygen, BOD, iron and manganese, and sedimentation. It is being systematically extended and is now with CE-QUAL the most widely used water quality model. However, the cost of the model available from the Water Research Centre, University of Western Australia, Nedlands, Western Australia, is high. A standard 1D, a 2D hydrodynamic (Hocking and Patterson 1988, 1991), and most recently a 3D water quality version (Romero and Imberger 1999) exist. An extension of the model to simulate the consequences of artificial destratification was formulated by Schladow and Hamilton (1995) as used, e.g., by Lindenschmidt and Chorus (1997) and Schlenkhof (1997).

(continued)

TABLE 2.3 (Continued)

Coupled Hydynamic–Ecological Models

- BEEKWAM is intended for nonstratified embankment reservoirs in the Netherlands. It combines an empirical equilibrium temperature model with differential equations for nutrients and algal growth. MATLAB® and Simulink® are used as simulation tools (Benoist et al. 1998).
- PROTECH-C (from Phytoplankton RespOnses To Environmental CHange) is a 2D model directed to detailed representation of species specific phytoplankton reactions on external effects, including the meteorological, hydrological, and human caused effects. The model from the Algal Modelling Unit, The Institute of Freshwater Ecology, Far Sawrey, Ambleside, Cumbria, UK, is based on extensive scientific phytoplankton studies by Reynolds (1987, 1992, 1998, 1999a, 1999b) and is used by a number of water quality agencies in the UK.

have used the modeling system SONCHES to apply AQUAMOD in Germany.

- PCLAKE, the shallow lake model designed specifically by Janse et al. (1993) for use on personal computers, is systematically being extended to increase its usefulness for managing shallow lakes, particularly with respect to eutrophication and attempts to treat it using biomanipulation (Janse et al. 1995). It is also being applied to the problem of two multiple stable states of shallow lakes—the phytoplankton state and the macrophyte state (Janse 1997; Janse et al. 1998).
- The BALATON model by Kutas and Herodek (1986) is a complex model specifically designed to aid in the management of the shallow Lake Balaton, which is characterized by high calcium content, low amounts of zooplankton, and high importance of bacterial processes.

2.4 Ecotoxicological Lake Models

During the past 25 years, models of toxic substances have emerged as a result of the increasing interest in the environmental management of toxic substances that can cause water pollution. Ecotoxicological models attempt to model the fate and/or effects of toxic substances in ecosystems. Toxic substance models are most often biogeochemical models because they attempt to describe the mass flows of the toxic substances being considered. There also are models of population dynamics, which include the influence of toxic substances on the birth rate, growth rate, and/or mortality and, therefore, should also be considered toxic substance models.

Toxic substance models have several characteristic properties (see also Chapter 15):

- The need for parameters to cover all possible toxic substance models is great. Thus, general estimation methods are widely used. Methods based on the chemical structure of chemical compounds are developed: the so-called quantitative structure activity research (QSAR) and structure activity research (SAR) methods.
- The safety margin should be high, for example, when expressed as the ratio between the actual concentration and the concentration that gives undesired effects.
- The possible inclusion of an effect component, which relates the output concentration to its effect. It is easy to include an effect component in the model. However, it often is a problem to find a well-examined relationship upon which to base it. Rather, it will require the use of good knowledge of the properties, particularly the toxicological properties for the component of concern.
- The possibility and need of simple models, based on points 1 and 2 above, and of our limited knowledge of the processes, parameters, sublethal effects, antagonistic and synergistic effects.

The decision regarding which model class to apply is based on the ecotoxicological problem that one wishes to solve with the model. It is often only the toxic substance concentration in one trophic level that is of concern. This includes the zero trophic level, which is understood to be the medium—lake water.

The acknowledgment of the uncertainty is of great importance for all models, but it is particularly important for ecotoxicological models. It may be taken into consideration either qualitatively or quantitatively. Another problem is: Where does one take the uncertainty into account? Should the economy or the environment benefit from the uncertainty? Unfortunately, most decision-makers up to now have used the uncertainty to the benefit of the economy. This is a completely unacceptable approach, since the same decision-makers would, for example, never consider whether uncertainty should be used for the benefit of the economy or the strength of a bridge in a civil engineering project. Table 2.4 gives an overview of a number of ecotoxicological models applied to lake and reservoir management.

Ecotoxicological models are widely used to perform environmental risk assessments (ERA). This is of particular interest for lakes and reservoirs used as drinking water supplies, or for which the fisheries are significant. ERA typically answers such questions as: What is the risk of contaminating drinking water or fish, for a given application of pesticides in an agricultural project adjacent to a lake or reservoir?

TABLE 2.4

Examples of Ecotoxicological Lake Models

Toxic Substance	Model Characteristic	Reference
Cadmium	Food chain as in an eutrophication model	Thomann et al. (1974)
Mercury	Six state variables: water, sediment, suspended matter, invertebrates, plant, and fish	Miller (1979)
Vinyl chloride	Chemical processes in water	Gillet (1974)
Heavy metals	Concentration factor, excretion, bioaccumulation	Aoyoma et al. (1978)
Lead	Hydrodynamics, precipitation, toxicity of ionic lead on algae, invertebrates, and fish	Lam and Simons (1976)
Radionuclides	Hydrodynamics, decay, uptake, and release by various aquatic surfaces	Gromiec and Gloyna (1973)
Polycyclic aromatic	Transport, degradation, bioaccumulation, hydrocarbons	Bartell et al. (1984)
Cadmium, PCB	Hydraulic overflow rate (settling), sediment interactions, steady state food chain submodel	Thomann (1984)
Hydrophobic organic	Gas exchange, sorption/desorption, hydrolysis compounds, photolysis, hydrodynamics	Schwarzenbach and Imboden (1984)
Mirex	Water–sediment exchange processes, adsorption, volatilization, bioaccumulation	Halfon (1984)
Toxins (aromatic hydrocarbons, Cd)	Hydrodynamics, deposition, resuspension, volatilization, photooxidation, decomposition, adsorption, complex formation (humic acid)	Harris et al. (1984)
Persistent organic chemicals	Fate, exposure, and human uptake	Paterson and Mackay (1989)
pH, calcium, and aluminum	Survival of fish populations	Breck et al. (1988)
Pesticides and surfactants	Fate in rice fields	Jørgensen et al. (1997)
Toxicants	Migration of dissolved toxicants	Monte (1998)
Growth promoters	Fate, agriculture	Jørgensen et al. (1998)
Toxicity	Effect on eutrophication	Legovic (1997)
Pesticides	Mineralization	Fomsgaard (1997)
Mecoprop (3)	Mineralization in soil or sediment	Fomsgaard and Kristensen (1999)

2.5 Lake Acidification Models

Lake acidification is caused by acid rain, which originates from emissions of sulfur and nitrogen oxides into the atmosphere (see Chapter 20). Thus, assessment of acidification for management purposes requires a chain of models, linking the emissions to the atmosphere to the effects of acid rain on soil chemistry in the lake's drainage area, and further to the pH changes in a lake. In this context, only the lake models and soil models will be mentioned. These latter models determine the composition of the drainage water flowing into lakes and reservoirs. Those interested in entire model chains are referred to Alcamo et al. (1990) and Jørgensen (1994).

Several models have been developed to translate emissions of sulfur and nitrogen compounds into changes in soil chemistry (in the first hand to changes in the pH of soil water). Kauppi et al. (1984) used knowledge about buffer capacity and velocity to relate the emissions to soil water pH. A critical pH value of 4.2 is usually applied to interpret the results.

From the results of the atmospheric deposition, it is possible to estimate the acid load in equivalents of hydrogen ions per square meter per year. In the European model, RAINS, the acid load is computed from the deposition, after accounting for forest filtering and atmospheric deposition of cations (for details, see Alcamo et al. 1990).

The Soil Map of the World classifies European soils into 80 soil types. The fraction of each soil type within each grid element is computerized to an accuracy of 5%. The resolution of the European RAINS model is such that each grid element includes one to seven soil types, with a mean number of 2.2. The model goal is to keep track of the development of soil pH and buffer capacity. Further development of soil models is still needed, as the soil model must be considered the weakest model in the chain.

It can be concluded that models of soil processes and chemical composition, including the concentrations of various ions and pH of the drainage water, are complex. It is not the intention here to present the models in detail—they are far too complex to do so—but rather, to mention the difficulties and a few of the basic ideas behind them. Further details can be found in the references given in this section.

Henriksen (1980) and Henriksen and Seip (1980) developed an empirical model for the relationship between lake water pH and sulfur load. They plotted related values of pH and sulfur loads for a given drainage area with specific sensitivities for acidification. The resulting curves look like a titration curve. These simple empirical approaches are used without consideration of the drainage water. Weathering processes, however, can be incorporated in the empirical approaches, although the results have little generality since they are based on regression analysis of local measurements. The presented modeling approach has been used on Scandinavian lakes with good results.

Most biological models of lake acidification focus on fish. On the basis of data from 719 lakes, Brown and Salder (1981) developed an empirical model relating the fish population to the pH of the water. For their study in lakes in southern Norway, they found that a 50% reduction of the sulfate emissions will give an average increase in pH of 0.2, which will only improve the fish populations in 9% of the lakes. However, a criticism of the model is that it underestimates the relationship between the reduction of sulfate emissions and pH.

Muniz and Seip (1982) developed another empirical model, in which they distinguished between lakes of different conductivity. Chen et al. (1982) developed a very comprehensive pH-effect model, which considers the effects on all levels in the food chain and the total effects on the ecosystem.

2.6 Lake Fisheries Models

Augmented exploitation of freshwater fish resources by sport and commercial fisheries and the deterioration of water quality has stimulated concern about the depletion of fish stocks in lakes. This reality has intensified the development models which take into account the effects of harvesting and water quality on the fish population. Models with a wide spectrum of complexity attempt to provide a management tool to assess an optimum fishery strategy.

For lakes with important commercial fisheries, landing records are often available for periods of multiple decades, and may open the possibilities developing a statistical fisheries model. However, a statistical model generally does not take into account a number of important factors, such as interactions among species, water quality, and changes in the concentrations of fish food. A statistical model would build on the assumption that present and past properties of the environment and the fish populations will be maintained. Statistical models will not be discussed further here, primarily because they do not consider the influences of water quality. It should be mentioned, however, that Pamolare (downloadable from www.unep.or.jp/ietc/pamolare), launched by the International Lake Environment Committee (ILEC) and the United Nations Environment Programme's International Environment Technology Centre (UNEP-IETC) the ILEC contains a simple eutrophication model with four state variables, which also computes the fish population with a simple regression equation relating primary production and the fish population.

The simplest approaches assume that the entire fish population is homogeneous, and does not consider the population dynamics and related age structure, which is essential for fishing policy. The more complex approaches consider the influences of water quality on the population dynamics and the age structure of the fish populations.

2.7 Structurally Dynamic Models Applied to Reveal the Ecosystem Properties of Lakes

It is important both in ecological research and in environmental management to understand the reactions of lakes as ecosystems to steadily changed conditions determined by the forcing functions. An example of lake reactions that we should be able to explain are the success and failure of biomanipulation and the hysteretic behavior of lakes. Success with biomanipulation varies considerably from case to case, depending on the specific circumstances. Thus, modeling seems an obvious tool to use in selecting biomanipulation as a control measure, since the primary task of management models is always to predict the results of alterations in an ecosystem. However, biomanipulation implies that the structure of the ecosystem (i.e., a lake) is changed, which is much more difficult to simulate by a model than a simple (and minor) change in state variables due to change in forcing functions. On the other hand, structural dynamic models (SDMs) are emerging (e.g., see Jørgensen 1992a, 1992b), and are increasingly used as an experimental tool that can sometimes also be used to explain ecosystem behavior (Jørgensen 1992b) associated, for example, with chaos and catastrophe theories.

Models based on case studies with structural changes are still not very frequent. However, we know that ecosystems have the ability to adapt to changed forcing functions, and to shift to species better suited to emerging conditions. Because radical changes are imposed on ecosystems, which will inevitably lead to the most pronounced changes in ecosystem structure, it is particularly interesting to apply models with dynamic structures, which are becoming increasingly important in environmental management. Thus, this type of model is discussed in this section in relation to a typical example of this problem, namely, biomanipulation applied for restoration of lakes and reservoirs.

Biomanipulation is based on changes in ecosystem structure imposed by intentional changes in fish populations. In this context, it is of interest to explain when and why top–down control measures are working (i.e., when the imposed structural changes can be predicted to work). This is also possible with the use of catastrophe theory on lake eutrophication models.

The short-term results of biomanipulation have been encouraging. It is unclear, however, whether a manipulated ecosystem will ultimately return to the initial eutrophic and turbid conditions. Some observations suggest that, if low nutrient concentrations are combined with a relatively high concentration of predatory fish, a stable steady state will be attained, whereas high nutrient concentration and high predatory concentration will lead to an unstable clear water state (see Hosper 1989; Van Donk et al. 1989). On the other hand, turbid conditions may prevail even at medium nutrient concentrations, provided that the predatory fish concentration is low. By introducing more predatory fish, however, the conditions may improve significantly, even at medium nutrient concentrations.

Willemsen (1980) distinguishes in the temperate region two possible conditions:

- A "bream state," characterized by turbid water, and a high degree of eutrophication relative to the nutrient concentration. Submerged vegetation is largely absent from such systems. Large amounts of bream are found, whereas pike are rarely found.
- A "pike state," characterized by clear water, and a low degree eutrophication relative to the nutrient level. Pike are abundant, whereas significantly fewer breams are found, compared with the "bream state." Willemsen's work shows that the pike/bream ratio is strongly correlated with water transparency, and that the separation between the two states is relatively distinct (see details in Jørgensen and Faith, 2011).

This behavior is clearly analogous to other examples described by catastrophe theory applied to biological systems (Jørgensen 1992a, 1994). Figure 2.2 shows the catastrophe fold, where a bream isocline is plotted in relation to the nutrient concentration, assuming that the pike population is in a steady state. The isocline consists of the stable lower or upper density. The arrow corresponds to the jumps between the two stable levels.

The discontinuous response to increasing and decreasing nutrient levels implies that decreased nutrient levels will not cause a significant decrease in lake or reservoir eutrophication, and a significant increase in water transparency, before a rather low level has been attained. It may be possible, however, to "push" the equilibrium from point 3 to point 1 with the addition of predatory fish. This modeling example illustrates that two different concentrations of planktivorous fish can coexist with a certain nutrient concentration, which explains the hysteresis reaction shown in Figure 2.2. The general modeling experience is that a given set of forcing functions will give a certain set of state variables, including the concentration of planktivorous fish. However, when the set of equations that describe the ecosystem has a formulation causing catastrophic behavior in the mathematical sense, the reactions described above are observed.

To further explain these observations, models have been used as an experimental tool in the sense that the model description is in accordance with well-working lake models, but for which simulations with forcing functions for which there are no data available to control the model output, are

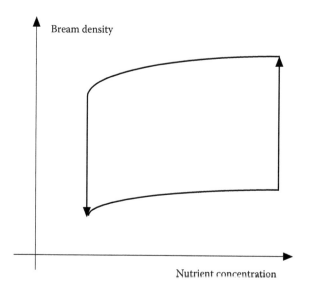

FIGURE 2.2
According to Willemsen, two stable bream densities exist. The upper level corresponds to a high eutrophication and the lower level to a low eutrophication. The jumps between the two levels take place at high nutrient concentration (from low to high bream density) and at low nutrient concentration (from high to low bream density).

carried out. The model used for these experiments is shown in Figure 2.3. Only phosphorus is considered as a nutrient in this case, although it is feasible to consider both nitrogen and phosphorus. The model encompasses the entire food chain. The modeling problem in the case of different inputs of nutrients, however, is that the phytoplankton, zooplankton, planktivorous fish, and carnivorous fish are all able to adjust their growth rates within certain ranges. Thus, it is necessary to test a series of simulations with different combinations of growth rates to find which combination can give the highest probability of survival for all four classes of species.

The results of these simulations can be summarized in the following points:

- Between a total phosphorus (TP) concentration of between 60 and 70 approximately up to between 120 and 140 µg/L, two levels of planktivorous fish give a stable situation, with a high probability of survival for all four classes of species. The level with the lowest level of planktivorous fish give the highest concentration of zooplankton and carnivorous fish, whereas the phytoplankton concentration is lowest. This interval between 60 and 70 approximately up to between 120 and 140 µg/L, is, of course, dependent on the model descriptions of the lake and must not be taken as very fixed values for all

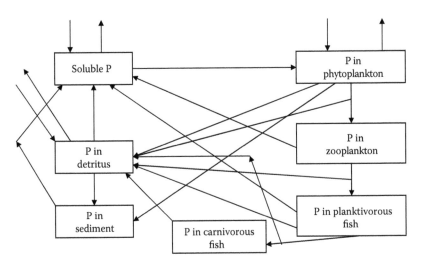

FIGURE 2.3
The model used as an experimental tool. For simulation results, see text. (See also Jørgensen and De Bernardi 1998.)

lakes. Rather, they only indicate that there is an interval corresponding approximately to mesotrophic conditions are approximately two stable situations.

- The stable situation with the lowest level of planktivorous fish corresponds to the lowest growth rate of zooplankton and phytoplankton, which normally implies species that are larger in size (see Peters 1983).
- Only one stable level is achieved at TP concentrations below 60–70 or above 120–140 µg/L. The zooplankton, relative to the level of phytoplankton, are present in highest levels at TP concentrations below 60–70 µg/L, and relatively low above 120–140 µg/L.
- When the planktivorous fish are present at low concentrations, the phytoplankton is controlled by the relatively high zooplankton level (i.e., the grazing pressure), whereas phytoplankton at a high planktivorous fish level is controlled by the nutrient concentration.

As already underlined, this exercise should only be considered qualitatively. However, the results suggest that the use of biomanipulation seems only to be successful at an intermediate nutrient level, which is consistent with the results of many biomanipulation experiments. The range at which biomanipulation can be used successfully is most probably dependent on the specific conditions in the considered lakes, which unfortunately makes it problematic to use results described in this section more quantitatively. The model example here uses a test of many combinations of growth rates to find the best combination of parameters from a survival perspective. This adjustment of the parameters corresponds to adaptation within some ranges, and for more radical changes of the parameters to a change in the species composition in the lake. If models are to be used as management tools in cases where significant changes in nutrient levels will occur, or where bio-manipulation is being considered, it is necessary to develop models that can account for structural dynamic changes. However, this model development is still in its infancy. Thus, only limited experience is available.

Our present models have generally rigid structures and a fixed set of parameters, indicating that no changes or replacements of the components are possible. However, it is necessary to introduce parameters (properties) that can change according to changing general conditions for the state variables (components). The current idea is to test if a change of the most crucial parameters produces a higher so-called goal function of the system and, if so, to use that set of parameters.

The type of models that can account for changes in species composition, as well as for the ability of the species (i.e., the biological components in the models) to change their properties (i.e., to adapt to the prevailing conditions imposed on the species), are as indicated in the introduction to this section, namely, the so-called SDMs (see Figure 2.4).

It can be argued that the ability of ecosystems to replace present species with better-fitted species can be considered by constructing models that encompass all the actual species for the entire period the model attempts to cover. However, this approach has two essential disadvantages. First, the model becomes very complex, since it will contain many state variables for each trophic level. This implies that the model will contain many more parameters that have to be calibrated and validated, and which will introduce a high uncertainty to the model predictions, rendering the application of the model very case-specific (Nielsen 1992a, 1992b). In addition, the model will still be rigid and not give the model the ecosystem property of having continuously changing parameters even without changing the species composition (Fontaine 1981).

SDMs can be developed by a current change of some of the most crucial parameters, reflecting the change of species. These changes may be based on expert knowledge, or may be determined with the use of a goal function as assumed in Figure 2.4. In the first case, the changes of parameters are made in accordance with good knowledge of the dominant species under different conditions (i.e., different combinations of forcing functions). Reynolds (1998) illustrates this approach. In the latter case, the parameters currently are changed according to goal functions. Several goal functions have been proposed. Straskraba (1979) used a maximization of biomass as the governing principle (i.e., the above-mentioned goal function). The model computes the biomass and adjusts one or more selected parameters to achieve the maximum biomass at every instance. The model has a computation program that computes the biomass for all possible combinations of parameters within a given realistic range. The combination that gives the maximum biomass is

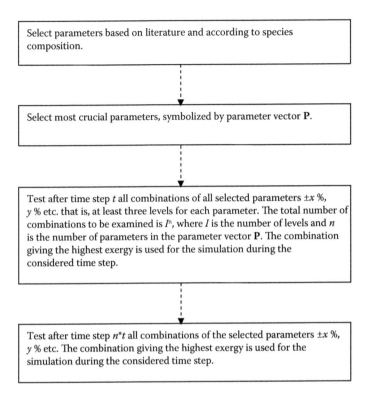

FIGURE 2.4
SDMs are developed by a procedure based on optimization of a goal function. In the case shown, exergy (eco-exergy) is used as goal function. More information about exergy, the most frequently applied goal function for development of SDMs, is given in the text.

selected for the next time step, etc. However, biomass can hardly be used in models with more trophic levels, and it will lead to biased results by, for example, adding together the biomass of fish and phytoplankton. The thermodynamic variable, exergy, has been used most widely as a goal function in ecological models.

The modeling experience shows that models can explain why biomanipulation works under some circumstances, and why it does not work under other circumstances. The applications of the catastrophe theory and/or SDMs are able to explain the appearance of hysteresis in the relations between nutrient level and eutrophication (see Figure 2.5). The results can be used to explain that this hysteresis relationship exists in an intermediate nutrient level, which explains why biomanipulation in this range of nutrient concentrations has worked properly (i.e., with long-term effects), but not above or below this intermediate range.

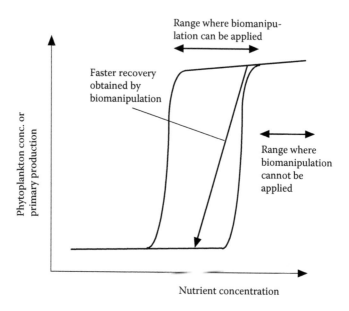

FIGURE 2.5
The hysteresis relation between nutrient level and eutrophication measured by the phyto-plankton concentration or the primary production. In a certain nutrient concentration range, it is possible to make a short cut between the two structures. (See also Jørgensen and De Bernardi 1998.)

Models have been an applicable tool in efforts to understand biomanipulation results. The results may be summarized in the following recommendations for using models for managing biomanipulation:

- Attempt either to use an SDM or the catastrophe theory to examine the possibilities of using biomanipulation.
- Consider the results only as approximations, unless very good data for applying SDMs are available.
- Consider intermediate nutrient levels, between approximately 60 and 120–140 μg/L of TP, or 7 times higher concentrations for nitrogen if it is the limiting nutrient, as the range within which biomanipulation will most probably work.

The results obtained by modeling and in-lake biomanipulation experiments may be explained by the presence of ecological buffer capacities and time lags in structural changes. When the nutrient level is increased (see Figure 2.5), the degree of eutrophication will initially not increase significantly. The lake ecosystem has an ecological buffer capacity, which can be explained by increased grazing and settling of the phytoplankton resulting from increased nutrient levels. At a certain nutrient level, however, the zooplankton can no longer control the phytoplankton concentration, which becomes limited only by the nutrient level. The result is that the eutrophication level increases considerably with increased nutrient levels. The entire structure of the ecosystem is changed because phytoplankton is now limited by nutrients, zooplankton by the predation from planktivorous fish rather than by the available food source (phytoplankton), and the planktivorous fish are now limited by the food source (zooplankton) rather than by the carnivorous fish.

This implies that other properties are selected (e.g., other growth rates). Furthermore, the carnivorous fish will be scarce since they mostly hunt by sight, with the increasing eutrophication causing an increased turbidity and, therefore, making hunting more difficult.

If nutrient levels are now decreased by, for example, discharging wastewater with significantly lower nutrient concentrations, will this "high eutrophication structure," characterized with a low concentration of carnivorous fish, high concentration of planktivorous fish, low concentration of zooplankton, and high concentration of phytoplankton, change immediately?

Thus, the low transparency will still make an increase in the carnivorous fish difficult, and grazing will still not be able to control the phytoplankton concentration. This is because they are present in relatively low concentrations, since they are still under stress from considerable predation pressure by planktivorous fish. In this situation, will biomanipulation—either by removal of planktivorous fish and/or introduction of more carnivorous fish—give a clear effect, as shown in Figure 2.5? Biomanipulation simply provides a faster recovery to the "low eutrophication structure," where grazing controls the phytoplankton concentration. Outside the intermediate nutrient level, the forcing functions (input of nutrients) determine the state of eutrophication.

Models are a useful tool for trying to understand the function of biomanipulation, including the fact that it sometimes gives positive results, and sometimes negative results. Models generally should be used as a tool to try to qualitatively understand the behavior of complex systems (e.g., lake ecosystems). When a good database is available, models can sometimes be used quantitatively in environmental management, an example being the control of eutrophication by wastewater management (e.g., see Jørgensen and Vollenweider 1988; Jørgensen 1986, 1994). The use of models in a quantitative sense for biomanipulation is possible by application of SDMs.

Several goal functions for the development of SDMs have been proposed. Straskraba (1979) used maximization of biomass as the governing principle (the goal function). The model adjusts one or more selected parameters in order to achieve maximum biomass at every instance. A modeling software routine is included that computes biomass for all possible combinations of parameters within a given realistic range. The combination that gives the maximum biomass is selected for the next time step, etc. However, the biomass can hardly be used in models with more trophic levels. To add the biomass of fish and phytoplankton together, for example, will lead to biased results.

The thermodynamic variable, exergy, has been used most widely as a goal function in ecological models. There are two pronounced advantages of exergy (also denoted as eco-exergy to underline its application in ecology and ecological modeling) as a goal function, compared to entropy and maximum power (Jørgensen 2002):

1. It is defined far from thermodynamic equilibrium.
2. It is related to state variables, which are easily determined or measured. However, because exergy is not a common thermodynamic function, we need first to describe it in this context.

Exergy expresses energy with a built-in measure of quality, quantifies natural resources, and can be considered as fuel for any system that converts energy and matter in a metabolic process (Schrödinger 1944). Thus, ecosystems convert exergy, and exergy-flow through the system is necessary to keep it functioning. Exergy (as the amount of work the system can perform when brought into equilibrium with its environment) measures the distance from the "inorganic soup" in energy terms, as will be explained further below. Thus, exergy is dependent both on the environment and the system, rather than the system alone. This means, for example, that it is not a state variable like free energy and entropy.

If we assume a reference environment representing an ecosystem at thermodynamic equilibrium, the definition illustrated in Figure 2.6 is valid. At

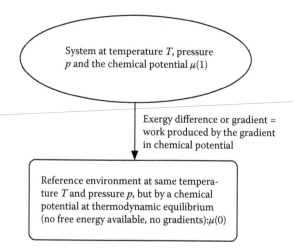

FIGURE 2.6

A reference system representing the same ecosystem at thermodynamic equilibrium and at the same temperature and pressure is assumed. At thermodynamic equilibrium, all components are inorganic, at the highest possible oxidation state, and homogeneously distributed in the system. Exergy, or eco-exergy to underline its application in ecological modeling, is the difference between the energy that can do work in the system and the reference state.

thermodynamic equilibrium all components are inorganic, at the highest possible oxidation state, and homogeneously distributed in the system. As the chemical energy embodied in the organic components and the biological structure contributes the most by far to the exergy content of the system, there seems to be no reason to assume a (minor) temperature and pressure difference between it and the reference environment. Under these circumstances, we can calculate the exergy content of the system provided entirely by chemical energy: $\sum_c(\mu_c - \mu_{ceq})N_i$. We can also determine its exergy, compared with that of a similar system at the same temperature and pressure, but in the form of an inorganic soup without any life, biological structure, informational, or organic molecules. Because $(\mu_c - \mu_{ceq})$ can be found from the definition of the chemical potential (replacing activities by concentrations), the following expression for the exergy can be obtained:

$$\text{Ex} = RT \sum_{i=0}^{i=n} C_i \ln C_i / C_{i,eq} \tag{2.1}$$

As shown, exergy measures the difference between free energy (given the same temperature and pressure) of an ecosystem, and that of the surrounding environment. If the system is in equilibrium with its surroundings, exergy is zero, with $n = 0$ accounting for inorganic compounds and $n = 1$ corresponding to detritus.

Since the only way to shift systems away from equilibrium is to perform work on them, and since the available work in a system is a measure of its ability to do so, we need to distinguish between the system and its environment (or thermodynamic equilibrium alias the inorganic soup). Thus, it is reasonable to use the available work (i.e., the exergy) as a measure of the distance from thermodynamic equilibrium.

Survival implies the maintenance of biomass, whereas growth means its increase. Exergy is needed to construct biomass, which then possesses exergy transferable to support other processes. Thus, survival and growth can be measured by use of the thermodynamic concept exergy, which may be understood as *the free energy relative to the environment* (see Equation 2.1).

Darwin's theory of natural selection, therefore, may be reformulated in thermodynamic terms and expanded to an ecosystem level as follows: *The prevailing conditions of an ecosystem steadily change. The system will continuously as a result of the changing conditions select those species which contribute most to the maintenance, or even to the growth of the exergy of that system.*

Note that the thermodynamic translation of Darwin's theory requires *populations* to possess properties of reproduction, inheritance, and variation. The selection of species that contribute most to the exergy of the system under the prevailing conditions requires that there are sufficient individuals with different properties for selection to take place. Reproduction and variation must be high and, once a change has taken place, it must be conveyed to the next generation via better adaptation. It is also noted also that a change in exergy is not necessarily ≥0, but rather depends on the resources of the ecosystem. However, the above proposition claims that an ecosystem attempts to reach the highest possible exergy level under the given circumstances with the available genetic pool (Jørgensen and Mejer 1977, 1979).

It is not possible to measure exergy directly. However, it is possible to compute it if the composition of the ecosystem by Equation 2.1 is known. C_i represents the ith component expressed in a suitable unit (e.g., for phytoplankton in a lake, C_i mg nutrient in the phytoplankton/liter of lake water) and $C_{i,eq}$ is the concentration of the ith component at thermodynamic equilibrium. For detritus, exergy can be found on basis of equilibrium constants, which give the ratio between the concentration of detritus in the ecosystem and that at thermodynamic equilibrium. The exergy content of detritus is approximately 18 kJ/g, which can be compared with the exergy (chemical energy) content of mineral oil (about 42 kJ/g).

By using this particular exergy based on the same system at thermodynamic equilibrium as reference, the eco-exergy becomes dependent only on the chemical potential of the numerous biochemical components.

It is possible to distinguish in Equation 2.1 between the contribution to the eco-exergy from the information and from the biomass. We define p_i as C_i/A, where

$$A = \sum_{i=1}^{n} C_i \tag{2.2}$$

is the total amount of matter density in the system. With introduction of this new variable, we get:

$$Ex = ART \sum_{i=1}^{n} p_i \ln p_i/p_{io} + A \ln A/A_o \tag{2.3}$$

As $A \approx A_o$, eco-exergy becomes a product of the total biomass A (multiplied by RT) and Kullback measure:

$$K = \sum_{i=1}^{n} p_i \ln\left(p_i/p_{io}\right) \tag{2.4}$$

where p_i and p_{io} are probability distributions, a posteriori and a priori to an observation of the molecular detail of the system. It means that K expresses the amount of information that is gained as a result of the observations. For different organisms that contribute to the eco-exergy of the ecosystem, the eco-exergy density becomes $cRT \ln(p_i/p_{io})$, where c is the concentration of the considered organism. $RT \ln(p_i/p_{io})$, denoted β, is found by calculation of the probability to form the considered organism at thermodynamic equilibrium, which would require that organic matter is formed and that the proteins (enzymes) controlling the life processes in the considered organism have the right amino acid sequence. These calculations can be seen in the work of Jørgensen and Svirezhev (2004). In the latter reference, the latest information about the β values for various organisms is presented (see Table A1 in Appendix 1). For humans, the β value is 2173, when the eco-exergy is expressed in detritus equivalent or 18.7 times as much or 40,635 kJ/g if the eco-exergy should be expressed as kJ and the concentration unit g/unit of volume or area. The β value has not surprisingly increased as a result of the evolution. To mention a few β values from Table 2.1: bacteria 8.5, protozoa 39, flatworms 120, ants 167, crustaceans 232, mollusks 310, fish 499, reptiles 833, birds 980, and mammals 2127. The evolution has, in other words, resulted in a more and more effective transfer of what we could call the classical work capacity to the work capacity of the information. A β value of 2.0 means that the eco-exergy embodied in the organic matter and the information are equal. As the β values (see above) are much larger than 2.0 (except for virus, where the β value is 1.01—slightly more than 1.0), the information eco-exergy is the most significant part of the eco-exergy of organisms.

In accordance with Equations 2.3 and 2.4 and the above presented interpretation of these equations, it now possible to find the eco-exergy density for a model as:

$$\text{Eco-exergy density} = \sum_{i=1}^{i=n} \beta_i C_i \tag{2.5}$$

The eco-exergy due to the "fuel" value of organic matter (chemical energy) is about 18.7 kJ/g (compare with coal, which is about 30 kJ/g, and crude oil, 42 kJ/g). It can be transferred to other energy forms for instance mechanical work directly, and be measured by bomb calorimetry, which requires destruction of the sample (organism), however. The information eco-exergy = $(\beta - 1) \times$ biomass or density of information eco-exergy = $(\beta - 1) \times$ concentration. A table with β values is presented in Appendix A. The information eco-exergy is taking care of the control and function of the many biochemical processes. The ability of the living system to do work is contingent upon its functioning as a living dissipative system. Without the information eco-exergy, the organic matter could only be used as fuel similar to fossil fuel. But due to the information eco-exergy, organisms are able to make a network of the sophisticated biochemical processes that characterize life. The eco-exergy (of which the major part is embodied in the information) is a measure of the organization (Jørgensen and Svirezhev 2004). This is the intimate relationship between energy and organization that Schrödinger (1944) was struggling to find.

The use of exergy calculations continuously to vary parameters has only been used in 21 case studies of biogeochemical modeling. One of these

(Søbygaard Lake) is described here as an illustration of what can be achieved with this approach. The results (Jeppesen et al. 1989) are particularly fitted to test its applicability to SDMs.

Søbygaard is a shallow lake (depth of 1 m) with a short retention time (15–20 days). The phosphorus nutrient load was significantly lowered in 1982, from 30 to 5 g P/m² yr. However, the decreased load did not result in reduced nutrient and chlorophyll concentrations during the period 1982–1985, because of internal loading of stored nutrients in the sediment (Jeppesen et al. 1989, 1990).

Radical changes were then observed during the period 1985–1988. Recruitment of planktivorous fish was significantly reduced during the interval 1984–1988 because of a very high pH. The zooplankton increased, and the phytoplankton decreased in concentration, with the average summer chlorophyll-*a* concentration reduced from 700 µg/L in 1985 to 150 µg/L in 1988. The phytoplankton population even collapsed during shorter periods because of extremely high zooplankton concentrations.

Simultaneously, phytoplankton species increased in size. Their growth rates declined and higher settling rates were observed. In other words, this case study illustrates that pronounced ecosystem structural changes were caused by biomanipulation-like events. However, the primary production was not higher in 1985 than in 1988, because of pronounced self-shading by smaller algae. Thus, it was very important to include a self-shading effect in the model. Simultaneously, sloppier feeding of zooplankton was observed, with a shift from *Bosmina* to *Daphnia* taking place.

The model contains six state variables, all of which represent forms of nitrogen, including fish, zooplankton, phytoplankton, detritus nitrogen, soluble nitrogen, and sedimentary nitrogen. The model equations are given in Table 2.5. Because nitrogen is the limiting nutrient for eutrophication in this particular case, it may be sufficient to only include this element in the model.

The aim of the study is to describe, by use of an SDM, the continuous changes in the most essential parameters, using the procedure shown in Figure 2.4. The data from 1984 to 1985 were used to calibrate the model. The two parameters which it was intended to change for the period 1985 to 1988 received the following values:

Maximum phytoplankton growth rate:	2.2/day
Phytoplankton settling rate:	0.15/day

The state variable, fish nitrogen, was kept constant at 6.0 during the calibration period. During the period 1985–1988, however, an increased fish mortality was introduced to reflect the increased pH. Thus, fish stock was reduced to 0.6 mg N/L—notice the equation "mort = 0.08 if fish > 6 (may be changed to 0.6) else almost 0." A time step of $t = 5$ days and $x\% = 10\%$ was applied (Figure 2.4). This means that nine runs were needed for each time step, in

TABLE 2.5

Mode Equations for Søbygaard Lake

---

fish = fish + dt * (–mort + predation)

INIT (fish) = 6

na = na + dt * (uptake – graz – outa – mortfa – settl – setnon)

INIT (na) = 2

nd = nd + dt * (–decom – outd + zoomo + mortfa)

INIT (nd) = 0.30

ns = ns + dt * (inflow – uptake + decom – outs + diff)

INIT (ns) = 2

nsed = nsed + dt * (settl – diff)

INIT (nsed) = 55

nz = nz + dt * (graz – zoomo – predation)

INIT (nz) = 0.07

decom = nd* (0.3)

diff = (0.015)*nsed

exergy = total_n*(Structura-exergy)

graz = (0.55)*na*nz/(0.4+na)

inflow = 6.8*qv

mort = IF fish > 6 THEN 0.08*fish ELSE 0.0001*fish

mortfa = (0.625)*na*nz/(0.4 + na)

outa = na*qv

outd = qv*nd

outs = qv*ns

pmax = uptake*7/9

predation = nz*fish*0.08/(1 + nz)

qv = 0.05

setnon = na*0.15*(0.12)

settl = (0.15)*0.88*na

Exergy density

= (nd + nsed/total_n)*(LOGN(nd + nsed/total_n) + 59) + (ns/total_n)*(LOGN(ns/total_n) – LOGN(total_n)) + (na/total_n)*(LOGN(na/total_n) + 60) + (nz/total_n)*(LOGN(nz/total_n) + 62) + (fish/total_n)*(LOGN(fish/total_n) + 64)

total_n = nd + ns + na + nz + fish + nsed

uptake = (2.0 – 2.0*(na/9))*ns*na/(0.4 + ns)

zoomo = 0.1 *nz

---

order to select the parameter combination that gives the highest exergy (eco-exergy). Changes in parameters from 1985 to 1988 (summer) are summarized in Table 2.6. It may be concluded that the proposed procedure (Figure 2.4) can approximately simulate the observed change in ecosystem structure.

The maximum phytoplankton growth rate is reduced by 50% from 2.2/day to 1.1/day, approximately in accordance with the increase in size. It was observed that the average size was increased from a few 100 μm^3 to

TABLE 2.6

Parameter Combination Giving the Highest Exery

Year	Max. Growth Rate, 1/24 h	Settling Rate, m/24 h
1985	2.0	0.15
1988	1.2	0.45

500–1000 µm³, a factor of 2–3 (Jeppesen et al. 1989). This would correspond to a specific growth reduction by a factor $f = 2^{2/3}$–$3^{2/3}$ (Jørgensen 1994; Peters 1983). Thus,

$$\text{Growth rate in 1988} = \text{Growth rate in 1985}/f, \qquad (2.6)$$

where f is between 1.58 and 2.08. In the above table, the value of 2.0 is found with the use of the structurally dynamic modeling approach. Kristensen and Jensen (1987) observed that settling was 0.2 m/day (range 0.02–0.4) during 1985, but 0.6 m/day (range 0.1–1.0) in 1988. Using the structural dynamic modeling approach, the increase was found to be 0.15 to 0.45 m/day, a slightly lower set of values. However, the same (3) phytoplankton concentration as chlorophyll-*a* was simultaneously reduced from 600 to 200 µg/L, approximately in accord with observations.

In this case, it may be concluded that structurally dynamic modeling gave an acceptable result. Validation of the model, and the procedure in relation to structural changes, was positive. Of course, the approach is never better than the model applied, and the model presented here may be criticized for being too simple, and not accounting for changes in zooplankton.

For further elucidation of the importance of introducing parameter shifts, an attempt was made to run data for 1985 with parameter combinations for 1988, and vice versa. These results (Table 2.7) show that it is of great importance to apply the appropriate parameter set to given conditions. If those for 1985 are used for 1988, significantly less exergy is obtained, and the model behaves chaotically. The parameters for 1988 used under 1985 conditions give significantly less exergy.

Experience mentioned previously in this chapter shows that models can be applied to explain why biomanipulation may work under some circumstances, and not in others. Qualitatively, the results can be used to explain that hysteresis exists over an intermediate range of nutrient loadings, so that biomanipulation has worked properly over this range, but not above or below it.

Another hysteresis behavior obtained and explained by the use of SDMs for lakes have recently been published (Zhang et al. 2003a, 2003b). It focuses on the structural change between a dominance of submerged vegetation and phytoplankton in shallow lakes. The model results show that between about 100 and 250 µg P/L, both structures can exist—they show hysteresis in this range. This result is in accordance with observations from many shallow

TABLE 2.7

Exergy and Stability by Different Combinations of Parameters and Conditions

	Conditions	
Parameter Set From	1985	1988
1985	75.0 stable	39.8 (average) oscillations, chaos
1988	38.7 stable	61.4 (average) minor oscillations

lakes (e.g., see Scheffer et al. 2000). SDMs are further illustrated in Chapters 12 and 14.

Ecosystems are very different from physical systems, due mainly to their enormous adaptability. Thus, it is crucial to develop models that are able to account for this property, in order to derive reliable model results. The use of exergy as goal functions to cover the concept of fitness seems to offer a good possibility for developing a new generation of models, which is able to consider the adaptability of ecosystems and to describe shifts in species composition. The latter advantage is probably the most important, because a description of the dominant species in an ecosystem is often more essential than assessing the level of the focal state variables.

The structurally dynamic approach has also recently been used to calibrate eutrophication models. It is known that the different phytoplankton and zooplankton species are dominant in different periods of the year. Thus, a calibration based on one parameter set for the entire year will not capture the succession that did take place over the year. By using exergy optimization to capture the succession (i.e., the parameter giving the best survival for phytoplankton and zooplankton over the year), it has been possible to improve the calibration results (see Jørgensen et al. 2002; Zhang 2003a).

The structurally dynamic modeling approach generally has been most widely applied in eutrophication models. A recent software package named Pamolare (downloadable from www.unep.or.jp/ietc/pamolare), and launched by the ILEC and the UNEP-IETC, also contains an SDM, in addition to a conventional two-layer model. A test of this model has shown that it is calibrated and validated faster than the conventional model and gives better results, understood as a smaller standard deviation.

In other words, models are an appropriate tool in our efforts to understand the results of structural changes to ecosystems. In addition to the use of a goal function, it also is possible to base the structural changes on knowledge, for example, of what conditions under which specific classes of phytoplankton are dominant. This knowledge can be used to select the correct combination of parameters, as well illustrated by Reynolds (1996). That the combined application of expert knowledge and the use of exergy as a goal function will offer the best solution to the problem of making models work more in accordance with the properties of real ecosystems, cannot be ruled out. Such combinations would draw upon the widest possible knowledge at this stage.

2.8 Latest Development in Lake and Reservoir Modeling

A number of lake model case studies can be found on the Internet. They represent mostly new applications of an already developed and published lake model from the seventies, eighties, or nineties. A few completely new lake models, including eutrophication models, that otherwise could have been covered in Chapter 12, will be presented in this section, which is partially based on Jørgensen (2010).

Zhang et al. (2008) published an interesting 2D eutrophication model of Lake Erie, characterized by

1. A 2D coupling between hydrodynamics and the ecological processes
2. Detailed submodels of zooplankton
3. Inclusion of the impact of zebra and quagga mussels

The model has 18 state variables and was acceptably validated. The model study shows the importance of having a detailed zooplankton model and how to include the impacts of mussels. If items (2) and (3) are of importance for the eutrophication, it can be recommended to apply the equations from this model study.

Bartleson et al. (2005) have developed a model that was supported by data from microcosm data. The model focuses on the competition between epithytic algae and submerged plants in the presence of grazers. The model results show that the initial concentration of submerged plants is very important for the final state of eutrophication. The results fit with the general experience of recovering shallow lakes by planting submerged vegetation and with the use of SDMs to describe the competition between phytoplankton and macrophytes (Zhang et al. 2003b). The competition between submerged plants and algae is important for the improvement of the water quality in shallow lakes. The model is recommended for use if a shallow lake is going to be restored by planting submerged plants. The model could probably be improved further by making it structurally dynamic by use of exergy as goal function, as shown by Zhang et al. (2003a).

Roth et al. (2007) have developed a model that looks into the role of woody habitat in lake food web. Trees fall into lakes from riparian forest and become habitat for aquatic organisms. The entire food web may thereby be changed. The production of benthic invertebrates prey is promoted and the fallen trees are refuge for prey fishes, which are consumed by piscivorous fishes. The presented model includes external processes, for instance, input of allochthonous energy sources that may lead to large shifts in the lake fish communities. The study demonstrates how these processes can be included in a model if it is relevant for a lake case study.

Jørgensen (2007) has shown that different aquatic ecosystems have a different carrying capacity. Lakes have a carrying capacity of about 6 kg detritus/m² yr or 112 MJ/m² yr.

Thapanand et al. (2007) used Ecopath to develop a model for the management of multispecies fishery. The model is useful for optimization of multispecies fishery, as well as showing the potential importance of the preservations of littoral zones for the fishery.

Zhao et al. (2008) have examined a number of improvements for eutrophication models:

1. Use of multiple nutrients cycles (P, N, Si, C, and O)

2. Multiple functional phytoplankton (diatoms, green algae, and cyanobacteria)

3. Inclusion of two zooplankton groups: copepods and cladocerans

4. Use of the recent advances in stoichiometric nutrient recycling theory, which allows one to examine the effect of food quality

The model results indicate that all four points offer significant model improvements, provided that the parameterization particularly of the zooplankton processes is reliable and robust.

Mieleitner and Reichert (2008) have modeled the functional groups of phytoplankton in three lakes of different trophic state. They consider four functional groups with different parameters in the model: F1 consists of small flagellates, small diatoms, and small green algae. F2 encompasses large diatoms, whereas F3 are large green algae, blue green algae, large chrysophyceae, and dinoflagellates. The last group, F4, consists of *Planktothrix rubescens*. The model results demonstrate that functional group modeling is useful for our understanding of the lake ecosystems and the characteristics that are decisive for the response to different nutrient impacts. The predictability of the model, however, is poor at the functional group level, which is in accordance with the discussion of structurally dynamic modeling by Jørgensen (2002).

Mieleitner and Reichert (2006) have examined the transferability of biogeochemical lake models and have shown that it is possible to transfer their model with the functional phytoplankton groups from one lake study to another by changing only three parameters based on a careful calibration using a good data set. The authors conclude that the attempt to make models more general seems to be achievable to some degree for biogeochemical models.

Hense and Beckmann (2006) have modeled the life stages of cyanobacteria. They consider four life stages: vegetative cells, vegetative cells with heterocysts, akines, and recruiting cells including germinates. The models assume that the life cycle is governed by the internal energy and nitrogen quotas of the cells. The study indicates that prediction of cyanobacteria blooms has to be based on a detailed knowledge of all stages of the life cycle. It is for instance insufficient only to take temperature and P/N ratio into account. A relatively complex equation is applied for the temperature dependence of the main cell processes of cyanobacteria. The model study is an important step toward an improved understanding and a reliable prediction of cyanobacteria blooms. As prediction and occurrence of harmful cyanobacteria bloom is an important task of environmental management, it is recommendable to pursue the ideas presented in this study.

Reid and Crout (2008) have developed an interesting model of freshwater Antarctic lake ice. Freshwater lake ice is affected more by air temperature than by any other variable, and is therefore a useful indicator of climate changes. The model performs well and can obviously be used to follow the climate changes and compare different climate change scenarios.

Bryhn and Blenckner (2007) have focused on nitrogen fixation in Lake Erken, and concluded that nitrogen fixation is unlikely ever to be limited by the nitrogen concentration in the lake water. Elliott et al. (2007) have developed a model for Lake Erken based on a coupling of a hydrophysical model and an ecological model. The main conclusion of their model study is that a coupling of a model focusing on the hydrology with a model focusing on the ecological processes shows a successful validation and can generally be recommended.

Rivera et al. (2007) have published an energy system language model of Broa Reservoir. The model study demonstrates how the energy system language developed by H. T. Odum can be used advantageously to obtain a more detailed overview of the state variables and their interactions in lake modeling.

Bruce et al. (2006) have demonstrated the importance of a good description of zooplankton processes to achieve a reliable lake eutrophication model. Both the grazing rate and the nutrients excretion rates are important parameters that should be determined on basis of good primary data, literature, and estimates of the nutrients cycling.

Hu et al. (2006) have developed a 3D model for Lake Taihu in China that is prepared for the use of the structurally dynamic approach.

Malmaeus et al. (2006) have developed a mechanistic phosphorus model, which has been applied on three sites. The model results have shown that the lakes have different sensitivity to climate changes, which is explained by the different water residence time.

Vladusic et al. (2006) have used what is called Q^2 learning (Qualitatively faithful Quantitative prediction) to interpret lake eutrophication data, and they claim that their approach is comparable to competing methods in terms of numerical accuracy and give good insight into domain phenomena.

Zhang et al. (2003a, 2003b) have developed an SDM that is able to explain the hysteresis in vegetation shifts for shallow lakes between phytoplankton dominated lakes and lakes dominated by the submerged vegetation. Such shifts are obviously a particular challenge for the structurally dynamic approach. The hysteresis in shifts between a zooplankton dominated structure and a structure dominated by phytoplankton and planktivorous fish was previously explained by the use of SDMs (Jørgensen and De Barnardi 1998). The result of the model is discussed in Section 2.7.

Zhang et al. (2004) have also examined the validation results that can be obtained by eutrophication models that were tailored from case to case and general models that were offered as ready-to-go software on the one side, and between general biogeochemical eutrophication and SDMs on the other side. They concluded that the structurally dynamic approach offers slightly better validation mainly due to the possibilities to consider seasonal changes in the calibration. They could, on the other hand, also conclude that the tailored model offers better validation results than the ready-to-go software models, although the difference was minor if the two models were using the same databases.

Gurkan et al. (2006) have applied the SDM on the restoration of a lake (Lake Fure, Denmark) by biomanipulation and aeration of the sediment—an obvious case for an SDM, because of major shifts in phytoplankton. Zooplankton and planktivorous fish could be foreseen. This model is interesting as a eutrophication model, and the results of the model are discussed in Chapter 19.

Fragoso et al. (2008) have developed a model that, by coupling of hydrodynamics with the ecological processes, is able to describe the spatial heterogeneity in a large shallow subtropical lake in Brazil. The model represents the advantages that are achieved by a better integration of the hydrodynamic and ecological components in lake modeling. The model was able to identify zones with a higher potential for eutrophication, which of course is crucial in environmental management.

Generally, it is more common to see models that integrate hydrodynamics and ecological processes today than a decade or two ago. Moreover, there has been tendency to integrate a watershed model with a lake model to be able to coordinate environmental strategies for the lake with environmental management for the entire watershed. For instance, Zhang and Jørgensen (2005) have, for the UNEP lake modeling software Pamolare, included a watershed model that is able to work together with a simple lake model or a structurally dynamic lake model with a medium to high complexity. They found it easy to apply the results from the watershed model to assess the nutrient loadings (the most crucial forcing functions) in the lake model.

Generally, the structurally dynamic modeling approach has been used for the development of several lake models, particularly during the past 5–6 years. All the papers conclude that this model type offers the possibilities to consider the structural changes—adaptation and shifts in species composition. The model type can be recommended to improve calibration, validation, and prognoses, when it is known that structurally dynamic changes take place. One of the models published recently has been able to combine a 3D model with an SDM. In general, 3D lake models are also more widely applied today than a few years ago.

A number of changes from the more classic eutrophication models have recently been tested to account for a number of processes that, in some case studies, may be of importance. It is recommended to use these changes whenever tailored models are developed for case studies where the processes have relevance. The following processes have been tested in modeling context:

1. Use of multiple nutrients cycles (P, N, Si, C, and O).
2. Multiple functional phytoplankton (diatoms, green algae, and cyanobacteria), although in many cases it is probable that better results can be achieved by using structurally dynamic modeling (see, e.g., the discussion in the work of Jørgensen 2002).
3. Inclusion of more zooplankton groups, for instance, copepods and cladocerans and other more complex zooplankton models.
4. Use of the recent advances in stoichiometric nutrient recycling theory, which allows researchers to examine the effect of food quality.
5. Competition between phytoplankton and macrophytes. Structurally dynamic modeling, in this case, also offers a good alternative since it is able to consider the hysteresis response of competition to changed conditions.
6. Woody habitats in lakes where fallen trees are important.
7. Cyanobacteria is a serious problem in many lakes, and it is therefore not surprising that there has been an increased interest in modeling this problem. It is recommended to use the detailed model of this problem proposed by Hense and Beckmann (2006). It would be beneficial for modeling this crucial problem to run wider tests of their model as it seems to offer a good modeling solution.
8. Nitrogen fixation has been modeled differently from the approach applied in classic eutrophication models, where it was considered a first-order reaction governed by the lack of nitrogen by nitrogen fixing species.
9. Mussels as important filter feeders in the lakes.
10. Use of 3D models and coupling between hydrodynamic models and ecological models or between watershed models and lake models.

Multi species fishery models are furthermore increasingly applied in lake models. The experience with these models is generally good and points toward wider application in the future, which is promising for the management of fishery.

The overall conclusion from a review of the papers on lake models is that the integration of hydrodynamic models and ecological models has continued. The latest published SDMs confirm that this model type offers a good solution to cases where adaptation and shifts in species composition are important. Moreover, there has been an increasing interest for modeling the impact of climate changes on lakes.

References

Alcamo, J., R. Shaw, and L. Hordijk, L., eds. 1990. *The Rains Model of Acidification.* Dordrecht: Kluwer Academic Publishers.

Ambrose, Jr., R. B., P. E. T. A. Wool, J. P. Connolly, and R. W. Schanz. 1988. *WASP4, A Hydrodynamic and Water Quality Model—Model Theory, User's Manual, and Programmer's Guide.* Environmental Research Laboratory, Office of Research and Development, U.S. Environmental Protection Agency, Athens, GA.

Aoyama, I., Y. Inoue, and Y. Inoue. 1978. Simulation analysis of the concentration process of trace heavy metals by aquatic organisms from the viewpoint of nutrition ecology. *Water Research* 12: 837–842.

Bartell, S. M., R. H. Gardner, and R. V. O'Neill. 1984. The fates of aromatics model. *Ecological Modelling* 22: 109–123.

Bartleson, R. D., W. M. Kemp, and J. C. Stevenson. 2005. Use of a simulation model to examine effects of nutrient loading and grazing on *Potamogeton perfoliatus* L. communities in microcosmoss. *Ecological Modelling* 185: 483–512.

Bedford, K. W., R. M. Sykes, and C. Libicki. 1983. Dynamic advective water quality model for rivers. *Journal of Environmental Engineering Division ASCE* 109: 535–554.

Benndorf, J., and F. Recknagel. 1982. Problems of application of the ecological model SALMO to lakes and reservoirs having various trophic status. *Ecological Modelling* 17: 129–145.

Benoist, A. P., A. G. Brinkman, P. M. J. S. van Diepenbeek, and J. M. J. Waals. 1998. BEKWAAM, a model fit for reservoir design and management. *Water Science & Technology* 37 (2): 269–276.

Bouron, T. 1991. COMMAS: A Communication and Environment Model for Multi-Agent Systems. In: *Modelling and Simulation 1991,* ed. E. Mosekilde, 220–225. European Simulation Multiconference, June 17–19, 1991. Copenhagen, Denmark: The Society for Computer Simulations International.

Breck, J. E., D. L. DeAngelis, W. Van Winke, and S. W. Christiansen. 1988. Potential importance of spatial and temporal heterogeneity in pH, Al and Ca in allowing survival of a fish population: A model demonstration. *Ecological Modelling* 41(1–2), 1–16.

Brown, D. J. A., and K. Salder. 1981. The chemistry and fishery status of acid lakes in Norway and their relationship to European sulfur emission. *Journal of Applied Ecology* 18: 434–441.

Bruce, L. C., D. Hamilton, J. Imberger, G. Gal, M. Gophen, T. Zohary, and K. D. Hambright. 2006. A numerical simulation of the role of zooplankton in C, N, and P cycling in Lake Kinneret, Israel. *Ecological Modelling* 193: 412–436.

Bryhn, A. C., and T. Blencker. 2007. Can nitrogen be deficient for nitrogen fixation in lakes? *Ecological Modelling* 202: 362–372.

Chang, N.-B., C. G. Wen, and Y. L. Chen. 1996. A grey fuzzy multiobjective programming approach for the optimal planning of a reservoir watershed. Part B: Application. *Water Research* 30 (10): 2335–2340.

Chen, C. W., J. D. Dean, S. A. Gherini, and R. A. Goldstein. 1982. Acid rain model: Hydrologic module. *Journal of Environmental Engineering Division ASCE* 108: 455–472.

DePinto, J. V., and P. W. Rodgers. 1994. Development of GEO-WAMS: A modeling support system for integrating GIS with watershed analysis models. *Lake and Reservoir Management* 9 (2): 68. Abstract of presentation.

Diogo, P. A., and A. C. Rodrigues. 1997. Two dimensional reservoir water quality modeling using CE-QUAL-W2. In: *Reservoir Management and Water Supply—An Integrated System*, Vol. 2, ed. P. Dolej and N. Kalousková, 41–44. Czech Republic: W&ET Team.

Elliott, J. A., I. Perrson, S. J. Thackeray, and T. Blencker. 2007. Phytoplankton modelling of Lake Erken, Sweden by linking the models PORBE and PROTECH. *Ecological Modelling* 202: 421–426.

Ellis, J. H. 1987. Stochastic water quality optimization using imbedded chance constraints. *Water Resources Research* 23: 2227–2238.

Fomsgaard, I. 1997. Modelling the mineralisation kinetics for low concentrations of pesticides in surface and subsurface soil. *Ecological Modelling* 102: 175–208.

Fomsgaard, I., and K. Kristensen. 1999. Influence of microbial activity, organic carbon content, soil texture and soil depth on mineralisation rates of low concentrations of 14-C mecoprop—development of a predictive model. *Ecological Modelling* 122: 45–68.

Fontaine, T. D. 1981. A self-designing model for testing hypotheses of ecosystem development. In: *Progress in Ecological Modelling*, ed. D. Dubois. Liege: Cebedoc.

Fragoso, C. R. et al. 2008. Modelling spatial heterogeneity of phytoplankton in Lake Mangueira, a large shallow subtropical lake in South Brazil. *Ecological Modelling* 219: 125–137.

Gillet, J. W. 1974. A conceptual model for the movement of pesticides through the environment. National Environmental Research Center U.S. Environmental Protection Agency, Corvallis or Report EPA 660/3-74-024.

Gnauck, A., E. Matthaus, M. Straskraba, and I. Affa. 1990. The use of SONCHES for aquatic ecosystem modelling. *Systems Analysis Modelling Simulation* 7: 439–458.

Gromiec, M. J., and E. F. Gloyna. 1973. Radioactivity transport in water. *Final Report No. 22 to US Atomic Energy Commission*, Contract AT (11-1)-490.

Gurkan, Z., J. Zhang, and S. E. Jørgensen. 2006. Development of a structurally dynamic model for forecasting the effects of restoration of lakes. *Ecological Modelling* 197: 89–103.

Halfon, E. 1984. Error analysis and simulation of Mirex behaviour in Lake Ontario. *Ecological Modelling* 22: 213–253.

Hamilton, D. 1999. Numerical modelling and reservoir management. Application of the DYRESM model. In: *Theoretical Reservoir Ecology and its Applications*, ed. J. G. Tundisi and M. Strakraba, 153–173. Leiden: International Institute of Ecology, Brazilian Academy of Sciences and Backhuys Publishers.

Hamilton, D. P., and B. T. de Stasio. 1998. Modelling phytoplankton-zooplankton interactions in Sparkling Lake. *Verhandlungen Internationale Vereinigung für Theoretische und Angewandte Limnologie* 26: 487–490.

Hamilton, D. P., and S. G. Schladow. 1997. Prediction of water quality in lakes and reservoirs: Part I. Model description. *Ecological Modelling* 96: 91–110.

Harris, J. R. W., A. J. Bale, B. L. Bayne, R. C. F. Mantoura, A. W. Morris, L. A. Nelson, P. J. Radford, R. J. Uncles, S. A. Weston, and J. Widdows. 1984. A preliminary model of the dispersal and biological effect of toxins in the Tamar estuary, England. *Ecological Modelling* 22: 253–285.

Henriksen, A. 1980. Acidification of freshwaters—a large scale titration. In: *Ecological Impact of Acid Precipitation*, ed. D. Drablřs and A. Tollan, 68–74. SNFS project.

Henriksen, A., and H. M. Seip. 1980. Strong and weak acids in surface waters of Southern Norway and Southwestern Scotland. *Water Research* 14: 809–813.

Hense, I., and A. Beckmann. 2006. Towards a model of cyanobacteria life cycle—effects of growing and resting stages on bloom formation of N_2-fixing species. *Ecological Modelling* 195: 205–218.

Hilborn, R. 1987. Living with uncertainty in resource management. *North American Journal of Fisheries Management* 7: 1–5.

Hocking, G. C., and J. Patterson. 1988. Two dimensional modelling of reservoir outflows. *Verhandlungen Internationale Vereinigung für Theoretische und Angewandte Limnologie* 23: 2226–2231.

Hocking, G. C., and J. Patterson. 1991. A quasi two dimensional reservoir simulation model. *Journal of Environmental Engineering Division ASCE* 117: 595–613.

Holling, C. S. 1978. *Adaptive Environmental Assessment and Management*. New York, NY: John Wiley and Sons.

Hosper, S. H. 1989. Biomanipulation, new perspectives for restoring shallow lakes in the Netherlands. *Hydrobiological Bulletin* 23: 5–11.

Hu, W., S. E. Jørgensen, and F. Zhang. 2006. A vertical-compressed three-dimensional ecological model in Lake Taihu, China. *Ecological Modelling* 190: 367–398.

Janse, J. H. 1997. A model of nutrient dynamics in shallow lakes in relation to multiple stable states. *Hydrobiologia* 342: 1–8.

Janse, J. H., E. Van Donk, and R. D. Gulati. 1995. Modelling nutrient cycles in relation to food web structure in a biomanipulated shallow lake. *Netherlands Journal of Aquatic Ecology* 29 (1): 67–79.

Janse, J. H., E. Van Donk, and T. Aldenberg. 1998. A model study on the stability of the macrophyte-dominated state as affected by biological factors. *Water Research* 32 (9): 2696–2706.

Janse, J. H., J. Van der Does, and J. C. Van der Vlugt. 1993. PCLAKE: Modelling eutrophication and its control measures in Reeuwijk lakes. In: *Proc. 5th Int. Conf. on the Conservation and Management of Lakes*, Stresa (Italy), ed. G. Giussani and C. Callieri, 117–120.

Jeppesen, E., E. Mortensen, O. Sortkjær, P. Kristensen, J. Bidstrup, M. Timmermann, J. P. Jensen, A.-M. Hansen, M. Søndergírd, J. P. Muller, H. Jerl Jensen, B. Riemann, C. Lindegírd Petersen, S. Bosselmann, K. Christoffersen, E. Dall, and J. M. Andersen. 1989. Restaurering af sřer ved indgreb i fiskebestanden. Status for igangvćrende undersřgelser. Del 2: Unsdersřgelser i Frederiksborg slotsø, Veng sø og Søbygård sø. Danmarks Miljøundersøgelser, 114 pp.

Jeppesen, E. J. et al. 1990. Fish manipulation as a lake restoration tool in shallow, eutrophic temperate lakes. Cross-analysis of three Danish case studies. *Hydrobiologia* 200/201: 205–218.

Jørgensen, S. E. 1986. Structural dynamic model. *Ecological Modelling* 31: 1–9.

Jørgensen, S. E. 1992a. Parameters, ecological constraints and exergy. *Ecological Modelling* 60: 160–170.

Jørgensen, S. E. 1992b. Development of models able to account for changes in species composition. *Ecological Modelling* 62: 195–208.

Jørgensen, S. E. 1994. *Fundamentals of Ecological Modelling*, 2nd ed. *Developments in Environmental Modelling*, 19. Amsterdam: Elsevier, 628 pp (3rd edition, G. Bendoricchio coauthor, 2001).

Jørgensen, S. E. 2002. *Integration of Ecosystem Theories: A Pattern*, 3rd revised ed. First edition 1992, second edition, 1997. Dordrecht: Kluwer Academic Publishers, 428 pp.

Jørgensen S. E. 2007. Description of aquatic ecosystem's development by eco-exergy and exergy destruction. *Ecological Modelling* 204: 22–28.

Jørgensen, S. E. 2010. A review of the recent development in lake modeling. *Ecological Modelling* 221: 689–692.

Jørgensen, S. E., and B. Fath. 2011. *Fundamentals of Ecological Modeling*. Amsterdam: Elsevier, 380 pp. (4th edition).

Jørgensen, S. E., and J. F. Mejer. 1977. Ecological buffer capacity. *Ecological Modelling* 3: 39–61.

Jørgensen, S. E., and H. F. Mejer, 1979. A holistic approach to ecological modelling. *Ecological Modelling* 7: 169–189.

Jørgensen, S. E., H. F. Mejer, and M. Friis. 1978. Examination of a lake model. *Ecological Modelling* 4: 253–278.

Jørgensen, S. E., and R. A. Vollenweider. (Eds.) 1988. *Principles of Lake Management. Guidelines of Lake Management, Vol. 1*. International Lake Environment Committee, Kusatsu, Japan.

Jørgensen, S. E., J. C. Marques, and P. M. Anastacio. 1997. Modelling the fate of surfactants and pesticides in a rice filed. *Ecological Modelling* 104: 205–214.

Jørgensen, S. E., and R. de Bernardi. 1998. The use of structural dynamic models to explain successes and failures of biomanipulation. *Hydrobiologia* 359: 1–12.

Jørgensen, S. E., H. Lützhøft, and B. Sørensen. 1998. Development of a model for environmental risk assessment of growth promoters. *Ecological Modelling* 107: 63–72.

Jørgensen, S. E., T. Chon, and F. Recknagel. (Eds.) 2009. *Handbook of Ecological Modelling and Informatics*. 432 pp.

Kalceva, R., J. Outrata, Z. Schindler, and M. Straskraba. 1982. An optimization model for the economic control of reservoir eutrophication. *Ecological Modelling* 17: 121–128.

Kauppi, P., M. Posch, E. Matzner, L. Kauppi, and J. Kdmdri. 1984. A model for predicting the acidification of forest soils: Application to deposition in Europe. *IIASA Research Report*.

Koponen, J., M. Virtanen, and J. Itkonen. 1998. Detailed validation of a 3D reservoir model with field measurements, laboratory tests and analytical solutions. In: *Proceedings of the 3rd International Conference on Reservoir Limnology and Water Quality*, ed. V. Strakrabová and J. Vrba, »eské Buddjovice, Czech Republic, August 11–15, 1997. *Internationale Revue Gesamten Hydrobiologie* 83: 697–704.

Kristensen, P., and P. Jensen. 1987. Sedimentation og resuspension i Søbygaard Lske. Special Rapport. Miljøstyrelsens Ferskvandslaboratorium & Botanisk Institut, Univ. Aarhus, 150 pp.

Kutas, T., and S. Herodek. 1986. A complex model for simulating the Lake Balaton ecosystem. In: *Modeling and Managing Shallow Lake Eutrophication*, ed. L. Somlyody and G. van Straten, 309–322. New York, NY: Springer-Verlag.

Lam, D. C. L., and T. J. Simons. 1976. Computer model for toxicant spills in Lake Ontario. In: *Metals Transfer and Ecological Mass Balances. Environmental Biochemistry*, Vol. 2, ed. J. O. Nriago, 537–549. Ann Arbor, MI: Ann Arbor Science.

Lavrik, V. I., A. N. Bilyk, and N. A. Nikiforovich. 1990. Multichamber simulation modelling of water quality in reservoir. *Systems Analysis Modelling Simulation* 7 (8): 625–635.

Legovic, T. 1997. Toxicity may affect predictability of eutrophication models in coastal sea. *Ecological Modelling* 99: 1–6.

Lindenschmidt, K.-E., and I. Chorus. 1997. The effect of aeration on stratification and phytoplankton populations in Lake Tegel, Berlin. *Archiv fuer Hydrobiologie* 139 (3): 317–346.

Malmaeus, J. M., T. Belnckner, H. Markensten, and I. Perrson. 2006. Lake phosphorus dynamics and climate warming: A mechanistic model approach. *Ecological Modelling* 190: 1–14.

Matsumura, T., and S. Yoshiuki. 1981. An optimization problem related to the regulation of influent nutrient in aquatic ecosystems. *International Journal of Systems Science* 12: 565–585.

Mejer, H. F. 1983. A menu driven lake model. *ISEM J.* 5: 45–50.

Mieleitner, J., and P. Reichert. 2006. Analysis of the transferability of a biogeochemical lake model to lakes of different trophic state. *Ecological Modelling* 194: 49–61.

Mieleitner, J., and P. Reichert. 2008. Modelling the functional groups of phytoplankton in three lakes of different trophic state. *Ecological Modelling* 211: 279–291.

Miller, D. R. 1979. Models for total transport. In: *Principles of Ecotoxicology Scope*, Vol. 12, ed. G. C. Butler, 71–90. New York, NY: Wiley.

Molen, D. T., F. J. van der Los, L. van Ballegooijen, and M. P. van der Vat. 1994. Mathematical modelling as a tool for management in eutrophication control of shallow lakes. *Hydrobiologia* 275/276: 479–492.

Monte, L. 1998. Predicting the migration of dissolved toxic substances from catchments by a collective model. *Ecological Modelling* 110: 269–280.

Muniz, I. P., and H. M. Seip. 1982. Possible effects of reduced Norwegian sulphur emissions on the fish populations in lakes in Southern Norway. *SI Report 8103*, 13-2.

Nielsen, S. N. 1992a. Application of maximum energy in structural dynamic models. PhD thesis. National Environmental Research Institute, Denmark.

Nielsen, S. N. 1992b. Strategies for structural-dynamic modelling. *Ecological Modelling* 63: 91–100.

Odum, H. T. 1983. *System Ecology*. New York, NY: Wiley Interscience, 510 pp.

Orlob, G. T. 1983. *Mathematical Modeling of Water Quality: Streams, Lakes, and Reservoirs*. Chichester: John Wiley & Sons.

Orlob, G. T. 1984. Mathematical models of lakes and reservoirs. In: *Lakes and Reservoirs. Ecosystems of the World 23*, ed. F. Taub, 43–62. Amsterdam: Elsevier.

Paterson, S., and D. Mackay. 1989. A model illustrating the environmental fate, exposure and human uptake of persistent organic-chemicals. *Ecological Modelling* 47 (1–2): 85–114.

Peters, R. H. 1983. *The Ecological Implications of Body Size*. Cambridge: Cambridge Univ. Press.

Recknagel, F., and J. Benndorf. 1995. SALMOSED. Internal report.

Recknagel, F., E. Beuschold, and U. Petersohn. 1991. DELAQUA—a prototype expert system for operational control and management of lake water quality. *Water Science & Technology* 24: 283–290.

Reid, T., and N. Crout. 2008. A thermodynamic model of freshwater Antarctic lake ice. *Ecological Modelling* 210: 231–241.

Reynolds, C. S. 1987. The response of phytoplankton communities to changing lake environments. *Schweizerische Zeitschrift für Hydrologie* 49 (2): 220–236.

Reynolds, C. S. 1992. Dynamics, selection and composition of phytoplankton in relation to vertical structure in lakes. *Archiv für Hydrobiologie–Beiheft Ergebnisse der Limnologie* 35: 13–31.

Reynolds, C. S. 1996. The plant life of the pelagic. *Proceedings of the International Association for Theoretical and Applied Limnology* 26: 97–113.

Reynolds, C. S. 1998. What factors influence the species composition of phytoplankton in lakes of different trophic status? *Hydrobiologia* 369/370: 11–26.

Reynolds, C. S. 1999a. Modelling phytoplankton dynamics and its application to lake management. In: *The Ecological Bases for Lake and Reservoir Management*, ed. D. M. Harper, B. Brierley, A. J. D. Ferguson, and G. Phillips, 123–131. The Netherlands: Kluwer Academic Publishers.

Reynolds, C. S. 1999b. Phytoplankton assemblages in reservoirs. In: *Theoretical Reservoir Ecology and its Applications*, ed. J. G. Tundisi and M. Strakraba, 439–456. Leiden: Brazilian Academy of Sciences, International Institute of Ecology, Backuys Publishers.

Riley, M. J., and H. G. Stefan. 1988. Development of the Minnesota lake water quality management model 'MINLAKE.' *Lake and Reservoir Management* 4 (2): 73–83.

Rivera, E. C., J. F. de Queiroz, J. M. Ferraz, and E. Ortega. 2006. System models to evaluate eutrophication in the Broa Reservoir, Sao Carlos, Brazil. *Ecological Modelling* 202: 518–526.

Romero, J. R., and J. Imberger. 1999. Seasonal horizontal gradients of dissolved oxygen in a temperate austral reservoir. In: *Theoretical Reservoir Ecology and its Applications*, ed. J. G. Tundisi and M. Strakraba, 211–226. Leiden: International Institute of Ecology, Brazilian Academy of Sciences and Backhuys Publishers.

Roth, B. M. et al. 2007. Linking terrestrial and aquatic ecosystems: The role of woody habitat in lake food webs. *Ecological Modelling* 203: 439–452.

Sale, M. J., E. D. Brill Jr., and E. E. Herricks. 1982. An approach to optimizing reservoir operation for downstream aquatic resources. *Water Resources Research* 18: 705–712.

Salençon, M. J. 1997. Study of the thermal dynamics of two dammed lakes (Pareloup and Rochebut, France), using the EOLE model. *Ecological Modelling* 104: 15–38.

Salençon, M. J., and J. M. Thébault. 1994a. Démarche de modélisation d'un écosystème lacustre: Application au Lac de Pareloup. *Hydroécologie Appliquée* 6: 315–327.

Salençon, M. J., and J. M. Thébault. 1994b. Modélisation de l'ecosysteme du lac de Pareloup avec les modeles ASTER et MELODIA. *Hydroécologie Appliquée* 6: 369–426.

Salençon, M. J., and J. M. Thébault. 1995. Simulation model of a mesotrophic reservoir (Lac de Pareloup, France): MELODIA, an ecosystem reservoir management model. *Ecological Modelling* 84: 163–187.

Scavia, D., and Chapra, S. C. 1977. Comparison of an ecological model of Lake Ontario and phosphorus loading models. *Journal Fisheries Research Board of Canada* 34: 286–290.

Scheffer, M., S. Carpenter, J. A. Foley, C. Folke, and B. Walker. 2000. Ecology of shallow lakes. *Nature* 413: 591–596.

Schindler, Z., and M. Straskraba. 1982. Optimalní Żizení eutrofizace údolních nadrůí. *VodohospodáŽsk_ Ěasopis SAV* 30: 536–548.

Schladow, S. G., and D. P. Hamilton. 1995. Effect of major flow diversion on sediment nutrient release in a stratified reservoir. *Marine and Freshwater Research* 46: 189–195.

Schladow, S. G., and D. P. Hamilton. 1997. Prediction of water quality in lakes and reservoirs. Part II—Model calibration, sensitivity analysis and application. *Ecological Modelling* 96: 111–123.

Schladow, S. G., D. P. Hamilton, and M. C. Burling. 1994. Modeling ecological impacts of destratification. In: *Hydraulic Engineering '94*, Vol. 2, ed. G. V. Cotroneo and R. R. Rumer, 1321–1325. New York, NY: American Society of Civil Engineers.

Schlenkhoff, A. U. 1997. Water quality management tool for the reservoir Wupper. In: *Reservoir Management and Water Supply—An Integrated System*, Vol. 2, ed. P. Dolej and N. Kalousková, 115–122. Czech Republic: W&ET Team.

Schrödinger, E. 1944. *What Is Life?* Cambridge: Cambridge University Press, 212 pp.

Schwarzenbach, R. P., and D. M. Imboden. 1984. Modeling concepts for hydrophobic pollutants in lakes. *Ecological Modelling* 22: 145–170.

Stefan, H. G., and J. J. Cardoni. 1982. *RESQUAL II: A Dynamic Water Quality Simulation Program for a Stratified Shallow Lake or Reservoir: Application to Lake Chicot, Arkansas.* Minneapolis, MN: Univ. of Minnesota.

Stefan, H. G., and X. Fang. 1994a. Dissolved oxygen model for regional lake analysis. *Ecological Modelling* 71: 37–68.

Stefan, H. G., and X. Fang. 1994b. Model simulations of dissolved oxygen characteristics in Minnesota lakes: Past and future. *Environmental Management* 18: 73–92.

Straskraba, M. 1976. Development of an analytical phytoplankton model with parameters empirically related to dominant controlling variables. *Abhandlungen der Akademie der Wissenshaften der DDR Jg* 1974: 33–65.

Straskraba, M. 1979. Natural control mechanisms in models of aquatic ecosystems. *Ecological Modelling* 6: 305–322.

Straskraba, M. 1998. Coupling of hydrobiology and hydrodynamics: Lakes and reservoirs. In: *Physical Limnology. Coastal and Estuarine Studies*, Vol. 54, ed. J. Imberger, 623–644. Washington, D.C.: American Geophysical Union.

Straskraba, M., and A. H. Gnauck. 1985. *Freshwater Ecosystems: Modelling and Simulation. Developments in Environmental Modelling 8.* Amsterdam: Elsevier.

Thapanand, T. et al. 2007. Toward possibly fishy management strategies in a newly impounded man-made lake in Thailand. *Ecological Modelling* 204: 143–155.

Thomann, R. V. 1984. Physico-chemical and ecological modeling the fate toxic substances in natural water system. *Ecological Modelling* 22: 145–170.

Thomann, R. V., D. Szumski, D. M. DiToro, and D. J. O'Connor. 1974. A food chain model of cadmium in western Lake Erie. *Water Research* 8: 841–851.

Thornton, J. A., W. Rast, M. M. Holland, G. Jolankai, and S. O. Ryding. 1999. *Assessment and Control of Nonpoint Source Pollution of Aquatic Ecosystems.* Paris: UNESCO and the Parhenon Publishing Group.

Tufford, D. L., H. N. McKellar Jr., J. R. V. Flora, and M. E. Meadows. 1998. A reservoir model for use in regional water resources management. *Journal of Lake Reserve Management* 15 (3): 220–230.

Van Donk, E., R. D. Gulati, and M. P. Grimm. 1989. Food web manipulation in Lake Zwemlust: Positive and negative effects during the first two years. *Hydrobiological Bulletin* 23: 19–35.

Varis, O. 1994. Water quality models: Tools for the analysis of data, knowledge, and decisions. *Water Science & Technology* 30 (2): 13–19.

Virtanen, M., J. Koponen, K. Dahlbo, and J. Sarkkula. 1986. Three-dimensional water quality transport model compared with field observations. *Ecological Modelling* 31: 185–199.

Vladusic, D., B. Kompare, and I. Bratko. 2006. Modelling Lake Glumsø with Q2 learning. *Ecological Modelling* 191: 33–46.

Willemsen, J. 1980. Fishery aspects of eutrophication. *Hydrobiological Bulletin* 14: 12–21.

Wlosinski, J. H., and C. D. Collins. 1985. Evaluation of a water quality model (CE-QUAL-RE) using data from a small Wisconsin reservoir. *Ecological Modelling* 29: 303.

Zhang, H., D. A. Culver, and L. Boegman. 2008. A two dimensional ecological model of Lake Erie: Application to estimate dreissenid impacts on large lake plankton population. *Ecological Modelling* 214: 219–240.

Zhang, J., S. E. Jørgensen, C. O. Tan, and M. Beklioglu. 2003a. A structurally dynamic modelling—Lake Mogan, Turkey as a case study. *Ecological Modelling* 164: 103–120.

Zhang, J., S. E. Jørgensen, C. O. Tan, and M. Beklioglu. 2003b. Hysteresis in vegetation shift—Lake Mogan prognoses. *Ecological Modelling* 164: 227–238.

Zhang, J., S. E. Jørgensen, and H. Mahler. 2004. Examination of structurally dynamic eutrophication models. *Ecological Modelling* 173: 213–333.

Zhang, J., and S. E. Jørgensen. 2005. Modelling of point and non-point nutrient loadings from a watershed. *Environmental Modelling Software* 20: 561–574.

Zhao, J., M. Ramin, V. Cheng, and G. B. Arhonditsis. 2008. Plankton community patterns across a trophic gradient: The role of zooplankton functional groups. *Ecological Modelling* 213: 417–436.

3

Estuary Ecosystem Models

Irene Martins and Joao C. Marques

CONTENTS

3.1 Introduction 63
3.2 An Estuarine-System Coastal Model: The Case Study of the
Xiangshan Gang (East China Sea)...67
    3.2.1 Problematic 67
    3.2.2 Aim 67
    3.2.3 Study System 68
    3.2.4 Conceptual Diagram, Model Structure, and Main
Equations..68
    3.2.5 Results 72
3.3 Brief Review of Some Recent Estuarine-Ecosystem Models 77
3.4 Models of Primary Producers in Estuaries and Other Coastal
Ecosystems..77
    3.4.1 Introduction...77
    3.4.2 Phytoplankton 84
    3.4.3 Macroalgae..95
    3.4.4 Microphytobenthos 108
    3.4.5 Seagrasses 113
3.5 Conclusions...122
References...124

3.1 Introduction

Estuaries worldwide are in severe decline, mainly as a result of pollution and the indirect impacts of climate change. This gives rise to an increasing awareness of the profound impact of humans on the functioning of estuarine ecosystems, and consequently to the need for approaches capable of sustaining those systems and where necessary, restoring them (Hughes et al. 2005). General environmental concern gave rise, approximately 2 decades ago, to the emergence of the idea of *sustainable development* (Pulselli et al. 2008), but researchers from different disciplines still attempt to understand and define more precisely the meaning of the term. Despite this, the most widely adopted definition has been "development that satisfies present

needs without compromising the possibility of future generations satisfy-
ing theirs" (Brundtland 1987). This is a rather vague nonoperational definition, which implies that the concept still requires a suitable quantification in
socioeconomic, cultural, and scientific terms (Böhringer and Jochem 2007;
Singh et al. 2009), taking into account (1) time, (2) relationships, and (3) biophysical limits (Pulselli et al. 2008).

Time is important as human society often does not evolve in accordance
with the environment's capacity to produce the resources required for
our development. Different living and nonliving natural resources are
needed by society, constituting what is called natural capital. Despite
the fact that sustainable development has become a key challenge for the
twenty-first century, the way human society interacts with that natural
capital is still controversial. In fact, there are two clearly opposite positions regarding the practical meaning of sustainability: weak and strong
sustainability. Weak sustainability implies that well-being must be maintained over intergenerational time scales, assuming that natural capital
and man-made capital are substitutes within specific production processes (Brand 2009). As a consequence, weak sustainability accepts that
the natural capital can be depleted, unless its requirement over time is
declining (Brand 2009). Conversely, strong sustainability states that natural capital and man-made capital have to be viewed as complementary.
As a consequence, human society must keep each type of capital intact
over time, and the whole stock of natural capital has to be preserved for
present and future generations in the long run (Brand 2009). In any case,
the recognition that humans, with their cultural diversity, are an integral
component of ecosystems, and the foreseeable threats represented by a
serious worldwide environmental degradation have put ecological sustainability in international agendas.

Moreover, independently from the conceptual approach adopted, in cases
in which uncertainties and change are key questions of environmental and
social organization, critical factors for sustainability are resilience, the capacity to cope and adapt, and the conservation of sources of innovation and
renewal (Lebel et al. 2006). The sustainable management of natural systems,
in which estuaries are included, may then be described as achieving a balance between delivering the economic goods and services provided by the
environment, which are required for societal health and functioning, while
maintaining and protecting the ecological goods and services required for
natural health and functioning.

Relationships therefore become crucial as the care of environment and
natural resources might be not compatible with the present economical
paradigm. In fact, economic instruments often appear to lack the criterion
of efficient allocation of resources, since they tend to consider only things
directly linked to the market (Pulselli et al. 2008). Such relationships imply
interdependencies but it is necessary to determine at what scale (regional,
national, etc.) different aspects are interdependent.

Finally, biophysical limits also need to be considered as each local human population can hardly meet its needs for materials, energy, land, waste sinks, and information from its own local resources. This is reflected in the concept of critical natural capital, which emerged between the "weak sustainability" and the "strong sustainability" positions, consisting of the part of the natural capital that performs important and irreplaceable environmental functions, that is, those ecosystem services that cannot be replaced by other types of capital (De Groot et al. 2003). In fact, it is widely accepted that the maintenance of such critical natural capital is essential to environmental sustainability and sustainable development (Ekins et al. 2003).

Sustainable management of estuaries can only be achieved if a multidisciplinary approach is undertaken; management actions within that approach are required to be environmentally and ecologically sustainable, economically viable, technologically feasible, socially desirable or at least socially tolerable, administratively achievable, legally permissible, and politically expedient (e.g., Elliott et al. 2006; Bunce et al. 2008; Mee et al. 2008; Ojeda-Martínez et al. 2009).

In estuaries, in particular, there is the need for environmental restoration involving dealing with problems such as: (1) losses of habitats and species diversity, as well as a decrease in habitats size and heterogeneity; (2) decrease in population size and changes in dynamics and distribution of many species; and (3) decrease in economically relevant services and goods naturally provided by the systems (e.g., Elliott et al. 2007). In this context, the search for estuaries ecologically sustainable management represents a great challenge, namely because if some ecological concepts are well understood, such as the nature of estuarine ecosystems structure and functioning, others such as resilience, carrying capacity, and estuarine ecosystem goods and services are in general still not well quantified (Marques et al. 2009).

The resilience concept, for instance, has suffered considerable changes in the past three decades (Walker et al. 2004). Recently, Elliott et al. (2007) attempted to remove the confusion regarding the term by taking the view that it referred the inherent ability of a system to return to a previous, or similar, state following disturbance. However, at least two other meanings for it can be distinguished. The first refers to dynamics close to equilibrium and is defined as the time required for a system to return to an equilibrium point following a disturbance event, or system's recovery. It is termed engineering resilience (Holling 1996; Folke 2006) and is largely equivalent to the stability property elasticity (Grimm and Wissel 1997), which can be seen as resistance to change (Levin and Lubchenco 2008). A further meaning refers to dynamics far from any equilibrium steady state and is defined as the capacity to absorb stress and yet still maintain "function"—this has been termed ecological resilience (Gunderson and Holling 2002; Folke 2006), that is the capacity to maintain functioning despite multiple stressors that affect a developing system (Levin and Lubchenco 2008). This is more related to renewal, regeneration, or reorganization following disturbance than to the system's recovery (Folke 2006). In this case, disturbance events and spatial heterogeneity cause each system's behavior to be unique, and the complexity of the system combined with unanticipated compounded effects can make recovery trajectories difficult or impossible to predict. A recovered system may look similar but it is not the same system, because like any living system it is continuously developing (Folke 2006). Elliott et al. (2007) termed the differences in degradation and recovery trajectories as hysteresis in the system. Ecological resilience has to be estimated via resilience surrogates (Carpenter et al. 2005), based on a comprehensive resilience analysis, including the identification of specific disturbance regimes and societal choices of the desired ecosystem services (Brand 2009).

Estuarine ecosystems and their change due to human stressors require to be analyzed according to their carrying capacity and its loss following stress. The concept of carrying capacity was originally an ecological construct, defined as the number of individuals of a population an environment can support without significant negative impacts to the given population and its environment (Elliott et al. 2007). This definition does not fully capture the multilayered processes of human–environment relationships, which have a fluid and nonequilibrium nature, and it may disregard the role of external forces in influencing environmental change (Moore et al. 2009). Because of this, several authors (MacLeod and Cooper 2005; Elliott et al. 2007) emphasize that carrying capacity also should relate to social and economic aspects of ecosystems, that is, what human activities and anthropogenic change can an estuarine ecosystem withstand before adverse change is experienced.

The linking between these ecological concepts and the management framework is also relatively recent and the concepts are now being integrated to provide a holistic approach not only to understand, but also to manipulate and manage the environment. Of particular importance with regard to estuarine management are ideas relating to the dynamics of ecosystems and the relations between biodiversity and ecosystem function, which have led toward the view that estuaries are particularly complex adaptive systems, characterized by historical dependency, nonlinear dynamics, threshold effects, multiple basins of attraction, and limited predictability (Folke et al. 2004; Duit and Galaz 2008; Moore et al. 2009).

Estuarine environmental problems are therefore intrinsically complex, and there is a clear need for solutions in a multiuse/multiuser/multisectorial system. To achieve this goal, the use of models becomes indispensable.

In this chapter, we provide a brief overview of recent models used in understanding estuarine processes, as well as in dealing with estuarine management problems. First, we present a case study of a model operating at the system level: the *multilayered model* for Xiangshan Gang bay (East China Sea), which was developed for the sustainable management of coastal ecosystems and used to simulate management scenarios that account for changes in multiple uses and enable assessment of cumulative impacts of coastal activities (Nobre et al. 2010). Then, we provide a summary, in the form of a table (Table 3.2), from other recent models developed for estuaries worldwide. Most of the presented models include a hydrodynamic submodel coupled to a biogeochemical one, whereas one of the presented models is an ECOPATH with ECOSIM model that analyzes the flows of energy and matter along estuarine food webs. Finally, because primary producers play quite relevant roles at estuaries and other coastal systems, we give particular insight into primary producer models from estuaries and coastal systems, contemplating all the four main different types of primary producers present at estuaries: phytoplankton, macroalgae or seaweeds, microphytobenthos, and seagrasses.

3.2 An Estuarine-System Coastal Model: The Case Study of the Xiangshan Gang (East China Sea)

3.2.1 Problematic

Coastal zones provide considerable benefits to society while, at the same time, human activities exert significant pressure on coastal ecosystems, therefore threatening those same benefits (Nobre 2009).

To promote the sustainable use of coastal zone resources, an ecosystem approach is required because it will contribute to the understanding of the causal relationships between environmental and socioeconomic components, plus the cumulative impacts of the range of activities developed in coastal ecosystems (Nobre and Ferreira 2009) and to manage coastal resources and biodiversity (Murawski et al. 2008).

Ecosystem modeling is a powerful tool that can contribute the required scientific grounding for the adoption of an Ecosystem-Based Management approach (Hardman-Mountford et al. 2005; Murawski 2007). Some of the improvements generated by this approach are to allow the examination of different development scenarios by altering variables of both the catchment and coastal systems and to provide insights for managers (Nobre et al. 2010).

3.2.2 Aim

The authors developed a multilayered catchment–coastal modeling approach to optimize the trade-offs presented above, through the use of a comprehensive set of models operating at different levels of complexity and geographical scales.

The specific objectives of the work were to (1) develop an integrated coastal management tool for decision-makers and (2) examine the outcomes of different development scenarios.

3.2.3 Study System

The Xiangshan Gang bay (volume of 3803×10^6 m³; area of 365 km²) located at the East China Sea (Figure 3.1).

3.2.4 Conceptual Diagram, Model Structure, and Main Equations

Multilayered ecosystem model. This work uses an integrated ecosystem modeling approach (Ferreira et al. 2008) to simulate the hydrodynamics, biogeochemistry, aquaculture production, and forcing functions, such as catchment loading within Xiangshan Gang (Figure 3.2).

The multilayered approach includes the coupling of several submodels, which were selected following the balance required in the choice of model complexity and structure (Jørgensen and Bendoricchio 2001). The considered key state variables and processes include: (1) production of multiple species in polyculture, (2) its effects on the coastal environment, and (3) impacts of other catchment–coastal system uses on the water quality and aquaculture resources.

Catchment submodel. The loading of substances into the Xiangshan Gang bay was simulated from the Soil and Water Assessment Tool (SWAT) model (Neitsch et al. 2002).

The main equations for catchment are the ones that describe the surface water balance and nutrient (for both N and P) export, respectively:

$$\frac{dSW}{dt} = PP_t - Q_{St} - Ea_t - W_{St} - Qgw_t \tag{3.1}$$

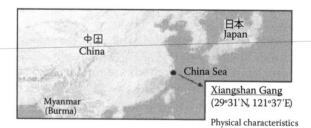

Volume: 3 803 10⁵m³
Surface area: 365 km²
Maximum depth: 45 m
Mean temperature: 24°C
Mean salinity: 24
Catchment area: 1 476 km²

FIGURE 3.1
Location of Xiangshan Gang and some of its physical characteristics. (Modified from Nobre et al. *Estuarine, Coastal and Shelf Science,* 87, 43–62, 2010. With permission.)

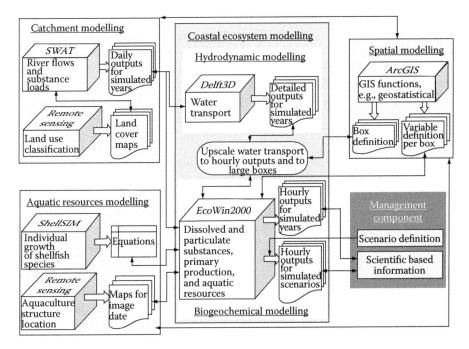

FIGURE 3.2
Conceptual diagram of the multilayered model. (Extracted from Nobre et al. *Estuarine, Coastal and Shelf Science*, 87, 43–62, 2010. With permission.)

where SW is the soil water content (mm^3 mm^{-2}), PP$_t$ is rainfall, Q$_{St}$ is the surface water runoff, Ea$_t$ is evapotranspiration, W$_{St}$ denotes the exchanges with the deep aquifer, Qgw$_t$ is the subsurface water runoff, and:

$$\frac{dN}{dt} = Fn_t + Rn_t + An_t - PUn_t - Qn_t - Ln_t - Vn_t - Dn_t \qquad (3.2)$$

where *N* is the soil nutrient concentration (kg ha^{-1}), Fn$_t$ is fertilization, Rn$_t$ is residue decomposition, An$_t$ is atmospheric fixation of N, PUn$_t$ is plant uptake, Qn$_t$ is lateral export, Ln$_t$ is leaching, Vn$_t$ volatilization of *N*, and Dn$_t$ is denitrification.

The model was calibrated against annual average discharge estimates for the most important rivers in the catchment, using a 30-year model run based on the 1961–1990 climatic normal built with the model's stochastic weather generator.

Model performance for water inputs was satisfactory as indicated by a significant correlation between simulated and observed values ($r^2 = 0.92$), low model bias (–5.3%) and high model efficiency (Nash–Sutcliffe efficiency index = 0.91). Additionally, the simulated annual nitrogen inputs from diffuse agricultural sources (960 tons year^{-1}) compared well with estimates based on export coefficients (900 tons year^{-1}; Huang et al. 2008a, 2008b).

The output from the catchment submodel was transformed into daily data series for offline coupling with the biogeochemical submodel.

The total nutrient load entering the bay was estimated to be about 11 tons day^{-1} of dissolved inorganic carbon (DIN) and 2 tons day^{-1} of phosphate. The point sources included untreated urban wastewater for ca. 600,000 inhabitants.

Hydrodynamic submodel. A three-dimensional hydrodynamic model (Delft3D-Flow-Delft Hydraulics 2006) was used to simulate the transport of substances among boxes and across the ocean boundary.

The calibration of the hydrodynamic model was achieved in two phases. First, the variations in tidal forcing were compared against measured water levels, followed by adjustment of bottom roughness to reproduce the water velocity characteristics (Huang et al. 2003). Overall, the model represented the amplitude of the main harmonic constituents well. In order to define the model boundary conditions, the salinity and temperature dataset was complemented with data from other authors (Hur et al. 1999; Isobe et al. 2004). In this second phase, the response of the system was gauged through existing knowledge of circulation as affected by tides and baroclinicity in tidal embayments (Fujiwara et al. 1997; Simpson 1997). Due to the lack of density and velocity data, the authors have used this procedure to tune the model within the theoretically acceptable boundaries for this type of system.

The model outputs provided a repeatable series of approximately 1 year of flows, which was used to force transport in the ecosystem model for calibration and validation years. The detailed flow fields were scaled up and converted into a data series of water fluxes between boxes and across the sea boundary with a 1-h time step and coupled offline with the biogeochemical submodel (see Ferreira et al. 2008).

Aquatic resource submodel. This submodel simulates the production of the Chinese oyster (*Ostrea plicatula*), the razor clam (*Sinonovacula constricta*), the Manila clam (*Tapes philippinarum*), and the muddy clam (*T. granosa*). The equations for shellfish aquaculture production were explicitly integrated into the ecosystem model with a four step approach: (1) use of a shellfish individual growth model (ShellSIM); (2) coupling of the individual growth model with a demographic model to simulate the population (Ferreira et al. 1997); (3) integration of the population growth model with an aquaculture practice model that implements the seeding of the population biomass and harvesting of the marketable cohorts for a given production cycle (Ferreira et al. 1997), and (4) use of a multiple-inheritance object-oriented approach (Nunes et al. 2003) to extend to multiple species in polyculture.

The individual growth model (ShellSIM) is driven by allometry, and it simulates feeding, metabolism, and individual growth of different shellfish species, under different environmental conditions. Shellfish growth depends on several environmental drivers, including salinity, temperature, and suspended particulate matter (SPM):

$$\eta = f(B) \cdot f(POM) \cdot f(SPM) \cdot f(L) \cdot f(T) \tag{3.3}$$

where η is the shellfish scope for growth, $f(B)$ is the function of phytoplankton, $f(POM)$ is the function of particulate organic detritus, $f(SPM)$ is the function of SPM, $f(L)$ is the function of salinity, and $f(T)$ is the function of water temperature.

For the Chinese oyster, razor clam, and muddy clam, the individual growth model was calibrated under local conditions (Ferreira et al. 2008). To simulate the growth of the Manila clam, the model used in Ferreira et al. (2007) was applied.

The population growth is simulated using a demographic model based on 10 weight classes, governed by the following equation:

$$\frac{dS(S,t)}{dt} = \frac{-d\left[S(s,t) \cdot \eta(s,t)\right]}{dS - \mu(S) \cdot S(s,t)} \tag{3.4}$$

where S is the number of individuals in each shellfish weight class, η is the shellfish scope for growth, and μ is the shellfish mortality rate.

The food resources (phytoplankton and detritus) removed by the population is scaled for each size class on the basis of the number of individuals in the class.

The aquaculture practice model (Ferreira et al. 1997) implements the seeding and harvesting strategies and interacts with the population model by respectively adding and subtracting individuals to the appropriate classes.

The model includes shrimp and fish production as forcing functions, contributing to dissolved and particulate waste (Ferreira et al. 2008).

Biogeochemical submodel. This submodel was developed to simulate the variation of salinity, dissolved nutrients, particulate matter, and phytoplankton. The main equations that describe the variations of these state variables are:

$$\frac{dB}{dt} = B \cdot \left(p_{max} \cdot f(I) \cdot f(NL) - r_b - e_b - m_b - S \cdot C_S\right) \tag{3.5}$$

where B is the phytoplankton biomass, p_{max} is the phytoplankton maximum gross photosynthetic rate, $f(I)$ is the Steele's equation for productivity with photoinhibition, $f(NL)$ is the Michaelis–Menten function for nutrient limitation, r_b is the respiration rate, m_b is the natural mortality rate, and C_S is shellfish grazing rate.

$$\frac{dN}{dt} = B \cdot \left(e_b + m_b\right) \cdot \alpha + S \cdot e_S + \text{POM} \cdot m_{\text{pom}} \cdot \varepsilon - B \cdot \left(p_{\max} \cdot f(I) \cdot f(\text{NL})\right) \cdot \alpha \tag{3.6}$$

where N denotes the dissolved inorganic nutrient, α is the conversion from phytoplankton carbon to nitrogen units, POM denotes particulate organic matter, ε is the conversion from POM dry weight to nitrogen units, m_{pom} is the POM mineralization rate, and e_S is the shellfish excretion rate.

$$\frac{d\text{POM}}{dt} = \text{POM} \cdot \left(e_{\text{pom}} - d_{\text{pom}}\right) + S \cdot f_S + B \cdot m_b \cdot \omega - \text{POM} \cdot \left(m_{\text{pom}} + p_{\text{pom}} \cdot S\right) \tag{3.7}$$

where POM denotes particulate organic matter, e_{pom} is the POM resuspension rate, d_{pom} is the POM deposition rate, f_S is the shellfish feces production, ω is the conversion from phytoplankton carbon to POM dry weight, and p_{pom} is the shellfish POM filtration rate.

$$\frac{d\text{SPM}}{dt} = \text{SPM} \cdot \left(e_{\text{spm}} - d_{\text{spm}}\right) + S \cdot f_S - \text{SPM} \cdot p_{\text{spm}} \cdot S \tag{3.8}$$

where SPM denotes suspended particulate matter, e_{spm} is the SPM resuspension rate, d_{spm} is the SPM deposition rate, and p_{spm} is the shellfish SPM uptake rate.

The parameterization of the model was in accordance with data obtained at the study site (Table 3.1).

The multilayered model was used to simulate coastal management options, which were based on scenario definition. The development scenarios were defined as a result of the participatory work among stakeholders, including modelers, local fishery and environmental managers, and aquaculture producers. The scenarios to be simulated by the multilayered ecosystem modeling framework were the following: (1) a reduction of fish cages corresponding to a 38% reduction in total fish production; (2) an extension of wastewater treatment to the entire population; and (3) a simultaneous reduction of fish cages and extended wastewater treatment.

The definition of scenarios is important for the evaluation of nutrient abatement strategies and to provide guidelines/grounding for future aquaculture policy and for eutrophication control.

3.2.5 Results

Ecosystem simulation. Simulations for the catchment submodel show two annual peaks for N input, in early spring and early summer, which can be related to both the fertilization of rice and the annual rainfall and runoff

TABLE 3.1

Ecosystem Model Parameters for the Standard Simulation

Shellfish Population	Number of Weight Classes		10
	Mortality, μ (% per day)	Oyster	0.40%
		Clam	0.56%
		Razor	0.20%
		Muddy	0.15%
Shellfish cultivation practice	Seed weight (gTFW ind^{-1})	Oyster	0.2
		Clam	0.5
		Razor	0.5
		Muddy	0.1
	Seeding period	Oyster	April–August
		Clam	May–June
		Razor	April–August
		Muddy	June–September
	Harvestable weight (gTFW ind^{-1})	Oyster	8
		Clam	14
		Razor	11
		Muddy	15
	Harvesting period	Oyster	December–March
		Clam	January–February
		Razor	October–February
		Muddy	November–March
	Aquaculture area (ha) and boxes cultivated	Oyster	2286 (Boxes 1–5, 8, 9, 11, 12)
		Clam	308 (Boxes 1–7,10)
		Razor	313 (Boxes 1–6)
		Muddy	187 (Boxes 1–3, 5, 6)
	Seedling density (tTFW ha^{-3})	Oyster	0.90
		Clam	0.45
		Razor	0.72
		Muddy	0.82
Phytoplankton growth	P_{max} (h^{-1})		0.2
	Lop (w m^{-2})		300
	Death loss – m_b day^{-1})		0.01
	KsDIN (μ mol L^{-1})		1
	KsPhosphate (μ mol L^{-1})		0.5
Suspended matter	POM mineralization rate (day^{-1})		0.02
	POM to nitrogen (DW to N)		0.0519
	POM to phosphorus (DW to P)		0.0074

Source: Nobre et al. *Estuarine, Coastal and Shelf Science,* 87, 43–62, 2010. With permission.

patterns. The same pattern was also found for particulate matter and P loads. Results indicate that the major sources of N are urban sewage discharges (56%), fertilization in rice crops (27%), and rangelands, mostly detritus decomposition from forests (17%). P followed a similar pattern, with 60% coming from urban sewage discharge and the rest from agricultural and natural sources.

Model outputs for DIN and phytoplankton fitted reasonably well with observed data (Figure 3.3), although in some points, the model overestimated DIN concentrations. Model outputs for SPM and POM in Box 10 (an outer box) did not represent the observed variability, but for an inner box (Box 3), there was a good agreement between predictions and observations. This may be related to the fact that the temporal resolution of SPM and POM being used to force the ocean boundary was not sufficient to represent the variability in the adjacent boxes. In the inner boxes, as the marine influence is lower and the catchment inputs of POM and SPM are more important, the daily inputs provided by the catchment model are realistic. Moreover, this limitation is not likely to significantly affect the simulation of aquaculture production, given that 83% of the bivalves are produced in the inner boxes (Boxes 1–5).

Simulations for total shellfish production were 2305 tons for an inner box (Box 3) and 741 tons (all oyster) for an outer box (Box 11). Overall, the outputs of harvested shellfish compared well with the landings data (Figure 3.4).

Development scenarios. In general, simulations indicate that the effects of changes implemented in the scenario simulations were mostly visible in the inner boxes. Variations of DIN, phosphate, phytoplankton, shellfish harvest, and productivity were less evident in the outer boxes, possibly due to exchanges with the ocean boundary.

In any of the scenarios, the reduction of nutrient loads resulted in very small changes in DIN concentration for any of the boxes, whereas phosphate concentrations were more sensitive to this. However, DIN and phosphate were still present in high concentrations for every scenario and in every box, which confirms the result of poor water quality in the Xiangshan Gang bay, even after scenario implementation.

For all scenarios, predictions indicate a decrease in shellfish productivity for all cultivated species, with more significant decreases in the inner boxes (e.g., less 286–864 tons year^{-1} in Box 3 compared to less 8–16 tons year^{-1} in Box 12).

Discussion. Model results indicate that the proposed scenarios will not achieve the management goals that they were designed to attain because (1) the modeled nutrient load reduction had no significant effect on the water quality of the Xiangshan Gang and (2) the decrease in phytoplankton biomass was limited to some areas of the bay.

Future work using this multilayered model includes the definition of further scenarios to assess how different land use and aquaculture practices may impact the bay.

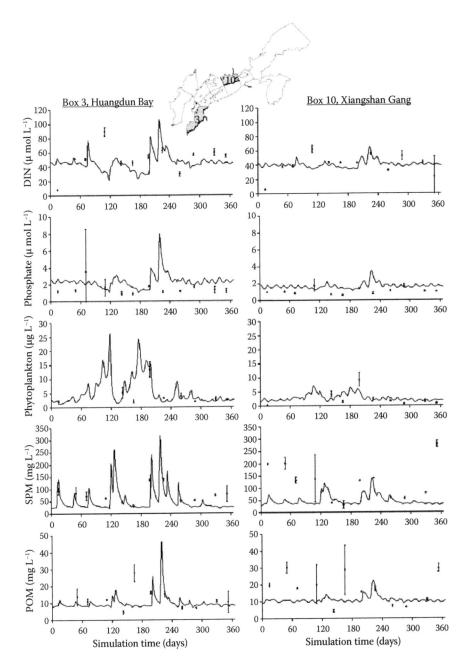

FIGURE 3.3
Standard simulation outputs for an inner box (Box 3) and a middle box (Box 10) plotted against average daily data ± standard deviation. Dissolved inorganic carbon (DIN), phytoplankton biomass, phosphate, suspended particulate matter (SPM), and particulate organic matter (POM). (From Nobre et al. *Estuarine, Coastal and Shelf Science*, 87, 43–62, 2010. With permission.)

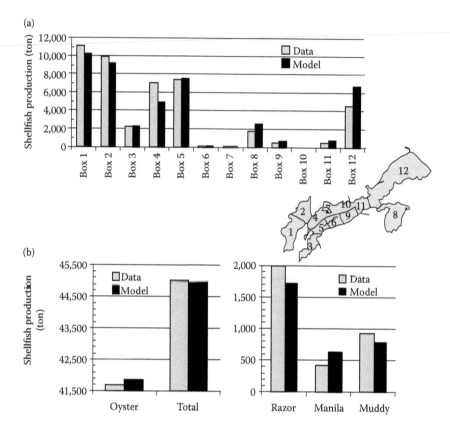

FIGURE 3.4
Standard simulation outputs and data for shellfish harvest (in tons year⁻¹).

The model outputs indicate that the nutrients and POM provided by fish cages and wastewater are sustaining shellfish growth in the inner boxes. In the scenarios that test a decrease of these substances, shellfish production also decreases, particularly in the inner boxes. The estimated total loss of harvested shellfish was between 4600 and 12,700 tons year⁻¹, corresponding to a loss of annual revenue between €555,000 and €1,500,000. Thus, the authors recommend reallocating part of the shellfish culture toward the mouth of the embayment.

Conclusions. This work clearly indicates that multilayered ecosystems models can play a key role for the adoption of ecosystem-based approaches to coastal and marine resource management. Furthermore, the integration of ecosystem-based tools can be used to fill data gaps, improve the temporal/spatial detail of setup datasets, and provide guidance to monitoring programs.

3.3 Brief Review of Some Recent Estuarine-Ecosystem Models

In this section, we present a summary in the form of a table, of some other estuarine-ecosystem models developed within the past decade (2000–2010), for a variety of estuaries worldwide (see Table 3.2), covering estuarine systems from Northern and Southern Europe, America, Asia, and Australia.

Most of the described models are hydrodynamic—coupled to biogeochemical—models (Das et al. 2010; Fulton et al. 2004; Timmermann et al. 2010; Trancoso et al. 2005; Blauw et al. 2009; García et al. 2010; Sohma et al. 2008), while one is a food web flow model (Patrício and Marques 2006). Some of the models were developed to perform as generic system-level models (e.g., Integrated Generic Bay Ecosystem Model [IGBEM], Fulton et al. 2004; or Generic Ecological Model [GEM] for estuaries, Blauw et al. 2009), whereas others were developed with specific aims for a given estuary.

3.4 Models of Primary Producers in Estuaries and Other Coastal Ecosystems

3.4.1 Introduction

Net primary production is defined as the amount of photosynthetically fixed carbon available to the first heterotrophic level and, as such, is the relevant metric for addressing environmental questions ranging from trophic energy transfer to the influence of biological processes on carbon cycling (Lindeman 1942).

Essentially, there are four types of primary producers that may have significant impacts on estuaries and coastal areas: phytoplankton, macroalgae, microphytobenthos, and seagrasses.

The biomass (B) variation of any primary producer over time (t) can be generically described by:

$$\frac{dB}{dt} = GP - R - E - L - M \tag{3.9}$$

where GP is the gross productivity, R is respiration, E is exudation, L is loss, and M is mortality.

Gross productivity is often described as a function of a maximum productivity rate (μ_{max}) and some limiting factors, usually, light (f_I), temperature (f_T), and nutrients (f_U):

$$GP - \mu_{max} \cdot f_I \cdot f_T \cdot f_U \tag{3.10}$$

TABLE 3.2

Summary of Some of the Estuarine Models Developed between 2000 and 2010

Study System	Aims	Software, Integration Algorithm, Background Models	Main Equations/Characteristics	Main Results	Reference
Mondego estuary (Portugal)	To assess food web flows and interactions along an eutrophication gradient.	ECOPATH with ECOSIM	$B_i \times (P/B)_i \times EE_i - \sum_{j=1}^{n} B_j \times (Q/B)_j \times DC_{ji}$ $- Y_i - BA_i - E_i = 0$ B_i, biomass of prey i; B_j, biomass of predator j; $(P/B)_i$, production/biomass ratio; EE_i, ecotrophic efficiency; Y_i, fisheries catch per unit area and time; $(Q/B)_j$, food consumption per unit biomass of j; DC_{ji}, fraction of prey i in the average diet of j; BA_i, the biomass accumulation rate for i; E_i, net migration of i	Total sum of flows, total system throughput, annual rate of net primary production highest in *Zostera* meadows, followed by the strongly eutrophic area and the intermediate eutrophic area.	Patrício and Marques (2006)
Barataria estuary (Gulf of Mexico)	To calculate the fluxes of water, nitrogen and carbon through the estuary. To estimate the importance of estuarine derived N and C for the overall C-budget and development of hypoxia.	Fourth-order Runge–Kutta with an integration step of 0.1 h.	Six box models. Mass balance equations for volumes in boxes: $\dfrac{\partial V_i}{\partial t} = F_i + P_i + R_i - E_i + Q_i$ V_i, segment volume; F_i, influx (or outflux) of water due to sea level variations; P_i, direct precipitation over the box area; R_i, runoff from the adjacent wetland areas; E_i, evaporation; Q_i, runoff from the Mississippi River diversions.	Annual TOC export from the estuary = 57 g Cm^2 yr^{-1}, corresponding to 2.7% of the riverine TOC. Not significant for the development of the Gulf's hypoxia	Das et al. (2010)

Port Phillip bay (Australia)	System-level description of temperate embayment ecosystems—IGBEM (Integrated Generic Bay Ecosystem Model).	Based on two existing ecosystem models: the Port Phillip Bay Integrated Model (PPBIM) and the European Regional Seas Ecosystem Model II (ERSEM II)	IGBEM integrates the biological and physical modules of PPBIM (Murray and Parslow 1997, 1999) and the biological modules from ERSEM II (Baretta et al. 1995; Baretta-Bekker and Baretta 1997)	The model behaves according to Sheldon Spectrum (Sheldon et al. 1972) and Monbet's relationship between chl. *a* and DIN (Monbet 1992), and reproduces spatial zonation and long-term cycles. It captures the system dynamics and allows exploration of the effects of ecological driving forces.	Fulton et al. (2004)
Horsens estuary (Denmark)	To assess the relative importance of nutrients supplied from freshwater outlets vs. nutrients entering the estuary from the adjacent sea. To detect which improvements	3D hydrodynamic model (Gustafsson and Bendtsen 2007-estuarymodel3) based on the COHERENS model (Luyten et al. 1999). The ecological component uses a modified	Numerous equations to define the 3D coupled hydrodynamic–ecological model. A numerical tracing technique (Menesguen et al. 2006) was used to assess the relative importance of nutrient discharges from streams, which requires a doubling of all equations dealing with state variables for nutrients. In this case, for simplicity, it was only applied to nitrogen tracking.	Results indicate that 40–90% of N in NO_3, NH_4 and detritus comes from streams, while only 15–50% of N in phytoplankton comes from streams; The impact of stream nitrogen reductions on NO_3, chl *a* and light attenuation was	Timmermann et al. (2010)

(continued)

TABLE 3.2 (Continued)

Summary of Some of the Estuarine Models Developed between 2000 and 2010

Study System	Aims	Software, Integration Algorithm, Background Models	Main Equations/Characteristics	Main Results	Reference
	can occur from different scenarios of nutrient load decrease.	version of the microplankton-detritus model (MPD model) (Tett 1998)		higher than the impact of open sea nitrogen reductions.	
Ria de Aveiro (Portugal)	To study the influence of benthic macroalgae on phytoplankton and nutrient concentrations.	MOHID modelling system, which integrates hydrodynamic, sediment transport and water quality elements (e.g., Braunschweig et al. 2003)	Primary production was described as $$\frac{\partial \Phi_X}{\partial t} = \left(\mu_X - r_X - ex_X - m_X - G_X\right)\Phi_z$$ t, time; Φ_X biomass (gC m^{-3} for phytoplankton; kgC m^{-2} for macralgae); μ_X gross growth rate (day^{-1}), r_X total respiration rate (day^{-1}); ex_X excretion rate (day^{-1}); m_X natural mortality rate (day^{-1}); G_X grazing rate (day^{-1})	Macroalgae significantly contribute to the system's primary production and to the nutrient limitation of phytoplankton.	Trancoso et al. (2005)
Case 1—Veerse Meer (The Netherlands) Case 2—North Sea Case 3—Venice Lagoon	To integrate biological and physical processes in simulations of basic ecosystem dynamics for a generic application to	Generic Ecological Model (GEM) for estuaries, an integrated model including physical, chemical and ecological processes, within	Numerous equations to describe phytoplankton processes (production, respiration, and mortality), extinction of light, decomposition of POM in the water and sediments, nitrification and denitrification, reaeration, settling, burial, filter-feeder processes (grazing, excretion, and respiration).	In general, the model results are accurate in both the spatial (horizontal and vertical) and the temporal dimension (seasonal and annual) for a variety of water systems.	Blauw et al. (2009)

| Case 4—Sea of Marmara (Turkey) | estuarine and coastal waters. | the Delft3D modeling suite (WL \| Delft Hydraulics 2003) | The "cost function" (OSPAR 1998; Radach and Moll 2006 in) was used to assess the "goodness-of-fit" between model outputs and measurements: $$C_x = \frac{\sum \left|M_{x,t} - D_{x,t}\right| / n}{sd_x} \times \left((1-c) + c(1-r_x)\right),$$ where C_x is the normalized deviation per station, annual value; $M_{x,t}$ is the mean value of the model results per station per month; $D_{x,t}$ is the mean value of the in situ data per station per month, and sd_x is the standard deviation of the annual mean based on the monthly means of the in situ data ($df = 11$), n is 12 months, c is 0.5, and r_x is the correlation over time between $M_{x,t}$ and $D_{x,t}$ (OSPAR 1998). | Model limitations are related to the application to shallow dynamic systems, with extensive tidal flats and periodic anoxia due to inappropriate formulations used in the sediment submodel. | |
| Urdaibai estuary (Basque Country) | To assess the impacts from hydrological inputs and wastewater loading. | 2D water quality model, including a hydrodynamic- and a water quality-submodel | For the water quality submodel, the following state variables were considered: ammonia-N, nitrate-N, phosphate-P, phytoplankton-C, dissolved BOD, suspended BOD, settled BOD and dissolved oxygen. | Pronounced decrease on phytoplankton concentration was obtained when decreased levels of nutrient discharge were coupled to highest river discharge. | García et al. (2010) |

(continued)

TABLE 3.2 (Continued)

Summary of Some of the Estuarine Models Developed between 2000 and 2010

Study System	Aims	Software, Integration Algorithm, Background Models	Main Equations/Characteristics	Main Results	Reference
Tokoy Bay (Japan)	To compare model outputs with the seasonal–daily dynamics of the benthic–pelagic compartments in the central bay and the tidal area.	Ecological Connectivity Hypoxia Model (ECOHYM)—a hydrodynamics–ecological model	General equation for the pelagic compartment: $$\frac{\partial C_W}{\partial t} = -(V_W \cdot \nabla)C_W + \nabla \cdot (K \cdot \nabla C_W) + \sum R$$ $$= -u_W \frac{\partial C_W}{\partial x} - v_W \frac{\partial C_W}{\partial y} - w_W \frac{\partial C_W}{\partial z}$$ $$+ \frac{\partial}{\partial x}\left(K_x \frac{\partial C_W}{\partial x}\right) + \frac{\partial}{\partial y}\left(K_y \frac{\partial C_W}{\partial y}\right)$$ $$+ \frac{\partial}{\partial z}\left(K_z \frac{\partial C_W}{\partial z}\right) + \sum R$$ where CW is the concentration of pelagic substances (phytoplankton, zooplankton, detritus, DOM, NH_4-N, NO_3-N, PC_4-P, dissolved oxygen); $V_W = (u_W, v_W, w_W)$ is the flow velocity; t is time; x, y, z are the spatial coordinates; $\sum R$ are the biochemical reactions and fluxes from outside the system; K is the eddy diffusion tensor.	The seasonal dynamics is more noticeable in the central bay area. The daily/tidal dynamics is determinant in the tidal flat area. Results reproduce accurately DO profiles during hypoxic events.	Sohma et al. (2008)

Macroalgal NPP can be estimated as (e.g., Duarte and Ferreira 1997):

$$\text{NPP} = \sum_{k=1}^{n} (B_{k,t+\Delta t} - B_{k,t} + M + Q)/\Delta t \tag{3.11}$$

where B is biomass, M represents losses through mortality, Q is biomass removed by frond breakage, k is the algal size class, n is the total number of classes, and Δt is the number of days between two consecutive samples.

In many models, the respiration loss term of algae is defined as an Arrhenius relationship with a reference temperature of 20°C (EPA 1985):

$$R = R_{\text{max20}} \cdot \theta^{(T-20)} \tag{3.12}$$

where R_{max20} is the maximum respiration rate at 20°C, θ is the empirical coefficient, and T denotes temperature.

Some models account for dark respiration and photorespiration (Trancoso et al. 2005), which is stimulated at high light intensities and dissolved oxygen concentrations (Carr et al. 1997):

$$R = k_{\text{re}} e^{0.069T} + k_{\text{rp}} \mu_A \tag{3.13}$$

where k_{re} is the endogenous respiration constant, k_{rp} is the fraction of actual photosynthesis rate oxidized by photorespiration, T is temperature, and μ_A is the algal growth rate.

In a very similar formulation, Blackford (2002) combines microphytobenthos activity respiration (r_{act}) as a fixed proportion of actual assimilation (Assim) and basal metabolism (r_{rest}) related to ambient temperature (f_t):

$$R = (r_{\text{rest}} f_T) + (\text{Assim} \cdot r_{\text{act}}) \tag{3.14}$$

Exudation (ex_A) may be described as a dependency on light (f_I) because of high photosynthate and photorespiratory compounds excretion at both low light levels and inhibitory high light levels (EPA 1985):

$$\text{ex}_A = K_e^A \mu_A \left(1 - f_I\right) \tag{3.15}$$

In some models, more complex formulations are used to describe algal excretion. Blackford (2002) describes three different processes of excretion. The excretion (e_a) of a fixed proportion (P_{ex}) of production (P):

$$e_a = P \cdot p_{\text{ex}} \tag{3.16}$$

Nutrient stress lysis (e_l) as a minimal rate (r_{lysis}) modified by the nutrient availability factor (i_N):

$$e_l = \frac{1.0}{i_N + 0.1} \cdot r_{lysis} \tag{3.17}$$

and nutrient stress excretion (e_n) of the potential assimilate ($1 - p_{ex}$):

$$e_n = P(1.0 - i_N) \cdot (1.0 - p_{ex}) \tag{3.18}$$

Mortality can be divided into grazing and nonpredatory mortality. Grazing on algae can be defined by simple formulations, which use a maximum grazing rate ($Graz_{max}$) and a temperature limitation function, reflecting the dependency of grazer's activity and abundance on temperature (f_{TGraz}) (Martins et al. 2007):

$$Graz = Graz_{max} \cdot f_{TGraz} \tag{3.19}$$

Or, by more complex formulations, which take into account the proportion of algae (microphytobenthos) available to the prey and different types of prey (meiobenthos and macrobenthos) (Blackford 2002):

$$Graz = \frac{r_{prey} \cdot p_A \cdot (p_A/(P_A + l))}{\sum prey + h} \tag{3.20}$$

where r_{prey} is the predator uptake rate, taking into account the uptake rate of microphytobenthos by deposit feeders, suspension feeders, and meiobenthos; p_A is the proportion of microphytobenthos available to each one of the predator groups; l is the feeding threshold for each one of the predator groups; and h is the half grazing rate constant for each one of the predator groups.

3.4.2 Phytoplankton

Phytoplankton constitutes the autotrophic component of the plankton with most species being microscopic unicellular organisms with a size range between 0.4 and 200 μm. Marine phytoplankton individuals live within the surface layers of the ocean, down to 200 m in the clearest waters (Simon et al. 2009). Although marine phytoplankters represent less than 1% of the Earth's photosynthetic biomass, they are responsible for more than 45% of the planet's annual net primary production (Field et al. 1998).

Estuarine phytoplankton is under strong temporal and spatial variations mostly due to tidal variations and seasonal freshwater runoff, which have significant effects on water column stability, residence time, light, and nutrient availability (Gameiro et al. 2007). Vertical mixing of the water column is also greatly influenced by other factors, such as wind-driven water mixing.

Since phytoplankton is the basis of many coastal and oceanic trophic chains, quantifying its productivity and understanding the factors that control phytoplankton dynamics is essential both from the ecological and management point of view. Additionally, detailed studies on phytoplankton taxonomy and functional groups are also of interest, as ecosystem eutrophication indexes may be derived from it, for example, the shift from diatom-dominated communities to phytoflagellate-dominated ones (Cloern 1996).

One of the most commonly used types of phytoplankton model is a time-integrated NPP (net primary productivity) phytoplankton model (Behrenfeld and Falkowski 1997), which can be defined by:

$$\sum NPP = \int_{z=0}^{zeu} P^b(z) \times PAR(z) \times DL \times Chl(z) \, dz \qquad (3.21)$$

where z denotes depth, z_{eu} is the euphotic depth, Pb(z) is the photoadaptive variable at depth z, PAR is the photosynthetic available radiation, DL denotes day length, and Chl(z) is the chlorophyll concentration at depth z.

A common application of productivity models is the calculation of global annual phytoplankton primary production (PPannu) (Behrenfeld and Falkowski 1997) and values range from 27.1 (Eppley and Peterson 1979) and 50.2 pg C year^{-1} (Longhurst et al. 1995). According to Behrenfeld and Falkowski (1997), such differences are mainly attributed to changes in depth integrated biomass (i.e., Csat × Zeu) and to spatial (i.e., horizontal) variability in the photoadaptive variable (Pb opt), which is defined as the maximum chlorophyll-specific carbon fixation rate observed within a water column measured under conditions of variable irradiance during incubations typically spanning several hours (Behrenfeld and Falkowski 1997).

Primary productivity models can be either empirical formulations of the P–E relationships, capable of mathematically describing the observed results but not describing physiological processes (e.g., Webb et al. 1974; Franks and Marra 1994) or of a mechanistic type, derived from known sequences of metabolic transformations (e.g., Eilers and Peters 1988, 1993; Han 2001a, 2001b; Rubio et al. 2003). Some primary productivity models assume a maximum threshold (saturation models) (e.g., Jassby and Platt 1976; Franks and Marra 1994), whereas others include photoinhibition (e.g., Steele 1962; Parker 1974; Pahl-Wostl and Imboden 1990). Macedo and Duarte (2006) have investigated the differences between static and dynamic photoinhibition models in three marine systems (turbid estuary, coastal area, and an open ocean ecosystem), and their results indicate that when photoinhibition development time is considered (dynamic model), the primary production estimates are always higher than when calculated with the static model. The authors also suggest that the quantitative importance of this underestimation appears to be more important in coastal areas and estuaries (21–72%) than in oceanic waters (10%).

3.4.2.1 Some Recent Phytoplankton Models

3.4.2.1.1 Nitrogen Isotope

3.4.2.1.1.1 Problematic   In the past three decades, nitrogen transformations in the marine environment have been actively investigated using ^{15}N techniques alone or in combination with other chemical techniques. Results from these studies, together with data on nutrients, particulate organic nitrogen, dissolved organic nitrogen, and biomass have improved the general understanding of nitrogen dynamics in marine ecosystems. Some models also describe the kinetics of nitrogen transformations and the factors that regulate such processes. However, the validation of transformation processes by only concentration data is difficult and limited. Furthermore, the incorporation of rates estimated by incubation experiments does not necessarily provide actual fluxes and concentrations. Since each approach has temporal and/or spatial limitations, combination studies provide a more comprehensive understanding of nitrogen dynamics in marine ecosystems (Sugimoto et al. 2010).

In recent years, ecosystem models incorporating δ^{15}N have proven to be useful tools for elucidating nitrogen dynamics in both the past and modern ocean (Giraud et al. 2000, 2003).

3.4.2.1.1.2 Aim   The aim of this model was to assess the quantitative contribution of different nitrogen sources (river, ocean, and regeneration) to primary production in Ise Bay (Japan) by using a combination of different approaches. For that, the authors have developed a three-dimensional physical-ecosystem model coupled to δ^{15}N, which was used to understand the flow of nitrogen in the coastal system. The authors highlight the idea that a combination of methods (models, concentration data obtained in the field, incubation experiments) provides a more comprehensive understanding of nitrogen cycling in marine systems.

The ability to detect small differences in the ^{15}N/^{14}N ratio of different pools of nitrogen combined with knowledge of kinetic isotope discrimination during chemical and biological reactions may potentially provide new ways to monitor nitrogen fluxes in marine systems on various temporal and spatial scales (Sugimoto et al. 2010).

The ^{15}N/^{14}N signal of phytoplankton may be used to estimate the extent and/or source of nitrogen used by phytoplankton.

3.4.2.1.1.3 Study System   The Ise Bay (Japan).

3.4.2.1.1.4 Model Structure and Main Equations   Five N-based state variables were included: NO_3^-, NH_4^+, phytoplankton, zooplankton, and detritus

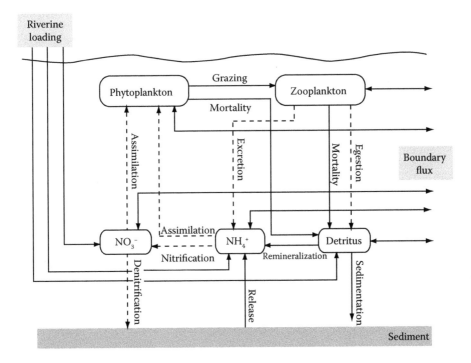

FIGURE 3.5
Conceptual diagram of the ecosystem model. Dashed and solid arrows indicate nitrogen flows with and without isotope fractionation, respectively. (From Sugimoto et al. *Estuarine, Coastal and Shelf Science*, 86, 450–466, 2010. With permission.)

(Figure 3.5). Besides photosynthesis, excretion, release, remineralization, sinking, mortality, grazing, egestion, and sedimentation, the processes of nitrification and denitrification are also included in the model.

Nitrification rates can be influenced by many factors, such as NH_4^+ concentration, oxygen, light, SPM, pH, and salinity (Herbert 1999). However, the seasonal rates of nitrification are more strongly correlated with temperature than with other factors (Berounsky and Nixon 1990), and the temperature coefficient of nitrification in the water column is higher than in the sediments (Sugimoto et al. 2008).

Thus, the authors described nitrification in the aphotic water column as:

$$\text{Nitrification} = V_{\text{Nit}}\exp(k_{\text{Nit}}T)[\text{NH}_4] \qquad (3.22)$$

where V_{Nit} is the nitrification rate at 0°C, k_{Nit} is the temperature coefficient of nitrification, and T is temperature.

Denitrification rate in the anoxic sediments was assumed to be dependent on temperature (T) and NO_3^- concentration in the water overlying the sediments (Sugimoto et al. 2008):

$$\text{Denitrification} = V_{\text{Denit}}\exp(k_{\text{Denit}}T)[NO_3] \qquad (3.23)$$

The fractionation of nitrogen isotopes in the model occurs in the processes of assimilation (photosynthesis) of NO_3^- and NH_4^+ by phytoplankton, excretion, and egestion by zooplankton and nitrification and denitrification (dashed arrows in Figure 3.6). During photosynthesis, the lighter isotope (^{14}N) is more readily incorporated by phytoplankton and, thus, the $\delta^{15}N$ values of NO_3^- and NH_4^+ increase as phytoplankton take up these nutrients (e.g., Waser et al. 1998a, 1998b).

The model includes δ values of nitrification in the water column and denitrification in the sediments obtained from field measurements (Sugimoto et al. 2008), and excretion and egestion by zooplankton were obtained from the literature (Yoshikawa et al. 2005), whereas the δ values of other assimilation processes were calibrated against measured values (Sugimoto et al. 2010).

The isotopic fractioning coefficient α_1 was described by:

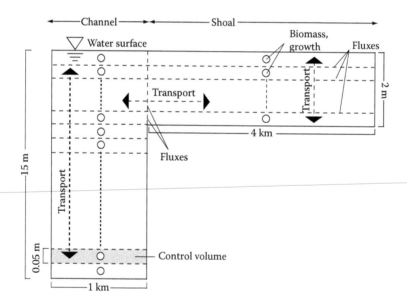

FIGURE 3.6
Schematic layout of computational grid of the phytoplankton-hydrodynamic model, representing a view of adjacent deep and shallow estuarine water columns. The deep compartment (left side) represents a channel; the shallow compartment (right side) represents a shoal. Dashed lines represent computational cell interfaces across which fluxes are calculated; solid lines represent domain boundaries. Circles represent the control volumes where phytoplankton biomass and growth-related quantities are calculated. (From Lucas et al. *Journal of Marine Systems*, 75, 70–86, 2009. With permission.)

$$\alpha_1 = \exp\left(\frac{-\varepsilon_1}{1000}\right) \tag{3.24}$$

where ε_1 is the isotopic discrimination for the considered biogeochemical processes.

The prognostic variables for the ^{15}N cycle are calculated as:

$$\frac{\partial\left[{}^{15}N_{NO_3}\right]}{\partial t} = -(\text{Photosynthesis}) \times F_{new} \times R_{NO_3} \times \alpha_1 + (\text{Nitrification}) \times R_{NH_4}$$
$$\times \alpha_5 - (\text{Denitrification}) \times R_{NO_3} \times \alpha_6 \tag{3.25}$$

$$\frac{\partial\left[{}^{15}N_{NH_4}\right]}{\partial t} = -(\text{Photosynthesis}) \times (1 - F_{new}) \times R_{NH_4} \times \alpha_2 + (\text{Excretion})$$
$$\times R_{Zoo} \times \alpha_3 + (\text{Decomposition}) \times R_{Det} - (\text{Nitrification})$$
$$\times R_{NH_4} \times \alpha_5 + (\text{Release}) \times R_{Sediment} \tag{3.26}$$

$$\frac{\partial\left[{}^{15}N_{Phy}\right]}{\partial t} = (\text{Photosynthesis}) \times F_{new} \times R_{NO_3} \times \alpha_1 + (\text{Photosynthesis})$$
$$\times (1 - F_{new}) \times R_{NH_4} \times \alpha_2 - (\text{Mortality}_{Zoo}) \times R_{Phy} - (\text{Grazing})$$
$$\times R_{Phy} + (\text{Sinking}_{Phy}) \times R_{Phy} \tag{3.27}$$

$$\frac{\partial\left[{}^{15}N_{Zoo}\right]}{\partial t} = (\text{Grazing}) \times R_{Phy} - (\text{Mortality}_{Zoo}) \times R_{Zoo} - (\text{Egestion}) \times R_{Zoo} \times \alpha_4$$
$$- (\text{Excretion}) \times R_{Zoo} \times \alpha_3 \tag{3.28}$$

$$\frac{\partial\left[{}^{15}N_{DET}\right]}{\partial t} = (\text{Mortality}_{Phy}) \times R_{Phy} + (\text{Mortality}_{Zoo}) \times R_{Zoo} + (\text{Egestion})$$
$$\times R_{Zoo} \times \alpha_4 - (\text{Decomposition}) \times R_{Det} + (\text{Sinking}_{Det}) \times R_{Det}$$
$$+ (\text{Sedimentation}) \times R_{Det} \tag{3.29}$$

where R_C is the $^{15}N/^{14}N$ ratio for a specific compartment C. The $\delta^{15}N$ values for a specific compartment C are calculated by:

$$\delta^{15}N = \left\{ \frac{\left(\dfrac{\left[^{15}N\right]}{\left[N\right]-\left[^{15}N\right]} \right)_C}{\left(\dfrac{\left[^{15}N\right]}{\left[N\right]-\left[^{15}N\right]} \right)_{Atmospheric}} - 1 \right\} \times 1000 \qquad (3.30)$$

where Atmospheric N_2 is defined as the standard sample including $[^{15}N]$ and equals 0.365%.

3.4.2.1.1.5 Results and Discussion　　Model results revealed that DIN (ammonium + nitrate) consumption by phytoplankton exceeds DIN supply from the rivers and ocean, indicating that a significant amount of phytoplankton production in Ise Bay depends on DIN regeneration within the bay. However, the ratio of consumption to external supply differs seasonally. The distribution of simulated $\delta^{15}N$ clearly showed the source of nitrogen incorporated by phytoplankton. The model reproduces well the decrease of $\delta^{15}N_{NO3}$ by the isotope effect of nitrification in spring.

　　The authors claim that further observations of $\delta^{15}N$ values of NH_4^+ are required, as well as sensitivity analysis, in order to estimate more accurate values. Additionally, further monitoring of river discharges are needed, including flood periods, to assess more precisely the effects of river loadings on the coastal environment.

3.4.2.1.2 Deep Channels and Broad Shallow

3.4.2.1.2.1 Problematic　　Many shallow estuaries around the world have one or more deep channels, laterally bounded by broad shallow regions (Lucas et al. 2009).

3.4.2.1.2.2 Aim　　The authors developed a model to explore the potential significance of hydrodynamic connectivity between a channel and a shoal and whether lateral transport can allow physical or biological processes (e.g., stratification, benthic grazing, light attenuation) in one subregion to control phytoplankton biomass and bloom development in the adjacent subregion.

3.4.2.1.2.3 Study System　　The South San Francisco Bay (USA) (SSFB).

3.4.2.1.2.4 Model Structure and Main Equations　　The developed model consists of two one-dimensional, vertically resolved water columns (one deep, one shallow) that exchange mass horizontally. Each water column is assumed to be laterally homogeneous. Lateral mass transport between the deep and shallow compartments is modeled as a diffusive process.

The two vertical subgrids are composed of several "boxes" or "control volumes" (Figure 3.6).

Within each compartment, the vertical transport, sources, and sinks of phytoplankton are represented by the advection–diffusion equation:

$$\frac{\partial B}{\partial t} = (\mu_{\text{net}} - ZP)B + \frac{\partial}{\partial z}\left(K_z \frac{\partial B}{\partial z}\right) - \frac{\partial}{\partial z}(W_s B) - \frac{\partial}{\partial z}(BG \cdot B) \qquad (3.31)$$

where B is the phytoplankton biomass concentration as chlorophyll a (mg chl a m^{-3}), z is the vertical location (m), μ_{net} is the phytoplankton net growth rate (day^{-1}), ZP is the zooplankton grazing rate (day^{-1}), K_z is the vertical turbulent diffusivity (m^2 day^{-1}), and BG is the benthic grazing rate (m^3 m^{-2} day^{-1}).

The lateral diffusion is given by:

$$\frac{\partial B}{\partial t} = \frac{\partial}{\partial y}\left(K_y \frac{\partial B}{\partial y}\right) \qquad (3.32)$$

where y is the transverse location (m) and K_y is the effective lateral diffusivity (m^2 s^{-1}).

3.4.2.1.2.5 Results and Discussion   Results indicated that (1) lateral transport from a productive shoal can result in phytoplankton biomass accumulation in an adjacent deep, unproductive channel; (2) turbidity and benthic grazing in the shoal can control the occurrence of a bloom system-wide; (3) turbidity, benthic grazing, and vertical density stratification in the channel are likely to only control local bloom occurrence or modify systemwide bloom magnitude.

The authors recommend that for other systems, which present a more gradual shallow–deep transition than SSFB, more compartments may be required to represent the range of functionalities present along the lateral bathymetric gradient. According to Lucas et al. (1999a, 1999b), transitional locations of moderate depth can be important ecologically, each offering a certain balance between light limitation and benthic grazing.

This study has a broader application for estuaries, in general, because it highlights the patchiness of process rates and the connectivity between subregions, indicating that kinetically distinct subenvironments can develop within an estuary and the hydrodynamic processes (e.g., tides, wind, river flow, density gradients) can drive the exchange of algal biomass between them.

3.4.2.1.3 Phytoplankton Motility

3.4.2.1.3.1 Problematic   To understand plankton communities and dynamics it is important to know the interplay between different phytoplankton species and the physical environment. Actually, the most successful approach in the prediction of harmful algal blooms focuses on the prediction of the physical environments in which blooms are known to occur, rather than an explicit prediction of the bloom itself (Ross and Sharples 2008).

It is known that about 90% of toxic phytoplankton species are dinoflagellates or flagellates (Smayda 1997) and that they seem to have significant physiological disadvantages compared to diatoms (e.g., lower photosynthetic rates and higher metabolic costs—Furnas 1990; Tang 1995; Smayda 1997), yet the two groups are often seen to coexist.

3.4.2.1.3.2 Aim   The focus is on a role for vertical migration in stratified coastal water, where the swimming speed is generally significantly less than the typical turbulent velocities in a tidally mixed bottom layer.

The specific aim is to address the question of whether motility is of any use in balancing resource requirements when a significant fraction of the water column is strongly turbulent.

3.4.2.1.3.3 Study System   Not applicable.

3.4.2.1.3.4 Model Structure and Main Equations   The model consists of three components: a k–ε turbulence model to describe the physics, a Lagrangian particle tracking model to describe the vertical displacement of cells in response to turbulence and cell motility, and a biological submodel to account for the swimming behavior of phytoplankton in relation to light and nutrient requirements.

The turbulence model must be able to capture the temporal and spatial variability of the tidally driven mixing in the bottom layer, which was achieved through a two-equation model of the k–ε variety (for details, refer to Canuto et al. 2001; Sharples et al. 2006).

The model accounts for tidal height variations through oscillating the sea surface slope at the period of the major lunar constituent of 12.42 h (M2). For the study area (Western English Channel), the current amplitudes are about 0.75 m s^{-1}, corresponding to a near-spring condition.

Vertical turbulence is controlled by the k–ε turbulence closure scheme and a parameterization for internal wave mixing at the thermocline is included following Large et al. (1994) (for details, see Sharples et al. 2006).

The model accounts for a total water depth of 80 m. A homogeneous surface layer (20 m) is separated from a 50-m-thick bottom mixed layer (BML) by a linear thermocline between 50 and 60 m above seabed. A constant temperature difference of 3.5°C is kept across the thermocline to allow easier analysis of thermocline reaccess by particles.

Model simulation produces two diffusivity maxima in the BML corresponding to the ebb and flood flows, which produce turbulent diffusivities in excess of 10^{-2} m^2 s^{-1}. The turbulence decreases rapidly while approaching the thermocline and in the stratified region the diffusivity K drops below 10^{-5} m^2 s^{-1}.

The following random walk equation is used to track a total of 10,000 individual particles through the eddy diffusivities produced by the k–ε model:

$$z_{n+1} = z_n + w_p \Delta t + K'(z_n)\Delta t + R \left[\frac{2K\left(z_n + \frac{1}{2}K'(z_n)\Delta t\right)\Delta t}{r} \right]^{1/2} \tag{3.33}$$

The second term of the equation accounts for motility, the third is a deterministic term, and the fourth is a random term.

Z_n is the vertical position of the particle after n iterations, w_p is the vertical swimming velocity, K is the turbulent eddy diffusivity with the abbreviation $K' = dK/dz$, Δt is the time step for the iteration and, R is a random process (for details, see Ross and Sharples 2008).

The lower end of the observed range of phytoplankton swimming speeds was used ($w_p = 0.1$ mm s^{-1}), to assess if even the lowest swimming velocities provide a measurable advantage to the cells.

The potential effectiveness of motility in a turbulent environment is the Péclet number, Pe, given by:

$$Pe = (\text{mixing time scale}/\text{swimming time scale}) = \frac{L^2/K}{L/w_p} = \frac{w_p L}{K} \tag{3.34}$$

where L is the mixed layer depth.

Model simulations indicate that only toward the base of the thermocline region should the swimming efforts of the cells be able to overcome the tidal turbulence.

3.4.2.1.4 *The Phytoplankton Growth Model*

The position and variation of the nutrient gradient within the thermocline is driven primarily by the uptake of nutrients by the cells within the subsurface chlorophyll maximum. In addition, the swimming strategies of motile cells also seem to depend on light and nutrient requirements (e.g., MacIntyre et al. 1997).

Phytoplankton production is described by a P–I (production–irradiance) curve following Denman and Marra (1986):

$$P^{d/l} = P_m^{d/l}\left[1 - \exp\left(-I/I_{d/l}\right)\right] \tag{3.35}$$

$P_m^{d/l}$ is the maximum dark-/light-acclimated production, $I_{d/l}$ is saturation irradiance, and I is irradiance in PAR at a certain depth.

PAR at a certain depth was calculated through a slightly modified Beer–Lambert equation:

$$I(z) = I_0 \exp[-(k + k_s(z))(H - z)] \tag{3.36}$$

where I_0 is the irradiance level at the sea surface (45% of which is assumed to be PAR), H is the total depth of the water column, k is light absorption coefficient, and $k_s(z)$ is the depth-dependent light absorption due to self-shading.

The model includes the limitation of dissolved inorganic nitrogen on phytoplankton growth. The uptake of DIN by phytoplankton was described through:

$$U = U_m \left(1 - \frac{Q}{Q_{max}}\right) \frac{N}{k_N + N} \tag{3.37}$$

Q is the cellular nitrogen/carbon ratio that needs to remain above a subsistence quota Q_{min} and below a maximum storage quota Q_{max}, U_m is the maximum uptake rate, and k_N is the half-saturation constant for nitrogen.

A Droop-type function was used to calculate the cellular increase in carbon:

$$\frac{dC}{dt} = \left[P\left(1 - \frac{Q_{min}}{Q}\right) - r - G - L_r \right] C \tag{3.38}$$

where R is the respiration rate, G is the grazing rate, C is the cellular carbon, and P is the instantaneous production.

The model takes into account a lower carbon threshold (C_{starve}), at which cells begin to die, and an upper level ($C_{fission}$) at which cells begin to divide (and the number of cells per Lagrangian particle doubles) (Broekhuizen 1999). Losses due to grazing and mortality reduce the number of cells within a Lagrangian particle. L_r represents this carbon-dependent mortality rate and was defined as:

$$L_r = 0.1 \text{ day}^{-1} \text{ if } C < C_{starve}, \text{ else } L_r = 0$$

The variation in the nitrogen concentration on a Eulerian grid was defined by:

$$\frac{dN_i}{dt} = \frac{d}{dz}\left(K\frac{dN_i}{dz}\right) - UC_i + \left[r + \beta G\right]_i N_i + \beta L_r N_i^{starve} \tag{3.39}$$

N_i denotes the total amount of cellular nitrogen in the ith element, N_{starve} is the cellular nitrogen content of cells that have $C < C_{starve}$, and β is the fraction of the grazed and starved nitrogen that is released back into the water column. The reminder is assumed to sink to the seabed. A fixed value of DINbed = 70 mg N m^{-3} is held in the model depth element at the seabed to include the resupply of DIN to the water column following remineralization processes in the sediments.

The light-nutrient driven swimming strategy of motile cells was expressed through (e.g., Broekhuizen et al. 2003):

$$w_p = \begin{cases} -w_p, & \text{if } Q < f_1 Q_{max} \\ +w_p, & \text{if } Q > f_2 Q_{max} \text{ and } I_{(z)} > I_c \\ 0, & \text{if } Q > f_2 Q_{max} \text{ and } I_{(z)} < I_c \end{cases} \tag{3.40}$$

The cellular N/C ratio (Q) increases through uptake of N and decreases as a result of carbon assimilation. If Q falls below a certain fraction f_1 of Q_{max}, the cell starts being nutrient limited and swims toward higher nutrient concentrations, that is, downward. As the cells encounter higher nutrient concentrations, Q starts to increase and when $Q > f_2 Q_{max}$, the cell gets nutrient replete and begins to swim toward higher light intensities (upward) to minimize light limitation. If light is below a certain threshold (I_c), a nutrient-replete cell simply becomes neutrally buoyant.

3.4.2.1.5 *Results and Discussion*

Results from this model show that tidal turbulence in the BML helps both motile and neutrally buoyant cells by periodically pushing them into the base of the thermocline. Motile cells have the additional advantage that they can swim further into the thermocline toward higher light, which will reduce the chance of being remixed back into the BML.

3.4.3 Macroalgae

Macroalgae tend to proliferate in nutrient-enriched, shallow, photic costal areas (e.g., Brush and Nixon 2010). Macrophyte algae are not only important because of their productivity, but also because of their associated roles in the ecosystem nutrient dynamics and as a sink for conservative pollutants such as heavy metals (Ferreira and Ramos 1989).

Traditionally, models of macroalgal populations may be divided into two groups:

(1) Demographic models, that simulate the numerical density of a population divided into size classes or different life history phases (Chapman 1993)

(2) Models that simulate the biomass density dynamics (Duarte and Ferreira 1993)

Duarte and Ferreira (1997) have developed a third type of model, resulting from the combination of types (1) and (2) and simulating simultaneously biomass and numerical density of predefined size classes as a function of type. The population was divided in four size classes: <5, 5–10, 10–15, and >15 cm.

To preserve model stability, asymptotic biomass values were assumed for each class, which were chosen according to observed values. When the biomass of a class is greater than its asymptotic value, the excess biomass and corresponding individuals are transferred randomly to the next class. Model forcing functions were only temperature and radiation. Nutrients were not included as limiting factors.

Results indicated that biomass dynamics and productivity are more sensitive to the light extinction coefficient than to the initial biomass conditions, biomasses losses due to respiration and exudation are comparable to those resulting from mortality, and frond breakage and different parameters should be used in productivity–irradiance curves according to seasonal variations (Duarte and Ferreira 1997).

3.4.3.1 Recent Macroalgal Models and Processes Taken into Account

3.4.3.1.1 Turbidity at the Intertidal Area

3.4.3.1.1.1 Problematic
In shallow waters, primary productivity is strongly dependent on the regulation of underwater light climate by SPM (e.g., Schild and Prochnow 2001) and, in case of nutrient surplus, light availability will be the key limiting factor for primary production. In intertidal areas, the combination of shallow waters and strong tidal currents creates complex patterns of SPM transport, deposition, and resuspension dynamics. In addition, the formation of isolated water pools on tidal flats at low tide, where SPM deposits quite fast, increases the complexity of these systems (Alvera-Azcárate et al. 2003).

3.4.3.1.1.2 Aim
The aim of the model was to develop a more mechanistic approach to model seaweed productivity, by improving the description of the underwater light climate in intertidal areas throughout the complete tidal cycle.

3.4.3.1.1.3 Study System
The model was applied to Tagus estuary (Portugal), a large mesotidal system, by combining field and experimental data, small spatial scale modeling, and geographical information systems (GIS).

3.4.3.1.1.4 Model Structure and Main Equations
Here, we will focus on the descriptions used by the authors for tide, velocity, and sediment dynamics. Details on the seaweed model can be found in the work of Alvera-Azcárate et al. (2003).

Interpolation method versus erosion–deposition method. Tidal variation was described from the basic harmonic constituents, considering a specific origin in time (January 1, 1980). The water velocity was simulated as a sinusoidal function:

$$u = A\sin(vt + p) \tag{3.41}$$

where A is the amplitude of the water current (m) and v is the wave velocity (m s^{-1}).

Velocity variation depends on the tide: the velocity was considered zero at high and low water, and reaches maximum values at mid-tide.

Interpolation. To describe the relationship between tidal range and the mass of suspended sediments in the water, the following equation was used:

$$SPM_t = \frac{z_t - z_{min}}{z_{max} - z_{min}} \cdot SPM_{lw} + \frac{z_{max} - z_t}{z_{max} - z_{min}} \cdot SPM_{hw} \tag{3.42}$$

SPM_t is the suspended particulate matter at time t; SPM_{hw} and SPM_{lw} are the SPM concentrations for high and low water conditions, respectively; z_t is the tidal height at time t; and z_{max} and z_{min} are high water and low water depths, respectively.

Erosion–deposition. The sediment modeling approach was improved by accounting for the sediment dynamics of tidal pools. The shear stress exerted over the bottom sediments depends mainly on water velocity and tidal height, and defines the rates of sediment erosion and deposition.

Deposition is given by Krone's equation:

$$D = \begin{cases} W_{fall}SPM\left(1 - \left(\frac{\tau}{\tau_d}\right)\right) & \text{if } \tau \leq \tau_d \\ 0 \text{ if } \tau > \tau_d \end{cases} \tag{3.43}$$

where the fall velocity of particles was described through Stoke's law:

$$W_{fall} = \frac{2}{9}g\frac{(\rho_P - \rho_W)}{\mu}r^2 \tag{3.44}$$

where ρ_w and ρ_p are the water and particle density, respectively; g is the acceleration gravity; μ is the water viscosity, and r is the particle radius.

τ_d is the critical shear stress and τ is the bed-shear stress, expressed by:

$$\tau = \rho_p c_b u^2 \tag{3.45}$$

where u is the water velocity and c_b the bed-shear coefficient:

$$c_{\text{b}} = \frac{gn^2}{h^{1/3}}$$ (3.46)

where n is the Manning roughness coefficient and h is the water column depth.

Erosion was calculated according to Partheniades (1965):

$$E = \begin{cases} M\left(\dfrac{\tau}{\tau_{\text{e}}} - 1\right) \text{ if } \tau \ge \tau_{\text{e}} \\[2mm] 0 \text{ if } \tau < \tau_{\text{e}} \end{cases}$$ (3.47)

where M is the erosion rate coefficient and τ_{e} is the critical shear stress for erosion. Erosion occurs when the bed shear stress is greater than the critical value.

During ebb, when tidal pools are formed, the deposition and erosion are different and must be treated separately from the rest of the tidal cycle (Alvera-Azcárate et al. 2003). In such circumstances, deposition is considered to be maximal, since there is no water turbulence; thus, the condition $\tau \le \tau_{\text{d}}$ is always true and deposition may be described by:

$$D = w_{\text{fall}}\text{SPM}$$ (3.48)

Bottom erosion is zero because there is no bed-shear stress, and the condition $\tau > \tau_{\text{c}}$ is always true.

Erosion (E) and deposition (D) depend on the water velocity, and deposition rate also depends on the total SPM in the water column; therefore, at each time step SPM is recalculated as:

$$\frac{\text{dSPM}}{\text{d}t} = E - D$$ (3.49)

The seaweed species considered for the model occupy an area of 16.5 km², with 36% coverage of *Fucus vesiculosus*, 46% of *Ulva lactuca*, and 18% of *Gracilaria verrucosa*. It was assumed that 50% of this area is occupied by tidal pools.

The "interpolation" model was upscaled over the total area of 16.5 km² and the "erosion–deposition" model was upscaled over 50% of this area, corresponding to tidal pools, and the remaining area was upscaled with the "interpolation" model. The two results were combined to determine the overall production.

Results upscalling were achieved through remote sensing and GIS (for details, see Alvera-Azcárate et al. 2003).

3.4.3.1.1.5 Results and Discussion   Results indicate that, in mesotidal or macrotidal turbid estuaries, the role of benthic autotrophs in cycling carbon and nutrients is quite significant. According to results, within the studied system (Tagus estuary), the annual carbon fixation by intertidal seaweeds corresponds to 21% of the total carbon fixed by all primary producers, and the nitrogen removal by seaweeds alone may be equivalent to the total loading of 490,000 people.

The authors argue that, in this type of estuary, where turbidity depends mainly on SPM resuspension rather than on phytoplankton, increased nutrient loading may not result in increased phytoplankton blooms, given the role of light as a limiting factor for pelagic production. Changes are more likely to occur as shifts in phytoplankton species composition and in alterations to benthic primary production.

The accurate description of the erosion–deposition processes used in this model has shown that estimates of carbon fixation and nutrient removal through uniform SPM dynamics may significantly underestimate the role of intertidal seaweeds and, thus, have important consequences to the understanding of carbon and nutrient cycles in mesotidal and macrotidal estuaries.

3.4.3.1.2 Macroalgal Life Stages (Adults and Spores) and Spatial Discrimination

3.4.3.1.2.1 Problematic   Macroalgal recruitment processes and factors affecting early life stages determine the development and dominance patterns of macroalgal blooms (Lotze and Worm 2000). Furthermore, the development of green macroalgal mats (frequently during spring) is either initiated by overwintering and regrowth of adult plants or by the formation of small propagules, such as vegetative fragments, zoospores, or zygotes (Schories et al. 2000). Thus, it is considered that macroalgal dynamics can only be fully understood if both adult individuals and early life stages are taken into account (Martins et al. 2007).

3.4.3.1.2.2 Aim   The aim of this work is to develop a model coupled to GIS, which is able to estimate algal productivity at the system-scale, taking into account the productivity of spores and adults, and predict the impacts of nutrient loading on the system and the adjacent coastal area (Figure 3.7).

For the sake of accuracy, productivity simulations take into account the spatial discrimination related to light and temperature conditions.

3.4.3.1.2.3 Study System   The Mondego estuary (Portugal).

3.4.3.1.2.4 Model Structure and Main Equations   *Processes affecting macroalgal spores.* Here, we will focus on the processes affecting the productivity of the initial life stages of macroalgae. A detailed description of the processes that control adult growth and productivity can be found in the work of Martins et al. (2007).

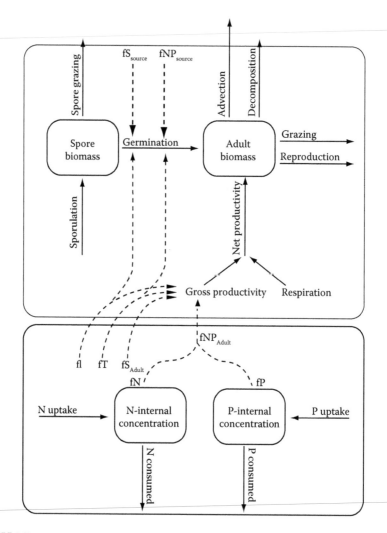

FIGURE 3.7
Simplified conceptual diagram of the spore and adult macroalgae productivity model.

The biomass of spores was considered dependent on three basic processes: sporulation (Sp), germination assumed as germination and growth to the next life stage (G) and grazing by macroinvertebrates (Zs):

$$\frac{dS}{dt} = Sp - (G + Zs) \qquad (3.50)$$

Sporulation is the release of spores by adult macroalgae (Santelices 1990). In the present work, sporulation was defined as the concentration of spores present in the water column, able to attach to any hard substrate, and it was experimentally obtained (Sousa et al. 2007).

The germination and growth of spores was described as:

$$G = G_{max} \cdot f(T) \cdot f(I) \cdot f(S_S) \cdot f(NP_S) \tag{3.51}$$

where G_{max} is the maximum germination rate and the f functions are limiting factors for temperature, light, salinity, and nutrients, respectively (for details, refer to Martins et al. 2007).

According to experimental evidence, the maximum grazing rate on spores (Z_{maxS}) was set at 0.14 day^{-1} (Lotze and Worm 2000) and the grazing rate was considered temperature-dependent:

$$Z_S = Z_{maxS} \cdot f(T_{Z_S}) \tag{3.52}$$

$$f(T_{Z_S}) = \exp\left[-2.3\left(\frac{T - T_{optZ}}{T_{xZ} - T_{optZ}}\right)^2\right] \tag{3.53}$$

where $T_{xZ} = T_{minZ}$ for $T \leq T_{optZ}$ and $T_{xZ} = T_{maxZ}$ for $T > T_{optZ}$; T_{optZ} is the optimum temperature for grazing activity, T_{minZ} is the lower temperature limit, below which grazing activity ceases, and T_{maxZ} is the upper temperature limit, above which growth ceases.

Light climate and temperature: accounting for the effect of spatial discrimination on algal productivity. Light intensity at surface (I_0) was estimated through the Brock model (1981), assuming a mean cloud cover of 0.41 (Martins et al. 2007). Photon flux density (PFD) was estimated from I_0, assuming that 42% of the overall energy is available for photosynthesis (Ferreira and Ramos 1989). PFD at depth z was calculated according to the Lambert–Beer equation, which assumes an exponential decrease of light with depth within the water column. The light extinction coefficient assumes different values according to local hydrodynamics variation, which accounts for turbidity, and adult macroalgal biomass, which accounts for self-shading (Martins et al. 2001, 2007).

Tidal height was simulated using the basic harmonic constituents and the bathymetry of the system (Mondego estuary, Portugal) was taken into account. Thus, the depth of algae depends on time and space:

$$\text{Algae depth} = \text{Tidal height} - \text{Bathymetry} \tag{3.54}$$

Desiccation stress has been pointed out as a mechanism related to the summer decline of macroalgal populations (Rivers and Peckol 1995; Hernández et al. 1997), particularly of those, located at latitudes where spring and summer temperature and irradiance values may be very high, such as the Mondego estuary (Martins and Marques 2002). Therefore, the model accounts for the seasonal and daily effect of desiccation on algae by assuming that, from April to September and from 11 A.M. until 4 P.M., emerged algae will exceed the air temperature in 5°C. According to Bell (1995), desiccated thalli have no evaporative water and can exceed air temperature up to 20°C.

The model was run at 33 different bathymetries (from −0.9 m [below sea level] to 2.30 m [above sea level] with a discrimination of 0.1 m). Model results were obtained at a scale of 1 m² and subsequently upscaled to the system using a bathymetric map and GIS and, taking into account, the patchy distribution of adult macroalgae within the system (Lopes et al. 2006).

3.4.3.1.2.5 Results and Discussion   Results indicate that there are significant spatial variations within the estuarine system regarding macroalgal growth. In this model, spatial variability is accounted for a dependency of light and temperature with depth, which determines that during winter and autumn, macroalgae growth is favored at lower depths (between +2.1 and +2.3 m above sea level) because it benefits from higher light availability, whereas in summer, macroalgae biomass is higher at deeper depths (between +1.8 and +2.2 m above sea level), where immersion periods are longer and desiccation stress is more limited.

Furthermore, the authors discuss that to account for a more accurate spatial variability of macroalgal standing stocks and growth at intertidal areas, other factors should be considered, such as the type of substrate (which affects the attachment conditions of algal spores and adults), the organic matter content of the underlying sediment, the presence or absence of rooted macrophytes, and the grazing pressure.

Calculations at the system level suggest that, in estuaries where hydrodynamics plays a major role, macroalgal biomass and, consequently, the amount of carbon, nitrogen, and phosphorus bounded to algal tissue, show enormous variations between years (240–21,205 tons macroalgal dry weight year⁻¹). Such variations will have significant impacts both in estuarine systems and on the adjacent coastal areas.

In years which do not favor macroalgal growth, there may be a potential nutrient surplus to the adjacent coastal areas because, in these circumstances, the role of macroalgae as nutrient sink is not significant. In years with intensive macroalgal growth, its internal requirements in nutrients may largely exceed the domestic loads of N and P to the Mondego estuary, indicating the significant role of opportunistic algae as a nutrient sink within this system. Nevertheless, in these years, up to 89 times more macroalgal tissue is likely to be exported to the adjacent coastal areas compared to years with limited algal growth. However, remineralization must occur before these nutrients are available to other primary producers, namely, coastal phytoplankton.

Regarding the impacts of spores on the dynamics of adult macroalgae, results suggest that in favorable years (mild winters and low rainfall) (Martins et al. 2001), the spring blooms occur at the expense of growth and reproduction of overwintering adults. However, during years with adverse conditions (Martins et al. 2001), where high rainfall in winter and spring prevent the occurrence of significant biomass of overwintering adults, the onset of the spring bloom may depend on the biomass of viable dormant spores (Martins et al. 2008).

3.4.3.1.3 The Mechanistic Approach

3.4.3.1.3.1 Problematic The use of empirical approaches in data-poor estuaries requires extrapolation from estuaries with possibly different physical environments and biota. Alternatively, it is possible to replace some empirical descriptions of ecological processes with more mechanistic descriptions (Baird and Emsley 1999). Concerning ecological processes, mechanistic is used to refer to the method of equation development. For example, using the process of diffusion to the cell surface to describe nutrient uptake is considered mechanistic because it uses a well-understood physical law (Fick's law of diffusion), and contains physically meaningful parameters such as the diffusion coefficient and the geometry of the cell (Pasciak and Gavis 1975). The use of mechanistic descriptions of ecological processes reduces the need for extrapolation of model to simulate ecological processes in data-poor estuaries (Baird et al. 2003).

3.4.3.1.3.2 Aim The aim was to develop a model of estuarine eutrophication built on mechanistic descriptions of a number of key ecological processes in estuaries.

3.4.3.1.3.3 Study Systems Several Australian estuaries.

3.4.3.1.3.4 Model Structure and Main Equations Photosynthetically available radiation (PAR) at the bottom of a layer of water (I_{bot}, mol photon m^{-2} s^{-1}) is given by:

$$I_{bot} = I_{top}e^{-K_d dz} \tag{3.55}$$

where I_{top} is the PAR at the top of the layer (mol photon m^{-2} s^{-1}), dz is the thickness of the layer (m), and K_d is the total light extinction coefficient (m^{-1}).
The equation that describes K_d:

$$K_d = k_w + n_{PS}\overline{aA}_{PS} + n_{PL}\overline{aA}_{PL} + n_{MPB}\overline{aA}_{MPB} + k_{other} \tag{3.56}$$

where k_w is the background attenuation coefficient of water (m^{-1}); n_{PS}, aA_{PS}, n_{PL}, aA_{PL}, and n_{MPB} and aA_{MPB} are the water concentration (cell m^{-3}) and absorption cross section (m^2 cell^{-1}) of cells of small phytoplankton, large phytoplankton, and microphytobenthos, respectively. K_{other} is the attenuation coefficient due to other components in the water column (e.g., dissolved organic matter or suspended organic matter, m^{-1}).

The average PAR in the layer, I_{av} (mol photon m^{-2} s^{-1}), is given by:

$$I_{av} = \frac{I_{top} - I_{bot}}{K_d dz} \tag{3.57}$$

In the model, it is assumed that light reaching the benthos is first attenuated by macroalgae and then seagrasses, which is described, respectively, by:

$$I_{below\ MA} = I_{bot} e^{-MA\,\overline{aA_{MA}}} \tag{3.58}$$

and

$$I_{below\ SG} = I_{below\ MA} e^{-SG\,\overline{aA_{SG}}} \tag{3.59}$$

where MA and SG are the biomass values (mg N m^{-2}) and aA_{MA} and aA_{SG} are the biomass-specific absorption cross sections (mg N^{-1} m^2) of the macroalgae and seagrass, respectively.

The remaining light passes through a thin layer of microphytobenthos at the surface of the sediment:

$$I_{below\ MPB} = I_{below\ SG} e^{-n_{MPB}\,\overline{aA_{MPB}dz}} \tag{3.60}$$

where n_{MPB} is the concentration of microphytobenthos (cell m^{-3}) with an absorption cross section of aA_{MPB} (m^2 $cell^{-1}$) in a sediment layer dz thick (m).

The average light flux available to microphytobenthos is given by:

$$I_{av} = \frac{I_{below\ SG} - I_{below\ MPB}}{n_{MPB}\,\overline{aA}_{MPB}dz} \tag{3.61}$$

Differently from other aquatic ecological models, autotroph absorption cross sections were used to parameterize both the dependence of autotroph growth rate on light availability and light attenuation in the water column.

In the Monod or Michaelis–Menten-type growth functions used in many ecological models, empirically determined half-saturation constants are required to determine the shape of the curve. In this model, the growth rate of autotrophs is determined from a functional form specifying the interaction of the maximum supply rates of nutrients and light, and the maximum growth rate.

Phytoplankton cells suspended in the water column obtain nutrients from the surrounding water, and light as a function of the average light in the considered layer. There is a physical limit to the rate at which a cell can absorb nutrients, which is given by the rate at which nutrient molecules can diffuse from the surrounding fluid to the cell surface and is known as the mass transfer limit (Pasciak and Gavis 1975). In this study, this parameter is used as the maximum rate of nutrient uptake (k_N, mol cell^{-1} s^{-1}) and is given by:

$$k_N = \psi DN \tag{3.62}$$

where ψ is the diffusion shape factor (m cell^{-1}), which for a sphere is $4\pi r$ and r is the cell radius (m); D is the molecular diffusivity of the nutrient (m^2 s^{-1}); and N is the concentration of the nutrient in the water column (mol m^{-3}).

In the case of light, the maximum supply rate (k_1, mol photon cell^{-1} s^{-1}) is given by:

$$k_1 = I_{av}\,\overline{aA}_{cell} \tag{3.63}$$

$\overline{aA}_{cell}$ is the absorption cross section of a cell (m^2 cell^{-1}) and I is the average PAR in the layer (mol photon m^{-2} s^{-1}).

The absorption cross section of a spherical cell is given by (Kirk 1975):

$$\overline{aA}_{cell} = \pi r^2 \left(1 - \frac{2(1 - (1 + 2\overline{y}_C r)e^{-2\overline{y}_C r})}{(2\overline{y}_C r)^2} \right) \tag{3.64}$$

y_C is the absorption coefficient (m^{-1}), which gives the rate at which light is attenuated.

For benthic macroalgae, the maximum rate of nutrient uptake, k_N (mol m^{-2} s^{-1}), can be calculated as a diffusion rate through an effective diffusive boundary layer thickness, δ (m):

$$k_N = \frac{D}{\delta} N \tag{3.65}$$

where D is the molecular diffusivity of the nutrient (m^2 s^{-1}) and N is the concentration of the nutrient in the water column (mol m^{-3}).

At typical hydrodynamic environments, the effective diffusive boundary layer thickness is of the order of 0.1 mm for typical benthic macrophytes. The effective thickness of the diffusive boundary layer is strongly influenced by shear stress at the water–benthos interface, which in turn is a function of surface roughness and water velocity (Hurd 2000).

Light capture by macroalgae, k_I (mol photon m^{-2} s^{-1}), is given by:

$$k_I = I_{bot}\left(1 - e^{-\overline{aA}_{MA}MA}\right)$$

(3.66)

where I_{bot} is the incident radiation at the top of the macroalgae (mol photon m^{-2} s^{-1}), MA is the biomass of macroalgae (mg N m^{-2}), and aA_{MA} is the nitrogen-specific absorption cross section of macroalgae (m^2 mg N^{-1}).

It is assumed that the projected area of the benthos is fully covered by macroalgae, but the macroalgae has varying thickness and, thus, absorbance depends on biomass.

The uptake rate of seagrasses is expressed as:

$$k_N = \frac{SG\mu_{max}}{K_{1/2}}N$$

(3.67)

where SG is the biomass of seagrass (mol N m^{-2}), μ_{max} is the maximum growth rate (s^{-1}), N is the sediment porewater nutrient concentration (mol m^{-3}), and $K_{1/2}$ is an experimentally determined half-saturation constant of nutrient-limited growth when fitted to the Monod growth equation (mol N m^{-3}).

The maximum uptake rate of benthic microalgae takes the same form as for suspended algal cells (Equation 3.62), although the diffusion coefficient is calculated from:

$$D_{porewater} = \frac{\vartheta D}{\theta}$$

(3.68)

where ϑ is the porosity and θ is the tortuosity (both dimensionless), which can be approximated by (Boudreau 1996):

$$\theta = 1 - \ln(\vartheta^2)$$

Light capture by microphytobenthos is modeled in the same manner as for algal cells suspended in the water column.

The authors have used a new empirical growth model scheme, which allows the logical use of mechanistically determined maximum uptake rates and maximum growth rates.

For planktonic autotrophs and assuming a light- and a single nutrient-limited system,

$$N = k_N\left(\frac{R_N^{max} - R_N}{R_N^{max}}\right) \text{ mol N cell}^{-1}\,\text{s}^{-1}$$

(3.69)

and

$$I = k_I \left(\frac{R_I^{max} - R_I}{R_I^{max}} \right) \text{mol photon cell}^{-1} \text{ s}^{-1} \tag{3.70}$$

where R denotes the reserves of nutrient and light available for growth (energy from light is assumed to be stored as fixed carbon) and R_{max} is the maximum value of R.

The growth rate, μ, is given by:

$$\mu = \mu^{max} \frac{R_N}{R_N^{max}} \frac{R_I}{R_I^{max}} \tag{3.71}$$

where μ_{max} is the maximum growth rate (s^{-1}).

To determine growth at any combination of k_I, k_N, and μ_{max}, a steady-state solution is obtained through:

$$k_N \left(\frac{R_N^{max} - R_N}{R_N^{max}} \right) = \mu_{max} (m_N + R_N) \frac{R_N}{R_N^{max}} \frac{R_I}{R_I^{max}} \tag{3.72}$$

$$k_I \left(\frac{R_I^{max} - R_I}{R_I^{max}} \right) = \mu_{max} (m_I + R_I) \frac{R_N}{R_N^{max}} \frac{R_I}{R_I^{max}} \tag{3.73}$$

where m_N and m_I are the stoichiometry coefficients specifying the number of moles of N and photons per cell.

3.4.3.1.3.5 Results and Discussion   The authors found significant differences for describing autotrophic growth between the mechanistic and empirical approaches. In many models, which follow empirical approaches (e.g., Madden and Kemp 1996), the growth rate of macroalgae is an exponential function of the biomass of the autotroph under all nutrient and light levels. In the mechanistic formulation, the rate of biomass increase varies from being an exponential function of biomass at nutrient and light-saturating conditions, to being independent of biomass at nutrient-limiting conditions, to having a complex exponential relationship with biomass at light-limiting conditions.

In the mechanistic approach, the half-saturation constants, which are often both poorly constrained and can impact significantly on model output, are replaced by parameters with very specific physical interpretations. These parameters can often be measured accurately and, in many cases, will have similar values in all estuaries.

However, although empirical models are constrained by observations, mechanistic approaches have the underlying assumption that all important processes have been captured. Consequently, as long as the empirical model behavior does not deviate significantly from the calibration data set, the model should perform well. In contrast, the mechanistic approach relies on the modeler capturing the important processes.

The authors conclude that the use of mechanistic descriptions in models of estuarine eutrophication may provide predictive capabilities beyond those of models with empirical descriptions.

3.4.4 Microphytobenthos

Microphytobenthos are major primary producers in estuarine mudflats systems (MacIntyre et al. 1996; Underwood and Kromkamp 1999). They are composed of different microalgae groups (e.g., diatoms, euglenoids, green algae, cyanobacteria) that inhabit various types of intertidal systems, from fine silt and mud to sand (MacIntyre et al. 1996; Paterson and Hagerthey 2001; Jesus et al. 2006).

Although models of microphytobenthos are less abundant than models of phytoplankton or macroalgae, they can also be found in literature with a somehow more recent appearance. For instance, Guarini et al. (2008) have developed a "hybrid dynamic model" for microphytobenthos, where continuous dynamic processes (i.e., production and mortality) coexist with discrete events (i.e., resuspension of the biofilm at the surface of the mudflat). Their model takes into account specific microphytobenthos behavior. On intertidal mudflats, photosynthetic active readiations often saturate the production of mycrophytobenthos (Guarini et al. 2000). When saturation is reached, the production rate varies mainly with temperature (Guarini et al. 1998), salinity, and pH. A general thermo-inhibition can be observed in temperate intertidal mudflats during summer (Guarini et al. 1998), which may lead to an inversion of the production–loss balance, with a consequent decrease in mycrophytobenthos biomass (Guarini et al. 1998).

In another model of microphytobenthos (Guarini et al. 2002), the authors combine deterministic processes (water elevation, light–dark cycle) and stochastic processes (carbon assimilation of the microalgal community) to estimate the potential productivity of microphytobenthos at the system level. Results indicate that microphytobenthos productivity exhibits significant spatial heterogeneity, which was attributed to different photosynthetic competence of microphytobenthos in different subbasins and spatial differences in the phase shifts between the tidal and solar cycles controlling the exposure of intertidal areas to sunlight.

3.4.4.1 Some Recent Microphytobenthos Models

3.4.4.1.1 Microphytobenthos Productivity

3.4.4.1.1.1 *Problematic* In the oceans, primary production is generally dominated by pelagic flora, but in shallow basins, primary production associated with the benthos may be locally significant (e.g., Barranguet et al. 1996). The annual production levels of intertidal benthic diatoms are reported to be significant and approaching the estimates of pelagic production in neighboring neritic habitats (e.g., McIntyre and Cullen 1995). In fact, several studies have shown that microphytobenthos contributes with a significant proportion of total microphytoplankton biomass and production, sometimes exceeding 50% (e.g., Cahoon and Cooke 1992). However, in many coastal systems worldwide, there is still a lack of studies that evaluate the contribution of microphytobenthos productivity to the system's dynamics and total production.

3.4.4.1.1.2 *Aim* The aim of the study was to assess the potential influence of the unicellular forms of microphytobenthos, principally episammic and epipelic diatoms, within the context of a dynamic intertidal ecosystem and related to carbon fixation, aerobic layer activity, light versus nutrient limitation, and nutrient regeneration.

3.4.4.1.1.3 *Study System* Northern Adriatic ecosystem.

3.4.4.1.1.4 *Model Structure and Main Equations* *The phytobenthic model.* The phytobenthic model (Figure 3.8) considers only diatoms, and their biomass variation is described by:

$$\frac{\partial A}{\partial t} = \text{production} - \text{respiration} - \text{loss} - \text{predation} \qquad (3.74)$$

Gross production rate (P) is calculated as:

$$P = r_{max} \cdot f_t \cdot f_i \cdot f_s \qquad (3.75)$$

where r_{max} is the maximum CO_2 uptake rate, and f_t, f_i, and f_s are the temperature-, light- and silicate limitation factors, respectively.

The temperature factor uses a standard Q10 formulation based on ambient temperature (T):

$$f_t = q_{10}^{0.1(T-10)} \qquad (3.76)$$

And the light limiting factor is described by:

$$f_i = \frac{l \cdot I \cdot P_{act}}{I_{opt}} \cdot e^{-\sigma D}$$

(3.77)

where l is day length, σ is the extinction coefficient, D is depth, I is irradiance, P_{act} is the proportion of light that is photosynthetically active, and I_{opt} is the optimum irradiance.

The adaptation of the optimal light parameter is described by:

$$\frac{d}{dt} I_{opt} = r_1 \cdot \left(I \cdot P_{act} - I_{opt} \right)$$

(3.78)

where r_1 is the relaxation process time constant. I_{opt} is constrained by a minimum value I_{min}.

The silicate limitation factor depends on the ambient silicate concentration and a constant:

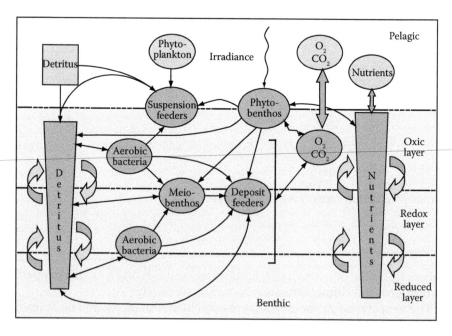

FIGURE 3.8
Conceptual diagram of the microphytobenthic model and the ERSEM benthic model showing state variables and fluxes. (From Blackford, J. C., *Estuarine, Coastal and Shelf Science*, 55, 109–123, 2002. With permission.)

$$f_s = \frac{[Si]}{h_S} \qquad (3.79)$$

The nutrient limitation factor (i_N) employs Droop kinetics:

$$i_N = \lim_{0 \to 1} \left[\frac{p_{cell} - p_{min}}{p_{max} - p_{min}} \cdot \frac{n_{cell} - n_{min}}{n_{max} - n_{min}} \right]^{0.5} \qquad (3.80)$$

where n_{cell} and p_{cell} are the actual internal cell C/N and C/P ratios, the maximum ratios (n_{max}, p_{max}), and the structural content of the cell (n_{min}, p_{min}).

Nutrient uptake (N and P) is defined by:

$$uptake = min(P \cdot N_{max} + A \cdot N_{max} - A_N, N_{aff} \cdot [N] \cdot A) \qquad (3.81)$$

which depends on the shortfall in existing nutrient content ($A \cdot N_{max} - A_N$), the uptake required to balance the calculated carbon production in the current time step ($P \cdot N_{max}$), and the cells physiological affinity to the nutrient and the available nutrient pool ($N_{aff} \cdot [N] \cdot A$).

The phytobenthic model was coupled to ERSEM (Baretta et al. 1995), a biomass-based model, which determines the fluxes of carbon and nutrients between several functional groups.

The physical model used in this work is the vertical diffusion submodel of the Princeton Ocean Model (POM) (Blumberg and Mellor 1980) coupled to a turbulence closure scheme model (Mellor and Yamada 1982). This model was previously applied in the study area by Vichi et al. (1998) and Allen et al. (1998).

Model simulations were performed for several simulations, from 5 to 25 m, along an offshore transect.

3.4.4.1.1.5 Results and Discussion   Predictions indicate that mycrophytobenthos can contribute in excess of 50% of primary production in shallow depths (<5 m), where light availability is the key factor limiting mycrophytobenthos, with self-shading dominating this effect in established populations. Grazing is also shown to be a significant control on microphytobenthic populations. Nutrient stress is not a major limiting factor of microphytobenthos due to benthic nutrient pools. Although nutrient uptake and cycling through microphytobenthos are not significant compared with regional inputs, microphytobenthic populations have significant impacts on nutrient regeneration rates, aerobic layer processes, and benthic trophic dynamics. The effects on the pelagic ecosystem are minimal.

3.4.4.1.2 Hydrobia Ulvae, Microphytobenthos, and Mudflat's Erosion

3.4.4.1.2.1 Problematic   Several studies have shown that the stability of inter-tidal mudflats is highly dependent on the biological activity and community structure (e.g., Herman et al. 2001). Benthic biology seems to influence the critical shear stress for erosion and the erosion rate significantly (Widdows et al. 2000; Amos et al. 2004).

Macroalgal mats, which occur more often during spring, summer, or early autumn, tend to stabilize the sediment bed, thus making mudflats more resistant to erosion (Paterson 1989). On the contrary, deposit-feeders such as *Hydrobia ulvae* may have a destabilizing action to the sediment bed, while moving and feeding on benthic algae (Andersen 2001).

Because it seems that there exists a strong positive feedback between mud deposition on mudflats and the functioning of the benthic food webs (Herman et al. 2001), it becomes important to estimate the overall net effect of biological processes on the sediment deposition rates at intertidal areas.

3.4.4.1.2.2 Aim   The authors have developed a numerical sediment trans-port model to describe how the presence of *H. ulvae* and microphytobenthos (mainly diatoms) affect the erosion and deposition patterns on the mudflat of a Danish estuary.

3.4.4.1.2.3 Study System   Lister Dyb tidal area at the Wadden Sea (between Germany and Denmark).

3.4.4.1.2.4 Model Structure and Main Equations   The authors have used an expo-nential erosion formulation (Parchure and Mehta 1985) appropriated to the dominant sediment type at the study area, that is, soft unconsolidated mud:

$$S_E = Ee\left[\alpha\sqrt{\tau_b - \tau_{ce}}\right] \tag{3.82}$$

where S_E is the erosion rate (kg m^{-2} s^{-1}), E is the floc erosion rate (kg m^{-2} s^{-1}), α is a coefficient (m N$^{-0.5}$), τ_b is the bed shear stress (N m^{-2}), and τ_{ce} is the critical bed shear stress for erosion (N m^{-2}).

Deposition is calculated through the settling flux formulation (Krone 1962; Einstein and Krone 1962):

$$S_D = w_s c_b p_d \tag{3.83}$$

where S_D is the deposition rate (kg m^{-2} s^{-1}), w_s is the settling velocity (m s^{-1}), and c_b is the near bed concentration (kg m^{-3}):

$$c_b = 1 + \frac{Pe}{1.25 + 4.75(p_d^{2.5})} \tag{3.84}$$

where *Pe* is the Peclet number, defined by:

$$Pe = \frac{w_s h}{D_z} \tag{3.85}$$

where D_z is the eddy diffusivity (m² s⁻¹) and p_d is the probability of deposition:

$$p_d = 1 - \left(\frac{\tau_b}{\tau_{cd}} \right) \tag{3.86}$$

where τ_{cd} is the critical bed shear stress for deposition (N m⁻²).

The sediment settling velocity is calculated through the exponential relationship:

$$w_s = kc^m \tag{3.87}$$

where *K* and *m* are time- and site-specific parameters.

The authors have run the model with four different sets of combinations covering the general range of erosion thresholds, erodibilities, and settling velocities observed in the study site. Only parameters related to the processes affected by the considered biological communities (microphytobenthos and *H. ulvae*) have been changed between different runs to predict and estimate the net effect of the varying biological parameters.

3.4.4.1.2.5 Results and Discussion   According to results, the net sedimentation on the intertidal mudflat is controlled by sediment characteristics hydrography, meteorology, and also benthic biology. The presence of large numbers of destabilizing *H. ulvae* results in higher net accumulation, particularly, if sediment aggregation comes in the form of fecal pellets from *H. ulvae*. These pellets cause settling velocities to be so high that, in spite of the destabilizing activity of the snail at the bed sediment, the increase in settling velocity compensates for this and the net result is enhanced deposition. However, the newly deposited sediment is loose and thus highly mobile and prone to subsequent resuspension. In contrast, the stability of the mudflats is increased by the presence of biofilms.

3.4.5 Seagrasses

Salt marshes are coastal systems from temperate zones, mainly occupied by halophytic vegetation, commonly designated as seagrasses or salt marshes plants. Ecologically, salt marshes are characterized by their high primary productivity and species diversity, representing habitat for migratory water-fowl, transient fish species, and indigenous flora and fauna (Simas et al. 2001). These ecosystems also provide important commercial resources because they act as nursery grounds for several fish and crustacean fisheries (e.g. Van Dijkeman et al. 1990).

In addition, seagrasses help the stabilization of sediments and play an important role in the nutrient budget of coastal marine ecosystems, as they take up nutrients both through below- and aboveground biomass (Oshima et al. 1999; Elkalay et al. 2003). Seagrasses also provide a physical connection between the water column and sediments by transporting photosynthetic- and seawater-derived oxygen to their roots and rhizomes (e.g. Miller et al. 2007).

3.4.5.1 Recent Seagrass Models and Processes Taken into Account

3.4.5.1.1 Effects of Sea-Level Rise in Salt-Marsh Areas

3.4.5.1.1.1 Problematic   The enhanced greenhouse effect results from human activities that have caused a recent increase in the atmospheric concentrations of the main greenhouse gases: carbon dioxide, methane, and nitrous oxide (e.g., IPCC 1990; Titus and Narayanan 1995; Houghton 1999). According to the reports from the IPCC (1990), the global average temperature of the planet will rise by about 2.5°C, with a range of 1.5–4.5°C, depending on the model used (Berner and Berner 1996; Houghton 1999; IPCC 1999). In spite of all uncertainties linked to these predictions, a rise in sea level is another expected consequence of the enhanced greenhouse effect, both because of oceanic thermal expansion and melting of Arctic and Antarctic glaciers (Titus et al. 1991). According to the estimates of the IPCC, the total average sea-level rise is predicted to be about 12 cm by 2030 and 50 cm by 2100 (IPCC 1999).

Salt marshes are important coastal ecosystems developed in temperate zones and, particularly, vulnerable to sea-level rise—as the sea rises, the outer salt marsh boundary will erode and new salt marsh will form inland (Simas et al. 2001). Salt marshes seem able to react by a negative feedback mechanism to the sea-level rise: a small increase in sea level leads to greater sediment deposition due to extended submersion time, and to reduced soil compaction due to reduced decomposition of organic matter (Nyman et al. 1994; Allen 1994). However, rapid sea-level rise may counteract this effect with a positive feedback one that increases salt-marsh loss, increased submersion times reduce salt marsh production and, consequently, organogenic sedimentation (Nyman et al. 1994).

3.4.5.1.1.2 Aim   The authors aimed at examining the consequences of sea-level rise in salt marsh areas, using an approach that combines ecological modeling, remote sensing, and geographical information analysis (Simas et al. 2001) (Figure 3.9).

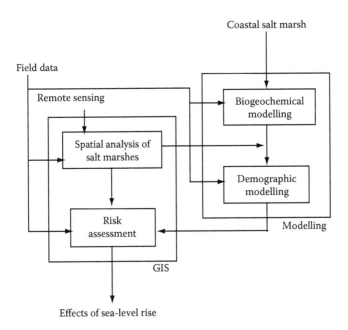

FIGURE 3.9
Methodology used to couple modelling and GIS approaches. (From Simas et al. *Ecological Modelling* 139, 1–15, 2001. With permission.)

3.4.5.1.1.3 Study System   The Tagus estuary (Portugal).

3.4.5.1.1.4 Model Structure and Main Equations   Salt marsh productivity was simulated by combining a biogeochemical model and a demographic model, which were developed for the aboveground vegetation. Because the biogeochemical model uses the same type of formulations previously described for other type of primary producers or described in the following section for another salt marsh plant (*Zostera noltii*), here we will focus in the demographic model. Details on biogeochemical models of seagrasses can be found in the work of Simas et al. (2001) and others (e.g. Simas and Ferreira 2007).
   The demographic model. Seagrass population dynamics was simulated through a class transition model, between weight classes, in order to describe plant population density per unit area. Four weight classes were considered and class transition was described by:

$$\frac{\partial n(s,t)}{\partial t} = -\frac{\partial\left[n(s,t)g(s,t)\right]}{\partial s} - \mu\left[(s)n(s,t)\right] \tag{3.88}$$

where t is time, s is weight class, n is the number of shoots, g is scope for growth, and μ is the mortality rate.

The number of shoots in each weight class depends on the individual shoot scope for growth and on allometric natural mortality rate. The model also takes into account that the erosion losses in the number of individuals are higher for the lower weight classes. Recruitment to class 1 is explicitly modeled as a source of new recruits to the population, which is accounted by the reproduction in weight class 4.

Sea-level rise scenarios. To simulate the gradual water elevation, a simple model was developed and coupled with the production module. The simple module for the water elevation assumed that the relative elevation to tidal datum of the surface of a salt marsh at a place (η) changes annually with "natural" sea-level rise due to mineral and organic accretion, which includes natural subsidence, soil compaction and depletion of groundwater levels, and sea-level rise due to climate change (Allen 1990a, 1990b, 1995, 1997):

$$\frac{\partial \eta}{\partial t} = \partial S_{min} + \partial S_{org} - \partial M - \partial P \qquad (3.89)$$

where t denotes time, S_{min} is the thickness of added minerogenic sediment, S_{org} is the thickness of added organogenic sediment, M is the change in relative sea level (positive upward), and P is change in the position of the marsh due to long-range compaction.

The water-level rise scenarios proposed by IPCC (1995) were chosen for the study. IPCC estimates a sea-level rise of 50 cm from 1999 to 2100, or 0.45 cm year^{-1}. The "high sea-level rise" scenario corresponds to a 95 cm rise, or 0.86 cm year^{-1} (IPCC 1995, 1999). A linear increase in this variable was considered during the simulation period.

3.4.5.1.1.5 Results and Discussion Biomass predictions for the lower and upper marshes fell within the range of observations (Catarino 1981; Catarino and Caçador 1981) (Figure 3.10).

Total plant production of the system was calculated by upscaling the ecological model using the GIS areas. Lower and upper marsh areas were simulated using the C_4 and C_3 models, respectively, because one C_4 species dominates the lower marsh (*Spartina maritima*), whereas two C_3 species dominate the upper marsh (*Halimione portucaloides* and *Arthrocnemum fruticosum*). According to the results, salt marsh production is responsible for the annual removal of about 1200 tons of nitrogen and 170 tons of phosphorus, which corresponds roughly to 270,000 and 170,000 inhabitants, respectively (Simas et al. 2001).

Further on, the model was used to simulate salt marsh production and biomass for both sea-level rise scenarios. In 2010, the model predicts losses in biomass and production of about 100% in the lower marsh, which suggests that by 2100, the current level of lower salt may potentially disappear. In the

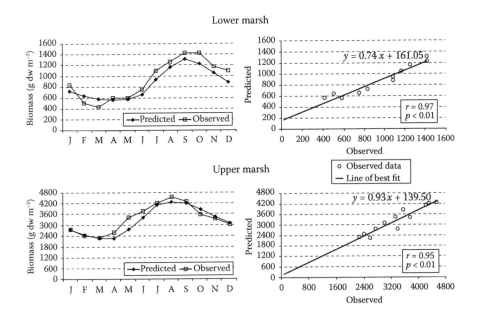

FIGURE 3.10
Observed and predicted biomass of salt marsh plants for the lower and upper marshes. The correlations between predictions and simulations are positive and significant ($P < 0.01$).

same scenario, the upper marsh biomass will decrease by 35% and production losses will reach 50% (Figure 3.11).

Finally, the proposed methodology was also used to carry out risk assessment in order to determine the susceptibility to submersion of different salt marsh areas in the estuary.

Results indicate that the eastern salt marsh area is the most resistant, because of its higher elevation, presenting a moderate to low risk of destruction. On the contrary, the lower salt marsh areas in the north and west of the estuary present high to very high risk.

3.4.5.1.2 *Z. noltii* and the Nitrogen Cycle

3.4.5.1.2.1 Problematic In some intertidal systems, rooted macrophytes show important standing stock values and cover significant areas (Gerbal and Verlaque 1995). The importance of their presence in those systems is related to their role in carbon fixation (e.g., Sfriso and Marcomini 1999), their impact on nutrient budgets (e.g., Touchette and Burkholder 2000) and their mitigation effect in eutrophication and anoxic episodes (e.g., Viaroli et al. 2001).

Z. noltii is an important macrophyte from the Thau lagoon system (Mediterranean coast of France), although few models of its growth or

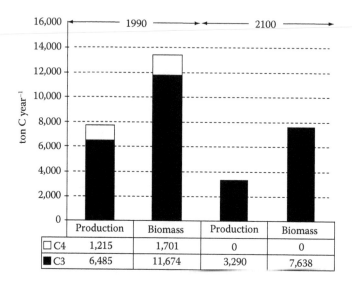

FIGURE 3.11
Simulation results of total biomass and production in lower (C4) and upper marsh (C3) for both
scenarios proposed marsh.

productivity have been developed compared to its more subtidal counter-
part *Zostera marina* (Plus et al. 2003).

3.4.5.1.2.2 Aim   The authors aimed at developing a *Z. noltii* ecosystem
model, able to simulate (1) the seasonal dynamics of biomass and nitrogen
content of the rooted macrophyte and (2) the contributions of different primary producers (seagrass, phytoplankton, and epiphytes) to the nitrogen
and oxygen cycle of the ecosystem.

3.4.5.1.2.3 Study System   The Thau lagoon at the French Mediterranean
coast.

3.4.5.1.2.4 Model Structure and Main Equations   The model takes into account
above- and belowground seagrass biomass, nitrogen quotas, and epiphytes.
The forcing functions are light intensity, wind speed, rainfall, and water
temperature (Figure 3.12).
    Z. noltii is represented by five state variables: aboveground biomass, below-
ground biomass, density of shoots, aboveground- and belowground-nitrogen
pools. In addition, the model includes a state variable representing epiphyte
biomass, and two state variables for dissolved inorganic nitrogen, ammonia,
and nitrate concentrations in the water column and in the sediment.
    The model takes into account a vertical structure, with a water box on top
(1.4 m depth), followed by three sediment boxes (0.5, 1.5, and 4.38 cm depth,

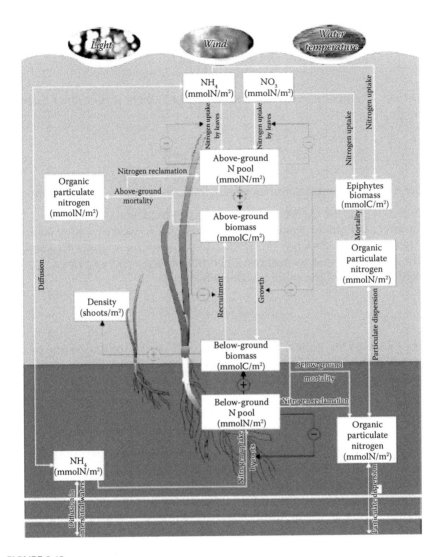

FIGURE 3.12
Conceptual diagram of the seagrass model. White rectangles are state variables, forcing functions are presented on the top of the diagram, white arrows represent fluxes between state variables and black arrows are limiting factors for processes. For simplication, processes related to the oxygen state variable are not shown (from Plus et al. *Ecological Modelling*, 161, 213 238, 2008. With permission.)

respectively). The vertical discrimination of the model allows describing the oxygen and nutrient fluxes at the water–sediment interface.

Seagrass photosynthesis is controlled by temperature and light according to the following equation:

$$P_{\text{tot}} = P_{\text{max}} \int_{z=0}^{\text{LAI}} \tanh\left(\frac{Q_{\text{can}} \times f_2(l)}{I_{k_1}}\right) dz \qquad (3.90)$$

The effect of temperature on the physiology of *Z. noltii* is described directly through:

$$P_{\text{max}} = \Theta_{P_{\text{max}}} \times t - P_{\text{max}}^{0°C} \qquad (3.91)$$

where $\Theta_{P_{\text{max}}}$ is the production increasing rate with temperature, t is temperature, and $P_{\text{max}}^{0°C}$ is the theoretical maximum production at 0°C.

The effect of light on seagrass productivity is described in a more complex manner because it depends on irradiance photosynthetic capacities and on shading effects caused by seagrass's bed morphology and epiphytes.

Light at the top of the canopy is described via:

$$Q_{\text{can}} = I \times f_1(E)e^{(-k_1 \times (D - 0.3))} \qquad (3.92)$$

where I denotes the light at the sea surface, $f_1(E)$ is the limitation due to epiphytes, D is the water column depth, and k_1 is the light extinction coefficient due to water.

$$f_1(E) = e^{(-k_2 \times (EB/LB))} \qquad (3.93)$$

where k_2 is the light extinction coefficient due to epiphytes, EB is the epiphyte biomass, and LB is the aboveground biomass.

The dependency of seagrass growth on nitrogen was described taking into account the uptake by leaves and rhizomes/roots and the storing capacity of the plant (Plus et al. 2003). Additionally, the authors have also accounted for the nutrient redistribution and reclamation capacity of *Z. noltii*, which tends to occur in nutrient-poor conditions. For the leaves and the rhizomes/roots, the part of nitrogen reclaimed was generically described through:

$$XN_{\text{rec}} = \begin{cases} \left[1 - \left(\dfrac{XN_{\text{sat}}}{S_{\text{rec}}}\right)^2\right] \times REC_{\text{max}} \\[2mm] 0, \text{ if } XN_{\text{sat}} > S_{\text{rec}} \end{cases} \qquad (3.94)$$

where XN_{sat} is the saturation level of leaf or rhizome/root nitrogen quota, S_{rec} is the internal reclamation threshold, and REC_{max} is the maximum reclamation rate.

Seagrass mortality includes two different processes, a temperature-dependent one ($f_5(t)$), which accounts for natural senescence and a physical stress-dependent mortality ($f_6(v)$), which describes the leaves or shoots sloughing due to wave motion:

$$LM = LMR_{20°C} \times f_5(t) + f_6(v) \tag{3.95}$$

where $LMR_{20°C}$ is the maximum leaf mortality at 20°C

$$f_5(t) = \Theta_{LM}^{t-20} \tag{3.96}$$

where Θ is coefficient for leaf mortality

$$f_6(v) = LMR_v \times \frac{1}{10}(V_{vent}) \times e^{-k_4 \times D} \tag{3.97}$$

where LMR_v denotes leaf sloughing due to wind, V_{vent} is the wind speed, D is depth, and k_4 is the wind effect attenuation with depth.

For the macrophyte recruitment, the model only takes into account vegetative reproduction because the sexual reproduction of *Z. noltii* in areas where the species is dominant is almost null (Laugier 1998).

Recruitment is considered as a transfer of matter from the rhizomes to the aboveground biomass. The initial biomass of new shoots is fixed at a constant weight and its apparition is controlled by temperature.

Epiphyte dynamics was accounted for and described through dependencies on light, temperature, and nitrogen.

3.4.5.1.2.5 Results and Discussion Simulations of the seasonal variations of *Z. noltii* biomass, shoot density, and nitrogen quotas followed the observed patterns. Although model validation was performed whenever possible, the authors point out the following weak points: (1) the model probably underestimates the seagrass rhizome and root biomass and (2) the absence of benthic detritus decomposition leads to an accumulation in the sediment.

Results also indicate that *Z. noltii* growth might suffer from nitrogen limitation during summer and that light competition with seagrasses is likely to take place. Epiphytes and phytoplankton also seem to compete for nitrogen.

Results from the coupling between the present model and another one developed for the same system (Chapelle et al. 2000) indicated that, within the Thau lagoon, seagrasses are more productive than phytoplankton and epiphytes and that such production increases oxygen concentrations in the water column. However, the impact of seagrasses on dissolved inorganic nitrogen in the water column is smaller than that of the phytoplankton's. Results also indicate that the exportation of seagrass leaves contributes more than phytoplankton or epiphyte mortality to the detritus compartment.

3.5 Conclusions

In the past decade, models have been developed for estuaries worldwide aiming at answering and clarifying some of the questions that arise related to these peculiar areas: to predict management scenarios that encompass both the great natural and socioeconomic value of estuaries, to quantify flows of energy and matter along food webs within estuarine areas with different disturbance levels; to understand the relationships between nutrient cycles and hypoxia; to quantify the relative proportion of nutrient input from freshwater and the adjacent sea; to ascertain the impacts from hydrological changes on estuarine communities; to assess if different areas within the estuarine area (e.g., central bay versus tidal area, deep channels versus broad shallow shoals) promote different nutrient- and primary producers-dynamics; and to assess the impacts of estuarine primary producers on the nutrient budget of the system.

It becomes evident that one of the most striking needs of models for estuaries and other coastal areas is related to developing integrated coastal management tools for decision-makers and examining the outcomes of different scenarios, particularly, those related with different nutrient loads, fish farming, and sea-level rise.

As ecologists are increasingly challenged to anticipate ecosystem change and emerging vulnerabilities (e.g., Clark 2005), reliable forecasts of the ecosystem state and services, and the natural capital are urgently required (Clark et al. 2001). This can be achieved with the availability of new data sets, the progress in computation and statistics and an engagement of scientists and decision-makers (Clark et al. 2001).

Ecosystem modeling seems to be the most suitable methodology to achieve the urgently required Ecosystem-Based Management approach, as it will contribute to give insights about ecological interactions within the ecosystem, to estimate the cumulative impacts of multiple activities operating at a given coastal system and to evaluate the susceptibility of the system to several stressors by means of scenario simulation (Nobre et al. 2010).

In spite of all the advances achieved in ecological modeling during the past decades, which were mostly related to progress in computation coupled to acquisition of large data sets, models still entail the dilemma of simplicity versus complexity, where the tradeoffs are well known (Clark 2005). If simple models with high generalization of processes may fail to capture essential ecosystem features and engage situation-specific and scale-dependent effects (Levin 1992; Carpenter 1996; Skelly 2002), complicated models require large datasets for setup and validation (Nobre et al. 2005), frequently leading to overfitting (Clark 2005) and generating outputs that are difficult to synthesize and interpret. The model complexity and, consequently, its dimensionality are decided upon what will be treated deterministically, what is assumed to be stochastic, and what can be ignored (Clark 2005). Stochasticity is central to the complexity dilemma, because it encompasses the elements that are uncertain and those that fluctuate due to factors that cannot be fully known or quantified (Clark 2005).

The ways or methodologies to overcome such dilemmas are still under discussion and few effective solutions have been presented. Among these options, there is the "Intermediate Approach," which includes different models running at different scales integrated to optimize the trade-offs between simple and complex models (Ferreira et al. 2008; Nobre et al. 2010). Another approach to this issue is the new hierarchical Bayes structure, which is based on the Bayesian approach and describes sampling-based algorithms for analysis of complex systems (Clark 2005).

One aspect that also needs further attention in models of natural systems is the need to assess how uncertainties in each component of the models reflect uncertainty in the final predictions (Borsuk et al. 2004). Again, Bayesian parameter estimation seems to be a suitable methodology to quantify uncertainty in model predictions, besides expressing model outputs as probability distributions, which are suited for stakeholders and policy makers when making decisions for sustainable environmental management (Arhonditsis et al. 2007).

In general and not surprisingly, estuarine models give particular emphasis to nutrients and primary producers, especially, phytoplankton and opportunistic macroalgae. In contrast, seagrasses and particularly microphytobenthos have been somehow neglected when considering modeling approaches of estuaries. However, microphytobenthos can contribute significantly to the productivity of shallow mudflats (McIntyre and Cullen 1995), whereas salt marsh plants play important roles in stabilizing sediments (Elkalay et al. 2003) or acting as sinks of nutrients (Simas and Ferreira 2007). Thus, estuarine models could potentially improve and describe more accurately the whole-system dynamics, if they would include the processes affecting the different types of primary producers inhabiting estuaries. The idea is not original and has been presented in several papers that have modeled, for instance, the competition between opportunistic macroalgae (*Ulva rigida*) and rooted macrophytes (*Z. marina*) (Coffaro and Bocci 1997), several different functional groups of macroalgae (Biber et al. 2004) or C3 and C4 salt marsh plants (Simas et al. 2001). We suggest that models from estuaries should include, as state variables, all existing types of primary producers at a certain system and not just one or two, as is frequently the case.

Overall, it is possible to conclude that during the past decade, estuarine models have became increasingly more accurate, complete, and robust due to an increasing knowledge of the estuarine intervenients and the processes affecting them, the acquisition and handling of large datasets, and the progress on computational and statistical fields. From all of these, we can reliably affirm that ecosystem modeling of estuaries is already an indispensable tool for the present and the future management and preservation of these extremely important ecosystems.

References

Allen, J. I., J. C. Blackford, and P. J. Radford. 1998. An 1-D vertically resolved modelling study of the ecosystem dynamics of the middle and southern Adriatic Sea. *Journal of Marine Systems* 18: 265–286.

Allen, J. R. L. 1990a. The formation of coastal peat marshes under an upward tendency of relative sea-level. *Journal of the Geological Society* 147: 743–745.

Allen, J. R. L. 1990b. Salt-marsh growth and stratification: A numerical model with special reference to the Severn Estuary, southwest Britain. *Marine Geology* 95: 77–96.

Allen, J. R. L. 1994. A continuity-based sedimentological model for temperate-zone tidal marshes. *Journal of the Geological Society* 151: 41–49.

Allen, J. R. L. 1995. Salt-marsh growth and fluctuating sea-level: Implications of a simulation model for Flandrian coastal stratigraphy and peat-based sea-level curves. *Sedimentary Geology* 100: 21–45.

Allen, J. R. L. 1997. Simulation models of salt-marsh morphodynamics: Some implications for high-intertidal sediment couplets related to sea-level change. *Sedimentary Geology* 113: 211–223.

Alvera-Azcárate A., J. G. Ferreira, and J. P. Nunes. 2003. Modelling eutrophication in mesotidal estuaries. The role of intertidal seaweeds. *Estuarine, Coastal and Shelf Science* 57: 715–724.

Amos, C. L., A. Bergamasco, G. Umgiesser, S. Cappucci, D. Cloutier, L. DeNat, M. Flindt, M. Bonardi, and S. Cristante. 2004. The stability of tidal flats in Venice Lagoon: The results of in-situ measurements using two benthic, annular flumes. *Journal of Marine Systems* 51: 211–241.

Andersen, T. J. 2001. Seasonal variation in erodibility of two temperate, microtidal mudflats. *Estuarine, Coastal and Shelf Science* 53: 1–12.

Arhonditsis, G. B., S. S. Qian, C. A. Stow, E. Conrad Lamon, and K. H. Reckhow. 2007. Eutrophication risk assessment using Bayesian calibration of process-based models: Application to a mesotrophic lake. *Ecological Modelling* 208: 215–229.

Baird, M. E., and S. M. Emsley. 1999. Towards a mechanistic model of plankton population dynamics. *Journal of Plankton Research* 21: 85–126.

Baird M. E., S. J. Walker, B. B. Wallace, I. T. Webster, and J. S. Parslow. 2003. The use of mechanistic descriptions of algal growth and zooplankton grazing in an estuarine eutrophication model. *Estuarine, Coastal and Shelf Science* 56: 685–695.

Baretta, J. W., W. Ebenhöh, and P. Ruardij. 1995. The European Regional Seas Ecosystem Model, a complex marine ecosystem model. *Netherlands Journal of Sea Research* 33: 233–246.

Baretta-Bekker, J. G., and J. W. Baretta, eds. 1997. Special Issue: European Regional Seas Ecosystem Model II. *Journal of Sea Research* 38 (3/4).

Barranguet, C., M. R. Plant-Cuny, and E. Alivon. 1996. Microphytobenthos production in the Gulf of Fos, French Mediterranean coast. *Hydrobiologia* 333: 181–193.

Bell, E. C. 1995. Environmental and morphological influences on thallus temperature and desiccation of the intertidal alga *Mastocarpus papillatus* Kützing. *Journal of Experimental Marine Biology and Ecology* 191: 29–55.

Behrenfeld, M. J., and P. G. Falkowski. 1997. A consumer's guide to phytoplankton primary productivity models. *Limnology and Oceanography* 42 (7): 1479–1491.

Berner, E. K., and R. A. Berner. 1996. *Global Environment: Water, Air and Geochemical Cycles.* Upper Saddle River, NJ: Prentice-Hall.

Berounsky, V. M., and S. W. Nixon. 1990. Temperature and the annual cycle of nitrification in waters of Narragansett Bay. *Limnology and Oceanography* 35: 1610–1617.

Biber, P. D., M. A. Harwell, and W. P. Cropper Jr. 2004. Modeling the dynamics of three functional groups of macroalgae in tropical seagrass habitats. *Ecological Modelling* 175: 25–54.

Blackford, J. C. 2002. The influence of microphytobenthos on the Northern Adriatic Ecosystem: A modelling study. *Estuarine Coastal and Shelf Science* 55: 109–123.

Blauw A. N., H. F. J. Los, M. Bokhorst, and P. L. A. Erftemeijer. 2009. GEM: A generic ecological model for estuaries and coastal waters. *Hydrobiologia* 618: 175–198.

Blumberg, A. F., and G. L. Mellor. 1980. A coastal ocean numerical model. In *Mathematical Modelling of Estuarine Physics. Proceedings of the International Symposium, Hamburg, August 1978*, eds. J. Sunderman and K. P. Holtz, 203–214. Berlin: Springer-Verlag.

Böhringer, C., and P. E. P. Jochem. 2007. Measuring the immeasurable—a survey of sustainability indices. *Ecological Economics* 63: 1–8.

Borsuk, M. E., C. A. Stow, and K. H. Reckhow. 2004. A Bayesian network of eutrophication models for synthesis, prediction, and uncertainty analysis. Ecological Modelling 173: 219–239.

Boudreau, B. P. 1996. The diffusive tortuosity of fine grained unlithified sediments. *Geochimica et Cosmochimica Acta* 60: 3139–3142.

Brand, F. 2009. Critical natural capital revisited: Ecological resilience and sustainable development. *Ecological Economics* 68: 605–612.

Braunschweig, F., F. Martins, P. Leitão, and R. Neves. 2003. A methodology to estimate renewal time scales in estuaries: The Tagus Estuary case. *Ocean Dynamics* 53 (3): 137–145.

Brock, T. D. 1981. Calculating solar radiation for ecological studies. *Ecological Modelling* 14: 1–19.

Broekhuizen, N. 1999. Simulating motile algae using a mixed Eulerian–Lagrangian approach: Does motility promote dinoflagelate persistence or co-existence with diatoms? *Journal of Plankton Research* 21 (7): 1191–1216.

Broekhuizen, N., J. Oldman, and J. Zeldis. 2003. Sub-grid-scale differences between individuals influence simulated phytoplankton production and biomass in a shelf-sea system. *Marine Ecology Progress Series* 252: 61–76.

Brundtland, G. H. Chair. 1987. Report on the World Commission on Environment and Development: Our common future. Transmitted to the General Assembly as an Annex to document A/42/427—Development and International Co-operation: Environment. UN Documents: Gathering a Body of Global Agreements. http://www.un-documents.net/wcedocf.htm.

Brush, M. J., and S. W. Nixon. 2010. Modelling the role of macroalgae in a shallow sub-estuary of Narragansett Bay, RI (USA). *Ecological Modelling* 221: 1065–1079.

Bunce, R. G. H., M. J. Metzger, R. H. G. Jongman, J. Brandt, G. De Blust, R. Elena-Rossello, G. B. Groom, L. Halada, G. Hofer, D. C. Howard, P. Kovar, C. A. Mücher, E. Padoa-Schioppa, D. Paelinx, A. Palo, M. Perez-Soba, I. L. Ramos, P. Roche, H. Skanes, and T. Wrbka. 2008. A standardized procedure for surveillance and monitoring of European habitats and provision of spatial data. *Landscape Ecology* 23 (1): 11–25.

Cahoon, L. B., and J. E. Cooke. 1992. Benthic microalgal production in Onslow Bay, North Carolina, USA. *Marine Ecology Progress Series* 84: 185–196.

Canuto, V. M., A. Howard, Y. Cheng, and M. S. Dubovikov, 2001. Ocean turbulence: Part I. One point closure model—momentum and heat vertical diffusivities. *Journal of Physical Oceanography* 31: 1413–1426.

Carpenter, S. R. 1996. Microcosm experiments have limited relevance for community and ecosystem ecology. *Ecology* 77: 677–680.

Carpenter, S. R., F. Westley, and M. G. Turner. 2005. Surrogates for resilience of social–ecological systems. *Ecosystems* 8: 941–944.

Carr, G., H. Duthie, and W. Taylor. 1997. Models of aquatic plant productivity: A review of the factor that influence growth. *Aquatic Botany* 59: 195–215.

Catarino, F. 1981. Papel das zonas húmidas do tipo sapal na descontaminação das água (in Portuguese). *Ciência* 1, 2 IV série.

Catarino, F. M., and M. I. Caçador. 1981. Produção de biomassa e estratégia do desenvolvimento em Spartina maritima e outros elementos da vegetação dos sapais do estuário do Tejo (in Portuguese). *Boletim da Sociedade Série* 2 (54): 387–403.

Chapelle, A., A. Ménesguen, J.-M. Deslous-Paoli, P. Souchu, N. Mazouni, A. Vaquer, and B. Millet. 2000. Modelling nitrogen, primary production and oxygen in a Mediterranean lagoon. Impact of oysters farming and inputs from the watershed. *Ecological Modelling* 127: 161–181.

Chapman, A. R. O. 1993. Hard data for matrix modelling of *Laminaria digitata* (Laminariales, Phaeophyta) populations. *Hydrobiologia* 260/261: 263–267.

Clark, J. S., S. R. Carpenter, M. Barber, S. Collins, A. Dobson, J. A. Foley, D. M. Lodge et al. 2001. Ecological forecasts: An emerging imperative. *Science* 239: 657–659.

Clark, J. S. 2005. Why environmental scientists are becoming Bayesians? *Ecology Letters* 8: 2–14.

Cloern, J. E. 1996. Phytoplankton bloom dynamics in coastal ecosystems: A review with some general lessons from sustained investigation of San Francisco Bay. *Reviews of Geophysics* 34: 127–168.

Coffaro, G., and M. Bocci. 1997. Resources competition between *Ulva rigida* and *Zostera marina*: A quantitative approach applied to the Lagoon of Venice. *Ecological Modelling* 102: 81–95.

Das A., D. Justić, and E. Swenson. 2010. Modeling estuarine-shelf exchanges in a deltaic estuary: Implications for coastal carbon budgets and hypoxia. *Ecological Modelling* 221: 978–985.

Delft Hydraulics. 2006. Delft3D-FLOW, a Simulation Program for Hydrodynamic Flow and Transports in 2 and 3 Dimensions. User Manual. Version 3.11.

De Groot, R., J. Van der Perk, A. Chiesura, and A. van Vliet. 2003. Importance and threat as determining factors for criticality of natural capital. *Ecological Economics* 44: 187–204.

Denman, K., and J. Marra. 1986. Modelling the time dependent photoadaptation of phytoplankton to fluctuating light. In *Marine Interfaces Ecohydrodynamics*, ed. J. Nihoul, 341–359. Amsterdam: Elsevier.

Duarte, P., and J. G. Ferreira. 1993. A methodology for parameter estimation in seaweed productivity modelling. *Hydrobiologia* 260–261: 183–189.

Duarte, P., and J. G. Ferreira. 1997. A model for the simulation of macroalgal population dynamics and productivity. *Ecological Modelling* 98: 199–214.

Duit, A., and V. Galaz. 2008. Governance and complexity—emerging issues for governance theory. *Governance. International Journal of Policy Administration, and Institutions* 21 (3): 311–335.

Eilers, P. H. C., and J. C. H. Peeters. 1988. A model for the relationship between light intensity and the rate of photosynthesis in phytoplankton. *Ecological Modelling* 42: 199–215.

Eilers, P. H. C., and J. C. H. Peeters. 1993. Dynamic behaviour of a model for photosynthesis and photoinhibition. *Ecological Modelling* 69: 113–133.

Einstein, H. A., and R. B. Krone. 1962. Experiments to determine modes of cohesive sediment transport in salt water. *Journal of Geophysical Research* 67: 1451–1461.

Ekins, P., S. Simon, L. Deutsch, C. Folke, and R. De Groot. 2003. A framework for the practical application of the concepts of critical natural capital and strong sustainability. *Ecological Economics* 44 (2–3): 165–185.

Elkalay, K., C. Frangoulis, N. Skliris, A. Goffart, S. Gobert, G. Lepoint, and J.-H. Hecq. 2003. A model of the seasonal dynamics of biomass and production of the seagrass *Posidonia oceanica* in the Bay of Calvi (Northwestern Mediterranean). *Ecological Modelling* 167: 1–18.

Elliott, M., S. J. Boyes, and D. Burdon. 2006. Integrated marine management and administration for an island state—the case for a new Marine Agency for the UK. *Marine Pollution Bulletin* 52 (5): 469–474.

Elliott, M., D. Burdon, K. L. Hemingway, and S. Apitz. 2007. Estuarine, coastal and marine ecosystem restoration: Confusing management and science—a revision of concepts. *Estuarine, Coastal and Shelf Science* 74: 349–366.

EPA. 1985. *Rates, Constants and Kinetics Formulations In Surface Water Quality Modelling*. 2nd ed. United States Environmental Prtotection Agency, Report EPA/600/3-85/040, 454 pp.

Eppley, R. W., and B. J. Peterson. 1979. Particulate organic matter flux and planktonic new production in the deep ocean. *Nature* 282: 677–680.

Ferreira, J. G., P. Duarte, and B. Ball. 1997. Trophic capacity of Carlingford Lough for oyster culture—analysis by ecological modelling. *Aquatic Ecology* 31 (4): 361–378.

Ferreira, J. G., and L. Ramos. 1989. A model for the estimation of annual production rates of macrophyte algae. *Aquatic Botany* 33: 53–70.

Ferreira, J. G., A. J. S. Hawkins, and S. B. Bricker. 2007. Management of productivity, environmental effects and profitability of shellfish aquaculture—the farm aquaculture resource management (FARM) model. *Aquaculture* 264: 160–174.

Ferreira, J. G., A. J. S. Hawkins, P. Monteiro, H. Moore, M. Service, P. L. Pascoe, L. Ramos, and A. Sequeira. 2008. Integrated assessment of ecosystem-scale carrying capacity in shellfish growing areas. *Aquaculture* 275: 138–151.

Field, C. B., M. J. Behrenfeld, J. T. Randerson, and P. G. Falkowski. 1998. Primary production of the biosphere: Integrating terrestrial and oceanic components. *Science* 281: 237–240.

Folke, C. 2006. Resilience: The emergence of a perspective for social–ecological systems analyses. *Global Environmental Change* 16: 253–267.

Folke, C., S. Carpenter, B. Walker, M. Scheffer, T. Elmqvist, L. Gunderson, and C. S. Holling. 2004. Regime shifts, resilience, and biodiversity in ecosystem management. *Annual Review of Ecology, Evolution, and Systematics* 35: 557–581.

Franks, P. J. S., and J. Marra, 1994. A simple new formulation for phytoplankton photoresponse and an application in a wind-driven mixed-layer model. *Marine Ecology Progress Series* 111: 143–153.

Fujiwara, T., L. P. Sanford, K. Nakatsuji, and Y. Sugiyama. 1997. Anti-cyclonic circulation driven by the estuarine circulation in a gulf type ROFI. *Journal of Marine Systems* 12: 1–4.

Fulton, E. A., A. D. M. Smith, and C. R. Johnson. 2004. Biogeochemical marine ecosystem models I: IGBMEM- a model of marine bay ecosystems. *Ecological Modelling* 174: 267–307.

Furnas, M. 1990. In situ growth rates of marine phytoplankton: Approaches to measurement, community and species growth rates. *Journal of Plankton Research* 12 (6): 1117–1151.

Gameiro, C., P. Cartaxana, and V. Brotas. 2007. Environmental drivers of phytoplankton distribution and composition in Tagus Estuary, Portugal. *Estuarine, Coastal and Shelf Science* 75: 21–34.

García A., J. A. Juanes, C. Álvarez, J. A. Revilla, and R. Medina. 2010. Assessment of the response of a shallow macrotidal estuary to changes in hydrological and wastewater inputs through numerical modelling. *Ecological Modelling* 221: 1194–1208.

Gerbal, M., and M. Verlaque. 1995. Macrophytobenthos de substrat meuble de l'étang de Thau (France, Méditerranée) et facteurs environnementaux associés (in French). *Oceanologica Acta* 18: 557–571.

Giraud, X., P. Bertand, V. Carcon, and I. Dadou. 2000. Modelling $\delta^{15}N$ evolution: First paleoceanographic applications in a coastal upwelling system. *Journal of Marine Systems* 58: 609–630.

Giraud, X., P. Bertand, V. Carcon, and I. Dadou, 2003. Interpretation of the nitrogen isotopic signal variations in the Mauritanian upwelling with a 2D physical-biological model. *Global Biogeochemical Cycles* 17: 1059. doi:10.1029/2002GB001951.

Grimm, V., and C. Wissel. 1997. Babel, or the ecological stability discussions: An inventory and analysis of terminology and a guide for avoiding confusion. *Oeacologia* 109: 323–334.

Guarini, J.-M., G. Blanchard, C. Bacher, P. Gros, P. Riera, P. Richard, D. Gouleau, R. Galois, J. Prou, and P. G. Sauriau. 1998. Dynamics of spatial patterns of microphytobenthic biomass: Inferences from geostatistical analysis of 2 comprehensive surveys in Marennes-Oléron Bay (France). *Marine Ecology Progress Series* 166: 131–141.

Guarini, J.-M., G. F. Blanchard, P. Gros, D. Gouleau, and C. Bacher. 2000. Dynamic model of the short-term variability of microphytobenthic biomass on temperate intertidal mudflats. *Marine Ecology Progress Series* 195: 291–303.

Guarini, J.-M., J. E. Cloern, J. Edmunds, and P. Gros. 2002. Microphytobenthic potential productivity estimated in three tidal embayments of the San Francisco Bay: A comparative study. *Estuaries* 25 (3): 409–417.

Gunderson, L. H., and C. S. Holling, eds. 2002. *Panarchy: Understanding Transformations in Human and Natural Systems*. Washington, DC: Island Press.

Gustafsson, K. E., and J. Bendtsen. 2007. Elucidating the dynamics and mixing agents of a shallow fjord through age tracer modelling. *Estuarine, Coastal and Shelf Science* 74: 641–654.

Han, B. P. 2001a. Photosynthesis–irradiance response at physiological level: A mechanistic model. *Journal of Theoretical Biology* 213: 121–127.

Han, B. P. 2001b. A mechanistic model of algal photoinhibition induced by photodamage to photosystem-II. *Journal of Theoretical Biology* 214: 519–527.

Hardman-Mountford, N. J., J. I. Allen, M. T. Frost, S. J. Hawkins, M. A. Kendall, N. Mieszkowska, K. A. Richardson, and P. J. Somerfield. 2005. Diagnostic monitoring of a changing environment: An alternative UK perspective. *Marine Pollution Bulletin* 50: 1463–1471.

Herman, P. M. J., J. J. Middelburg, and C. H. R. Heip. 2001. Benthic community structure and sediment processes on an intertidal flat: Results from the ECOFLAT project. *Continental Shelf Research* 21: 2055–2071.

Herbert, R. A. 1999. Nitrogen cycling in coastal marine ecosystems. *FEMS Microbiology Reviews* 23: 563–590.

Hernández, I., G. Peralta, J. L. Pérez-Lloréns, J. J. Vergara, and F. X. Niell. 1997. Biomass and dynamics of growth of *Ulva* species in Palmones river estuary. *Journal of Phycology* 33: 764–772.

Holling, C. S. 1996. Engineering resilience versus ecological resilience. In *Engineering within Ecological Constraints*, ed. P. C. Schulze, 31–44. Washington, DC: National Academy Press.

Houghton, J. 1999. *Global Warming: The Complete Briefing*, 2nd ed., 251. UK: Cambridge University Press.

Huang, S., H. Lou, Y. Xie, and J. Hu. 2003. Hydrodynamic environment and its effects in the Xiangshan Bay. In *International Conference on Estuaries and Coasts*, November 9–11, 2003. Hangzhou, China.

Huang, D. W., H. L. Wang, J. F. Feng, and Z. W. Zhu. 2008a. Modelling algal densities in harmful algal blooms (HAB) with stochastic dynamics. *Applied Mathematical Modelling* 32: 1318–1326.

Huang, X. Q., J. H. Wang, and X. S. Jiang. 2008b. *Marine Environmental Capacity and Pollution Control in Xiangshan Gang*, 348 pp. China Ocean Press.

Hughes, T. P., D. R. Bellwood, C. Folke, R. S. Steneck, and J. Wilson. 2005. New paradigms for supporting the resilience of marine ecosystems. *Trends in Ecology & Evolution* 20 (7): 380–386.

Hur, H. B., G. A. Jacobs, and W. J. Teague. 1999. Monthly variations of water masses in the Yellow and East China Sea. *Journal of Oceanography* 55: 171–184.

Hurd, C. L. 2000. Water motion, marine macroalgal physiology and production. *Journal of Phycology* 36: 453–472.

Isobe, A., E. Fujiware, P. Chang, K. Sugimatsu, M. Shimizu, T. Matsuno, and A. Manda. 2004. Intrusion of less saline shelf water into the Kuroshio subsurface layer in the East China Sea. *Journal of Oceanography* 60: 853–863.

Intergovernmental Panel on Climate Change (IPCC). 1990. Climate change. In *The IPCC Assessment*, eds. J. T. Houghton, G. J. Jenkins, and J. J. Ephraumus. UK: Cambridge University Press. http://www.ipcc.ch/pub/reports.htm.

Intergovernmental Panel on Climate Change (IPCC). 1995. Climate Change 1995. The IPCC Second Assessment Synthesis of Scientific-Technical Information Relevant to Interpreting Article 2 of the UN Framework Convention on Climate Change. Switzerland: IPCC. http://www.ipcc.ch/pub/reports.htm.

Intergovernmental Panel on Climate Change (IPCC). 1999. Aviation and the global atmosphere. A special report of working groups I and III of the Intergovernmental Panel on Climate Change. Summary for policymakers. Switzerland: IPCC. http://www.ipcc.ch/pub/reports.htm.

Jassby, A. D., and T. Platt. 1976. Mathematical formulation of the relationship between photosynthesis and light for phytoplankton. *Limnology and Oceanography* 21: 540–547.

Jesus, B., C. R. Mendes, V. Brotas, and D. M. Paterson. 2006. Effect of sediment type on microphytobenthos vertical distribution: Modelling the productive biomass and improving ground truth measurements. *Journal of Experimental Marine Biology and Ecology* 332: 60–74.

Jørgensen, S. E., and G. Bendoricchio. 2001. Fundamentals of ecological modelling. In *Developments in Environmental Modelling*, 3rd ed., vol. 21, 530 pp. Netherlands: Elsevier.

Kirk, J. T. 1975. A theoretical analysis of the contribution of algal cells to the attenuation of light within natural waters. II. Spherical cells. *New Phytologist* 75: 21–36.

Krone, R. B. 1962. Flume studies of the transport of sediment in estuarine shoaling processes. Final Report to San Francisco District U.S. Army Corps of Engineers, Washington, DC.

Large, W. G., J. C. McWilliams, and S. C. Doney. 1994. Oceanic vertical mixing: A review and model with a nonlocal boundary layer parameterization. *Reviews of Geophysics* 32 (1): 363 403.

Laugier, T. 1998. *Ecologie de deux phanérogames marines sympatriques—Zostera marina L. et Z. noltii Hornem. dans l'étang de Thau (Hérault, France)*, 162. France: Thèse de l'université de Montpellier II.

Lebel, L., J. M. Anderies, B. Campbell, C. Folke, S. Hatfield-Dodds, T. P. Hughes, and J. Wilson. 2006. Governance and the capacity to manage resilience in regional social–ecological systems. *Ecology and Society* 11 (1): 19.

Levin, S. A. 1992. The problem of pattern and scale in ecology. *Ecology* 73: 1943–1967.

Levin, S. A., and J. Lubchenco. 2008. Resilience, robustness, and marine ecosystem-based management. *Bioscience* 58 (1): 27–32.

Lindeman, R. L. 1942. The trophic–dynamic aspect of ecology. *Ecology* 23: 399–418.

Longhurst, A., S. Sathyendranath, T. Platt, and C. Caver-Hill. 1995. An estimate of global primary production in the ocean from satellite radiometer data. *Journal of Plankton Research* 17: 1245–1271.

Lopes, R. J., M. A. Pardal, T. Múrias, J. A. Cabral, and J. C. Marques. 2006. Influence of macroalgal mats on abundance and distribution of dunlin *Calidris alpina* in estuaries: A long-term approach. *Marine Ecology. Progress Series* 323: 11–20.

Lotze, H. K., and B. Worm. 2000. Variable and complementary effects of herbivores on different life stages of bloom-forming macroalgae. *Marine Ecology. Progress Series* 200: 167–175.

Lucas, L. V., J. R. Koseff, J. E. Cloern, S. G. Monismith, and J. K. Thompson. 1999a. Processes governing phytoplankton biomass blooms in estuaries: I. The local production–loss balance. *Marine Ecology. Progress Series* 187: 1–15.

Lucas, L. V., J. R. Koseff, J. E. Cloern, S. G. Monismith, and J. K. Thompson. 1999b. Processes governing phytoplankton biomass blooms in estuaries: II. The role of horizontal transport. *Marine Ecology. Progress Series*. 187: 17–30.

Lucas L. V., J. R. Koseff, S. G. Monismith, and J. K. Thompson. 2009. Shallow water processes govern system-wide phytoplankton bloom dynamics: A modelling study. *Journal of Marine Systems* 75: 70–86.

Lumborg, U., T. J. Andersen, and M. Pejrup. 2006. The effect of *Hydrobia ulvae* and microphytobenthos on cohesive sediment dynamics on an intertidal mudflat described by means of numerical modelling. *Estuarine, Coastal and Shelf Science* 68: 208–220.

Luyten, P., J. E. Jones, R. Proctor, A. Tabor, P. Tett, and K. Wild-Allen. 1999. COHERENS documentation: A coupled hydrodynamical–ecological model for regional and shelf seas: user documentation. MUMM internal document.

Macedo, M. F., and P. Duarte. 2006. Phytoplankton production modelling in three marine ecosystems—static versus dynamic approach. *Ecological Modelling* 190: 299–316.

MacIntyre, H. L., and J. J. Cullen. 1995. Fine-scale resolution of chlorophyll and photosynthetic parameters in shallow-water benthos. *Marine Ecology. Progress Series* 122: 227–237.

MacIntyre, H. L., R. J. Geider, and D. C. Miller. 1996. Microphytobenthos: the ecological role of the "secret garden" of unvegetated, shallow-water marine habitats: I. Distribution, abundance and primary production. *Estuaries* 19: 186–201.

MacIntyre, H. L., J. Cullen, and A. Cembella. 1997. Vertical migration, nutrition and toxicity of the dinoflagellate *Alexandrium tamarense*. *Marine Ecology. Progress Series* 148: 201–216.

MacLeod, M., and J. A. G. Cooper. 2005. Carrying capacity in coastal areas. In *Encyclopedia of Coastal Science*, ed. M. Schwartz, 226 pp. Heidelberg: Springer.

Madden, C. J., and W. M. Kemp. 1996. Ecosystem model of an estuarine submersed plant community: Calibration and simulation of eutrophication responses. *Estuaries* 19: 457–474.

Marques, J. C., A. Basset, T. Brey, and M. Elliot. 2009. The ecological sustainability trigon—a proposed conceptual framework 4 for creating and testing management scenarios. *Marine Pollution Bulletin* 58: 1773–1779.

Martins, I., M. A. Pardal, A. I. Lillebø, M. R. Flindt, and J. C. Marques. 2001. Hydrodynamics as a major factor controlling the occurrence of green macroalgal blooms in a eutrophic estuary: A case study on the influence of precipitation and river management. *Estuarine, Coastal Shelf Science* 52: 165–177.

Martins, I., and J. C. Marques. 2002. A model for the growth of opportunistic macroalgae (*Enteromorpha* sp.) in tidal estuaries. *Estuarine, Coastal Shelf Science* 55 (2): 247–257.

Martins I., R. J. Lopes, A. I. Lillebø, J. M. Neto, M. A. Pardal, J. G. Ferreira, and J. C. Marques. 2007. Significant variations in the productivity of green macroalgae in a mesotidal estuary: Implications to the nutrient loading of the system and the adjacent coastal area. *Marine Pollution Bulletin* 54: 678–690.

Martins, I., A. Marcotegui, and J. C. Marques. 2008. Impacts of macroalgal spores on the dynamics of adult macroalgae in a eutrophic estuary: High versus low hydrodynamic seasons and long-term simulations for global warming scenarios. *Marine Pollution Bulletin* 56: 984–998.

Mee, L. D., R. L. Jefferson, D. A. Laffoley, and M. Elliott. 2008. How good is good? Human values and Europe's proposed Marine Strategy Directive. *Marine Pollution Bulletin* 56: 187–204.

Mellor, G. L., and T. Yamada. 1982. Development of a turbulence closure model for geophysical fluid problems. *Reviews of Geophysics* 20, 851–875.

Menesguen, A., P. Cugier, and I. Leblond. 2006. A new numerical technique for tracking chemical species in a multisource, coastal ecosystem applied to nitrogen causing Ulva blooms in the Bay of Brest (France). *Limnology and Oceanography* 51: 591–601.

Miller III, H. L., C. Meile, and A. B. Burd. 2007. A novel 2D model of internal O₂ dynamics and H₂S intrusion in seagrasses. *Ecological Modelling* 205: 365–380.

Monbet, Y. 1992. Control of phytoplankton biomass in estuaries: A comparative analysis of microtidal and macrotidal estuaries. *Estuaries* 15: 563–571.

Moore, S. A., T. J. Wallington, R. J. Hobbs, P. R. Ehrlich, C. S. Holling, S. Levin, D. Lindenmayer, C. Pahl-Wostl, H. Possingham, M. G. Turner, and M. Westoby. 2009. Diversity in current ecological thinking: Implications for environmental management. *Environmental Management* 43: 17–27.

Murray, A., and J. Parslow. 1997. Port Phillip Bay Integrated Model: Final Report. Technical Report No. 44. Port Phillip Bay Environmental Study. CSIRO, Canberra, Australia.

Murray, A. G., and J. S. Parslow. 1999. Modelling of nutrient impacts in Port Phillip Bay—a semi-enclosed marine Australian ecosystem. *Marine and Freshwater Research* 50: 597–611.

Murawski, S. A. 2007. Ten myths concerning ecosystem approaches to marine resource management. *Marine Policy* 31: 681–690.

Murawski, S., N. Cyr, M. Davidson, Z. Hart, M. Noaa Balgos, K. Wowk, and B. Cicin-Sain. Global Forum 2008. Policy brief on achieving EBM and ICM by 2010 and progress indicators. In *4th Global Conference on Oceans, Coast, and Islands: Advancing Ecosystem Management and Integrated Coastal and Ocean Management in the Context of Climate Change*, April 7–11, 2008, p. 70, Hanoi, Vietnam.

Neitsch, S. I., J. G. Arnold, J. R. Kiniry, J. R. Williams, and K. W. Kiniry. 2002. Soil and Water Assessment Tool Theoretical Documentation. TWRI report TR-191. Texas Water Resources Institute, College Station.

Nobre, A. M., J. G. Ferreira, A. Newton, T. Simas, J. D. Icely, and R. Neves. 2005. Management of coastal eutrophication: integration of field data, ecosystem-scale simulations and screening models. *Journal of Marine Systems* 56: 375–390.

Nobre, A. M. 2009. An ecological and economic assessment methodology for coastal ecosystem management. *Environmental Management* 44 (1): 185–204.

Nobre, A. M., and J. G. Ferreira. 2009. Integration of ecosystem-based tools to support coastal zone management. *Journal of Coastal Research* SI 56: 1676–1680.

Nobre, A. M., J. G. Ferreira, J. P. Nunes, X. Yan, S. Bricker, R. Corner, S. Groom et al. 2010. Assessment of coastal management options by means of multilayered ecosystem models. *Estuarine, Coastal and Shelf Science* 87: 43–62.

Nunes, J. P., J. G. Ferreira, F. Gazeau, J. Lencart-Silva, X. L. Zhang, M. Y. Zhu, and J. G. Fang. 2003. A model for sustainable management of shellfish polyculture in coastal bays. *Aquaculture* 219: 257–277.

Nyman, J. A., R. D. DeLaune, H. H. Roberts, W. H. Patrick Jr. 1993. Relationship between vegetation and soil formation in a rapid submerging coastal salt marsh. *Marine Ecology. Progress Series* 96: 269–279.

Nyman, J. A., M. Carloss, R. D. DeLaune, and W. H. Patrick Jr. 1994. Erosion rather than plant dieback as the mechanism of marsh loss in an estuarine marsh. *Earth Surface Process and Landforms* 19: 69–84.

Ojeda-Martínez, C., F. G. Casalduero, J. T. Bayle-Sempere, C. B. Cebrián, C. Valle, J. L. Sanchez-Lizaso, A. Forcada et al. 2009. A conceptual framework for the integral management of marine protected areas. *Ocean & Coastal Management* 52 (2): 89–101.

OSPAR. 1998. Report of the ASMO modelling workshop on eutrophication issues, 5–8 November 1996, The Hague, The Netherlands. OSPAR Commission Report,

Netherlands Institute for Coastal and Marine Management/RIKZ, The Hague, The Netherlands.

Oshima, Y., M. J. Kishi, and T. Sugimoto. 1999. Evaluation of the nutrient budget in a seagrass bed. *Ecological Modelling* 115: 19–33.

Pahl-Wostl, C., and D. M. Imboden. 1990. DYPHORA—a dynamic model for the rate of photosynthesis of algae. *Journal of Plankton Research* 12: 1207–1221.

Parchure, T. M., and A. J. Mehta. 1985. Erosion of soft cohesive sediment deposits. *Journal of Hydraulic Engineering* 111: 1308–1326.

Partheniades, E. 1965. Erosion and deposition of cohesive soils. *Journal of the Hydraulics Division* 91: 105–139.

Pasciak, W. J., and J. Gavis, 1975. Transport limited nutrient uptake rates in *Dictylum brightwellii*. *Limnology and Oceanography* 20: 604–617.

Parker, R. A. 1974. Empirical functions relating metabolic processes in aquatic systems to environmental variables. *Journal of the Fisheries Research Board of Canada* 31: 1150–1552.

Paterson, D. M. 1989. Short-term changes in the erodibility of intertidal cohesive sediments related to the migratory behaviour of epipelic diatoms. *Limnology and Oceanography* 34: 223–234.

Paterson, D. M., and S. E. Hagerthey. 2001. Mycrophytobenthos in contrasting coastal ecosystems: Biology and dynamics. In *Ecological Comparisons of Sedimentary Shores*, ed. K. Reise, 105–125. Berlin: Ecolgical Studies. Springer.

Patrício, J., and J. C. Marques. 2006. Mass balanced models of the food web in three areas along a gradient of eutrophication symptoms in the south arm of the Mondego estuary (Portugal). *Ecological Modelling* 197: 21–34.

Plus, M., A. Chapelle, A. Ménesguen, J.-M. Deslous-Paoli, and I. Auby. 2003. Modelling seasonal dynamics of biomasses and nitrogen contents in a seagrass meadow (*Zostera noltii* Hornem): Application to the Thau lagoon (French Mediterranean coast). *Ecological Modelling* 161: 213–238.

Pulselli, F. N., S. Bastianoni, N. Marchettini, and E. Tiezzi. 2008. *The Road to Sustainability. GDP and Future Generations*, 197 pp. Southampton: WIT Press.

Radach, G., and A. Moll. 2006. Review of three-dimensional ecological modelling related to the North Sea shelf system: Part II. Model validation and data needs. *Oceanography and Marine Biology* 44: 1–60.

Rivers, J., and P. Peckol. 1995. Summer decline of *Ulva lactuca* (Chlorophyta) in a eutrophic embayment: Interactive effects of temperature and nitrogen availability? *Journal of Phycology* 31: 223–228.

Ross, O. N., and J. Sharples. 2008. Swimming for survival: A role of phytoplankton motility in a stratified turbulent environment. *Journal of Marine Systems* 70: 248–262.

Rubio, F. C., F. G. Camacho, J. M. F. Sevilla, Y. Chisti, and E. Molina. 2003. A mechanistic model of photosynthesis in microalgae. *Biotechnology and Bioengineering* 81: 459–473.

Santelices, B. 1990. Patterns of reproduction, dispersal and recruitment in seaweeds. *Oceanography and Marine Biology* 28: 177–276.

Sfriso, A., and A. Marcomini. 1999. Macrophyte production in a shallow coastal lagoon: Part II. Coupling with sediment, SPM and tissue carbon, nitrogen and phosphorus concentrations. *Marine Environmental Research* 47: 285–309.

Sharples, J., O. N. Ross, B. E. Scott, S. P. R. Greenstreet, H. Fraser. 2006. Inter-annual variability in the timing of stratification and the spring bloom in the northwestern North Sea. *Continental Shelf Research* 26 (6): 733–751.

Schild, R., and D. Prochnow. 2001. Coupling of biomass production and sedimentation of suspended sediments in eutrophic rivers. *Ecological Modelling* 145: 263–274.

Schories, D., J. Anibal, A. S. Chapman, E. Herre, I. Isaksson, A. I. Lillebø, L. Pihl, K. Reise, M. Sprung, and M. Thiel. 2000. Flagging greens: Hydrobiid snails as substrata for the development of green algal mats (*Enteromorpha* spp.) on tidal flats of North Atlantic coasts. *Marine Ecology. Progress Series* 199: 127–136.

Sheldon, R. W., A. Prakash, and W. H. Sutcliffe Jr. 1972. The size distribution of particles in the ocean. *Limnology and Oceanography* 17: 327–340.

Simas, T. C., and J. G. Ferreira. 2007. Nutrient enrichment and the role of salt marshes in the Tagus estuary (Portugal). *Estuarine, Coastal Shelf Science* 75: 393–407.

Simas, T., J. P. Nunes, and J. G. Ferreira. 2001. Effects of global climate change on coastal salt marshes. *Ecological Modelling* 139: 1–15.

Simon, N., A.-N. Cras, E. Foulon, and R. Lemée. 2009. Diversity and evolution of marine phytoplankton. *Comptes Rendus Biologies* 332: 159–170.

Simpson, J. H. 1997. Physical processes in the ROFI regime. *Journal of Marine Systems* 12: 1–4.

Singh, R. K., H. R. Murty, S. K. Gupta, and A. K. Dikshit. 2009. An overview of sustainability assessment methodologies. *Ecological Indicators* 9: 189–212.

Skelly, D. 2002. Experimental venue and estimation of interaction strength. *Ecology* 83: 2097–2101.

Smayda, T. J. 1997. Harmful algal blooms: Their ecophysiology and general relevance to phytoplankton blooms in the sea. *Limnology and Oceanography* 42 (5, part 2): 1137–1153.

Sohma A., Y. Sekiguchi, T. Kuwae, and Y. Nakamura. 2008. A benthic–pelagic coupled ecosystem model to estimate the hypoxic estuary including tidal flat—Model description and validation of seasonal/daily dynamics. *Ecological Modelling* 215: 10–39.

Sousa, A. I., I. Martins, A. I. Lillebø, M. R. Flindt, and M. A. Pardal. 2007. Influence of salinity, nutrients and light on the germination and growth of *Enteromorpha* sp. spores. *Journal of Experimental Marine Biology and Ecology* 341: 142–150.

Steele, J. H. 1962. Environmental control of photosynthesis in the sea. *Limnology and Oceanography* 7: 137–150.

Sugimoto, R., A. Kasai, T. Miyajima, and K. Fujita. 2008. Nitrogen isotopic discrimination by water column nitrification in a shallow coastal environment. *Journal of Oceanography* 64: 39–48.

Sugimoto, R., A. Kasai, T. Miyajima, and K. Fujita. 2010. Modeling phytoplankton production in Ise Bay, Japan: Use of nitrogen isotopes to identify dissolved inorganic nitrogen sources. *Estuarine, Coastal and Shelf Science* 86: 450–466.

Tang, Y. 1995. The allometry of algal growth rates. *Journal of Plankton Research* 17 (6): 1325–1335.

Tett, P. 1998. *Parameterising a Microplankton Model*. Edinburgh, UK: Napier University.

Timmermann, K., S. Markager, and K. E. Gustafsson. 2010. Streams or open sea? Tracing sources and effects of nutrient loadings in a shallow estuary with a 3D hydrodynamic-ecological model. *Journal of Marine Systems* 82: 111–121.

Titus, J. G., and V. K. Narayanan. 1995. *The Probability of Sea-Level Rise*. USA: Environmental Protection Agency. http://users.erols.com/ititus/Holding/NRJ.html.

Titus, J. G., R. A. Park, S. P. Leatherman, J. R. Weggel, M. S. Greene, P. W. Mausel, S. Brown, C. Gaunt, M. Trehan, and G. Yohe. 1991. Greenhouse effect and sea-level rise: The cost of holding back the sea. *Coastal Management* 19: 171–210.

Touchette, B. W., and J. M. Burkholder. 2000. Review of nitrogen and phosphorus metabolism in seagrasses. *Journal of Experimental Marine Biology and Ecology* 250: 133–167.

Trancoso, A. R., S. Saraiva, L. Fernandes, P. Pina, P. Leitão, and R. Neves. 2005. Modelling macroalgae using a 3D hydrodynamic-ecological model in a shallow, temperate estuary. *Ecological Modelling* 187: 232–246.

Underwood, G. J. C., and J. Kromkamp. 1999. Primary production by phytoplankton and microphytobenthos in estuaries. *Advances in Ecological Research* 29: 93–153.

Van Dijkemann, K. S., J. H. Bossinade, P. Bouwema, and R. J. Glopper. 1990. Salt marshes in the Netherlands Wadden Sea: Rising high-tide levels and accretion enhancement. In *Expected Effects of Climatic Change on Marine Coastal Ecosystems*, 173–188. The Netherlands: Kluwer Academic Publishers.

Viaroli, P., R. Azzoni, M. Bartoli, G. Giordani, and L. Tajè. 2001. Evolution of the trophic conditions and dystrophic outbreaks in the Sacca di Goro lagoon (Northern Adriatic Sea). In *Mediterranean Ecosystems: Structures and Processes*, eds. E. M. Faranda, L. Guglielmo, and G. Spezie, 443–451. Italy: Springer-Verlag.

Vichi, M., N. Pinardi, M. Zavatarelli, G. Matteucci, M. Marcaccio, M. C. Bergamini, and F. Frascari. 1998. One dimensional ecosystem model results in the Po prodelta area (northern Adriatic Sea). *Environmental Modelling and Software* 13: 471–481.

Walker, B., C. S. Hollin, S. R. Carpenter, and A. Kinzig. 2004. Resilience, adaptability and transformability in social-ecological systems. *Ecology and Society* 9 (2): 5.

Waser, N. A., P. J. Harrison, B. Nielsen, S. E. Calvert, and D. H. Turpin. 1998a. Nitrogen isotope fractionation during uptake and assimilation of nitrate, nitrite, ammonium and urea by a marine diatom. *Limnology and Oceanography* 43: 215–224.

Waser, N. A., K. Yin, Z. Yu, K. Tada, P. J. Harrison, D. H. Turpin, and S. E. Calvert. 1998b. Nitrogen isotope fractionation during uptake and assimilation of nitrate, nitrite, ammonium and urea by marine diatoms and coccolithophores under various conditions of N availability. *Marine Ecology Progress Series* 169: 29–41.

Webb, W. L., M. Newton, and D. Starr. 1974. Carbon dioxide exchange of *Alnus rubra*: A mathematical model. *Oecologia* 17: 281–291.

Widdows, J., S. Brown, M. D. Brinsley, P. N. Salkeld, and M. Elliot. 2000. Temporal changes in intertidal sediment erodability; influence of biological and climatic factors. *Continental Shelf Research* 20: 1275–1289.

WL Delft Hydraulics. 2003. Delft3D-WAQ User's Manual. WL Delft Hydraulics, Delft, The Netherlands.

Yoshikawa, C., Y. Yamanaka, and T. Nakatsuka. 2005. An ecosystem model including nitrogen isotopes: Perspectives on a study of the marginal nitrogen cycle. *Journal of Oceanography* 61: 921–942.

4

Models of Rivers and Streams

Sven Erik Jørgensen

CONTENTS

4.1 Introduction... 137
4.2 Overview of River and Stream Models 138
4.3 A Detailed River Model of Medium Complexity................................... 140
References.. 151

4.1 Introduction

Streams and river have two energy sources:

1. Sunlight, the fuel of the primary production
2. Plant litter from riparian vegetation

The greater the plant litter inputs, the greater the limitation of light reaching the water surface, which implies limitation of algae and vascular plant growth. The processes determining the energy budget are ingestion, assimilation, production, egestion, and respiration. Sixty-six percent of the ingested energy is in average egested and 34% is assimilated, of which 74% in average is respired and 26% is used for production (Cummins 1974). The energy budget is often an important guideline for river model development.

Rivers and streams are characterized by interactions of a number of very dynamic processes: hydrological, geomorphic, aquatic, chemical, and ecological (biological) processes. The last processes are controlled by many taxonomic groups characteristic for running waters: algae, macrophytes, benthic macroinvertebrates, and fishes. The key question for the development of river and streams models is: "Which processes are crucial for the focal problem?" The following problems have been considered in models of rivers and streams:

A. Oxygen depletion (the oxygen is consumed by oxidation of organic matter and by nitrification)
B. Acidification

C. Eutrophication by algae and macrophytes

D. Elevated temperature due to discharge of cooling water

E. Turbidity due to suspended matter

F. Impacts due to climate change

G. Toxic substances—heavy metals and toxic organic matter

H. Decreased biodiversity

A mathematical representation of most of the processes included in models focusing on problems A to H are presented in Section II of this book, "Environmental Management Problem." The equations used for the processes determining the eutrophication, for instance, in rivers and streams, are in principle not different from the same processes in other aquatic ecosystems. Similarly, the processes of toxic substances in river and streams can be described by the same equations that are used for other aquatic ecosystems. A few of the equations will be repeated in the next section, which will present an overview of models of rivers and streams, including an overview of the processes included in a wide spectrum of applicable river and streams models.

The following section presents a further development of the simple Streeter–Phelps model, which includes discussion of nitrification and denitrification, settling, sediment oxygen demand (SOD), and the development of coliform bacteria. Eutrophication and photosynthesis could easily have been added, too, but these processes are presented in detail in Chapters 3 and 12. It was therefore deemed more important to present the other important processes that are included in several river and stream models, and show how these processes could easily be added even to a simple river and stream model such as the Streeter–Phelps model.

4.2 Overview of River and Stream Models

The first river model was developed by Streeter and Phelps 90 years ago. It was two differential equations covering the dynamic of oxygen and organic matter. The model included the following processes:

- Oxygen was consumed by degradation of organic matter, which was described as a first-order reaction and oxygen transfer from the atmosphere to the river water by reaeration.
- The Streeter–Phelps model can be solved analytically.
- The model covered the two most important processes for problem A (see list in Section 4.1).

Oxygen depletion is probably the most important environmental problem in rivers and streams, and therefore models that included more processes of significance for the oxygen concentration in water have been developed, particularly after 1970, when a massive development of ecological models began and more computer power became available. Table 4.1 gives an overview of the processes that have been included in a few characteristic and illustrative river and stream models that focused mainly on oxygen depletion, *in addition* to reaeration, degradation of organic matter, and nitrification. All three processes could be included in various versions of the Streeter–Phelps model that was already in use during the 1920s. It is clear from this short overview that river and stream models with a wide spectrum of complexity are available, and that attempts had been made to include all processes that are relevant for the oxygen depletion problem in at least one model, but often in more models. Several models have included the impact of photosynthesis, which could, of course, contribute considerably to the oxygen balance. The settling of organic matter and oxygen consumption by the organic matter in the sediment have been likewise included in several river and stream models. Photosynthesis is covered by equations similar to those applied in eutrophication models in general (see Chapters 3 and 12). The settling of particulate organic matter (POM) and oxygen consumption by sediments are mostly covered by first-order reactions in the river and stream models. It is noticeable that river and streams models developed to include relevant processes for oxygen depletion inevitably have to include processes that are relevant for the eutrophication of rivers and streams.

TABLE 4.1

Overview of Processes Included in a Number of Characteristic River and Stream Models

Processes	Model Name and Reference
Inflows, mixing, primary production, nutrient dynamics	PRAM, Brown and Barnell (1987)
Sediment processes, benthic organisms	MBM, Cazelles (1987); Cazelles, Fontvieille and Chau (1991)
Settling, coliform growth, photosynthesis of epiphytes and macrophytes, sediment oxygen demand, dispersion	Park and Uchrim (1988) STREAM, Kangwon National University 1993
Photosynthesis, grazing, coliform growth, denitrification	RMB10, Bowler at al. (1992)
Coliform growth, benthic oxygen demand, radioactive material, photosynthesis, and nutrient dynamics	QUAL, the model is in various editions by Texas Water Development Board
N-cycling processes in water and sediment, growth of benthic algae and cyanobacteria, POM	Loucks and van Beek (2005)
Benthic oxygen demand, coliform growth, spatial and temporal flow, and temperature variations	Armstrong (1977)

A number of river models focused on other problems than oxygen depletion or eutrophication. A few illustrative examples are as follows:

- ASRAM, developed by Korman et al. (1994), focuses on acidification of river and streams and how a change in pH influence the salmonoid life.
- EXWAT, developed by Paasivirta (1994), focuses on the fate of chlorophenoles and tetraethyl lead.
- OHIO RIVER OIL SPILL MODEL, developed by Mitsch (1990) (see Cronk, Mitsch and Sykes 1990). The model is based on an oil spill in a river. Oil spill loss is via sedimentation, whereas high flow conditions may cause resuspension of oil from the sediment.

The influence of a number of factors—pH, total solid, alkalinity, hardness, chloride, phosphate, potassium, sodium, ammonium-N, nitrate-N, chemical oxygen demand (COD)—on the biological oxygen demand/dissolved oxygen (BOD/DO) balance has been examined by Singh et al. (2009) using an artificial neural network (ANN) model. They found that all of these factors are important (with a relative importance ranging from 5% to 12%), which implies that knowledge of these factors would enable an ANN model to predict with good approximations the BOD/DO balance.

4.3 A Detailed River Model of Medium Complexity

The model is presented by the use of the software STELLA. The components used for the conceptual diagram by the software are shown in Figure 4.1. The presented example illustrates a river model with the following state variables:

1. Organic matter with a typical average decomposition rate; the concentration is expressed as BOD (mg/L)
2. POM (mg/L)
3. Coliform microorganisms as indicator bacteria [mg/L (dry matter)]
4. Ammonium-N (mg/L)
5. Nitrate-N (mg/L)
6. Oxygen (mg/L)

The state variables characteristic for eutrophication are not included, because they are presented in detail in Chapters 3 and 12. The equations used for the various processes are presented in detail below. A STELLA

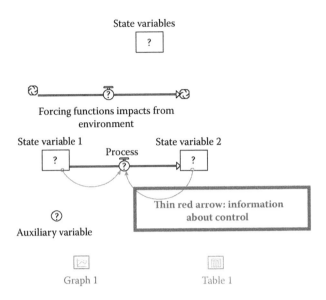

State variables

Forcing functions impacts from environment

State variable 1

State variable 2

Process

Thin red arrow: information about control

Auxiliary variable

Graph 1   Table 1

FIGURE 4.1
Components of the conceptual diagram used in STELLA.

model that considers only the decomposition of organic matter, the reaeration, and the nitrification—in principle, the Streeter–Phelps model—can be found in the work of Jørgensen (2009).

Organic matter is decomposed with good approximations by a first-order reaction. If L is the concentration of organic matter (mg/L) and k_1 is the rate coefficient for the decomposition, the following differential equation is valid:

$$dL/dt = - k_1 L \qquad (4.1)$$

The equation has the following analytical solution:

$$L_t = L_0 e^{-k_1 t} \qquad (4.2)$$

where L_t is the concentration at time t and L_0 is the initial concentration.

L is expressed as BOD_5—this denotes the mg/L oxygen consumption during a period of 5 days. If it is expressed as mg/L (average) organic matter or detritus, the concentration should—according to the stoichiometry of the chemical decomposition equation—be multiplied by 1.39. It means, 1 g detritus or organic matter requires 1.39 g of oxygen to be decomposed into inorganic components in the highest oxidation step (mainly carbon dioxide and water) during a period of 5 days (which will be close to 100% decomposition).

Nitrification of ammonium also causes oxygen depletion. If we denote ammonium concentration as NC (mg N/L in the form of ammonium) and it is presumed that nitrification follows a first-order reaction, the following differential equation is valid:

$$dNC/dt = -k_N NC \qquad (4.3)$$

where k_N is the rate coefficient for the nitrification. Equation 4.3 has the following analytical solution:

$$NC_t = NC_0 e^{-k_N t} \qquad (4.4)$$

where NC_t is the concentration at time t and NC_0 is the initial concentration.

Note that NC is the concentration in mg ammonium-N/L, and the corresponding oxygen consumption is found from the chemical equation for the nitrification:

$$NH_4^+ + 2O_2 \rightarrow NO_3^- + H_2O + H^+ \qquad (4.5)$$

This means that 1 g of ammonium-N requires $2 \times 32/14 = 4.6$ g of oxygen, which will be included in the model, when the nitrification is "translated" to oxygen depletion. The factor, however, is in practice 4.3 and not 4.6, because of the bacterial assimilation of ammonium by the nitrifying microorganisms. It is, however, better to apply a Michaelis–Menten equation instead of a first-order expression (see Jørgensen and Bendorichio 2001). According to the Michaelis–Menten expressions, Equation 4.3 should be multiplied by

$$\min([NC]/(k_{ma} + [NC]), [Ox]/(k_{ao} + [Ox])) \qquad (4.6)$$

to account for the influence of both ammonium and oxygen as possible limiting factors of decomposition. The decomposition of organic matter and nitrification is, of course, temperature-dependent. A simple Arrhenius expression may be applied:

The rate coefficient at temperature (°C) $T =$

$$\text{rate coefficient at } 20°C * K^{(T-20)} \qquad (4.7)$$

K is, with good approximation, 1.05 for decomposition of organic matter, whereas nitrification is more sensitive to temperature changes, and K for this process is therefore 1.07–1.08 (see Jørgensen 2000).

Typical values for the rate coefficients and the initial concentrations for various sources of organic matter and ammonium are shown in Table 4.2.

If oxygen concentration is below the saturation concentration, that is, water is in equilibrium with the atmosphere, reaeration from the atmosphere takes

TABLE 4.2

Characteristic Values for k_1, k_N (1/24 h) and Initial Concentrations (mg/L) for Various Sources of Oxygen Depletion in Streams and Rivers

Source	k_1	k_N	L_0	NC_0
Municipal wastewater	0.35–0.40	0.15–0.25	180–300	20–45
Mechanically treated wastewater	0.32–0.36	0.10–0.15	100–200	18–35
Biologically treated wastewater	0.10–0.25	0.05–0.20	10–40	15–32
Potable water	0.05–0.10	0.03–0.06	0–2	0–1
River water (average)	0.05–0.15	0.04–0.10	1–4	0–2
Agricultural drainage water	0.08–0.20	0.04–0.12	5–25	0–10
Wastewater, food industry	0.4–0.5	0.1–0.25	200–5000	20–200

place. The equilibrium concentration can be found via Henry's law. The saturation concentration is dependent on the temperature and the salinity of the water. In Table 4.3, the equilibrium concentration of oxygen can be found as a function of temperature and salinity.

Aeration is proportional to the difference between the oxygen concentration at saturation, Ox_{sat}, and the actual oxygen concentration, $[Ox]$. The driving force for the aeration is this difference. Ox_{sat} is easily determined as a function of temperature and salinity (see Table 4.3), and is expressed by the following equation:

$$\text{Reaeration} = d[Ox]/dt = K_a(Ox_{sat} - [Ox]) \tag{4.8}$$

where K_a is the reaeration coefficient (1/24 h).

K_a is dependent on the temperature and the flow rate of water. Finally, aeration is also proportional to the surface area in relation to the volume—which means that it is inversely proportional to the water depth. There are several hundred empirical equations that can be used to estimate the reaeration or the reaeration coefficient, considering also additional factors such as frictions and turbulence flows. An often-applied equation is:

$$K_a = 2.26 \, v \times \exp(0.024(T - 20)/d), \tag{4.9}$$

where v is the water flow rate (m/s), T is the temperature (°), and d is the depth of the stream or river (m).

Oxygen concentration is determined by the difference between consumption and reaeration. If we consider only decomposition of organic matter and nitrification, oxygen concentration is determined using the following differential equation:

$$d[Ox]/dt = K_a(Ox_{sat} - [Ox]) - k_1L - k_NNC \tag{4.10}$$

TABLE 4.3

Dissolved Oxygen (ppm, mg/L) in Fresh, Brackish, and Sea Water at Different Temperatures, and at Different Chlorinities (%)

Temperature	0%	0.2%	0.4%	0.6%	0.8%	1.0%	1.2%	1.4%	1.6%	1.8%	2.0%
1	14.24	13.87	13.54	13.22	12.91	12.58	12.25	11.99	11.70	11.42	11.15
2	13.74	13.50	13.18	12.88	12.56	12.26	11.98	11.69	11.40	11.13	10.86
3	13.45	13.14	12.84	12.55	12.25	11.96	11.68	11.39	11.12	10.85	10.59
4	13.09	12.79	12.51	12.22	11.93	11.65	11.38	11.10	10.83	10.59	10.34
5	12.75	12.45	12.17	11.91	11.63	11.36	11.09	10.83	10.57	10.33	10.10
6	12.44	12.15	11.86	11.60	11.33	11.07	10.82	10.56	10.32	10.09	9.86
7	12.13	11.85	11.58	11.32	11.06	10.82	10.56	10.32	10.07	9.84	9.63
8	11.85	11.56	11.29	11.05	10.80	10.56	10.32	10.07	9.84	9.61	9.40
9	11.56	11.29	11.02	10.77	10.54	10.30	10.08	9.84	9.61	9.40	9.20
10	11.29	11.03	10.77	10.53	10.30	10.07	9.84	9.61	9.40	9.20	9.00
11	11.05	10.77	10.53	10.29	10.06	9.84	9.63	9.41	9.20	9.00	8.80
12	10.80	10.53	10.29	10.06	9.84	9.63	9.41	9.21	9.00	8.80	8.61
13	10.56	10.30	10.07	9.84	9.63	9.41	9.21	9.01	8.81	8.61	8.42
14	10.33	10.07	9.86	9.63	9.41	9.21	9.01	8.81	8.62	8.44	8.25
15	10.10	9.86	9.64	9.43	9.23	9.03	8.83	8.64	8.44	8.27	8.09

16	9.89	9.66	9.44	9.24	9.03	8.84	8.64	8.47	8.28	8.11	7.94
17	9.67	9.46	9.26	9.05	8.85	8.65	8.47	8.30	8.11	7.94	7.78
18	9.47	9.27	9.07	8.87	8.67	8.48	8.31	8.14	7.97	7.79	7.64
19	9.28	9.08	8.88	8.68	8.50	8.31	8.15	7.98	7.80	7.65	7.49
20	9.11	8.90	8.70	8.51	8.32	8.15	7.99	7.84	7.66	7.51	7.36
21	8.93	8.72	8.54	8.35	8.17	7.99	7.84	7.69	7.52	7.38	7.23
22	8.75	8.55	8.38	8.19	8.02	7.85	7.69	7.54	7.39	7.25	7.11
23	8.60	8.40	8.22	8.04	7.87	7.71	7.55	7.41	7.26	7.12	6.99
24	8.44	8.25	8.07	7.89	7.72	7.56	7.42	7.28	7.13	6.99	6.86
25	8.27	8.09	7.92	7.75	7.58	7.44	7.29	7.15	7.01	6.88	6.85
26	8.12	7.94	7.78	7.62	7.45	7.31	7.16	7.03	6.89	6.86	6.63
27	7.98	7.79	7.64	7.49	7.32	7.18	7.03	6.91	6.78	6.65	6.52
28	7.84	7.65	7.51	7.36	7.19	7.06	6.92	6.79	6.66	6.53	6.40
29	7.69	7.52	7.38	7.23	7.08	6.95	6.82	6.68	6.55	6.42	6.29
30	7.56	7.39	7.25	7.12	6.96	6.83	6.70	6.58	6.45	6.32	6.19

Note: Values are at saturation.

The solution of the differential equations for L and NC can be used in this differential equation to yield the following expression for the solution of [Ox] as function of time: (note that it is only valid if a first-order equation can be applied for the nitrification):

$$d[Ox]/dt = \left(K_a Ox_{sat} - [Ox]\right) - k_1 L_0 e^{-k_1 t} - k_N NC_0 e^{-k_N t} \qquad (4.11)$$

where k_N is presumed to contain the stoichiometric factor 4.3, or the expression for nitrification can be multiplied by 4.3. This equation can be solved analytically, but it is more easy to use a computer.

In the more elaborate model that we want to develop, the oxygen consumption (demand) of the sediment SOD and of the suspended POM should be added. SOD is expressed as g oxygen/m² 24 h. Table 4.4 presents SOD for different bottom types. SOD should, in other words, be expressed as a forcing function that changes with time or distance from the discharge point. Oxygen consumption in water is SOD/d. The relationship between the two independent variables are discussed below.

POM covers particulate organic matter that will be decomposed at a much slower rate than the dissolved organic matter, but following L as a first-order reaction:

$$d[POM]/dt = -k_p[POM] \qquad (4.12)$$

where k_p is the rate coefficient for the decomposition, which may be 10–100 times smaller than k_1. POM may also be removed from the water to the sediment by settling, which can also be described as a first-order reaction completely parallel to 4.12. A good average settling rate, s_r, is 0.2 m/24 h (range 0.05–0.5 m/24 h). The rate of POM removal by settling is therefore equal to 0.2/d in the unit 1/24 h. Equation 4.12 expresses the decomposition of POM, and the corresponding oxygen consumption is 1.39 times as much as this rate (see the discussion on Equation 4.2 above). Denitrification applies organic matter as a carbon source. The stoichiometry of the denitrification process

TABLE 4.4

Oxygen Demand (SOD) for Various Sediment Types (g oxygen/m² 24 h)

Bottom Type	Range	Average
Filamentous bacteria	5–10	7
Municipal sewage sludge	2–10	4
Downstream of outfall of sewage	1–2	1.5
Estuarine mud, high organic content	1–3	2.0
Sandy sediment	0.2–0.8	0.5
Mineral soil, stones, gravel	0.05–0.1	0.07

shows that the BOD$_5$ removal corresponds to 1.97 the mg nitrate-N/L denitrified. Denitrification is usually expressed by a Michaelis–Menten's equation:

$$d[NIT\text{-}N]/dt = -k_D[NIT\text{-}N]/(k_{md} + [NIT\text{-}N]) \qquad (4.13)$$

Nitrate-N is determined by the discharge of wastewater + the discharge of agricultural drainage water + nitrification according to Equation 4.6. Denitrification takes place under anaerobic conditions, which requires that SOD is at least 2 g oxygen/m^2 24 h. Moreover, denitrification is more inversely proportional to depth, d, as anaerobic conditions only occur in the bottom water. In average, it can be assumed that 10 cm of water close to the bottom may be anaerobic.

It is now possible to erect the differential equations for L and oxygen:

$$dL/dt = \text{discharge of waste} + \text{discharge of agricultural drainage}$$

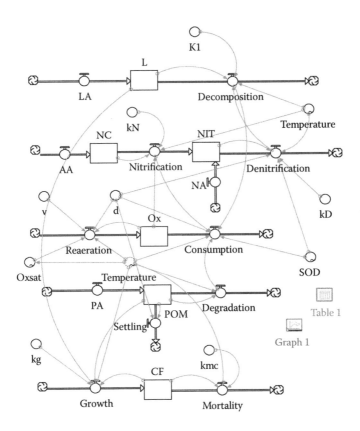

FIGURE 4.2
Conceptual diagram of the presented river model by application of the software STELLA.

water – decomposition according to 4.1 – denitrification
according to 4.13 × stoichiometric factor 1.97   (4.14)

$$d[Ox]/dt = K_a(Ox_{sat} - [Ox]) - k_1L - k_N\ 4.3[NC]\min([NC]/(k_{ma} + [NC]),$$
$$[Ox]/(k_{ao} + [Ox]) - 1.39k_p[POM] - SOD/d + \text{contributions of}$$
$$\text{oxygen from inflows}$$ (4.15)

The concentration of coliforms is determined by the discharge of waste-water and other drainage water + the growth and mortality. The growth is determined by the concentration of organic matter L in which the coliforms are feeding and of POM that adsorbs the coliforms on the surface. The mortality is a first-order reaction. The following differential equation can be used to express the dynamics of coliforms:

$$D[CF]/dt = k_g[CF]\min([L]/(k_{ml} + [L]), [POM]/(k_{pm} + [POM]) - k_{me}[CF]$$ (4.16)

The model could be made more comprehensive by considering different biodegradability, changed reaeration at turbulent flow, and so on. As always, when we develop models, the problem, the available data, and the system

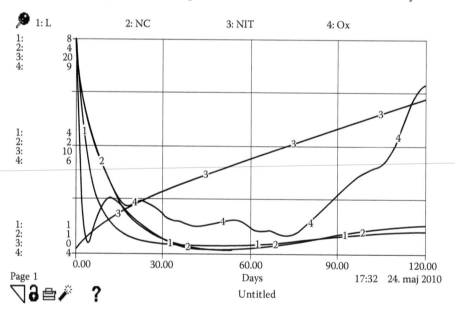

FIGURE 4.3
Results of the river model. Oxygen has a minimum rate of 4.2 mg oxygen/L after 4 days or 9.6 km according to the flow rate (see text). Ammonium-nitrogen decreases up to about day 50 and increases slightly after 50 days. Nitrate increases although at a lower rate. BOD$_5$ decreased during the first 430 days, then increased afterward because of the nonpoint pollution. The minimum BOD$_5$ is close to 1 mg/L at day 40 or after 96 km.

with a wide spectrum of possible processes should determine the complexity of the model. Let us, however, develop the model by using STELLA to illustrate how to use this software, and to illustrate how powerful even relatively medium complex models are.

The STELLA diagram for the model is shown in Figure 4.2. Note that the state variables are *L*, Ox, NC, NIT, POM, and concentration factor (CF.) Time is considered the independent variable, but it could also be the distance from the discharge for instance of wastewater. If the water flow rate for instance is 0.2 m/s, the time in days will correspond to 0.2 × 3600 × 24 m = 17,280 m. The model has a constant discharge, which can be considered as agricultural drainage water along the shoreline of the stream. SOD is described as a time variable forcing function. The point discharge of wastewater takes place at time 0, and will therefore correspond to the initial value of *L*, NC, NIT, POM, and CF. The dilution, of course, needs to be considered when the initial values are calculated. If, for instance, 1000 m^3 wastewater/h with 30 mg/L of BOD$_5$, 12 mg/L ammonium-N, 4 mg/L NIT-N, 6 mg/L POM and 0.06 mg/L CF is discharged to a river with 9000 m^3 of water flow h, then the dilution factor is 10. If the river water has 4 mg BOD$_5$/L and 2 mg ammonium-N/L and no NIT, POM, and CF, the mixture of river water and wastewater will have (4 × 9 + 30 × 1)/10 = 6.6 mg BOD$_5$/L, (2 × 9 + 12 × 1)/10 = 3 mg ammonium-

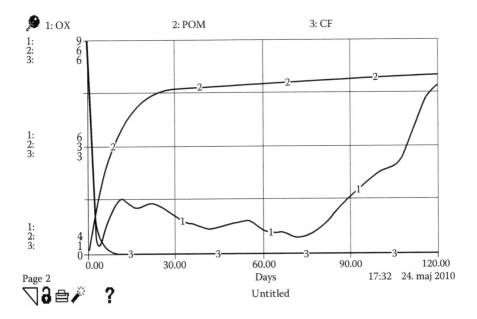

FIGURE 4.4
Results of the river model. The POM, CF, and Ox are shown. POM increased but after day 30, did so at a very moderate rate. The increase is attributed to the nonpoint pollution. CF decreases rapidly exponentially, as expected, and is almost at 0 after 15 days or 36 km.

CF(t) = CF(t - dt) + (growth - mortality) * dt
INIT CF = 6

INFLOWS:
growth = kg*CF*1.05^(temperature-20)*MIN(L/(L+2),POM/(POM+0.2))
OUTFLOWS:
mortality = kmc*CF*1.06^(temperature-20)
L(t) = L(t - dt) + (LA - decomposition) * dt
INIT L = 6.6

INFLOWS:
LA = 0.25
OUTFLOWS:
decomposition = K1*L*1.05^(temperature-20)+1.97*denitrification
NC(t) = NC(t - dt) + (AA - nitrification) * dt
INIT NC = 3

INFLOWS:
AA = 0.05
OUTFLOWS:
nitrification = kN*NC*1.08^(temperature-20)*MIN(NC/(NC+1),Ox/(Ox+2))
NIT(t) = NIT(t - dt) + (NA + nitrification - denitrification) * dt
INIT NIT = 0.4

INFLOWS:
NA = 0.05
nitrification = kN*NC*1.08^(temperature-20)*MIN(NC/(NC+1),Ox/(Ox+2))
OUTFLOWS:
denitrification = IF(SOD)>2.0 THEN(1.08^temperature-20)*(kD*0.1/d)*NIT/(NIT+1)ELSE(0)
Ox(t) = Ox(t - dt) + (reaeration - consumption) * dt
INIT Ox = 8.1

INFLOWS:
reaeration = (Oxsat-Ox)*2.26*v*EXP(0.024*(temperature-20/d))
OUTFLOWS:
consumption = decomposition+degradation+nitrification+SOD*1.05^(temperature-20)/d
POM(t) = POM(t - dt) + (PA - settling - degradation) * dt
INIT POM = 0.6

INFLOWS:
PA = 0.5
OUTFLOWS:
settling = 0.2*POM/d
degradation = 0.01*POM*1.05^(temperature-20)
d = 2
K1 = 0.3
kD = 4
kg = 0.1
kmc = 1
kN = 0.2
v = 0.2
Oxsat = GRAPH(temperature)
(5.00, 12.8), (7.50, 12.0), (10.0, 11.3), (12.5, 10.7), (15.0, 10.1), (17.5, 9.54), (20.0, 9.11),
(22.5, 8.68), (25.0, 8.27), (27.5, 7.90), (30.0, 7.56)

FIGURE 4.5
Equations of the river model.

SOD = GRAPH(TIME)
(0.00, 8.25), (5.00, 5.45), (10.0, 5.00), (15.0, 4.80), (20.0, 4.20), (25.0, 4.00), (30.0, 4.05),
(35.0, 3.75), (40.0, 3.75), (45.0, 3.80), (50.0, 3.95), (55.0, 4.10), (60.0, 4.80), (65.0, 5.15),
(70.0, 5.80), (75.0, 6.00), (80.0, 6.15), (85.0, 5.90), (90.0, 5.85), (95.0, 5.80), (100, 5.75),
(105, 5.70), (110, 4.70), (115, 3.60), (120, 3.50)
temperature = GRAPH(TIME)
(0.00, 13.2), (5.00, 13.9), (10.0, 15.9), (15.0, 18.4), (20.0, 20.1), (25.0, 21.4), (30.0, 22.8),
(35.0, 24.1), (40.0, 24.8), (45.0, 24.3), (50.0, 23.4), (55.0, 22.6), (60.0, 21.4), (65.0, 20.1),
(70.0, 18.8), (75.0, 17.7), (80.0, 16.5), (85.0, 15.8), (90.0, 15.0), (95.0, 14.1), (100, 13.7),
(105, 13.4), (110, 13.2), (115, 12.6), (120, 12.3)

FIGURE 4.5 (Continued)

N/L, 0.4 mg/L NIT, 0.6 mg/L POM, and 0.006 mg/L or 6 µg/L CF, which are applied as the initial concentrations.

The model should be able to give information on how these concentrations are changed over time. The concentration of oxygen in treated wastewater will almost always be close to 0 mg/L. If the river water has for instance 9 mg/L oxygen, the mixture of wastewater and river water will have an oxygen concentration of 8.1 mg/L, corresponding to a dilution of the wastewater by a factor 10. The constant input per unit of time from agricultural drainage water of dissolved organic matter, ammonium-N, nitrate-N, and POM are denoted in the STELLA diagram as LA, AA, NA, and PA, respectively.

Figures 4.3 and 4.4 show the results of running the model 120 days by using the following initial values:

For BOD_5, 6.6 mg/L

For ammonium-N, 3 mg/L

For oxygen, 8.1 mg/L

The other initial values are indicated above.

The equations are presented in Figure 4.5. Time can be translated to distance from the discharge point by the flow rate. If the cross-sectional area is 100m³, the water flow of 10,000 m³/h corresponds to 100 m/h. Twenty-four hours will therefore correspond to 2400 m and the 120 days to 288 km. The minimum oxygen concentration occurs after 8 days or 19.2 km.

References

Armstrong, N. E. 1977. Technical Report CRWR-145. Simulation of water quality in rivers and estuaries. Center for Research in Water Resources. The University of Texas at Austin.

Bowler et al. 1992. *Proceedings of the 1992 Annual Symposium*, Volume XXIV ISSN 1068–0381. Desert Fishes Council, Bishop, CA.

Brown, L. C., and T. O. Barnwell. 1987. QUAL2E and QUAL2E-Uncas. Documentation and user model. EPA-600-3-87-007.

Cazelles, B. 1987. Thesis, Centere de Bioinformatique, Paris, France. Modelisation d'un Ecosystem.

Cazelles, B., M. Fontvieille, and N. T. Chau. 1991. Self-purification in a lotic ecosystem: A model of dissolved organic carbon and benthic microorganisms dynamic. *Ecological Modelling* 58: 91–117.

Cronk, J. K., W. J. Mitsch, and R. M. Sykes. 1990. Effective modelling of a major inland oil spill on the Ohio River. *Ecological Modelling* 51: 161–192.

Cummins, K. W. 1974. Structure and function of stream ecosystems. *Bioscience* 24: 631–641.

Jørgensen, S. E. 2009. *Ecological Modelling, An Introduction*. Southampton, UK: WIT. 198 pp.

Jørgensen, S. E., and G. Bendorichio. 2001. *Fundamentals of Ecological Modelling*. Third edition. Amsterdam: Elsevier. 528 pp.

Jørgensen, L. A., S. E. Jørgensen, and S. N. Nielsen. 2000. *ECOTOX: Ecological Modelling and Ecotoxicology*. Amsterdam: Elsevier. CD-ROM.

Kangwon National University, 1993. Seok Park. Chuncheon 200-701 Korea.

Korman et al. 1994. Developments and evaluation of a biological model to assess regional scale effects of acidification on Atlantic salmon. *Canadian Journal of Fisheries and Aquatic Science* 51: 662–680.

Paasivirta, J. 1994. Environmental fat model in toxic risk estimation of a chemical spill. Research Centre of the Defence Forces, Finland. Publications A/4 1994, 11–21.

Park, S. S., and C. G. Uchrim. 1988. A numerical mixing zone model for water quality assessment in natural streams. *Ecological Modelling* 42: 233–244.

Singh, K. P., A. Basant, A. Malik., and G. Jain. 2009. Artificial neural network modelling of the river water quality. *Ecological Modelling* 220: 888–895.

5

Models of Coral Reefs

Sven Erik Jørgensen

CONTENTS

5.1 Introduction 153
5.2 Overview of Coral Reef Models 154
5.3 A Coral Reef Model 156
References 159

5.1 Introduction

Corals are invertebrates that build large living structures—where algae and corals through symbiosis form a very productive ecosystem—that give shelter to many species of plants and animals. Coral reefs are among the most biologically diverse ecosystems on Earth. In addition, they have a very high productivity rate, on average about 80 MJ/m² year or about the same as tropical fain forests. The calcification and growth of corals depends on the mutualism between corals and photosynthetic dinoflagellates, *Symbiodinium* spp. Corals receive more than 90% of their carbon from the photosynthesis of the dinoflagellates, whereas dinoflagellates acquire nutrients from the coral excretion products. Coral reefs are abundant in shallow tropical, clear, warm water coastlines. Therefore, coral reefs are very rarely found in areas above 30° latitude. Coral reefs cover approximately 250,000 km² of the Earth's surface.

Up until recently, only very few models have been developed for coral reefs, but the environmental focus on global warming and its impact on coral reefs has enormously increased the level of interest on these unique ecosystems. In general, the threat to coral reefs has been attributed to many stress factors, in addition to global warming, such as overfishing, intense land use of the coastal zone, and pollution by nutrients, toxic substances, and organic matter. In the Caribbean, the coral cover has decreased by 80% or more during the past couple of decades. Corals may bleach in response to elevated sea water temperature and increased ultraviolet radiation.

Models of coral reefs are of particular interest in environmental management because they focus on uniquely beautiful, productive, and diverse ecosystems that are sensitive to human impacts on global conditions. To a certain extent, coral reefs can be considered an indicator of the decrease in global sustainability due to human activities and, in particular, of global warming.

A short overview of the most characteristic coral reefs models will be presented in the next section, followed by a section where a coral reef model will be presented in more detail.

5.2 Overview of Coral Reef Models

Several models have been developed to be able to make bleaching predictions. Various mathematical methods are used to reveal the environmental influences on bleaching, for instance, a Bayesian approach (see Wooldridge and Done 2004; Renken and Mumby 2009) and clustered binomial regression (Yee, Santanvy, and Barron 2008). The results of these mathematical methods are applied to make predictions on the future influence of environmental factors such as temperature changes and changes in other meteorological variables such as wind, current, and solar radiation. The influence of environmental factors on coral reefs, however, is very complex; for instance, on the one hand temperature may exacerbate bleaching and on the other hand, it may enhance coral resistance to bleaching. At the same time, solar radiation, water depth, current, and water turbidity are important and interacting environmental factors. In addition, the influence of these factors may be different for different species, and since the diversity is high and at the same important for the development of coral reefs, it is necessary to consider at least a number of representative species.

Yee, Santavy, and Barron (2008) determined the probability of coral colony bleaching, p_b, as a function of the effective temperature, T_e, the effective solar radiation, S_e; when both additive effects and interactions are considered:

$$\ln(p_b/(1 - p_b)) = a_1 + a_2 \times T_e + a_3 \times S_e + a_4 \times (S_e \times T_e) \tag{5.1}$$

T_e was quantified as degree heating weeks (DHW), the average daily wind speed in the 12 weeks before the survey, W, by use of the following expression:

$$T_e = b_1 \times DHW + b_2 \times W + b_3 \times (DHW \times W) \tag{5.2}$$

S_e was determined by the following expression:

$$S_e = \ln(S_w \times \exp(-KD), \tag{5.3}$$

where S_w is the solar radiation at the water surface, K is the diffuse attenuation coefficient, and D is the depth of coral.

Different species have different bleaching susceptibility. Because coral community composition varies and different coral reefs have different relative abundance of species, it is necessary to apply a hybrid model by adding the best model of the influence of the relative abundance with the best models of the environmental factors considered in Equations 5.1–5.3. The relative abundance is considered by describing the probability as functions of relative abundance:

$$\ln(p_b / (1 = p_b)) = s_o \sum_{i=1}^{N} s_i D_i \tag{5.4}$$

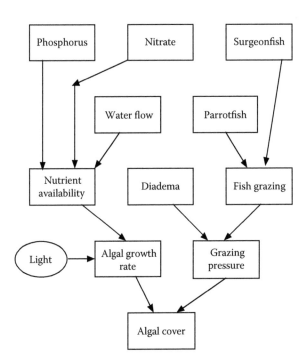

FIGURE 5.1

The BBN used in the described model is shown. Parrotfish and surgeonfish represent biomass of the fish grazers. *Diadema* denotes the abundance of the urchin. Nitrate and phosphorus are the nutrient concentrations. Water flow is the water motion on the reef; light represents photosynthetic active radiation, named PAR. Nutrient availability covers the interaction of nutrient and water flow. Algal growth is a function of nutrient availability and light, whereas grazing pressure represents the combined effect of fish grazers and *Diadema*. Algal growth and grazing pressure determine the algal cover.

where s_o and s_i are the parameters that fit best into the equation, and D_i is the abundance of the I taxa in a given survey (modeled coral reef).

The model has been used to predict coral bleaching resulting from climate changes in the future, although the model has unexplained variability in observed colony bleaching even when taxon-specific models were applied. The future application of the model will be able to assess the validation of such prognoses.

Renken and Mumby (2009) have developed a model for macro algae in coral reef. Their dynamics influence the resilience and thereby disturbances of coral reefs, but the physical, chemical, and ecological processes behind the dynamics of macro algae in coral reefs need to be revealed according to Renken and Murby. Thus, they applied a Bayesian belief network (BBN) to integrate these processes in a useful model. The proposed BBN is shown in Figure 5.1. The mathematical tool behind BBN is able to find the parameters to be applied for the BBN, based on available observations. As shown in Figure 5.1, the BBN has 12 nodes, of which seven are input nodes, four are intermediate nodes, and one is the final predicted node—the macro algal cover of a given species. The validation of the developed model showed that algal cover predicted by the model was correct 55% of the time, but that it was possible to raise this rate to 72% if the scale was subsumed to equal-sized categories of 10% interval. The model has demonstrated that BBN is a useful tool to structure the understanding of a complex dynamics as is characteristic for coral reefs. Development of the BBN, however, requires a good knowledge of the causality of the processes that are connected in the network.

5.3 A Coral Reef Model

The model presented in this section has been published by Riegl and Purkis (2009) in *Ecological Modelling*. The model is based on observations of mass mortality caused by temperature anomalies, after the climax was observed in 1995 (Riegle 1999). The model assumes three competitive species groups: *Acropora*, faviids, and *Porites*. *Acropora* (an aggressive group), dominates at equilibrium.

Small *Acropora* is modeled by:

$$dN_1/dt = R_1 \times N_2(K - N_1 - N_2 - N_4 - N_6)/K - N_1(G_1 + D_1 + n) + mA \quad (5.5)$$

where N_1 is the number of small species, N_2 is the number of large species, N_4 is the number of faviids, and N_6 is the number of *Porites*. R denotes the specific fertility; K is the carrying capacity; G denotes the loss of small species due to growth into the large size class, N_2; D is mortality; n is the specific emigration rate; m is a specific migration rate from a connected population A. N_2 is covered by the following equation:

$$dN_2/dt = G_1 \times N_1 - D_2 \times N_2 \qquad (5.6)$$

Faviids are massive coral with slower growth than *Acopora*. They are competitively subordinate, but dominate *Porites*. The number of small faviids, N_3, and large faviids, N_4, are covered by similar equations:

$$dN_3/dt = R_3 \times N_4(K - N_3 - N_1 - N_2 - N_4 - N_6)/$$
$$K - N_3(G_3 + D_3 + n + R_1 \times N_2/K) + mB \qquad (5.7)$$

$$dN_4/dt = G_3 \times N_3 - D_4 \times N_4 \qquad (5.8)$$

where m is a specific migration rate from a connected population B.

Small *Porides* face limitation due to space pre-emption by all big and small colonies and loose in encounters with recruits of other species. *Porides* small (N_5) and large (N_6) are covered by the following equations:

$$dN_5/dt = R_5 \times N_4(K - N_3 - N_1 - N_2 - N_3 - N_4 - N_5 - N_6)/$$
$$K - N_5(G_5 + D_5 + n + R_1 \times N_2/K) + mC \qquad (5.9)$$

$$dN_6/dt = G_5 \times N_5 - D_6 \times N_6 \qquad (5.10)$$

Parameters were determined by combining remote sensing with in situ ground measurements. Population sizes were obtained by assigning coral density per unit of area from digitized photo transects and ground control points to color-coded habitat classes on classified satellite imagery. The carrying capacity of the system, K, was estimated as the sum of all coral pixels including substratum onto which corals could recruit and form new coral patches. The intrinsic properties of the model are shown in Figure 5.2.

The parameters found using these methods (see the tables in Riegl and Purkis 2009) were used to simulate observed severe disturbances with 90% reduction of small and large *Acropora*, whereas small and large faviids were reduced by 25%. It was found—in accordance with the observations—that *Acropora* population strongly rebounded without outside larval input, as did faviids and *Porites* (see the simulation in Figure 5.3). A closely spaced 1996–2008–2002 disturbance triplet, however, did not allow sufficient time for population recovery, pushing all population to very low levels and *Acropora* to extinction. However, Faviids and *Porites* showed strong signs of recovery from 2002 to 2008. The model was able to follow the described changes as a result of the severe disturbances. It was therefore concluded by the authors

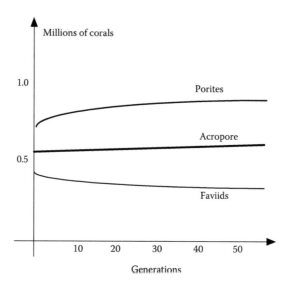

FIGURE 5.2
Intrinsic properties of the model. The species group reach carrying capacity and then remain stable. The figure shows the results of the use of the equations shown in the text. The parameters are calculated from values, where the three species coexist. With these parameters, the model allows coexistence, as shown on the graph. *Porites* will be most frequent, followed by *Acropora*, and then faviids.

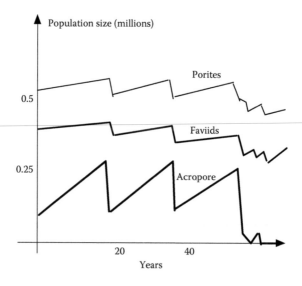

FIGURE 5.3
Model result of six disturbance cycles, suggesting the persistence of the coral assemblage. The results are for Ras Hasyan, a coral reef in the Persian Gulf. The shown result corresponds to $m = 0$, $R_1 = 0.87$, $D_1 = D_3 = D_5 = 0$, and 90% mortality of *Acropora* and 25% mortality of faviids and *Porites*. The observed recovery is clearly demonstrated.

that the model can be applied to describe the response to bleaching and mass mortality events that are predicted to increase in frequency with climate changes.

References

Renken, H., and P. J. Mumby. 2009. Modelling the dynamics of coral reef macroalgae using a Bayesian belief network approach. *Ecological Modelling* 220: 1305–1314.

Riegl, B. M. 1999. Coral communities in a non-reef settling in the Southern Arabian Gulf: Fauna and community structure in response to recurrent mass mortality. *Coral Reef* 18: 63–73.

Riegl, B. M., and S. J. Purkis. 2009. Model of coral population response to accelerated bleaching and mass mortality in a changing climate. *Ecological Modelling* 220: 192–208.

Wooldridge, S., and T. Done. 2004. Learning to predict large-scale coral bleaching from past events: A Bayesian approach using remotely sensed data in situ data and environmental proxies. *Coral Reef* 23: 96–108.

Yee, S. H., D. L. Santavy, and M. G. Barron. 2008. Comparing environmental influences on coral bleaching across and within species using clustered binomial regression. *Ecological Modelling* 208: 162–174.

6

Models of Forest Ecosystems

Sven Erik Jørgensen

CONTENTS

6.1   Introduction.. 161
6.2   Overview of the Characteristic Forest Models..................................... 163
6.3   A Model of Wood Quality and Forest Ecosystem Functioning.......... 167
6.4   A Thermodynamic Forest Growth Model... 169
References.. 171

6.1 Introduction

Numerous models of forest ecosystems have been developed to optimize forest management both from an environmental and an economic point of view. Models have a wide spectrum of applications in forestry. They may be used to assess the management strategy that gives the highest production of wood, to determine the response to environmental stress factors, to protect forests as a valuable natural ecosystem offering indispensable ecosystem services, or to assess the best feasible pollution control when a forest is threatened by various pollution sources.

Most forest models have been developed for temperate forests, whereas models for complex tropical rainforests tend to be rather simple or even lacking because of the lack of data. Fortunately, in the past decade there has been a tendency to expand the application of existing models to tropical forests, for which more data have recently been provided. In general, many new forest models have been developed in the past decade, although many of the applied core equations were already applied in forest models that were developed in 1980s.

The wide spectrum of forest models available today covers different climate conditions and the following seven relevant issues:

1. Growth, succession, production, and yield models of natural forests and plantations
2. Spatial distribution models of forest attributes
3. Landscape/forest models (see also Chapter 24)

4. Impact of climate changes on forest ecosystems
5. The role of soil properties, water, and nutrient acquisition for forest growth
6. The effects of pesticide pollution and air pollutants on forest ecosystems
7. The interactions of trees, birds, and insects

Several models cover two or more issues simultaneously. Recently, only a few class 6 models have been developed, although there has been a growing interest for class 4 models.

The applied equations for the ecological processes that are included in the forest models are, in principle, not different from the expression applied in other ecosystem models. The growth equations, for instance, are not in principle different from the equations applied for plant growth and photosynthesis, including the influence of limiting factors such as phosphorus, nitrogen, and water, applied in models of agricultural systems, grassland, rivers, lakes, and other ecosystems. Growth off trees, however, is slightly more complex than growth of macrophytes and phytoplankton because we have to distinguish between leaves, stems, total biomass, roots, and soil processes. In addition, these different tree components are interacting, which is of great importance for the growth of trees. These special features of forest models compared with other ecosystem models will be shown and discussed in Section 6.3.

The forest models focusing on spatial distribution are to a large extent based on remote sensing, and the results are most often represented by the use of geographic information system (GIS). In many ways, forest models are similar to landscape models. Because the growth of trees involves the diameter, the total biomass, the stem, and the height, many forest models apply allometric principles (see, e.g., Fleming and Bruns 2003). The influence of temperature on the process rates, the description and growth of parasites, and the destruction of plants are also normally covered by the same equations as we are using for the processes and influences in other ecological models.

The next section gives a short overview of the most characteristic forest models that can be found in the modeling literature. Tree growth is a core process in many forest models applied in forestry. Section 6.3 presents the most characteristic features and equations of an up-to-date growth model, ANAFORE (ANalysis of FORest Ecosystems) (see Deckmyn et al. 2008). The conceptual model will be presented in addition to the most important equations and the conceptual considerations behind the model development. Section 6.4 describes a forest growth model based on the thermodynamic theory. In the chapter, we also include a more holistic modeling approach, which is applicable to make "short cuts" in the modeling of ecosystems and ecological processes. The presentation of this "creative" approach may inspire others to apply the same idea for the development of other ecosystem models.

6.2 Overview of the Characteristic Forest Models

Numerous forest models have been developed thus far, and although it is not possible here to give an overview of all forest models, an attempt is made to cover a characteristic spectrum of the available models. This move will hopefully give the reader a sufficient impression of the spectrum of forest models to facilitate the choice of a proper and workable forest model for new case studies.

Table 6.1 gives an overview of the models that have been selected to represent the spectrum of forest models. The most characteristic features of the models, the authors, and the model class (1 to 7; see Section 6.1) are summarized in the table. Note that several of the listed models cover more than one model class (issue). Additional information on some of the models, which are not covered in this table, is discussed below.

The model GROMAP gives the suitability of any perennial plant taxon for cultivation across Africa. The monthly values of maximum temperature, minimum temperature, precipitation, evaporation, and solar radiation are estimated for each location. Information describing the climatic requirements of selected plants is entered into the model. Color maps of the edaphic, climatic, and overall limitation for the focal plants are produced by the model. The program contains a database for 10,000 locations.

The model PnEt is able to predict the effects of changes in physical and chemical climate on water quality and quantity and forest production. The model is based on a carbon and water balance and two principal relationships: (1) maximum photosynthetic rate as a function of foliar nitrogen concentration and (2) stomatal conductance as a function of realized photosynthetic rate. These relationships are combined with standard equations describing the light attenuation in canopies and photosynthetic response to radiation intensity, along with the effect of soil water stress and vapor pressure deficit.

Formix is a model of the growth dynamics of natural tropical forests under various management conditions and silvicultural treatments. The model is based on an energy balance in the form of carbon, and accounts for the mutual interactions of the forest growth and light conditions that cause vertical and horizontal differentiation in the natural forest mosaic. A special version of the model, denoted Formix 2, incorporates the multistory canopy layer structure of the natural forest.

The model named Treedyn 3 describes the growth dynamics of forests under the influence of various pollutants. It is based on carbon and nitrogen balance. The state variables are the number of trees, base diameters, tree height, leaf mass, root mass, fruit biomass, carbon, and nitrogen in litter. Carbon and nitrogen I organic matter in soil and inorganic carbon and nitrogen in soil available for the plants. The model is particularly fitted to examine the effect of climate, site, silvicultural management strategy, and effects of pollutants. The model is furthermore an excellent teaching tool in forestry classes.

TABLE 6.1

List of Models Representing the Spectrum of Forest Models

Model Name	Characteristics	Class	Reference
GROMAP	Mapping climatic and other limitations for plant growth	2	Booth (1991)
PnEt	Effect on climate on forest ecosystems	4	Aber and Federer (1994)
FORMIX	Forest growth/management strategies	1	Bossel and Krieger (1991)
Treedyn 3	Growth, carbon, and nitrogen dynamics	5	Bossel and Schæfer (1989)
Fsyl-mod	C and water budget of beech forest	1	Stickan et al. (1994)
CARDYN	C-dynamic of deciduous forest	1	Veroustraete (1994)
FAGUS	Growth as f(N, water, soil, temp)	5	Hoffmann (1995)
FED	Influence of climate changes (spatial)	4	Levine et al. (1993)
HYBRID	Growth, N, soil,water, light	5	Friend et al. (1993)
Hobo model of predation	Two tree species and foliage-feeding larvae	7	Lhotka (1994)
Ecolecon	Spatial, IBM, landscape, animals	3+7	Liu (1993)
GENDEC	Cycling of C and N, including detritus	5	Moorhead and Reynolds (1991)
FORET	Growth and mortality of trees	1	Dale et al. (1988)
	Tropical forest, spatial	2+5	Boersma et al. (1991)
TEMFES	Growth as f(C, N, water, tree physiology)	5	Nikolov and Fox (1994)
	Physiology of individual trees	1+5	Chen and Gomez (1988)
FOREST-BGC	Cycling of C, N, and water	1	Kremer (1994)
GLOBTREE	Growth and competition	1+5	Soares and Tome (2003)
CAPSIS	Different management strategies	1	Colygny et al. (2003)
TRIPLEX	"Classical" forest model	1(+5)	Zhang et al. (2008)
	Alpine line dynamics	4	Wallentin et al. (2008)
	Growth of shelterbelt	1	Zhou et al. (2007)
MOFM	Multiple Objective Forest Management	1	Gardingen (2003)
	Woodland landscape dynamics	3	Gillet (2008)
	Spatial + hierarchical interactions	2	Grabarnik and Särkkä (2009)
	Spruce, bark beetles, climate changes	4+7	Ogris and Jurc (2010)
	Changing frost regimes/forest	4	Ramming et al. (2010)
	Mixed forest stand, spatial variations of humus	2	Wælder et al. (2008)
	Parameter estimation for tree growth	1	Colbert et al. (2004)

(continued)

TABLE 6.1 (Continued)

List of Models Representing the Spectrum of Forest Models

Model Name	Characteristics	Class	Reference
	Nitrogen cycling in forest ecosystems	5	Corbeels et al. (2005)
	Dispersal model	2	Wagner et al. (2004)
	Mapping composition and diversity	2	Foody and Cutler (2006)
FVS	Forest vegetation for main species	2	Lacerte et al. (2006)
ZELIG	Forest succession	1	Larocque et al. (2006)
	Management of tropical forest	1	Tietjen and Huth (2006)
	Vegetation structure	2	Kucharik et al. (2006)
	Effect of pollutants on the tree growth	6	Ford and Kiester (1990)

Cardyn is a model that aggregates the behavior of several deciduous species. Four compartments are included in the model: the green compartment, the nongreen compartment, the litter compartment, and the atmosphere, which is considered an infinite source and sink. The relationship between the leaf area index (LAI) and leaf carbon is presented by using a conversion factor, which entails averaging the area to the mass relation of light- and shade-adapted leaves.

The model FED provides a flexible framework for modeling a forest ecosystem, where naturally and anthropogenically induced changes in climate is altering the soil properties, the vegetation, the temperature, and the radiation. The model also considers how these factors will modify each other and their spatial distribution. The results are also given spatially, including the soil profile. The model entails the following processes: (1) fluxes of water, heat, and trace gases; (2) ion exchange and leaching in soil; (3) formation and distribution of coatings; (4) pH changes of soil; (5) decomposition of detritus; (6) changes in density of clay minerals.

The model Ecolecon is based on a spatially explicit, individual-based, and object-oriented program. It is able to simulate the dynamics of animal populations and economic revenues. The model is linked with GIS to run simulations on real forest landscapes. It can predict population dynamics changes, changes in spatial distributions, future landscape structure, and income from timber harvesting. As seen from this description, Ecolecon is a typical forest management tool.

The model developed by Boersma et al. (1991) (see Table 6.1) is a spatial model for tropical forests. It covers 300 patches with each patch large enough to contain one mature tree. The growth of each tree is followed as the size compared with the maximum size, and determined by the availability of light and nutrients. The amounts of nutrients in the soil are state variables. State variables also include the height and the diameter of the tree.

The model TEMFES couples water, carbon, nitrogen, and population dynamics of tree species for forest ecosystems. The equations are based on mass and energy balances. The model has five modules: (1) physiologically based tree growth module; (2) a biophysical canopy module; (3) a soil, water, and heat module; (4) a module to follow the decomposition of organic matter; and (5) a regeneration and mortality module.

The model developed by Chen and Gomez (1990) has many state variables to account for the different forms of environmental stresses. These state variables include LAI, leaf mass and age classes, stem wood and sapwood, coarse and fine roots, calcium, magnesium, potassium, carbon, nitrogen, and phosphorus. Environmental stresses are covered by modification of the rate of the physiological processes. For instance, ozone and water deficit cause closure of stomata, resulting in reduced carbon dioxide uptake and photosynthetic rate. A cumulative dose of ozone amplifies the mortality and therefore also the litterfall.

The model called Forest BGC considers the cycling of carbon, water, and nitrogen in forest ecosystems. The model includes canopy interception, evaporation, transpiration, photosynthesis, tree growth, tree respiration, allocation of carbon, litterfall, decomposition of detritus, nitrogen mineralization, and primary production.

Soares and Tome (2003) have developed an individual tree model, named GLOBTREE. The model is characterized by the use of many variables describing the individual tree and its growth: tree crown ratio, diameter at 1.3 m of tree height (in cm), quadric mean diameter, maximum tree diameter, total height, age (years), and total volume (m³). Equations for the change of each of these variables are applied in the model together with equations regulating the right ratios between diameter and height and between maximum and mean diameter. Competition from other trees is considered, and a submodel focuses on the probability of survival. The model is probably the most complete model for the growth and development of an individual tree. A proper coupling with an individual based model could probably lead to a good and workable model of the trees in a forest.

The model TRIPLEX has been applied to simulate forest growth and biomass production in subtropical forest. The model considers the following core variables: tree density, tree height, tree diameter at breast height, litter, total biomass, and above surface biomass. The model considers the influence of temperature, solar radiation, and precipitation on growth. The model has yielded acceptable validation results and represents an up-to-date classical forest model (see Zhang et al. 2008).

Models for different tree species have been developed.

The most important ones are (see Amaro, Reed, and Soares 2003):

1. Eucalyptus trees
2. Cork oak

3. Pine tree

4. Black spruce

Wallentin et al. (2008) have developed a model of the alpine tree line dynamics. It considers the effects of climate change and land use change. This individual based model has produced acceptable validation results.

Malanson, Wang, and Kupfer (2007) have developed a model that follows the ecological consequences of deforestation. The model is able to follow the spatial patterns before, during, and after simulated deforestation.

A Bayesian model can be generally applied to describe carbon, water, and heat fluxes in boreal forests with acceptable calibration results and indication of standard deviation for gross primary production, total ecosystem respiration, and net ecosystem productivity. Bayesian models are widely applied in forest management models, and can generally be recommended as a powerful modeling tool.

6.3 A Model of Wood Quality and Forest Ecosystem Functioning

The model (see Deckmyn et al. 2008) named ANAFORE simulates the relevant processes on a wide range of spatial and temporal scales, as shown in Figure 6.1. The forcing functions are radiation, temperature, vapor pressure, wind, and daily precipitation. The state variables are carbon, water, nitrogen, and tissue of the leaves, tree, tree stand, and carbon, water, nitrogen, and organic matter in the soil. The model is characterized by having a detailed description of the processes in leaves, tree, tree stand, and soil, and the interaction between these levels. For the leaf processes, it uses a half-an-hour time step, for the trees 24-h time step, and for the stand biomass and the soil processes time steps of 1 year. The time steps are selected in accordance with the dynamics of the focal processes. The leaf layer includes the following processes: light interception, stomatal conduntance, photosynthesis, transpiration, and stem sap flow.

The results of these processes are used to determine the daily net leaf photosynthesis and the daily maintenance of respiration for the entire tree. Based on these results, the daily C gain is calculated for the entire tree, and it is distributed over the different biomass pools according to the basic phonological patterns. The daily net gain of each pool is thus the result of C gain from the allocation minus the loss through mortality. Single trees are used as representatives of a limited number of tree cohorts (currently 10), which make up the tree stand. Tree cohorts and trees within cohorts compete for light, water, and nitrogen, and are linked to a daily running soil submodel.

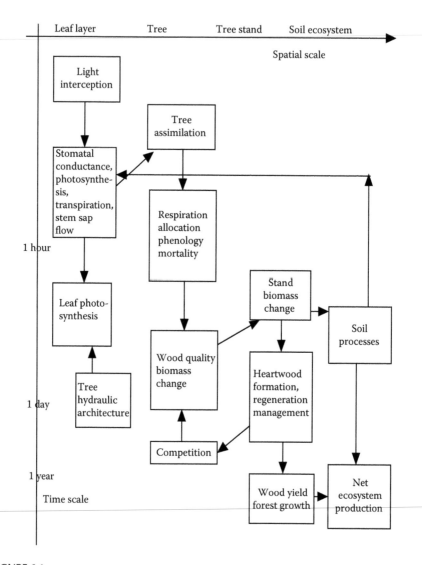

FIGURE 6.1
The ANAFORE model. The flowchart shows the submodels and processes included in the model, and how the submodels are interconnected.

ANAFORE has two detailed submodels covering the wind profile considering the canopy structure and the crown extinction coefficient for direct sunlight considering the leaf angle distribution in the crown and the solar elevation profile. The soil and the tree stands make up the forest. The model outputs are produced at three time levels: half-hourly, daily, and yearly. The model is very complex, and the model presentation has a five-page list of symbols (variables, forcing functions, universal constant, and parameters).

Bayesian statistics for the parameter estimation was implemented, which gave a good result of the calibration of the model. Several equations are, for instance, used to cover the allocation of the biomass to the various pools. The model, however, is a very good tool if one wishes to study the interactions between different species' parameters as well as the impact of current and future climate on the forest ecosystem. The acceptable calibration and validation results are probably rooted in the three time levels—half hour, day, and year—which are in accordance with the dynamics of the leaf, tree, and ecosystem processes. The model can be recommended to develop a holistic analysis of forest ecosystems with particular emphasis on the interactions between different tree species.

6.4 A Thermodynamic Forest Growth Model

Forest growth is normally modeled by a logistic model (see, e.g., Fukuda 2003; Zeide 2004) or by the use of mass balance equations (see, e.g., Huth and Ditzer 2000; Komarov et al. 2003). Both types of models work very well in describing the observed growth but they are not designed to predict future growth. Alexandrov (2008) has developed a thermodynamic model that can predict the future growth of a forest based on the present state. The ecological law of thermodynamics (ELT) (further details are given by Jørgensen 2002, 2008, 2009; Jørgensen et al. 2007; Jørgensen and Svirezhev 2004; also discussed in Section 2.7) has provided a hint as to which variables can be used to calculate forest growth. ELT is applied to develop the so-called structurally dynamic models, and an example is presented in Chapters 19 and 22. ELT postulates that an ecosystem uses the incoming energy of solar radiation to move as far away from thermodynamic equilibrium as possible and thereby achieves more information and organization. This implies self-organization of canopies—which means that the randomness of foliage distribution is reduced (Alexandrov and Oikawa 1997).

The basal area of the stand, S, can be shown (Alexandrov and Oikawa 1997) based on ELT to be proportional to $A^{(4/5)}$, where A is the age of the stand. Since biomass, B, is proportional to S, B is also proportional to $A^{(4/5)}$. Exergy, Ex, is proportional to the crown density R, which implies that E/R is a constant. It means that the similarity of canopy patterns in stands of different ages can be assumed, and that the future growth of S and B can be predicted from its current value. This conditions furthermore implies that age-dependent productivity $p(A)$ times the ratio between biomass increment and productivity, $f(A)$, will be proportional to dB/dA or proportional to $A^{(-1/5)}$, since

$$dB/dA = (4/5)A^{(-1/5)} \tag{6.1}$$

and

$$dB/dA = P_p(a)f(A) \tag{6.2}$$

where P is the potential productivity.

We can therefore obtain, by using these equations and by denoting $b = (5/4)p(A)f(A)A^{(1/5)}$, that

$$B = bPA(4/5) \tag{6.3}$$

where b is the potential productivity of a given species and b is a parameter. This is in accordance with the Osnabruck biosphere model (Esser 1987). This model has yielded 6.3 empirically with $b = 0.59181$ and the exponent = 0.79216, which is very close to 4/5.

Testing of this 4/5-law against the data proposes a minor revision of Equation 6.3 to:

$$B = bP(A - u)^{(4/5)} \tag{6.4}$$

Figure 6.2 shows how the biomass is predicted by using this equation as a function of age. The 4/5 law, however, is only applicable where the processes of growth and thinning are naturally balanced to maintain a constant E/R.

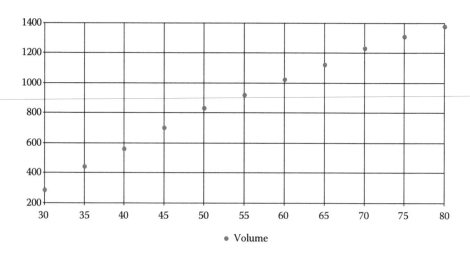

• Volume

FIGURE 6.2
The biomass as volume m³/ha (y-axis) is plotted versus the stand age in years (x-axis). The biomass volumes are calculated based on the 4/5 law. The obtained values are in good (approximate) agreement with measurements and the Chapman–Richards model, but only in the shown range.

Similar equations have been proposed by Tsoularis (2001) and, as mentioned above, by Esser (1987). This model, however, is not based on empirical data, but on a fundamental thermodynamics hypothesis, the ELT.

Alexandrov (2008) writes as conclusion of the presented model the following: This theoretical proposition reduces the variety of forest growth curves that can be derived from logistic and mass-balance models, and allows us to "see the forest behind the trees." The huge number of microscopic descriptions of individual patterns of growth, which is found in most forest models, has been replaced by a few macroscopic parameters. It demonstrates a generalized pattern, which enables modelers to make prediction about forest growth.

References

Aber, J. D., and C. A. Federer. 1993. A generalized, lumped parameter model of photosynthesis, evaporation and net primary production in temperate and boreal forest ecosystems. *Oecologia* 92: 463–474.

Alexandrov, G. A. 2008. Forest growth in the light of thermodynamic theory of ecological systems. *Ecological Modelling* 216: 102–106.

Alexandrov, G. A., and T. Oikawa. 1997. Contemporary variations of terrestrial net primary production: The use of satellite data in the light of extremal principle. *Ecological Modelling* 95: 113–118.

Amaro, A., D. Reed, and P. Soares. 2003. *Modelling Forest Systems*, 400 pp. CABI Publishing.

Boersma, M. et al. 1991. Nutrient gradients and spatial structure in tropical forest: A model study. *Ecological Modelling* 55: 219–240.

Booth, T. H. 1991. A climatic/edaphic data base and plant growth index prediction system for Africa. *Ecological Modelling* 56: 127–134.

Bossel, H., and H. Krieger. 1991. Simulation model of natural tropical forest dynamics. *Ecological Modelling* 59: 37–71.

Bossel, H., and H. Schæfer. 1989. Generic simulation model of a forest growth, carbon and nitrogen dynamics, and application to tropical acacia and European spruce. *Ecological Modelling* 48: 221–265.

Chen, C. W., and L. E. Gomez. 1988. Application of the ILWAS Model. Report to Southern California Edison Company, Palo Alto, CA, USA.

Colbert, J. J. et al. 2004. Individual tree basal-area growth parameter estimates for four models. *Ecological Modelling* 174: 115–126.

Colygny, F. de et al. 2003. CAPSIS: Computer-aided projection for strategies in silviculture: Advantages of a shared forest-modelling platform. In: *Modelling Forest Systems*, eds. A. Amaro, D. Reed, and P. Soares, 400 pp. CABI Publishing.

Corbeels, M. et al. 2005. A process-based model of nitrogen cycling in forest plantation. Part I and Part II. *Ecological Modelling* 187: 426–474.

Dale, V. H. et al. 1988. Using sensitivity and uncertainty analysis to improve predictions of broad-scale forest development. *Ecological Modelling* 42: 165–178.

Deckmyn, G. et al. 2008. ANAFORE: A standscale process-based forest model that includes wood tissue development and labile carbon storage in trees. *Ecological Modelling* 215: 337–344

Esser, G. 1987. Sensitivity of global carbon pools and fluxes to human and potential climate impacts. *Tellus Series B Chemical and Physical Meteorology* 39: 245–260.

Hoffmann, F. 1995. FAGUS—a model for growth and development of beech. *Ecological Modelling* 83: 327–348.

Foody, G. M., and M. E. J. Cutler. 2006. Mapping the species richness and composition of tropical forest from remotely sensed data with neural network. *Ecological Modelling* 195: 37–42

Ford, E. D., and A. R. Kiestter. 1990. Modelling the effects of pollutants on the processes of growth. In: *Process Modelling of Forest Growth Responses to Environmental Stress*, ed. R. K. Dixon et al., 440 pp. Portland, OR: Timber Press.

Friend, A. D. et al. 1993. A physiology-based gap model of forest dynamics. *Ecology* 79: 792–797.

Gillet, F. 2008. Modelling vegetation dynamics in heterogeneous pasture-woodland landscapes. *Ecological Modelling* 217: 1–18.

Grabarnik, P., and A. Särkkä. 2009. Modelling the spatial structure of forest stands by multivariate point processes with hierarchical interactions. *Ecological Modelling* 220: 1232–1240.

Huth, A., and T. Ditzer. 2000. Simulation of growth of a lowland Dipterocarp rain forest with FORMIX 3. *Ecological Modelling* 134: 1–25.

Jørgensen, S. E. 2002. *Integration of Ecosystem Theories: A Pattern*, 3rd ed., 428 pp. Dordrecht: Kluwer Academic Publishers.

Jørgensen, S. E. 2008. *Evolutionary Essays*. Amsterdam: Elsevier, 230 pp.

Jørgensen, S. E. 2009. The application of structurally dynamic models in ecology and ecotoxicology. In *Ecotoxicological Modeling*, ed. J. Devillers, 377–394. Springer Verlag.

Jørgensen, S. E., and Y. M. Svirezhev. 2004. *Towards a Thermodynamic Theory for Ecological Systems*. Oxford: Elsevier.

Jørgensen, S. E., B. Fath, S. Bastiononi, M. Marques, F. Müller, S. N. Nielsen, B. C. Patten, E. Tiezzi, and R. Ulanowicz. 2008. *A New Ecology. Systems Perspectives*, 288 pp. Amsterdam: Elsevier.

Kucharik, C. et al. 2006. A multiyear evaluation of a Dynamo Global Vegetation Model at three AmeriFlux forest sites: Vegetation structure, phenology, soil temperature, carbon dioxide and water vapor exchange. *Ecological Modelling* 196: 1–32.

Komarov, A. S. et al. 2003. EFIMOD 2—a model of growth and cycling of elements in boreal forest ecosystems. *Ecological Modelling* 170: 373–392.

Kremer, R. G. 1994. Model documentation by Kremer. University of Montana School of Forestry, Missoula, MT 59812.

Lacerte, V. et al. 2006. Calibration of the forest vegetation simulator (FSV) model for the main forest species of Ontario, Canada. *Ecological Modelling* 199: 336–349.

Larocque G. R. et al. 2006. Modelling forest succession in two southeastern Canadian mixedwood ecosystem types using the ZELIG model. *Ecological Modelling* 199: 350–362.

Levine, E. et al. 1993. Forest ecosystem dynamics: Linking forest succession, soil processes and radiation models. *Ecological Modelling* 65: 199–219.

Lhotka, L. 1994. Implementation of individual-oriented models in aquatic ecology. *Ecological Modelling* 74: 47–62.

Liu, J. 1993. Ecolecon, model for species conservations in complexed forest landscapes. *Ecological Modelling* 70: 63–88.

Malanson, G. P., Q. Wang, and J. Kupfer. 2007. Ecological processes and spatial patterns, before, during and after simulated deforestation. *Ecological Modelling* 202: 397–409.

Moorhead, D. L., and Reynolds, J. F. 1991. A general model of litter decomposition in the northern Chihuahuan Desert. *Ecological Modelling* 59: 197–219.

Nikolov, N. T., and D. G. Fox. 1994. A coupled carbon–water–energy–vegetation model to assess responses of temperate forest ecosystems in climate and atmospheric carbon dioxide: Part 1. Model concepts. *Environmental Pollution* 83: 251–262.

Ogris, N., and M. Jurc. 2010. Sanitary felling of Norway spruce due to spurge beetles in Slovenia: A model and projections for various climate change scenarios. *Ecological Modelling* 221: 290–302.

Ramming, A. et al. 2010. Impacts of changing forest regimes on Swedish forests: Incorporating cold hardiness in a regional ecosystem model. *Ecological Modelling* 221: 303–313.

Soares, P., and M. Tome. 2003. GLOBTREE: An individual tree growth model for *Eucalpytus globulus* in Portugal. In *Modelling Forest Systems*, eds. A. Amaro, D. Reed, and P. Soares, 400 pp. CABI Publishing.

Stickan, W. et al. 1994. Modelling the influence of climatic variability on carbon and water budgets of trees using field data for beech saplings. *Ecological Modelling* 75–76: 331–344.

Tietjen, B., and A. Huth. 2006. Modelling dynamics of managed tropical rainforests—an aggregated approach. *Ecological Modelling* 199: 421–431.

Tsoularis, A. 2001. Analysis of logistic growth models. *Research Letters in Information and Mathematical Sciences* 2: 23–46.

Veroustraete, F. 1994. On the use of a simple deciduous forest model for the interpretation of climate change effects at the level of carbon dynamics. *Ecological Modelling* 75–76: 221–238.

Wagner, S. et al. 2004. Directionality in fruit dispersal models for anemochorous forest trees. *Ecological Modelling* 179: 487–498.

Wallentin, G. et al. 2008. Understanding the Alpine tree line dynamics: An individual based model. *Ecological Modelling* 218: 235–246.

Wælder, K. et al. 2008. Analysis of O_F-layer humus mass variation in a mixed stand of European beech and Norway spruce: An application of structural equation modelling. *Ecological Modelling* 213: 319–330.

Zeide, B. 2004. Intrinsic units in growth modeling. *Ecological Modelling* 175: 249–259.

Zhang, J. et al. 2008. TRIPLEX model testing and application for predicting forest growth and biomass production in the subtropical forest zone of China's Zhejiang Province. *Ecological Modelling* 219: 264–275.

Zhang, Y. Z., and X. Yu. 2006. Measurement and evaluation of interactions in complex urban ecosystem. *Ecological Modelling* 196: 77–89.

Zhou, X. et al. 2007. Developing above-ground woody biomass equations for open-grown, multiple-stemmed tree species: Shelterbelt-grown Russian-olive. *Ecological Modelling* 202: 311–323.

7

Grassland Simulation Models: A Synthesis of Current Models and Future Challenges

Debra P. C. Peters

CONTENTS

7.1 Introduction 175
7.2 Overview of Available Models 176
    7.2.1 Demographic Models 177
    7.2.2 Physiological Models 182
    7.2.3 Physical Models 184
    7.2.4 Biogeochemical Models 186
    7.2.5 Dynamic Global Vegetation Models 188
7.3 Recommendations on Model Selection 191
7.4 Future Challenges in Grassland Modeling 192
References 194

7.1 Introduction

Grasslands occur on all continents, except Antarctica, in areas that are transitional between deserts that are drier and forests that are wetter. Historically, grasslands comprised almost 42% of the world's plant cover (Anderson 2006). However, many perennial grasslands have converted to other system states (degraded shrublands or shrub steppes, savannas, woodlands, cultivated fields) or have been susceptible to invasion by nonnative species over the past several centuries. For example, North American grasslands are now listed as critically endangered with declines in spatial extent as high as 98% in some locations (Noss et al. 1995). Consequences of these vegetation conversions are consistent across grasslands globally: local ecosystem properties are modified, including primary production, biodiversity, and rates and patterns of nutrient cycling (Schlesinger et al. 1990; Ricketts et al. 1999; Huenneke et al. 2002; Briggs et al. 2005; Knapp et al. 2008). Regional to global processes are altered, including transport of dust to the atmosphere, redistribution of water to the oceans and groundwater reserves, and feedbacks to weather (Jaffe et al. 2003; McKergow et al. 2005; Pielke et al. 2007).

Aside from conversion to cultivation, historic shifts in grass species composition and dominance are related to changes in environmental drivers, in particular climate (i.e., periodic drought), fire, and grazing animals (Oesterheld et al. 1999); the relative importance of these drivers to ecosystem dynamics varies by grassland type (Pieper 2005). Dry grasslands change to more desertlike systems as xerophytic shrubs increase in density and cover with livestock overgrazing in times of drought (Schlesinger et al. 1990; Archer 1994). Mesic grasslands shift to more forest-like systems as trees increase in density and cover with a reduction in grazing and fire frequency, particularly in extended wet periods (Briggs et al. 2005). Shifts in species composition can occur in grasslands containing mixtures of C_3 and C_4 species when temperature and water availability are modified to favor one physiology over another (Steuter 1987; Paruelo and Lauenroth 1996). Shifts in dominance between species that are grazing- or fire-adapted occur with changes in grazing management or fire frequency (Knapp et al. 1998). Invasion by nonnative species depends on the availability of their propagules in the presence of disturbance under appropriate environmental conditions that favor their recruitment and expansion (Sheley and Petroff 1999).

Although many studies have been conducted in grasslands, it is often difficult experimentally to quantify and distinguish the role of each driver to grassland dynamics, and the consequences of changing species composition and dominance to other ecosystem processes (Peters et al. 2006). In addition, the environmental drivers continue to change, thus making predictions about future states increasingly difficult (Smith et al. 2009). Simulation models provide a powerful approach to improving our understanding of historic dynamics, and to synthesizing the importance of changing drivers to future dynamics.

7.2 Overview of Available Models

Five major classes of simulation models are commonly used to study grasslands (Tables 7.1 to 7.5). Each class is discussed below in terms of its drivers and key response variables relative to the types of questions typically addressed by that class of model. Specific models are described for each class that differ in the drivers (climate, grazing, fire), input parameters, unit of response, processes simulated, and response or output variables that determine the questions that can be addressed. An exhaustive description of all possible grassland models is not intended, but rather specific models are given as examples. In addition, a representation of each model is given rather than all possible applications that may, in some cases, include different plot sizes, functional types, input parameters, and response variables. The discussion is limited to simulation models such that theoretical and mathematical models are not discussed here (e.g., Rietkerk et al. 1997; HilleRisLambers et al. 2001).

7.2.1 Demographic Models

Demographic models include both individual-based models (IBMs), models that simulate an aggregate of individuals, and cellular automata (CA) models (Table 7.1). IBMs simulate the recruitment, growth, and mortality of individual plants, and how these demographic processes are affected by competition for resources with neighboring plants. The models are often deterministic in competition for resources that affect growth, and have stochastic elements for recruitment and mortality. IBMs have a long history of use in forests where individual trees compete primarily for light, although more recent applications include simple submodels of water and nitrogen (Botkin et al. 1972; Shugart 1984; Smith and Urban 1988; Pacala et al. 1996; Moorcroft et al. 2001). The forest modeling paradigm was adapted to grasslands where plants compete for belowground resources (water, nitrogen) (Coffin and Lauenroth 1990; Peters 2002).

IBMs are often gap models where individual plants are simulated on a plot scaled to the resource space associated with a full-size plant of the dominant species. Forest models use light gaps associated with the canopy of a full-size dominant tree (Shugart 1984), whereas grassland models define a gap as the belowground resource space (i.e., active rooting volume) associated with a full-size grass of the dominant species (Coffin and Lauenroth 1990). Because grass plants are small, the plot size of grassland models is much smaller (0.1 to 1 m²) compared with forest gap models (>100 m²). Time step is typically annual for incrementing plant growth, although other dynamics, such as soil water, may be simulated at a finer resolution (daily, weekly, monthly), and this information is aggregated to obtain an annual amount of water available for growth by each plant. Grids of plots are used to simulate landscapes, either with or without spatial processes such as seed dispersal, that connect plots (Coffin and Lauenroth 1989; Rastetter et al. 2003).

Because of the challenges and uncertainties in explicitly simulating belowground resources and rooting distributions at the temporal and spatial resolution required for individual grasses, IBMs have not been used as extensively in grasslands compared to the number and variety of forest models (Bugmann 2001; Perry and Enright 2006). One approach to simulating competition for water is to link an IBM to a multilayer soil water model that simulates water dynamics (i.e., interception, evaporation, transpiration, infiltration, and deep drainage) at a fine-scale resolution in time (daily) and space (single plot with detailed soil layers). The approach is to compare the distribution of simulated soil water content with the distribution of active root biomass for each individual relative to total root biomass on the plot to determine the amount

TABLE 7.1

Demographic Models: Examples of Specific Models and Their Key Characteristics

Model Type	Model Name	Ecosystem Type	Unit of Response	Plot Size	Time Step	Drivers	Plot-Scale Processes	Spatially Explicit Processes	Response Variables	References
IBM	STEPPE STEPPE + SOILWAT	Semiarid grasslands	Individual	0.12 m²	Annual Annual, daily	PPT, temp. disturbance	Competition for water – rule based – dynamic	Dispersal Dispersal	Production biomass for herbaceous plants individuals, populations, communities	Coffin and Lauenroth (1989, 1990) Coffin et al. (1993)
IBM	Ecotone	Arid to semiarid grasslands	Individual	0.1 to 1 m²	Annual	PPT, temp. disturbance	Competition for water (dynamic)	Dispersal	Grass–shrub state changes individuals, populations, communities	Peters (2002) Peters and Herrick (2002) Rastetter et al. (2003)
IBM	COIRON	Grass steppes of Patagonia (*Festuca pallescens*)	Individual	0.09 m²	Annual	PPT grazing (intensity, seasonality)	Competition for water (rule-based)	Lateral water movement grazing	Density of individuals Species ANPP	Paruelo et al. (2008)
PFT	FATE LAMOS	Semiarid grasslands (Sweden)	Groups of individuals of four life stages	20 × 20 m	Annual	Grazing, fragmentation	Competition for light	Grazing frequency, intensity; seed dispersal	Biomass by group	Moore and Noble (1990) Cousins et al. (2003)
Cellular automata		Artificial landscapes	Pops. PFT	100s of meters	Annual	Weather, fragmentation	Competition (rule-based)	Seed dispersal	Population size, turnover rate, portion of occupied habitats	Körner and Jeltsch (2007)

Note: ANPP, aboveground net primary production; Pops, populations; PPT, precipitation.

of water allocated to a plant (Coffin et al. 1993; Peters 2002). Soil water content per layer is simulated daily by a soil water model using information on daily precipitation and temperature and soil properties. Active root biomass simulated by the vegetation model is determined for each plant based on the temperature of that day relative to optimum temperature for growth. Water is allocated proportionally to each plant daily based on its proportion of the total active root biomass in that layer. Total amount of water available to each plant in each year (cm water/year) is found by summing across layers and days. Water-use efficiency values (g production/cm water) are then used to convert water (cm) to plant production (grams) per year. Root to shoot ratios are used to allocate total production to above- and belowground. Root production by layer is distributed based on the proportion of water available to a plant in each layer relative to the total water in the profile. This approach to competition for soil water has been used successfully to link the SOILWAT model (Parton 1978) to two grassland models (STEPPE, Coffin et al. 1993; Ecotone, Peters 2002).

Grassland IBMs have been used to examine the role of different drivers and key processes on species dominance and composition, primarily in arid and semiarid grasslands where competition for light can be ignored (Table 7.1). STEPPE was developed to examine the importance of local (competition for water) and spatial (seed dispersal) processes to successional dynamics of semiarid grasslands (Coffin and Lauenroth 1989, 1990). Although the model successfully represented successional patterns for a range of disturbance sizes (Coffin and Lauenroth 1989, 1994), the rate of recovery was too slow as disturbance size increased to include abandoned agricultural fields (Coffin et al. 1993). The model assumes that recovery is dependent solely on wind dispersal of seeds from the undisturbed edge of a field. Results suggest that additional processes besides wind are operating to disperse seeds over long distances to result in faster recovery times than simulated (Coffin et al. 1993).

The Ecotone IBM (Figure 7.1) was developed to examine shifts in species dominance between grasses and shrubs in arid and semiarid grasslands, either with (Rastetter et al. 2003) or without seed dispersal (Peters 2002). This model explicitly includes competitive interactions among grasses and shrubs, and the drivers that promote shrub expansion or grass dominance. Results showed that soil properties can overwhelm climatic conditions that favor shrub expansion to allow grass persistence (Peters 2002), and that seed dispersal can limit the suitable microsites for grass dominance in a shrub-dominated landscape (Rastetter et al. 2003). Ecotone has also been used to simulate conditions that limit or promote the spread of herbaceous invasive species (Goslee et al. 2001, 2006), and to identify thresholds in disturbance frequency where dominance shifts from perennial grasses to shrubs (Peters and Herrick 2002). The COIRON model was developed to examine effects of precipitation and livestock grazing on density of perennial grasses in South America (Paruelo et al. 2008). Interannual variability in precipitation

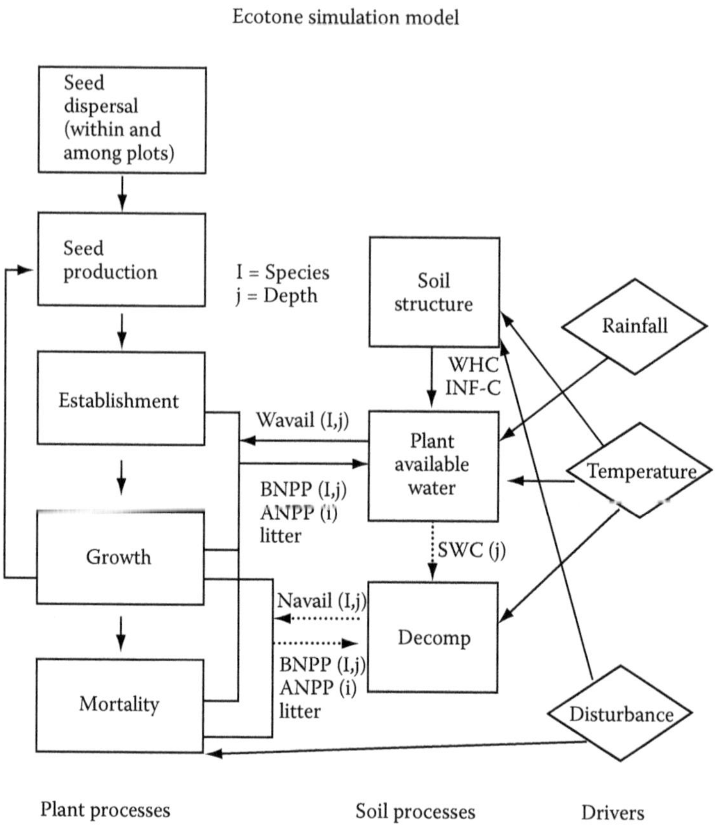

Ecotone simulation model

FIGURE 7.1
The Ecotone IBM for grasslands simulates plant (recruitment, competition, mortality), water, and soil processes within a plot as well as seed dispersal among plots. (From Peters, D.P.C., *Ecological Modelling* 152, 5–32, 2002. With permission.) Competition for soil water is determined using water availability and rooting distribution by depth for each plant relative to all plats on a plot. Time scales vary from days to years depending on the process simulated. Spatial scales range from individual plants to groups of plants (patches), landscape units, and geomorphic units. Environmental drivers include climate (precipitation, temperature) and disturbances, such as human and animal activities, that kill plants and modify soil structure.

was found to be more important than grazing regime in explaining plant density.

Two models similar to IBMs deserve mention here. First, the FATE model simulates groups of individuals in similar life stages rather than simulating individual plants (Moore and Noble 1990). Linking the model to LAMOS allowed simulation of the effects of grazing frequency and intensity on plant species persistence at the landscape scale, and the relative effects of grassland size and pattern (Cousins et al. 2003). Results showed that continuous, low intensity grazing is more favorable to grasslands than discontinuous, high intensity grazing. Second, cellular automata (CA) models have been used to simulate plant populations across landscapes using rules for population dynamics in each cell. In CA models, each particular cell is affected by its neighbors in a simple, rule-based manner. A CA model for grasslands showed that plant functional groups selected to represent populations with different traits have different vulnerabilities to fragmentation (Körner and Jeltsch 2007).

Advantages. Demographic models, and in particular, IBMs, are relevant to many key questions in grasslands that deal with invasion by woody plants or herbaceous, nonnative species. A focus on individual plant dynamics, including recruitment and dispersal, allows the simulation of invasion and recovery dynamics by populations and species. In addition, IBMs allow an examination of how system properties emerge from the behavior of individuals, and how system dynamics affect individuals. Variation among individuals, local interactions such as competition for limiting resources, complete life cycles, and individual responses to changing environmental conditions can all be studied within the context of multiple levels of organization (populations, communities, ecosystems) (Grimm et al. 2006). IBMs are intuitively appealing because of the use of individuals as the fundamental unit, and of life cycle stages with stochastic elements (recruitment, mortality) that can often be parameterized based on natural history information for each species.

Limitations. IBMS can be difficult to parameterize for grasslands given that competition among individuals is simulated, and little is known about how plants compete for belowground resources. IBMs are computationally intensive as spatial extent increases because many small plots with many individuals on each plot need to be simulated. Thus, simulations can become prohibitive for large landscapes and regions. IBMs require many data for calibration and validation, although simple functional relationships are often possible for some parameters, in particular those related to recruitment and mortality when they are assumed to be stochastic processes. The models can not simulate conservation of mass and energy, and can become complex if many processes are included. Developing IBMs for mesic grasslands where both above- and belowground competition need to be explicitly simulated is a challenge for the future.

Approaches to address limitations. Approaches have been developed to address some of these limitations in forest models; these approaches are also expected to be applicable to grassland models. For example, the HYBRID model replaces species-specific functions with physiological and biogeochemical relationships similar to ecosystem models (e.g., Century described below). This approach reduces the number of parameters required, yet maintains mechanistic relationships. To scale up, each 1° Global Climate Model (GCM) grid cell is assumed to be homogeneous, and is simulated using the ensemble average of 10 tree-size plots (Friend et al. 1997). Hundreds or thousands of tree-size plots would be required to account for within GCM grid cell heterogeneity. More recent approaches have focused on developing equations that govern the ensemble average of an IBM directly from fine-scale processes without simulating each plant (Moorcroft et al. 2001). More tractable macroscopic equations have been developed to predict mean densities and size structures for each species using only individual-level parameter values and functional forms in an IBM (Strigul et al. 2008). These equations can be solved analytically to significantly reduce computational time for large spatial extents.

Meta-modeling is a different approach where many simulations of an IBM are used to develop relationships between state variables and drivers or to create transition probabilities from one state to another (Urban 2005). These relationships or transition probabilities can then be used to simulate large spatial extents, by using a CA model, a semi-Markov model, or a stage-based transition model, depending on the level of detail included in the processes in the original IBM (Urban et al. 1999).

7.2.2 Physiological Models

Physiological models simulate carbon assimilation, allocation, and growth by plant functional types (PFTs) or species (Table 7.2). Models typically include additional processes, such as local water and nitrogen dynamics, and spatial processes associated with water redistribution and seed dispersal. The time step is daily to weekly, and key response variables are biomass and production by PFTs or key species. Plot size is sufficiently large to assume homogeneity within a patch (tens to hundreds of meters). PALS is an example of these models that have been used to explore effects of rainfall variability on primary production and plant–soil water dynamics (Reynolds et al. 2000), and decomposition and nutrient cycling (Kemp et al. 2003). MALS is a spatially explicit version of PALS that includes water redistribution and seed dispersal among plots (Gao and Reynolds 2003). Results generally support the hypothesis that wetter winters and drier summers decrease grasses and increase shrubs, but the model was unable to reproduce major shifts in state from grasslands to shrublands.

The SAVANNA model is included here, although this model also includes simple demographic processes of plants as well as detailed livestock grazing processes at local, landscape, and regional scales (Coughenour 1993). In this model, flows of biomass, nitrogen, and organisms are simulated weekly, and vegetation and ungulate production are related to climate, soils, and topography. SAVANNA was originally developed for studies of African pastoralism, and has been applied to western U.S. and Canadian national parks, and to savannas in Australia as an ecosystem management tool (Ludwig et al. 2001; Weisberg and Coughenour 2003). SAVANNA has also been used to examine the effects of climate change (precipitation, temperature, CO_2) on grazing system sustainability in steppe grasslands of Inner Mongolia (Christensen et al. 2004). Large increases in precipitation, temperature, and CO_2 left the

TABLE 7.2

Physiological Models: Examples of Specific Models and Their Key Characteristics

Model Type	Model Name	Ecosystem Type	Unit of Response	Plot Size	Time Step	Drivers	Plot-Scale Processes	Spatially Explicit Processes	Response Variables	Citations
PFT	PALS	Chihuahuan Desert	PFT	10 × 30 m	Day	Weather	Carbon assimilation allocation growth	None	Biomass, ANPP, nutrient cycling by PFT	Reynolds et al. (2000) Kemp et al. (2003)
PFT	MALS	Chihuahuan Desert	PFT	10 × 30 m	Day	Weather	Carbon assimilation allocation growth	Water runon–runoff Seed dispersal	Biomass by PFT	Gao and Reynolds (2003)
PFT	SAVANNA	Arid and semiarid grasslands	PFT (trees, shrubs, grasses) Animal FT	Scaled to landsc	Week	Weather, soil properties, elevation, aspect plus CO_2	Carbon assimilation allocation growth Competition for water, N, light, demog	Seed dispersal, grazing, water runon–runoff	Plant biomass, production by PFT number of animals	Coughenour (1993) Boone et al. (2004) Christensen et al. (2004)
PFT	STEP STEP-RT	Annuals in Sahel	PFT (C_3/C_4; proportion of dicots/grasses)		Day	Weather (rainfall daily, temp., PET, radiation decade), soil depth, texture Radiative transfer functions	Carbon assimilation allocation growth	None	Biomass, LAI Canopy height, fraction green vegetation All above plus land surface reflectance	Mougin et al. (1995) Tracol et al. (2006) Jarlan et al. (2008)

Note: Animal FT, animal functional type; ANPP, aboveground net primary production; demog, demography; landsc, landscape; N, nitrogen; PET, potential evapotranspiration; PFT, plant functional type; PPT, precipitation.

simulated system vulnerable to shrub invasion when grazed. SAVANNA has also been used to guide management decisions in Africa where simulated production was improved if livestock stocking was reduced before a forecasted drought (Boone et al. 2004).

A set of physiologically based models have been developed for use with remotely sensed images. The STEP model simulates daily water fluxes, photosynthesis, respiration, growth, and senescence driven by weather and global radiation (Mougin et al. 1995). STEP was coupled with radiative transfer models in the optical (Lo Seen et al. 1995) and active/passive microwave domains (Frison et al. 1998) to interpret temporal variations of satellite observations over the Sahel. Normalized Difference Vegetation Index (NDVI) values were also assimilated into STEP to estimate grassland production in Mali (Jarlan et al. 2008). Values simulated by a canopy process model were also used in canopy radiative transfer equations to determine effects of vegetation change on energy balance with feedbacks to the atmosphere (Cayrol et al. 2000).

Advantages. Physiological models account for fine-scale temporal resolution in carbon, nitrogen, and water dynamics (daily), and are used for questions related to controls on carbon assimilation and allocation as related to photosynthesis. The focus is on biomass and production of PFTs or key species, thus the spatial resolution assumes homogeneity within plots sufficiently large to support multiple PFTs (e.g., SAVANNA: trees, shrubs, grasses) or to reduce heterogeneous responses (PALS). Detailed vegetation–herbivore interactions can be simulated at fine temporal scales needed to understand the effects of livestock grazing on grassland production.

Limitations. Because demographic processes are not included, physiological models are limited in their ability to simulate shifts in species dominance or state changes from grasslands to woody plant dominance. The short time steps (daily, weekly) require intensive parameterization of plant physiological processes.

7.2.3 Physical Models

Physical models simulate physical processes, typically soil water dynamics and other environmental variables, then rules or functional relationships are used to calculate vegetation responses (Table 7.3). One model, the SOILWAT model of soil water dynamics, has been used extensively to calculate the probability of germination and establishment for perennial grasses in North America. SOILWAT is a daily time step, multilayer soil water dynamics model of interception, evaporation, transpiration, infiltration, and deep storage (percolation) (Parton 1978). Combining this model of soil water dynamics with detailed growth chamber studies of the sequence of microenvironmental conditions required for germination and establishment has allowed recruitment probabilities to be calculated for a dominant perennial grass in the shortgrass steppe for different soil properties (Lauenroth

TABLE 7.3

Physical Models: Examples of Specific Models and Their Key Characteristics

Model Type	Model Name	Ecosystem Type	Unit of Response	Plot Size	Time Step	Drivers	Plot-Scale Processes	Spatially Explicit Processes	Response Variables	References
Tipping bucket	SOILWAT	Shortgrass steppe SGS–CD CD	Plot	1 × 1 m	Day	Biomass, weather, soil properties	Soil water dynamics	None	Soil water content Estab. from rules	Lauenroth et al. (1994) Minnick and Coffin (1999) Peters (2000) Peters et al. (2010)
Membership functions		Mediterranean: semiarid	Plot	25 × 25 m	Day	Weather, topographysoil hydraulic properties, rock cover, radiation flux	Soil water dynamics	Runon–runoff	Germination and production calculated from rules	Svoray et al. (2008)
Niche-based	Biomod	Mostly forests	Species	0.5° grid	Month	Temp. PPT PET	Statistical relationships	Migration rates	Species presence/absence	Thuiller (2003) Morin and Thueller (2009)

Note: CD, Chihuahuan Desert; estab, establishment; PET, potential evapotranspiration; PPT, precipitation; SGS, shortgrass steppe; Temp, air temperature.

et al. 1994). Results showed that infrequent recruitment (1/10 to >1/100 years) can explain, at least in part, the inability of this species to recover following disturbance. The model was also extended to a dominant species in the Chihuahuan Desert grasslands, and was used to explore the importance of recruitment to the geographic distributions of dominant grass in the short-grass steppe and Chihuahuan Desert grasslands (Minnick and Coffin 1999). Effects of seasonality and decadal patterns in rainfall on establishment of grasses were examined at an ecotonal site between the two grassland types (Peters 2000), and more recently to examine effects of soil properties, climate, and historical shifts from grass to shrub dominance on grass establishment in the Chihuahuan Desert (Peters et al. 2010).

A similar approach was used for germination and production of annuals in Israel (Svoray et al. 2008). TOPMODEL was used to simulate soil moisture dynamics within and among topographic units, and relationships for other environmental variables were used in a fuzzy logic (rule based) approach to determine combined effects on germination and production of annuals in the Negev. Results showed that water redistribution and climate are of similar importance in explaining variation in ANPP (Svoray et al. 2008). Another type of physical model is a niche-based model where the establishment of statistical or theoretical relationships between environmental drivers and observed species distributions are used to predict future distributions as the drivers change. Niche-based models have been used to predict effects of climate and land use change on species distributions globally, including grasslands (e.g., Thomas et al. 2004; Thuiller 2003).

Advantages. Physical models are relatively easy to parameterize depending on the complexity of the underlying physical processes and the relationships with biotic responses. Niche-based models are relatively simple, yet can project modeled niches of many species given distribution data.

Limitations. No biology is explicitly included, thus many assumptions are needed for these models to work well. These models do not include biotic interactions, mortality, or growth, and rely on observed patterns based on historical drivers that may change nonlinearly in the future (Hampe 2004).

7.2.4 Biogeochemical Models

Biogeochemical models simulate changes in the cycling of carbon, water, and nutrients with fixed vegetation types (Table 7.4). Consequences of changes in vegetation type can be examined by imposing a change in vegetation or management regime. Although similar drivers (climate, soils) are used, the specific models within this class of models use very different approaches that can generate different results at the grid scale for the same input parameters, yet all models examined in one analysis converged on continental scale total values for NPP and total carbon storage (Schimel et al. 1997). In many cases, the models have been used for both global simulations as well as local, site-

TABLE 7.4

Biogeochemical Models: Examples of Specific Models and Their Key Characteristics

Model Type	Model Name	Ecosystem Type	Unit of Response	Plot Size	Time Step	Drivers	Plot-Scale Processes	Spatially Explicit Processes	Response Variables	Citations
Nutrient cycling	CENTURY	C_3 C_4 Grasslands globally (also forests, shrublands)	Ecosystem	Plot to 0.5° grid	Month	Maximum, minimum ATEMP PPT Soil texture/Depth CO_2	Soil carbon pools/fluxes Nitrogen, phosphorus, Sulfur	None	ANPP, SOC, net N min, E-T	Parton et al. (1987, 1993)
	DAYCENT	C_3 C_4 grasslands	Ecosystem	Plot to 0.5° grid	Day	Same as above and manage. type, timing	Same as above	None	ANPP, SOC, trace gas flux	Parton et al. (1998)
Nutrient cycling	TEM	Grasslands globally (also forests, shrublands, tundra)	Ecosystem	Plot to 0.5° grid	Month	Maximum, minimum Temp PPT Soil texture/ Depth CO_2	E-T photosyn. decomp. Soil N turnover	None	ANPP, SOC net N min, E-T	McGuire et al. (1993) Clein et al. (2000)
Nutrient cycling	BIOME-BGC	C_3 C_4 Grasslands globally (also forests, shrublands)	Ecosystem	Plot to 0.5° grid	Day/ annual	Maximum, minimum Temp PPT Soil texture/ Depth CO_2 N dep	E-T photosyn. decomp. Soil N	None	ANPP, SOC net N min, E-T, LAI	Running and Hunt (1993) White et al. (2000) Thornton and Rosenbloom (2005)

Note: ANPP, aboveground net primary production; decomp, decomposition; E-T, evapotranspiration; LAI, leaf area index; N, nitrogen; net N min, net nitrogen mineralization; photosyn, photosynthesis; PPT, precipitation; SOC, soil organic carbon.

specific conditions. A large number of these models have been developed that allow model comparisons (e.g., Cramer et al. 1999); only a few are shown in Table 7.4 that include grasslands as part of one of many vegetation types simulated globally. One model was developed specifically for grasslands (CENTURY) that is described in more detail here.

The CENTURY model was originally developed to simulate Great Plains grasslands of North America (Parton et al. 1987). This model simulates carbon, nitrogen, and phosphorus cycling as well as plant production at a monthly time step. Multiple plant, litter, and carbon pools are simulated. Soil respiration is computed from decomposition of litter, and SOM is regulated by soil temperature and moisture. Runoff from a plot is calculated from ecosystem water balance. CENTURY has been used extensively to simulate grasslands both in the United States (Burke et al. 1991; Schimel et al. 2000) and globally (Parton et al. 1993; Cramer et al. 1999). DAYCENT is a more recent formulation that uses a daily time step, and more detailed submodels for simulating soil moisture, temperature, and nitrogen, trace gas flux, and soil organic matter (Parton et al. 1998; Gerten et al. 2008).

Advantages. Biogeochemical models are based on conservation principles. These models are easy to parameterize, apply to new systems, and use for predictions. Models have been developed for individual sites and ecosystem types as well as applied globally.

Limitations. These models cannot represent population or spatial variability, or dynamic state changes (grasses to shrubs). Although multiple PFTs can be simulated, species composition cannot be simulated. Proportional biomass by C_3 and C_4 functional groups has been used to represent total biomass and production.

7.2.5 Dynamic Global Vegetation Models

Dynamic global vegetation models (DGVMs) were designed to simulate vegetation functional and structural dynamics at the global scale, thus they are not specific to grassland dynamics. However, grasslands are one of several PFTs simulated globally, and some models have been used at the landscape scale to simulate grassland–forest ecotones. Thus, this class of models is briefly described here with particular reference to the models relevant to grasslands (Table 7.5).

DGVMs simulate vegetation dynamics within a coarse resolution grid cell (typically 0.5° × 0.5°, 1° × 1°) as a fractional coverage of populations of different PFTs. The general structure of DGVMs is similar, although the level of detail varies among models (Cramer et al. 2001). Fast processes (e.g., energy and gas exchange at the canopy–atmosphere interface, photosynthesis, plant–soil water exchange) are simulated hourly or daily, seasonal dynamics (e.g., plant phenology, growth and soil organic matter dynamics) are simulated monthly, and vegetation dynamics are simulated annually (Prentice et al.

TABLE 7.5

Dynamic Global Vegetation Models: Examples of Specific Models and Their Key Characteristics

Model Type	Model Name	Ecosystem Type	Unit of Response	Plot Size	Time Step	Drivers	Plot-Scale Processes	Spatially Explicit Processes	Response Variables	Citations
Biogeography–biogeochemistry	LPJ	Global	PFT Average individual	0.5° grid	Day Month Annual	PPT Temp CO$_2$ Soil properties	Establishment Growth Mortality Above, belowground carbon, water	None	ANPP, carbon, water fluxes	Sitch et al. (2003) Gerten et al. (2004)
+ Land surface model	LPJ + LSM	Global	PFT Average individual	3° grid	20 min Day Month Annual	Same as above	Above plus energy, moisture, momentum fluxes between land and atmosphere, hydrologic cycle, soil temperature	None	ANPP, cover, biomass, respiration, net ecosystem production	Bonan et al. (2003)
Biogeography + biogeochemistry	MAPSS	Grasslands	PFT	200 m 10 km 0.5° grid	Month	PPT Temp. Soils	Plant production Water, energy	Fire	Production Biomass LAI	Neilson (1995)
	MC1 (MAPSS + CENTURY + fire)	C$_3$, C$_4$ grasslands	PFT	50 m 0.5° grid	Month Annual	PPT Temp. soils	Above plus Soil organic matter decomp. nutrient cycling grazing		Production Biomass LAI	Bachelet et al. (2000) Daly et al. (2000)

Note: ANPP, aboveground net primary production; decomp, decomposition; LAI, leaf area index; PFT, plant functional type; PPT, precipitation; Temp, air temperature.

2007). Some models provide more detail to plant physiological processes (e.g., HYBRID; Friend et al. 1997). Other models simulate energy and water fluxes needed by atmospheric circulation models (e.g., IBIS; Foley et al. 1996) or were designed as vegetation dynamics models and include scaling of individual-level processes to the grid cell (e.g., LPJ; Sitch et al. 2003). DGVMs have also been coupled interactively with climate models to allow vegetation feedbacks to climate (e.g., Foley et al. 1998; Bonan et al. 2003). A number of studies have compared DGVMs and have shown uncertainties in the way ecosystem responses to climate are simulated, in particular for water-limited systems (Cramer et al. 2001; Bachelet et al. 2003). These uncertainties led to models where biogeochemistry and water dynamics are explicitly simulated using ecosystem models such as Century (Daly et al. 2000).

The use of DGVMs in grasslands has been primarily to include grassland types as one or more PFTs in global simulations. Grasslands have been distinguished either based on physiology and photosynthetic pathway (C_3 and C_4; Bonan et al. 2003), vegetation structure (tall and short grasses; Hickler et al. 2006) or drought and fire tolerance (Hély et al. 2006). More detailed representations of water fluxes in the soil profile and wildfire effects are likely needed in global models to accurately distinguish grassland types (Hickler et al. 2006). In some cases, DGVMs have been used at the landscape scale to simulate effects of fire, grazing, and climate on grassland–forest ecotones (Bachelet et al. 2000; Daly et al. 2000). Woody encroachment was enhanced by grazing and limited by frequent fires, similar to observed patterns in the field (Bachelet et al. 2000). A warmer and slightly wetter future simulated climate increased the extent of grasslands, and reduced the spatial distributions of forests and savannas (Daly et al. 2000). In addition, model results were sensitive to rooting distributions suggesting that large-scale models will require more detailed accurate belowground representations before shifts in life form can be accurately simulated.

Advantages. Broad spatial extents, from landscapes to the globe, can be simulated for different PFTs, including grassland types. Effects of drivers on dynamics of groups of species can be simulated for large areas. Landscapes and regions can be simulated at finer resolution of PFTs for a greater range of drivers, including fire spread. Broad-scale changes in carbon and nitrogen pools and cycling as a result of interactions between the terrestrial biome and the atmosphere can be simulated as well as patterns in potential natural vegetation.

Limitations. DGVMS assume PFTs migrate rapidly and remain in equilibrium with the climate. At the global scale, the models do not include dispersal, disturbance, or human activities, although there are exceptions at the continental (Lenihan et al. 2008) and landscape scales that include these processes, either within or among grid cells (e.g., Bachelet et al. 2000). Range distribution shifts and extinction for particular species cannot be simulated without detailed knowledge of potential migration rates (Midgley et al. 2007). Detailed parameters for many processes are needed at multiple temporal scales. There is no variation within PFT responses, for example, species-specific variability in growth responses, age, and phenological patterns are not accounted for. These models also do not currently include human activities such as logging, habitat fragmentation, and introduction of invasive species.

7.3 Recommendations on Model Selection

In general, the unit of response is an important determinant of the question that can be addressed by each class of model, and can be used with the needed response variables to guide model class selection. Because demographic models simulate individual plants or populations, they can be used to address controls on shifts in species composition and invasive species dynamics as well as state changes between grasses and woody plants. Size and age class distributions of individuals as well as population density can be output from these models. Because many questions related to future grassland dynamics require an understanding of shifts in dominance, more detail is provided below on demographic models. Physiological models simulate carbon assimilation, allocation, and growth differences among PFTs in response to drivers, and are restricted to dynamics of broad categories of plants (grasses, woody plants) rather than individual plants or populations. Physical models simulate physical properties of a system, such as soil water dynamics, and then use rules to generate biological responses; the resolution of the rules determines the details of the response. Biogeochemical models simulate carbon, water, and nitrogen cycling to generate plant production and soil organic carbon as response variables. DGVMs are described in terms of their ability to simulate grasslands within a broader context of multiple ecosystem types across regions, continents, and the globe.

The selection of a particular model within each class often depends on the way processes are represented relative to the specific question or system being simulated. Model comparisons have been used very effectively to highlight variability in responses driven by different processes included in each model (Cramer et al. 2001; Gerten et al. 2008). These comparisons involve the use of the same input parameters and generation of the same output responses; thus the differences in response depend on which processes are included, assumptions about these processes and their relationships with the drivers, and how the processes are represented (Schimel et al. 1997; Cramer et al. 1999). Similar responses across multiple models can be used to generate testable hypotheses about key processes (Luo et al. 2008), and to provide greater certainty on modeled estimates, such as the annual carbon sink or species range shifts (Schimel et al. 2000; Morin and Thuiller 2009).

7.4 Future Challenges in Grassland Modeling

Many of the future science questions to be addressed in perennial grasslands globally will revolve around the conditions that shift these grasslands to alternative states, dominated by woody plants, by nonnative herbaceous species, or by novel assemblages of species, and the consequences of these state changes to ecosystem services, including biodiversity and primary production, rates of nutrient cycling and carbon storage, and air and water quality and quantity. In many cases, these state changes are "ecological surprises" in that they are observed and confirmed after they occur. These surprises result from an inability to understand the full suite of mechanisms and interactions occurring across spatial and temporal scales that act to drive and maintain these shifts (Peters et al. 2004). In addition, there is increasing interest in identifying the weather–vegetation–soil conditions that may allow a reversal of these states back to perennial grasslands under global changes in drivers (e.g., Holmgren and Scheffer 2001; Allington and Valone 2010).

New modeling approaches will be needed to improve understanding of these mechanisms in order to detect, predict or promote state changes, in particular those that impact the delivery of goods and services to human populations (Peters et al. 2009). Modeling approaches will be needed that enhance our understanding of cross-scale interactions and elucidate the role of these interactions in determining ecosystem thresholds (the level or magnitude of an ecosystem process that results in a sudden or rapid change in ecosystem state). Critical thresholds are often crossed during or following a state change such that a return to the original state is difficult or seemingly impossible (Bestelmeyer 2006). Thresholds can occur either in the environmental driver, the rate of a process, or a state variable. Thresholds indicate that a change in a dominant process has occurred and that distinct exogenous drivers or endogenous positive feedbacks are governing rates of change (Peters et al. 2004). Feedbacks tend to maintain a state, and it is often the change in these feedbacks and the resultant alteration in pattern–process relationships that differentiate a regime shift from a reversible ecosystem change that is not maintained through time (Carpenter 2003). For example, shifts from grasslands to woodlands can be maintained for hundreds of years by positive feedbacks between woody plants and soil properties. In some cases, state changes are driven by processes at one spatial or temporal scale interacting with processes at another scale (Carpenter and Turner 2000; Peters et al. 2007).

Modeling approaches that explicitly account for thresholds and cross-scale interactions are expected to improve our understanding of the mechanisms driving state changes in grasslands, and to allow more informed predictions of impending changes (Bestelmeyer et al. 2006). Most studies of cross-scale interactions have documented changing patterns in vegetation through time and across space, and then assumed changing patterns resulted from changing ecological processes (Peters et al. 2004). However, an approach that combines pattern analyses with experimental manipulation of processes and simulation modeling of rates of ecosystem change under different drivers is needed to tease apart the role of role of drivers and processes in determining patterns at different scales (Peters et al. 2009).

In order to address these science questions, thresholds, feedbacks, and cross-scale interactions will need to be included in models that simulate shifts from grasslands to dominance by other species or assemblages. In some cases, this can be accomplished by linking models of different types

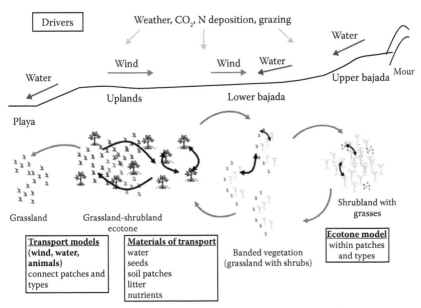

Transport within (black arrows) and among (gray arrows) grassland types

FIGURE 7.2
The ENSEMBLE model for grasslands is being developed to link existing models of vegetation, water, and soil processes at the scale of individual plants (Ecotone) with transport models of wind, water, and animals. Transfer of materials is simulated both within (black arrows) and among spatial units (gray arrows) across a range of scales to represent different grassland types that typically occur along an elevation gradient in the Basin and Range Province of North America. A stylized landscape for the Jornada Basin USDA-LTER site in the northern Chihuahuan Desert is shown as an example. The relative importance of each driver (weather, CO_2, nitrogen deposition, grazing, wind, and water) varies along this gradient as a result of interactions among patterns in atmospheric circulation, elevation, soils, vegetation cover, composition, and spatial distribution of vegetated and bare soil gaps. (From Peters et al., in Miao et al., eds., *Real World Ecology: Large-Scale and Long-Term Case Studies and Methods*, Springer, New York, 2009. With permission.) These interactions across scales can generate nonlinear responses and threshold dynamics in the conversion of grasslands to other states (from Peters et al., *Proceedings of the National Academy Sciences* 101: 15130–15135, 2004; with permission), and are hypothesized to be important to grass recovery. Both native woody plants and nonnative herbaceous plants can be simulated as invasive species that shift dominance to alternative states of the system with consequences for species composition and cycling of carbon, water, and nutrients.

identified in Tables 7.1–7.5, and by adding processes and functional relation-ships to account for nonlinear dynamics and interactions. At a minimum, models will need to maintain species parameters and within-plot responses to drivers, and include biogeochemical processes that feedback to the vegeta-tion. For example, IBMs have been linked with nutrient cycling models or incorporated nutrient cycling routines from these models to simulate land-scapes (e.g., Ecotone with SOILWAT and Century routines) (Figure 7.1) or regions (MAPSS + Century, Bachelet et al. 2000; Daly et al. 2000). Linking veg-etation dynamics models with landscape-scale physical models (e.g., SWEMO of wind erosion–deposition, Okin 2008) or mesoscale climate models (e.g., RAMS, Beltrán-Przekurat et al. 2008) will allow more realistic simulations of additional processes known to be important to grassland–shrubland dynam-ics. More detailed soil development routines will be needed with wind and water movement of materials as soil is added or removed. Using IBMs in these linked models will require a reduction in complexity through statistical or numerical approaches to maintain detailed species information on many small plots, yet simulate large spatial extents (described in Section 7.2.1).

Under conditions where spatial processes are important to connect spatial units, transport of materials will also need to be simulated. Seed dispersal, spread of fire, and water runoff are the most commonly simulated spatial processes (e.g., MALS, Gao and Reynolds 2003; SAVANNA, Coughenour 1993). However, natural landscapes may include multiple drivers occurring across a range of spatial and temporal scales that act to transport materials both within and among spatial units corresponding to different assemblages of grasslands and shrublands (Figure 7.2). Simulating realistic landscapes will require more complicated approaches that include multiple spatial and temporal scales as well as vegetation–soil feedbacks. Scaling this landscape-scale approach to broader spatial extents, such as continents containing con-nected ecosystem types, is possible with a robust conceptual framework (Peters et al. 2008).

References

Allington, G. R. H., and T. J. Valone. 2010. Reversal of desertification: The role of phys-ical and chemical soil properties. *Journal of Arid Environments*. doi:10.1016/j.jaridenv.2009.12.005.

Anderson, R. C. 2006. Evolution and origin of the central grassland of North America: Climate, fire, and mammalian grazers. *Journal of the Torrey Botanical Society* 133: 626–647.

Archer, S. 1994. Woody plant encroachment into south-western grasslands and savan-nas: rates, patterns and proximate causes. In *Ecological Implications of Livestock Herbivory in the West*, eds. M. Vavra and W. Laycock, and R. Pieper, 13–68. Denver, CO: Society of Rangeland Management.

Bachelet, D., J. M. Lenihan, C. Daly, and R. P. Neilson. 2000. Interactions between fire, grazing and climate change at Wind Cave National Park, SD. *Ecological Modelling* 134: 229–244.

Bachelet, D., R. P. Neilson, T. Hickler, R. J. Drapek, J. M. Lenihan, M. T. Sykes, B. Smith, S. Sitch, K. Thonicke. 2003. Simulating past and future dynamics of natural ecosystems in the United States. *Global Biogeochemical Cycles* 17: 1045. doi:10.1029/2001GB001508.

Beltrán-Przekurat, A., R. A. Pielke Sr, D. P. C. Peters, K. A. Snyder, and A. Rango. 2008. Modelling the effects of historical vegetation change on near-surface atmosphere in the northern Chihuahuan Desert. *Journal of Arid Environments* 72: 1897–1910.

Bestelmeyer, B. T. 2006. Threshold concepts and their use in rangeland management and restoration: The good, the bad, and the insidious. *Restoration Ecology* 14: 325–329.

Bestelmeyer, B. T., D. A. Trujillo, A. J. Tugel, and K. M. Havstad. 2006. A multi-scale classification of vegetation dynamics in arid lands: What is the right scale for models, monitoring, and restoration? *Journal of Arid Environments* 65: 296–318.

Bonan, G. B., S. Levis, S. Sitch, M. Vertenstein, and K. W. Oleson. 2003. A dynamic global vegetation model for use with climate models: Concepts and description of simulated vegetation dynamics. *Global Change Biology* 9: 1543–1566.

Boone, R. B., K. A. Galvin, M. B. Coughenour, J. W. Hudson, P. J. Weisberg, C. H. Vogel, and J. E. Ellis. 2004. Ecosystem modeling adds value to a South African climate forecast. *Climatic Change* 64: 317–340.

Botkin, D. B., J. F. Janak, and J. R. Wallis. 1972. Some ecological consequences of a computer model of forest growth. *Journal of Ecology* 60: 849–872.

Briggs, J. M., A. K. Knapp, J. M. Blair, J. L. Heisler, G. A. Hoch, M. S. Lett, and J. K. McCarron. 2005. An ecosystem in transition: Causes and consequences of the conversion of mesic grassland to shrubland. *Bioscience* 55: 243–254.

Bugmann, H. 2001. A review of forest gap models. *Journal of Climate Change* 51: 259–305.

Carpenter, S. R. 2003. *Regime Shifts in Lake Ecosystems: Pattern and Variation.* 21385 Oldendorf/Luhe, Germany: International Ecology Institute.

Carpenter, S. R., and M. G. Turner. 2000. Hares and tortoises: Interactions of fast and slow variables in ecosystems. *Ecosystems* 3: 495–497.

Cayrol, P., A. Chehbouni, L. Kergoat, G. Dedieu, P. Mordelet, and Y. Nouvellon. 2000. Grassland modeling and monitoring with SPOT-4 VEGETATION instrument during the 1997–1999 SALSA experiment. *Agricultural and Forest Meteorology* 105: 91–115.

Burke, I. C., T. G. F. Kittel, W. K. Lauenroth, P. Snook, C. M. Yonker, and W. J. Parton. 1991. Regional analysis of the Central Great Plains: Sensitivity to climate variability. *Bioscience* 41: 685–692.

Christensen, L., M. B. Coughenour, J. S. Ellis, and Z. Chen. 2004. Vulnerability of the Asian typical steppe to grazing and climate change. *Climatic Change* 63: 351–368.

Clein, J. S., B. L. Kwiatkowski, A. D. McGuire, J. E. Hobbie, E. B. Rastetter, J. M. Melillo, and D. W. Kicklighter. 2000. Modelling carbon responses of tundra ecosystems to historical and projected climate: A comparison of a plot- and a global-scale ecosystem model to identify process-based uncertainties. *Global Change Biology* 6 (Suppl. 1): 127–140.

Coffin, D. P., and W. K. Lauenroth. 1989. Disturbances and gap dynamics in a semiarid grassland: A landscape-level approach. *Landscape Ecology* 3 (1): 19–27.

Coffin, D. P., and W. K. Lauenroth. 1990. A gap dynamics simulation model of succession in the shortgrass steppe. *Ecological Modelling* 49: 229–266.

Coffin, D. P., and W. K. Lauenroth. 1994. Successional dynamics of a semiarid grassland: Effects of soil texture and disturbance size. *Vegetatio* 110: 67–82.

Coffin, D. P., W. K. Lauenroth, and I. C. Burke. 1993. Spatial dynamics in recovery of shortgrass steppe ecosystems. In *Theoretical Approaches to Predicting Spatial Effects in Ecological Systems. Lectures on Mathematics in the Life Sciences*, ed. R. Gardner, vol. 23, 75–108.

Coughenour, M. B. 1993. *SAVANNA—Landscape and Regional Ecosystem Model. User Manual*. Fort Collins, CO, USA: Colorado State University.

Cousins, S. A. O., S. Lavorel, and I. Davies. 2003. Modelling the effects of landscape pattern and grazing regimes on the persistence of plant species with high conservation value in grasslands in south-eastern Sweden. *Landscape Ecology* 18: 315–332.

Cramer, W., A. Bondeau, F. I. Woodward, I. C. Prentice et al. 2001. Global response of terrestrial ecosystem structure and function to CO_2 and climate change: Results from six dynamic global vegetation models. *Global Change Biology* 7: 357–373.

Cramer, W., D. W. Kicklighter, A. Bondeau, B. Moore III et al. 1999. Comparing global models of terrestrial net primary productivity (NPP): Overview and key results. *Global Change Biology* 5 (Suppl): 1–15.

Daly, C., D. Bachelet, J. M. Lenihan, R. P. Neilson, W. J. Parton, and D. S. Ojima. 2000. Dynamic simulation of tree–grass interactions for global change studies. *Ecological Applications* 10: 449–469.

Foley, J. A., S. Levis, I. C. Prentice, D. Pollard, and S. L. Thompson. 1998. Coupling dynamic models of climate and vegetation. *Global Change Biology* 4: 561–579.

Foley, J. A., C. I. Prentice, N. Ramankutty, S. Levis, D. Pollard, S. Sitch, and A. Haxeltine. 1996. An integrated biosphere model of land surface processes, terrestrial carbon balance, and vegetation dynamics. *Global Biogeochemical Cycles* 10: 603–628.

Friend, A. D., A. K. Stevens, R. G. Knox, and M. G. R. Cannell. 1997. A process-based, terrestrial biosphere model of ecosystem dynamics (Hybrid v. 3.0). *Ecological Modelling* 95: 249–287.

Frison, P. L., E. Mougin, and P. Hiernaux. 1998. Observations and interpretation of ERS windscatterometer data over northern Sahel (Mali). *Remote Sensing of the Environment* 63: 233–242.

Gao, Q., and J. F. Reynolds. 2003. Historical grass-shrub transitions in the northern Chihuahuan Desert: Modeling the effects of shifting rainfall seasonality and event size along a landscape gradient. *Global Change Biology* 9: 1475–1493.

Gerten, D., Y. Luo, G. Le Maire, W. J. Parton, C. Keough et al. 2008. Modelled effects of precipitation on ecosystem carbon and water dynamics in different climatic zones. *Global Change Biology* 14: 2365–2379.

Gerten, D., S. Schaphoff, U. Haberlandt, W. Lucht, and S. Sitch. 2004. Terrestrial vegetation and water balance: Hydrological evaluation of a dynamic global vegetation model. *Journal of Hydrology* 286: 249–270.

Goslee, S. C., D. P. C. Peters, and K. G. Beck. 2001. Modeling invasive weeds in grasslands: The role of allelopathy in *Acroptilon repens* invasion. *Ecological Modelling* 139: 31–45.

Goslee, S. C., D. P. C. Peters, K. G. Beck. 2006. Spatial prediction of invasion success across heterogeneous landscapes using an individual-based model. *Biological Invasions* 8: 193–200.

Grimm, V., U. Berger, F. Bastiansen, S. Eliassen et al. 2006. A standard protocol for describing individual-based and agent-based models. *Ecological Modelling* 198: 115–126.

Hampe, A. 2004. Bioclimate envelope models: what they detect and what they hide. *Global Ecology and Biogeography* 13: 469–476.

Hély, C., L. Bremond, S. Alleaume, B. Smith, M. T. Sykes, and J. Guiot. 2006. Sensitivity of African biomes to changes in the precipitation regime. *Global Ecology and Biogeography* 15: 258–270.

Hickler, T., I. C. Prentice, B. Smith, M. T. Sykes, and S. Zaehle. 2006. Implementing plant hydraulic architecture within the LPJ Dynamic Global Vegetation Model. *Global Ecology and Biogeography* 15: 567–577.

HilleRisLambers, R., M. Rietkerek, F. van den Bosch, H. H. T. Prins, and H. de kroon. 2001. Vegetation pattern formation in semi-arid grazing systems. *Ecology* 82: 50–61.

Holmgren, M., and M. Scheffer 2001. El Niño as a window of opportunity for the restoration of degraded arid ecosystems. *Ecosystems* 4: 151–159.

Huenneke, L. F., J. P. Anderson, M. Remmenga, and W. H. Schlesinger. 2002. Desertification alters patterns of aboveground net primary production in Chihuahuan ecosystems. *Global Change Biology* 8: 247–264.

Jaffe, D., I. McKendry, T. Anderson, and H. Price. 2003. Six "new" episodes of trans-Pacific transport of air pollutants. *Atmospheric Environment* 37: 391–404.

Jarlan, L., E. Mangiarotti, P. Mougin, E. Mazzega, P. Hiernaux, and V. Le Dantec. 2008. Assimilation of SPOT/VEGETATION NDVI data into a Sahelian vegetation dynamics model. *Remote Sensing of Environment* 112: 1381–1394.

Kemp, P. R., J. F. Reynolds, R. A. Virginia et al. 2003. Decomposition of leaf and root litter of Chihuahuan Desert shrubs: Effects of three years of summer drought. *Journal of Arid Environments* 53: 21–39.

Knapp, A. K., J. M. Briggs, S. L. Collins, S. R. Archer, M. S. Bret-Harte, B. E. Ewers, D. P. C. Peters et al. 2008. Shrub encroachment in North American grasslands: Shifts in growth form dominance rapidly alters control of ecosystem carbon inputs. *Global Change Biology* 14: 615–623.

Knapp, A. K., J. M. Briggs, D. C. Hartnett, and S. L. Collins. 1998. *Grassland Dynamics: Long-Term Ecological Research in Tallgrass Prairie*. New York, NY: Oxford University Press.

Körner, K., and F. Jeltsch. 2007. Detecting general plant functional type responses in fragmented landscapes using spatially-explicit simulations. *Ecological Modelling* 210: 287–300.

Lauenroth, W. K., O. E. Sala, D. P. Coffin, and T. B. Kirchner. 1994. The importance of soil water in the recruitment of *Bouteloua gracilis* in the shortgrass steppe. *Ecological Applications* 4: 741–749.

Lenihan, J. M., D. Bachelet, R. P. Neilson, and R. Drapek. 2008. Simulated response of conterminous United States ecosystems to climate change at different levels of fire suppression, CO_2 emission rate, and growth response to CO_2. *Global and Planetary Change* 64: 16–25.

Lo Seen, D., E. Mougin, S. Rambal, A. Gaston, and P. Hiernauz. 1995. A regional Sahelian grassland model to be coupled with multispectral satellite data: 2. Towards its control by remotely sensed indices. *Remote Sensing of Environment* 52: 194–206.

Ludwig, J. A., M. B. Coughenour, A. C. Liedloff, and R. Dyer. 2001. Modelling the resilience of Australian savanna systems to grazing. *Environment International* 27: 167–172.

Luo, Y. Q., D. Gerten, G. le Maire, W. J. Parton et al. 2008. Modelled interactive effects of precipitation, temperature, and CO2 on ecosystem carbon and water dynamics in different climatic zones. *Global Change Biology* 14: 1986–1999.

McGuire, A. D., L. A. Joyce, D. W. Kicklighter, J. M. Melillo, G. Esser, and C. J. Vorosmarty. 1993. Productivity response of climax temperate forests to elevated temperature and carbon dioxide: A North American comparison between two global models. *Climatic Change* 24: 287–310.

McKergow, L. A., I. P. Prosser, A. O. Huges, and J. Brodie. 2005. Sources of sediment to the Great Barrier Reef World Heritage Area. *Marine Pollution Bulletin* 51: 200–211.

Midgley, G. F., W. Thuiller, and S. I. Higgens. 2007. Plant species migration as a key uncertainty in predicting future impacts of climate change on ecosystems: Progress and challenges. In *Terrestrial Ecosystems in a Changing World*, eds. J. G. Canadell, D. E. Pataki, and L. F. Pitelka, 129–137. New York: Springer.

Minnick, T. J., and D. P. Coffin. 1999. Geographic patterns of simulated establishment of two *Bouteloua* species: Implications for distributions of dominants and ecotones. *Journal of Vegetation Science* 10: 343–356.

Moorcroft, P. R., G. C. Hurtt, and S. W. Pacala. 2001. A method for scaling vegetation dynamics: The ecosystem demography model (ED). *Ecological Monographs* 71: 557–586.

Moore, A. D., and I. R. Noble. 1990. An individualistic model of vegetation stand dynamics. *Journal of Environmental Management* 31: 61–81.

Morin, X., and W. Thuiller. 2009. Comparing niche- and process-based models to reduce prediction uncertainty in species range shifts under climate change. *Ecology* 90: 1301–1313.

Mougin, E., D. Lo Seen, S. Rambal, A. Gaston, and P. Hiernaux. 1995. A regional Sahelian grassland model to be coupled with multispectral satellite data: 1. Description and validation. *Remote Sensing of the Environment* 52: 181–193.

Neilson, R. P. 1995. A model for predicting continental-scale vegetation distribution and water balance. *Ecological Applications* 5: 362–385.

Noss, R. F., E. T. La Roe III, Scott. 1995. *Endangered ecosystems of the United States: a preliminary assessment of loss and degradation, Biological Report* 28, 59 pp. Washington, DC: U.S. Department of Interior, National Biological Service.

Oesterheld, M., J. Loreti, M. Semmartin, and J. M. Paruelo. 1999. Grazing, fire, and climate effects on primary productivity of grasslands and savanna. In *Ecosystems of the World 16. Ecosystems of Disturbed Ground*, ed. L. R. Walker, 287–306. New York, NY: Elsevier.

Okin, G. S. 2008. A new model of wind erosion in the presence of vegetation. *Journal of Geophysical Research* 113: F02S10, doi: 10.1029/2007JF000758.

Pacala, S. W., C. D. Canham, J. Saponara, J. A. Silander, R. K. Kobe, and E. Ribbens. 1996. Forest models defined by field measurements: estimation, error analysis and dynamics. *Ecological Monographs* 66: 1–43.

Parton, W. J. 1978. Abiotic section of ELM. In *Grassland Simulation Model*, ed. G. S. Innis, 31–53. New York, NY: Springer-Verlag.

Parton, W. J., M. Hartman, D. Ojima, and D. Schimel. 1998. DayCent and its land surface submodel-description and testing. *Global and Planetary Change* 19: 35–48.

Parton, W. J., D. S. Schimel, C. V. Cole, D. S. Ojima. 1987. Analysis of factors controlling soil organic matter levels in Great Plains grasslands. *Soil Science Society of American Journal* 51: 1173–1179.

Parton, W. J., J. M. O. Scurlock, D. S. Ojima et al. 1993. Observations and modeling of biomass and soil organic matter dynamics for the grassland biome worldwide. *Global Biogeochemical Cycles* 7: 785–809.

Paruelo, J. M., and W. K. Lauenroth. 1996. Relative abundance of plant functional types in grasslands and shrublands of North America. *Ecological Applications* 6: 1212–1224.

Paruelo, J. M., S. Pütz, G. Weber, M. Bertiller, R. A. Golluscio, M. R. Aguiar, and T. Wiegand. 2008. Long-term dynamics of a semiarid grass steppe under stochastic climate and different grazing regimes: A simulation analysis. *Journal of Arid Environments* 72: 2211–2231.

Perry, G. L. W., and N. J. Enright. 2006. Spatial modelling of vegetation change in dynamic landscapes: A review of methods and applications. *Progress in Physical Geography* 30: 47–72.

Peters, D. P. C. 2002. Plant species dominance at a grassland–shrubland ecotone: An individual-based gap dynamics model of herbaceous and woody species. *Ecological Modelling* 152: 5–32.

Peters, D. P. C. 2000. Climatic variation and simulated patterns in seedling establishment of two dominant grasses at a semiarid-arid grassland ecotone. *Journal of Vegetation Science* 11: 493–504.

Peters, D. P. C., B. T. Bestelmeyer, J. E. Herrick, H. C. Monger, E. Fredrickson, and K. M. Havstad. 2006. Disentangling complex landscapes: New insights to forecasting arid and semiarid system dynamics. *Bioscience* 56: 491–501.

Peters, D. P. C., B. T. Bestelmeyer, A. K. Knapp, J. E. Herrick, H. C. Monger, K. M. Havstad. 2009. Approaches to predicting broad-scale regime shifts using changing pattern–process relationships across scales. In *Real World Ecology: Large-Scale and Long-Term Case Studies and Methods*, eds. S. Miao, S. Carstenn, and M. Nungesser, 47–72. New York: Springer.

Peters, D. P. C., B. T. Bestelmeyer, and M. G. Turner. 2007. Cross-scale interactions and changing pattern-process relationships: Consequences for system dynamics. *Ecosystems* 10: 790–796.

Peters, D. P. C., P. M. Groffman, K. J. Nadelhoffer, N. B. Grimm, S. L. Collins, and W. K. Michener. 2008. Living in an increasingly connected world: A framework for continental-scale environmental science. *Frontiers in Ecology and the Environment* 5: 229–237.

Peters, D. P. C., and J. E. Herrick. 2002. Modelling vegetation change and land degradation in semiarid and arid ecosystems: An integrated hierarchical approach. In *Land Degradation in Drylands: Current Science and Future Prospects. Advances in Environmental Monitoring and Modelling*, eds. J. Wainwright and J. Thornes, vol. 1(2). http://www.kcl.ac.uk/advances.

Peters, D. P. C., J. E. Herrick, H. C. Monger, and H. Huang. 2010. Soil–vegetation–climate interactions in arid landscapes: Effects of the North American monsoon on grass recruitment. *Journal of Arid Environments* 74: 618–623.

Peters, D. P. C., R. A. Pielke Sr, B. T. Bestelmeyer, C. D. Allen, S. Munson-McGee, and K. M. Havstad. 2004. Cross scale interactions, nonlinearities, and forecasting catastrophic events. *Proceedings of the National Academy Sciences* 101: 15130–15135.

Pielke Sr, R. A., J. Adegoke, A. Beltrán-Przekurat, C. A. Hiemstra, J. Lin, U. S. Nari, D. Niyogi, T. E. Nobis. 2007. An overview of regional land-use and land-cover impacts on rainfall. *Tellus*. doi: 10.1111/j.1600-0889.2007.00251.x.

Pieper, R. D. 2005. Grasslands of central North America. In *Grasslands of the World*, eds. J. M. Suttle, S. G. Reynolds, C. Batello, 221–263. Rome, Italy: Food and Agriculture Organization of the United Nations.

Prentice, I. C., A. Bondeau, W. Cramer, S. P. Harrison, T. Hickler, W. Lucht, S. Sitch, B. Smith, and M. Sykes. 2007. Dynamic global vegetation modeling: Quantifying terrestrial ecosystem responses to large-scale environmental change. In *Terrestrial Ecosystems in a Changing World*, eds. J. G. Canadell, D. E. Pataki, and L. F. Pitelka, 175–192. New York: Springer.

Rastetter, E. B., J. D. Aber, D. P. C. Peters, and D. S. Ojima. 2003. Using mechanistic models to scale ecological processes across space and time. *Bioscience* 53: 1–9.

Reynolds, J. F., P. R. Kemp, and J. D. Tenhunen. 2000. Effects of long-term rainfall variability on evapotranspiration and soil water distribution in the Chihuhuan Desert: A modeling analysis. *Plant Ecology* 150: 145–159.

Ricketts, T. R., E. Dinerstein, D. M. Olson, and C. J. Loucks et al. 1999. *Terrestrial Ecoregions of North America: A Conservation Assessment*. Washington, DC: Island Press.

Rietkerk, M., F. van den Bosch, and J. van de Koppel. 1997. Site-specific properties and irreversible vegetation changes in semi-arid grazing systems. *Oikos* 80: 241–252.

Running, S. W., and E. R. Hunt Jr. 1993. Generalization of a forest ecosystem model for other biomes, BIOME-BGC, and an application for global-scale models. In *Scaling Physiological Processes: Leaf to Globe*, eds. J. R. Ehleringer, and C. B. Field, 141–158. San Diego, CA: Academic Press.

Schlesinger, W. H., J. F. Reynolds, G. L. Cunningham, L. F. Huenneke, W. H. Jarrell, R. A. Virginia, and W. G. Whitford. 1990. Biological feedbacks to global desertification. *Science* 247: 1043–1048.

Schimel, D., J. Melillo, H. Tian, A. D. McGuire, D. Kicklighter, T. Kittel et al. 2000. Contribution of increasing CO_2 and climate to carbon storage by ecosystems in the United States. *Science* 287: 2004–2006.

Schimel, D. S., VEMAP participants, and B. H. Braswell. 1997. Continental scale variability in ecosystem processes: Models, data, and the role of disturbance. *Ecological Monographs* 67: 251–271.

Sheley, R. L., and J. K. Petroff. 1999. *Biology and Management of Noxious Rangeland Weeds*, 438 pp. Corvallis, OR, USA: Oregon State University Press.

Shugart, H. H. 1984. *A Theory of Forest Dynamics*. New York, NY: Springer-Verlag.

Sitch, S., B. Smith, I. C. Prentice et al. 2003. Evaluation of ecosystem dynamics, plant geography and terrestrial carbon cycling in the LPJ dynamic global vegetation model. *Global Change Biology* 9: 161–185.

Smith, M. D., A. K. Knapp, and S. L. Collins. 2009. Global change and chronic resource alterations: Moving beyond disturbance as a primary driver of contemporary ecological dynamics. *Ecology* 90: 3279–3289.

Smith, T. M., and D. L. Urban. 1988. Scale and resolution of forest structural pattern. *Vegetatio* 74: 143–150.

Steuter, A. 1987. C3/C4 production shift on seasonal burns—northern mixed prairie. *Journal of Range Management* 40: 27–31.

Strigul, N., D. Pristinski, D. Purves, J. Dushoff, and S. Pacala. 2008. Scaling from trees to forests: Tractable macroscopic equations for forest dynamics. *Ecological Monographs* 78: 523–545.

Svoray, T., R. Shafran-Nathan, Z. Henkin, and A. Perevolotsky. 2008. Spatially and temporally explicit modeling of conditions for primary production of annuals in dry environments. *Ecological Modelling* 218: 339–353.

Thomas, C. D., A. Cameron, R. E. Green, M. Bakkenes et al. 2004. Extinction risk from climate change. *Nature* 427: 145–148.

Thornton, P. E., and N. A. Rosenbloom. 2005. Ecosystem model spin-up: Estimating steady state conditions in a coupled terrestrial carbon and nitrogen cycle model. *Ecological Modelling* 189: 23–48.

Thuiller, W. 2003. BIOMOD—optimizing predictions of species distributions and projecting potential future shifts under global change. *Global Change Biology* 9: 1353–1362.

Tracol, C. J., E. Mougin, P. Hiernaux, and L. Jarlan 2006. Testing a sahelian grassland model against herbage mass measurements. *Ecological Modelling* 193: 437–446.

Urban, D. L. 2005. Modeling ecological processes across scales. *Ecology* 86: 1996–2006.

Urban, D. L., M. F. Acevedo, and S. L. Garman. 1999. Scaling fine-scale processes to large-scale patters using models derived from models: Meta-models. In *Spatial Modeling of Forest Landscape Change: Approaches and Applications*, eds. D. Mladenoff and W. Baker, 70–98. Cambridge, UK: Cambridge University Press.

Weisberg, P., and M. B. Coughenour. 2003. Model-based assessment of aspen responses to elk herbivory in Rocky Mountain Park, USA. *Environmental Management* 32: 152–169.

White, M. A., P. E. Thornton, S. W. Running, and R. R. Nemani. 2000. Parameterization and sensitivity analysis of the BIOME-BGC terrestrial ecosystem model: Net primary production controls. *Earth Interactions* 4: 1–85.

8

Agricultural Systems

Søren Hansen, Per Abrahamsen, and Merete Styczen

CONTENTS

8.1 Introduction 204
8.2 Basic Steps in Modeling a Field 205
8.3 Overview of the DAISY-Model Code 212
    8.3.1 Water Balance 213
    8.3.2 Heat Balance 214
    8.3.3 Carbon Balance 214
        8.3.3.1 Crop Growth 215
        8.3.3.2 Turnover of SOM 216
    8.3.4 Nitrogen Balance 217
        8.3.4.1 Organic Nitrogen 218
        8.3.4.2 Plant Organic Nitrogen 218
        8.3.4.3 Soil Organic Nitrogen (Mineralization/
                   Immobilization Model) 219
        8.3.4.4 Ammonium Nitrogen 219
        8.3.4.5 Nitrate Nitrogen 219
    8.3.5 Validation 220
8.4 Variables/Parameters, Sensitivity, and Good Modeling Practice 221
    8.4.1 Weather and Nitrogen Deposition 221
    8.4.2 Soil and Soil System 223
    8.4.3 Crop 228
        8.4.3.1 Reduction of the Photosynthesis 228
        8.4.3.2 Reduction of LAI 229
        8.4.3.3 Adjusting Uptake of Nitrogen 229
    8.4.4 Agricultural Management and Crop Rotation 230
8.5 Analysis of Simulation Results 231
8.6 Discussion and Conclusion 231
References 233

8.1 Introduction

For centuries, the most important objective of agricultural technology has been to increase the agricultural production. However, in recent decades awareness of the adverse effects of agricultural production on the environment has increased. In particular, leaching of nitrate and pesticides or surface runoff containing sediment, phosphorus, and pesticides are important environmental problems. Hence today we face two often contradicting objectives when we are planning agricultural production: (1) increase agricultural production and (2) minimize agricultural pollution.

Many fields of science contribute to our understanding of the agricultural system. Furthermore, during the past century our knowledge within these scientific disciplines has increased tremendously. The complexity of agricultural systems and the processes taking place within them make it beyond our minds' ability to fully comprehend system response to environmental conditions and human activity. Mathematical modeling is an extension of our brain and helps us integrate aspects of many scientific disciplines, and is therefore a key element in the modern system approach.

Numerous model codes* exist that describe elements of environmental problems related to agriculture, for example, PESTLA (Van den Berg and Boesten 1998), PRZM (Carsel et al. 1998), or MACRO (Larsbo and Jarvis 2003), which describe pesticide transport in soil, or WEPP (2009; Flanagan et al. 1995) and GUEST (Misra and Rose 1999; Rose et al. 1997), which describe soil erosion. None of these model codes simulate crop growth dynamically. In this chapter, focus will be on the management of water, carbon, and nitrogen within agricultural systems as it is of paramount importance to agricultural productivity and to the ecological impact of agriculture on the surrounding environment.

Several model codes simulate agricultural production, and water, carbon, and nitrogen dynamics and balances in agricultural systems, for example, ANIMO (2009; Groenendijk et al. 2005), APSIM (2009; see homepage), CERES (Jones and Kiniry 1986; Littleboy et al. 1992), COUPmodel (Jansson and Karlberg 2009), Daisy (Hansen et al. 1991a, 1991b, 1991c; Abrahamsen and Hansen 2000), EPIC (Williams 1995), NCSOIL (Molina et al. 1983), NLEAP (Shaffer et al. 1991), RISK-N (Gusman and Mariño 1999), SOIL-N (Eckersten et al. 1998; Johnsson et al. 1987), SUNDIAL (2009; see homepage), and RZWQM (Ahuja et al. 2000). In general, these model codes have a lot in common because they simulate many of the same processes. However, they

---

* The following definitions from Refsgaard and Henriksen (2004) are used in this chapter: A model code is a generic software program, which can be used for different study areas without modifying the source code. A model is a site application of a code to a particular study area, including input data and parameter values.

may differ in the manner in which they formulate these processes. They may also differ in their objectives and scopes. The CENTURY model code (2009; Parton et al. 1987), for example, predicts long-term changes in the organic pool, whereas the model codes mentioned above focus on shorter-term changes. Some model codes work at field scale, other model codes at farm scale, and yet other model codes work at catchment scale. In this hierarchy of scales, the field scale is the most fundamental one, and in the following discussion we restrict ourselves to that scale. Hence, we consider the soil–plant–atmosphere system at field scale in an agricultural context. Another concept used synonymously with the soil–plant–atmosphere system is the soil–plant–atmosphere continuum (SPAC). The latter emphasize that when modeling the soil–plant–atmosphere system, we assume that the system can be considered as a continuum. From a mathematical point of view, this assumption has considerable advantages.

8.2 Basic Steps in Modeling a Field

The first step in applying a SPAC model code is to identify the field by delineating system boundaries. In an agricultural context, a field is an area with the same vegetation (crop) and management. However, in a modeling context, a field is an area characterized by homogeneous weather, soil, vegetation, and management conditions. Weather, soil, and vegetation exhibit spatial variation. Weather data are normally obtained from a nearby weather station or a network of weather stations, and it is assumed that these weather data are representative for the considered field. In general, it is soil that exhibits the largest spatial variability. For most applications it is sufficient to consider a field homogeneous with respect to soil if the soil within the field belongs to the same mapping unit. A common assumption in SPAC modeling is that the field can be considered a 1-D system, that is, the model code only considers vertical fluxes. Some model codes do also take artificial drainage (e.g., tile drains) into account, thus describing one horizontal flux. A typical delineation is:

1. Upper boundary: the atmosphere
2. Sideward boundaries: no flux except for drainage flow and subsequent transport of solutes
3. Lower boundary: a groundwater table or a free drainage condition

In addition to the boundaries, the discretization of the model is important. How thick layers or how large units can we consider homogeneous within the soil column? Fine discretization is required where conditions change rapidly, for example, close to the soil surface, and if good descriptions of threshold values is required, for example, for initiation of macropore flow.

As stated above, SPAC model codes in general have a lot in common and therefore the same type of information is needed when applying this type of model code. The required information pertains to:

1. Weather
2. Soil and soil system
3. Crop and crop growth
4. Agricultural management practices

In the following, we shall exemplify this by applying the DAISY model code, and compare some of the functionalities to APSIM, RZWQM, COUPmodel, and MACRO. An overview of key functionalities in these four model codes are presented in Table 8.1.

DAISY was originally designed to simulate nitrate leaching from agricultural systems (Hansen et al. 1990, 1991a, 1991b, 1991c). Since then, the model code has been further developed. For instance, it can now also simulate the full carbon balance and the fate of pesticides in soil. Furthermore, the software supports alternative process descriptions, which makes it possible to choose the process description on the basis of the required accuracy, the available data, or available resources in terms of computer time, and to compare different process descriptions (Abrahamsen and Hansen 2000).

Model applications comprise both scientific studies and management-related studies aimed at decision support. The latter pose special problems. In 1995 we participated in a comparative test of SPAC models. One of the conclusions from the workshop was: "The simulation results obtained from 19 participants of the workshop 'Validation of Agro-ecosystem Models' were compared and discussed. Although all models were applied to the same data set, the results differ significantly. From the results it can be concluded that the experience of a scientist applying a model is as important as the difference between various model approaches" (Diekkrüger et al. 1995). We believe that this conclusion still holds true. This is a serious issue especially when model results have legal implications, for example, when a farmer applies for a permit to increase his livestock production (in parts of the world—e.g., in Denmark—agriculture are regulated by law and such permits are required). In such a case, it is of utmost importance that model application complies with good modeling practice. In Denmark, we have been working with this issue (Styczen et al. 2004). The presentation in this chapter is based on this work, but first, we will give a short overview of the DAISY-model code.

TABLE 8.1

Overview of Key Functionalities of RZWQM, APSIM, COUPmodel, and MACRO

	RZWQM	APSIM	COUPMODEL	MACRO
Flow description	Green-Ampt equation to simulate infiltration, and Richards' equation to simulate water redistribution	Simple overflow Water cannot leave the bottom of the soil column	Richards equation	Richards equation
Richards equation parameters	Brooks–Corey but a different approach close to saturation		Brooks–Corey, van Genuchten modified at the end with log-lin functions	Van Genuchten combined with a linear approach close to saturation
Frost/thaw, water in soil	Snow, yes. Frozen ice in soil does not influence infiltration	No	Snow, frozen soil, including related shrinking and swelling	Snow, yes. Ice in the soil does not influence infiltration
Macropores	Yes	No	Included as simple bypass	Yes
Hysteresis	No	No	Included	No
Water vapor flow	Yes, in development version	No	Included	No
Evaporation	Shuttleworth and Wallace model, which is an extension of the Penman–Monteith concept that extends the concept to partial canopy conditions (soil evap and crop transp components). Further extended to include surface residue, described by Farahani and Ahuja (1996)	(Priestly and Taylor 1972) Soil evaporation is not a function of suction and K but an empirical function Input data radiation, max and min temp, rain	Penman–Moneteith	As input or Penman–Monteith

(continued)

TABLE 8.1 (Continued)

Overview of Key Functionalities of RZWQM, APSIM, COUPmodel, and MACRO

	RZWQM	APSIM	COUPMODEL	MACRO
Runoff creation	Difference between rainfall and infiltration. No storage Can be coupled to kinematic wave calculations. Depression storage included. Splash and flow detachment equations are added	Curve number, based on daily rainfall	Throughfall > infiltration + storage	Not clear, the documentation is not consistent. Older versions did not allow runoff as all excess water ended up in macropores. Presently, it seems also macropores have a maximum infiltration, meaning that it must be possible to generate surface ponding if not runoff
Erosion	As addition. Sediment and solutes can be moved	Yes, from PERFECT (Littleboy et al. 1989), a modified USLE equation Freebairn and Wockner (1986) and the model by Rose (1985)	No	No
Drainage	Tile drain flow is simulated using Hooghoudt's steady-state equation	Pipe drains not mentioned	Four options: empirical, linear, Hcoghoud's equation, and Ernst	Flux from sat. layers above drainage depth: $q_{d(tot)} = A_f E$ (A_f is a shape factor, depending on distance between drains and E is seepage potential, which is a function of hydraulic conductivity and water table height). Hooghoud's equation is used for layers below drainage depth

Heat	Heat convective–dispersive equation	Heat convective–dispersive equation	Heat convective–dispersive equation, but also effect of heat pump	Heat convective–dispersive equation
Organic pools	Fast residue pool; slow residue pool; fast humus pool; transition humus pool; stable humus pool; aerobic heterotrophs pool; anaerobic heterotrophs pool; autotrophs pool; initial urea-N; initial NO_3-N; initial NH_4-N; able to produce methane	Biomass and humus Fresh organic matter pool C-inert pool	C(humus), C(litter), C(litter Surface), C(faeces), optional pool of DOC. C(microbes is optional)	No, organic matter is an input
Plants	Dynamic plant growth Simulated growth stage development is influenced by probabilities which again depend on environmental fitness (temp, N, water, other factors) Photosynthesis is simulated	Dynamic plant growth	1. Static or simulated, simplest approach: is a logistic growth function, where potential growth is a function of time 2. Growth is estimated from a water use efficiency parameter and from the simulated transpiration 3. Light use efficiency can be used to estimate potential growth rate, limited by unfavorable temperature, water and nitrogen conditions. A biochemical model after Farquhar et al. (1980) can be used if hourly values of photosynthesis and transpiration is of interest	Static

(continued)

TABLE 8.1 (Continued)

Overview of Key Functionalities of RZWQM, APSIM, COUPmodel, and MACRO

	RZWQM	APSIM	COUPMODEL	MACRO
Biomass increase	Photosynthesis is a function of incoming radiation. This includes light use efficiency coefficient	In APSIM, the biomass increase is calculated differently, but it is still dependent on water and nitrogen stress and influenced indirectly by temperatures and senescence	Response function for soil water stress $f(Eta/Etp)$ is multiplied with the stomatal resistance to account for stomatal closure due to plant water stress. Includes salinity stress, temperature stress, and N stress	Na
Calibration parameters:	μ_1 max active N uptake daily resp as f of photosynthate (Φ), biomass to leaf area conversion coef (C_{LA}), age effect for plants in propagule dev. Stage (A_p), and age effect in seed dev stage (A_s)	APSIM builds on CERES-crop growth model. In CERES, parameters controlling leaf and root growth, plant height, tissue N concentration and partitioning of dry matter are built into the code of the CERES models, rather than being user-specified inputs, so it is implicitly assumed that none of these parameters vary among cultivars	ε_L is the radiation use efficiency and η is a conversion factor from biomass to carbon	

N uptake	Passive uptake in proportion to plant transpiration + active uptake (Michaelis–Menten) + fixation if elevant	Nitrogen uptake maximum = the nitrogen uptake required to bring all plant part N contents to the maximum allowable concentration. Nitrogen supply is the sum of nitrogen available via mass flow (Equation 8.9) and by diffusion + fixation if rel.	Demand and supply calculation	N.A.
N-fix by microorganisms	Rhizobia, not other	Rhizobia, not other	Yes	No
Ammonia sorption	Modeled as ion exchange, balanced with Na^{2+}	Not described in documentation	Defined as a fraction	No
Soil P	No	Yes	No	No
Tillage	Yes, changes hydraulic properties and incorporates organic material	Yes	Yes	Yes, changes hydraulic properties
Crop rotation	Yes	Yes	No	No
Economics	No	Simple economic module	No	No

8.3 Overview of the DAISY-Model Code

This overview comprises the version of the model code that is used in management-related studies aimed at decision support.

Figure 8.1 gives a schematic overview of the model code. It is noted that the code comprises three main modules: a bioclimate, a vegetation, and a soil component. Driving variables are weather data, which are read from a special weather file, and management information, which can be rather complex—field activities (e.g., irrigation, fertilizing, soil tillage)—and which may depend on the state of the system. Management information is coded in a simple computer language, and is supplied to the model in a text file. Initialization and parameterization is provided by setup files. All files are read by a special parser component.

The input system is very flexible; it allows for storing full or partial parameterizations of selected process models in separate library files. The information stored in the library file is included in the setup by a simple reference to the library file (Abrahamsen and Hansen 2000). For example, this facility is used for storing crop parameterizations in a crop library, and by including

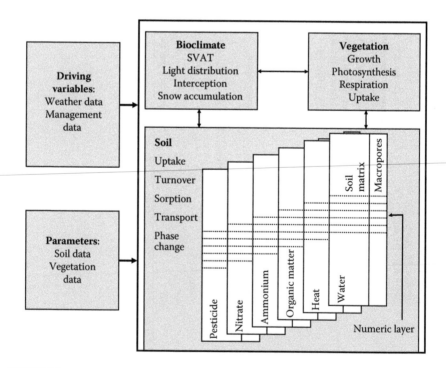

FIGURE 8.1

Schematic representation of the SPAC model code DAISY. The code comprises three main modules: a bioclimate, a vegetation, and a soil component.

the crop library or crop libraries, the various crop parameterizations are available just by referring to the name under which the parameterization is stored. Similarly, soil information required to parameterize a soil horizon can be stored in a soil horizon library and referred to by name.

If more than one process model description is implemented, the selection between these implementations is done in the parameterization. Furthermore, the parser allows for the allocation of default values to model parameters. This facility is applied to allocate default values to model parameters whenever it is feasible. Parameters, which are not given a specific value, need not appear in the input files, and in the following it is assumed that default parameters are used if not otherwise specified. In the recommendations of Styczen et al. (2004), it is argued that good modeling practice requires that when we in an application deviate from default parameters, it should be clearly stated. This practice simplifies the documentation of the application. Certain parameters may be estimated by transfer functions if they are not provided with input files.

The output system is also very flexible, and allows the user to define his output, which can be very detailed or contain only major simulation results. Output files are readily read by common spreadsheet software.

APSIM, COUPmodel, and RZWQM are also built in a modular fashion, whereas in MACRO, which is aimed specifically, and macropore flow and pesticide transport are one functional unit.

8.3.1 Water Balance

The water balance model comprises the following storages: water intercepted by the canopy, snow storage, soil surface storage, and soil water. Intercepted water is fed by precipitation and overhead irrigation, and is lost by evaporation and drip off. Water in snow storage is fed by precipitation, and overhead irrigation is lost by evaporation/sublimation and percolation of liquid water out of the snow pack. Freezing and melting take place within the snowpack. The soil surface storage water is fed by precipitation, overhead irrigation, and drip off of intercepted water, and is lost by evaporation and infiltration. Soil water is fed by infiltration of rain and irrigation water and water coming from drip off and percolation out of the snow pack; in addition, water can be gained from raising groundwater and capillary rise. Soil water is lost by evaporation from the surface, by water taken up by the plant and henceforth transpired, and by deep percolation. All evaporative fluxes are driven by a potential evapotranspiration, which are provided by the bioclimate module. Internal distribution of potential evapotranspiration is governed by leaf area index (LAI) simulated by the vegetation model.

The internal distribution of soil water is described by a numeric solution to Richards' equation, and soil water flow including deep percolation is calculated by Darcy's law. The soil water model requires site-specific information on soil water retention and soil water conductivity. The water uptake by plants is modeled by the single root concept (see, e.g., Hansen and Abrahamsen 2009). Furthermore, calculation of root uptake also requires information on root length density. This information is supplied by the vegetation model. In addition, the lower boundary condition should be specified. If tile drains are present a simple pseudo-two-dimensional (2-D) solution based on Hooghoudt's equation is used.

Some of the typical differences between SPAC model codes are:

1. Flow description with Richards' equation or as "tipping bucket"
2. Macropore inclusion and details in the description of the macropore flow
3. Exact method used for calculation of reference evaporation
4. Inclusion of snow/frost/thaw processes

APSIM uses the tipping bucket method and does not include macropores. The other model codes are based on Richards' equation and include macropores. The COUPmodel has a very detailed description of snow and frozen soils, hysteresis, and water vapor flow, whereas MACRO and RZWQM include snow. For calculation of reference evaporation, MACRO and COUPmodel use Penman–Monteith, RZWQM a modified Penman–Monteith, and APSIM a method by Priestly and Taylor (1972), but some model codes may allow other options, too.

8.3.2 Heat Balance

Soil temperature strongly influences biological processes such as root growth and microbial transformation of carbon and nitrogen in soil. The present soil temperature model is based on the 1-D heat flow equation, which takes into account heat flow due to conduction and convection. The thermal parameters of soil (heat capacity and thermal conductivity, which depend on soil water content) are calculated on the basis of the composition of the soil and the properties of the individual soil constituents. Air temperature is used as the upper boundary condition, and average temperature variation at the location is used in estimating the lower boundary condition. All the model codes compared here handle heat in a similar fashion.

8.3.3 Carbon Balance

Input to the field carbon balance includes organic fertilizers and photosynthesis. Losses constitute plant and soil respiration. Transport of dissolved organic matter in the soil profile can in DAISY only be simulated with considerable uncertainty (Gjettermann et al. 2008), and is neglected in the version recommended for decision support. The COUPmodel also allows DOC calculations.

8.3.3.1 Crop Growth

Photosynthesis and plant respiration are integral parts of the crop growth model. Figure 8.2 shows the carbon flow and main crop growth processes included in the model code. It is noted that the crop comprises four compartments: storage organ, leaf, stem, and root. Storage organ is a common denomination for the ear in a small grain cereal, the cub in maize, or the tuber in potato crop, etc.

Photosynthesis depends on absorbed photosynthetically active radiation (assumed to be a fixed fraction of the global radiation), temperature, canopy structure, senescence, and stress factors. The canopy structure is defined by the LAI and the leaf area distribution. Senescence is assumed to be a function of the development stage of the crop (DS; DS = 0 at emergence; DS = 1 at flowering; DS = 2 at maturity). In the vegetative stage ($0 \leq DS \leq 1$), the increase in DS is determined by temperature and the length of the photoperiod. In the reproductive stage ($1 < DS \leq 2$), the increase in DS is determined by temperature only. Stress factors comprise water stress and nitrogen stress only. Other stress factors are neglected. The photosynthesis produces assimilates that are transferred to the assimilate pool. Part of the assimilate is used for maintenance respiration, which depends on storage organ, leaf, stem, and root weight and on temperature. The rest of the assimilate is used for crop growth. The released assimilate is partitioned among the main compartments. It is

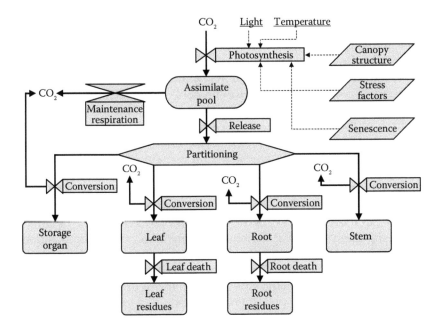

FIGURE 8.2
Overview of the carbon flow in the default crop model code included in DAISY.

converted to structural dry matter, and a part of the carbon is lost as growth respiration. Leaf and root death depend on DS. Dead leaf can defoliate and is then transferred to the soil surface. Dead roots are transferred directly to the soil organic matter (SOM). At harvest, organic carbon is removed and residuals are left on the surface.

LAI depends on leaf weight (leaf compartment) and DS and leaf area distribution depends on DS. Root density distribution depends on root weight (root compartment) and rooting depth. Rooting depth depends on soil temperature and textural composition of the soil.

Crop model parameters are plant species–specific. Parameterization of a new crop is a job for a specialist and requires detailed information. Crop parameters are available in the DAISY crop parameter library.

Plant dry matter and carbon content are considered equivalent in the model.

Some model codes, for example, MACRO, operate with a prespecified growth curve, meaning that plant growth do not react to stress. APSIM, COUPmodel, and RZWQM are all able to simulate crop growth. However, the exact description of processes may differ.

8.3.3.2 Turnover of SOM

Input to the SOM model comprises plant residuals and organic fertilizers. Root residues are transferred directly to the SOM model. Plant residues left on the surface are transferred by either tillage or by a bio-incorporation model. Organic fertilizers can at application be incorporated or left at the surface and subsequently be incorporated by tillage. Carbon loss is due to respiration of the soil microbial biomass (SMB).

The turnover of SOM is described by a multi compartment model, Figure 8.3. The model code considers three main pools: dead native SOM, SMB, and added organic matter (AOM), for example, organic fertilizers and plant residues. Each of these distinct pools is considered to contain a continuum of substrate qualities, but in order to facilitate the description of all turnover processes by first-order kinetics each of these main pools has been divided into two subpools: one with a slow turnover (e.g., SOM1, SMB1, and AOM1) and one with a faster turnover (e.g., SOM2, SMB2, and AOM2). SOM3 in a pool has a very slow decomposition rate that the decomposition can be neglected (an inert pool). Decomposition rates are assumed to be affected by soil temperature and soil moisture. Furthermore, the decomposition of SOM1, SOM2, and SMB1 is affected by the clay content mimicking physical protection from degradation. Carbon flow and partitioning are indicated by arrows and partitioning coefficients shown in the figure.

Carbon loss, that is, the production of carbon dioxide, results from all C fluxes into the microbial biomass (SMB) pools (substrate utilization efficiencies being less than unity). Furthermore, microbial maintenance respiration produces carbon dioxide. Maintenance respiration is described by a first-

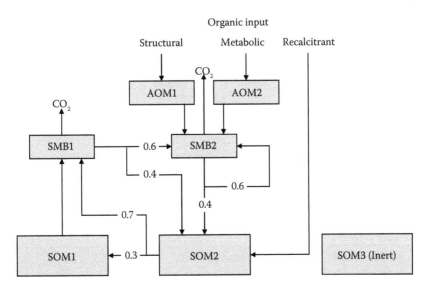

FIGURE 8.3
Pools and subpools (1 and 2) of organic matter and related partitioning of organic C flow between pools. AOM, added organic matter; SMB, soil microbial biomass; SOM, soil organic matter.

order process and is affected by abiotic factors in the same fashion as the decomposition rates.

The soil carbon turnover model is closely linked to nitrogen mineralization/immobilization, and low availability of nitrogen exhibits feedback to the carbon turnover model and reduces turnover rates.

Information on the general parameterization of the turnover model can be found in the work of Jensen et al. (2001).

When calibrating the model, it is a common procedure to apply a warm-up period. During this period, the size of AOM1 and AOM2 pools and the SMB2 pool will adjust themselves to field management strategy (fertilization practice, plant residue management, etc.). Furthermore, it is assumed that the SMB1 pool is in equilibrium with the SOM1 and SOM2 pool. However, in general, it is not a valid assumption to assume that the size of SOM1 and SOM2 is in equilibrium to the prevailing field management strategy, as this would take 100 years or more of similar management. Hence, the main calibration parameters of the model are the initial distribution of the SOM between the SOM pools (Table 8.2).

8.3.4 Nitrogen Balance

The model code considers organic nitrogen, ammonium, and nitrate.

TABLE 8.2

Overview of Organic Pools in the Selected Model Codes

	RZWQM	APSIM	COUPMODEL	DAISY
Fresh organic matter	Fast residue pool, slow residue pool	Fresh organic matter pool	C(litter), C(litter surface), C(faeces)	AOM1 AOM2
DOC			Optional pool of DOC	DOC
Microbial biomass	Aerobic heterotrophs pool, anaerobic heterotrophs pool, autotrophs pool	Biomass	C(microbes is optional)	SMB1 SMB2
Humus pools	Fast humus pool, transition humus pool, stable humus pool	Humus	C(humus)	SOM1 SOM2
Inert C-pool		C-inert-pool		SOM3

Note: For MACRO, the organic matter content is input.

8.3.4.1 Organic Nitrogen

Organic nitrogen is closely linked to organic carbon as an integral part of organic matter, which is characterized by a C/N ratio, and is present in plant material characterized by N percentages for the different crop components. Organic fertilizers are characterized by both a carbon and nitrogen content.

8.3.4.2 Plant Organic Nitrogen

Plant nitrogen is formed when plants take up ammonium and nitrate from the soil solution. The nitrogen uptake model is based on the single root concept and is closely linked to the water uptake model (see Hansen and Abrahamsen 2008). Two mechanisms are assumed active: mass flow and diffusion. It is assumed that the plant take up ammonium in preference to nitrate; however, as ammonium adsorbs soil constituents, it is less mobile than nitrate and hence the uptake of ammonium in general is less than that of nitrate. The driving force is the plant demand, which is the difference between the potential plant N content and the actual plant N content. The potential plant N content depends on the size of the crop compartments (storage organ, leaf, stem, and root) and corresponding N concentrations, which depend on DS. If the actual plant N content gets below a certain critical level, nitrogen stress will occur and result in a feedback to the photosynthesis model. The critical N content is calculated in a similar manner as the potential N content, by just substituting the potential N concentrations with corresponding critical ones. For further details, see Hansen et al. (1991a, 1991b, 1991c). At harvest, organic nitrogen is removed, residuals are left on the surface, and root organic nitrogen is momentarily transferred to the mineralization/immobilization model (see soil organic N). At tillage, organic N stored at the surface is transferred to the mineralization/immobilization model.

8.3.4.3 Soil Organic Nitrogen (Mineralization/Immobilization Model)

Soil organic nitrogen turnover is an integral part of the soil organic carbon turnover model. Figure 8.3 shows the carbon pools considered by the model code and the related flows of carbon. Each of these pools is characterized by a C/N and consequently by a content of soil organic nitrogen. Nitrogen flows are associated with corresponding carbon flows, and it is assumed that the C/N in the nitrogen flow is determined by the C/N in the pool where it originates. A carbon balance of the individual pool determines whether the size of the pool increases or decreases in terms of carbon content and due to the characterizing C/N also in terms of nitrogen. The nitrogen balances of all pools determine whether the total system is in surplus or deficit of nitrogen. If a surplus exists, it is released as mineralization (as ammonium). If a deficit exists, immobilization will occur and mineral nitrogen (ammonium and nitrate) will be extracted from the soil solution and turned into organic nitrogen. It is assumed that ammonium is immobilized in preference to nitrate. Furthermore, it is assumed that the maximum immobilization rates are proportional to the content of ammonium or nitrate in the soil solution, respectively. If the maximum immobilization rate cannot fulfill the demand of the turnover model, the turnover rates of nitrogen demanding processes are reduced correspondingly. In agricultural soils, it is AOM pools with high C/N that are responsible for the immobilization.

8.3.4.4 Ammonium Nitrogen

Atmospheric deposition, fertilization, and mineralization of organic matter are input to the ammonium balance. Losses comprise plant uptake, immobilization, nitrification, autotrophic denitrification, and leaching.

Atmospheric deposition and fertilizer ammonium can be stored at the surface and is transferred to the soil at the first infiltration or tillage event. Furthermore, fertilizer ammonium can be directly incorporated into the soil. Movement of ammonium in the soil is described by a numeric solution to the advection–dispersion equation assuming instantaneous sorption/desorption of ammonium. Leaching is a consequence of the movement. Due to the sorption/desorption, the contribution of ammonium leaching is in general negligible. Nitrification is an important process converting ammonium to nitrate. During the nitrification process, a small part of the converted ammonium is lost to the atmosphere as N_2O (autotrophic denitrification). The nitrification rate is described by Michaelis–Menten kinetics. Rate coefficients are affected by soil temperature and soil moisture.

8.3.4.5 Nitrate Nitrogen

Input to the nitrate balance includes atmospheric deposition, fertilization, and nitrification. Losses comprise plant uptake, immobilization, heterotrophic denitrification, and leaching.

Atmospheric deposition and fertilizer nitrate can be stored at the surface, and is transferred to the soil at the first infiltration or tillage event. Furthermore, fertilizer nitrate can be directly incorporated into the soil. Movement of nitrate in the soil is described in a numeric solution to the advection–dispersion equation. Leaching is a consequence of the movement. Heterotrophic denitrification converts nitrate to gaseous nitrogen compounds. The heterotrophic denitrification is based on the concept of potential denitrification (anoxic conditions), which is assumed to be proportional to the CO_2 evolution simulated by the carbon turnover model. The oxygen status of the soil system is mimicked by a function of the soil water content. The actual denitrification is then based on the potential denitrification rate, the oxygen status of the soil, and the availability of nitrate. The latter is simulated in the mineralization/ immobilization model.

APSIM, COUPmodel, and RZWQM contain the same compounds, but particularly the organic fractions differ to suit the C pools of the respective model codes. MACRO, which is a flow and pesticide model code, does not deal with N. APSIM deals with both N and P.

8.3.5 Validation

In practice, a comprehensive model code like DAISY cannot be validated in the strict sense of the word. It can only be validated at a given site with a given set of parameters. However, the system is so complex that it is virtually impossible to collect a dataset that gives a thoroughly comprehensive description of the system; furthermore, the model's demand for parameters are so huge that parameter assessment virtually always includes guesstimates and some degree of calibration. Nevertheless, some kind of comparison with reality is needed in order for the model to gain credibility. Hence, we will use the word validation referring to successful application of the model code at a given site, that is, when the model has shown itself able to reproduce observation data with accuracy sufficient for the actual purpose. This is a very weak use of the word. In this sense, DAISY is a well-validated model. We list a number of studies reported internationally: Vereecken et al. (1991), Hansen et al. (1991a, 1991b, 1991c), Willigen (1991), Jensen et al. (1992, 1993), Styczen and Storm (1993a, 1993b), Hansen et al. (1994), Jensen et al. (1994a, 1994b), Diekkrüger et al. (1995), Petersen et al. (1995), Svendsen et al. (1995), Muller et al. (1996), Jensen et al. (1996, 1997), Smith et al. (1997), Djurhuus et al. (1999), Hansen et al. (1999), Refsgaard et al. (1999), Børgesen et al. (2001), Hansen et al. (2001), Jensen et al. (2001), Thorsen et al. (2001), Van der Keur et al. (2001), Bruun and Jensen (2002), De Neergaard et al. (2002), Bruun et al. (2003), Muller et al. (2003), Boegh et al. (2004), Christiansen et al. (2004), Olesen et al. (2004), Bruun et al. (2006), Hansen et al. (2006), Müller et al. (2006), Veihe et al. (2006), Hansen et al. (2007), Olesen et al. (2007), Pedersen et al. (2007), Heidmann et al. (2008), Gjettermann et al. (2008), Boegh et al. (2009), Hansen et al. (2009), Pedersen et al. (2009), Van der Keur et al. (2008).

We do not claim that the list comprises all internationally reported applications, but it covers a wide range of applications.

An overview of RZWQM evaluations can be found at http://s1004 .okstate.edu/S1004/Regional-Bulletins/Modeling-Bulletin/RZWQM2- word.html#1000000028. Similar information on APSIM can be found on its homepage.

8.4 Variables/Parameters, Sensitivity, and Good Modeling Practice

8.4.1 Weather and Nitrogen Deposition

As a minimum, the model requires daily values of global radiation, air temperature, and precipitation. These data are obtained from a nearby weather station. If global radiation is not available, it can be calculated with the Angstrom formula, which relates solar radiation to extraterrestrial radiation and relative sunshine duration:

$$S_i = \left(a_s + b_s \frac{n}{N} \right) S_e \tag{8.1}$$

where S_i is the global or shortwave radiation (MJ m^{-2} day^{-1}), n is the actual duration of sunshine (h), N is the maximum possible duration of sunshine or daylight hours (h), S_e is the extraterrestrial radiation (MJ m^{-2} day^{-1}), a_s is the regression constant, expressing the fraction of extraterrestrial radiation reaching the earth on overcast days ($n = 0$), and $a_s + b_s$ is the fraction of extraterrestrial radiation reaching the earth on clear days ($n = N$). The Angstrom parameters a_s and b_s are site-specific. Where no calibration has been carried out for improved a_s and b_s parameters, the values $a_s = 0.25$ and $b_s = 0.50$ are recommended by the Food and Agriculture Organization (FAO).

If average daily temperature, T_a (°C), is not available, it can be estimated from observations of daily maximum, T_{max} (°C) and minimum temperatures, T_{min} (°C):

$$T_a = \left(\frac{T_{min} + T_{max}}{2} \right) \tag{8.2}$$

Precipitation is probably the most site-specific of the required weather data. Hence, it is important that it is measured at the site. If this is not the case, a considerable modeling uncertainty will be associated with model predictions (growth, N leaching, etc.). Precipitation is often measured at 1.5 m elevation.

In this case, observations of precipitation are subject to systematic errors. Typical errors are the aerodynamic error caused by turbulence generated by the rain gauge, evaporation from the collector, and the wetting error caused by evaporation of water intercepted by the funnel conducting the precipitation to the collector. For Danish conditions, Allerup et al. (1998) has proposed the corrections given in Table 8.3. It is noted that the corrections depend on shelter conditions and time of the year. The shelter conditions are important because the aerodynamic error is the dominant error source. The very high corrections in wintertime are caused by snow; hence, the values cannot be directly transferred to other regions.

If the only weather data supplied to the model comprise global radiation, air temperature, and precipitation, then the reference evapotranspiration will be estimated by a version of the Makkink equation adapted to Danish conditions (Makkink 1957; Hansen 1984):

$$E_r = 0.7 \frac{s}{s+\gamma} \frac{S_i}{L_v} \tag{8.3}$$

where E_r is the reference evapotranspiration (mm/day), s is the slope of the saturation vapor pressure curve versus temperature at air temperature, γ is the psychrometer coefficient (66.7 Pa/K), S_i is the global radiation (MJ m^{-2} day^{-1}), and L_v is latent heat of volatilization (2.45 MJ/kg). This simple approach has given reliable results in the Danish climate (humid northwest European climate). However, in other climates other methods may be preferable. A generally applicable method is the FAO–Penman–Monteith method. FAO–Penman–Monteith reference evapotranspiration calculation procedures taking different data availability into account can be found at the homepage of FAO. If reference evapotranspiration is supplied with the weather data, these data are used by the model.

Crop-specific potential evapotranspiration (corresponding to crop evapotranspiration under standard conditions in the FAO terminology) is estimated based on the reference evapotranspiration:

TABLE 8.3

Monthly and Annual Correction of Precipitation at Different Shelter Classes (SC) in Percent

S.C.[a]	J	F	M	A	M	J	J	A	S	O	N	D	Y
A	29	30	26	19	11	9	8	8	9	10	17	26	16
B	41	42	35	24	13	11	10	10	11	14	23	37	21
C	53	53	45	29	16	13	12	12	13	17	29	48	27

Source: Allerup et al. Standardværdier (1961–90) af nedbørskorrektioner. DMI. Teknisk rapport 98-10, 1998. With permission.

[a] Shelter class: A, vertical angle 19–30°; B, vertical angle 5–19°; C, vertical angle 0–5°.

$$E_{p,c} = K_c E_r \tag{8.4}$$

where $E_{p,c}$ is the crop-specific potential evapotranspiration (mm/day) and K_c is a crop coefficient. In DAISY, K_c is estimated as:

$$K_c = K_{ca}\left(1 - e^{-0.5L_{ai}}\right) + K_s e^{-0.5L_{ai}} \tag{8.5}$$

where K_{ca} is a crop canopy coefficient, K_s is a soil coefficient, and L_{ai} is the leaf area index. K_{ca} would often resemble the mid-season K_c (full cover) suggested by FAO, and correspondingly K_s would resemble the initial value of K_c (FAO).

Considering weather data uncertainty associated with precipitation is the largest contributor to the modeling uncertainty under Danish conditions (see Table 8.4).

DAISY considers wet and dry deposition of NH_y and NO_x, respectively. Often only the total deposition is known. For Danish conditions, this in general amounts to between 12 and 28 kg N ha^{-1} year^{-1}; however, close to large sources of emission the values can be much higher. We often assume that dry deposition constitutes 40% of the deposition and wet deposition constitutes the remaining 60%. For dry deposition, we assume that 60% is as NH_y and 40% is as NO_x. For wet deposition we assume equal distribution, and based on the annual precipitation we calculate the average concentration in the precipitation. The contribution by this procedure to the total modeling uncertainty is small.

Dry and wet deposition require paramerization in all the evaluated model codes to suit the site of simulation.

8.4.2 Soil and Soil System

The soil system is defined by a number soil horizons and a lower boundary characterizing the soil hydrology. The soil horizons (typically three horizons) are characterized by soil texture, dry bulk density, soil organic matter content, C/N in SOM, and soil hydraulic properties, that is, soil water characteristic and soil hydraulic conductivity. If soil bulk density is not supplied with the input data, it is estimated based on information on porosity or saturated water content in mineral soil.

The soil water characteristic can be described by common retention models, viz. the Brooks and Corey model (Brooks and Corey 1964), the Campbell model (Campbell 1974), and the van Genuchten model (van Genuchten 1980) (Table 8.5). The retention models can be combined with the theory of Burdine (1952) or Mualem (1976) to yield the soil hydraulic conductivity (Table 8.6). In general, soil hydraulic properties are not readily available. Hence, it is customary to use pedo-transfer functions to estimate these properties. DAISY has built-in pedo-transfer functions of Cosby et al. (1984) and HYPRES (Wösten et

TABLE 8.4

Comparison of Differences between Meteorological Data Sources (Station and Grid Data, 10 km for Precipitation and 20 km for Global Radiation and Temperature) for Three Locations and the DAISY-Calculated Percolation

		Jyndevad		Aarslev		Taastrup		%-Wise Deviations		
Soil	Vegetation	Station	Grid	Station	Grid	Station	Grid	Station	Grid	Taastrup
				Precipitation						
		1143	1031	855	878	650	749	9.8	-2.7	-15.2
				Evaporation						
		591	595	609	612	591	605	-0.7	-0.5	-2.4
				Percolation						
JB1	Spring crops	715	616	471	487	297	372	13.8	-3.4	-25.3
<5% clay	Winter crops	697	596	453	467	278	349	14.5	-3.1	-25.5
	Grass	659	563	421	437	248	318	14.6	-3.8	-28.2
JB4	Spring crops	609	502	350	367	181	258	17.6	-4.9	-42.5
5–10%	Winter crops	594	486	331	346	159	232	18.2	-4.5	-45.9
Clay	Grass	579	479	339	354	167	234	17.3	-4.4	-40.1
JB6	Spring crops	674	581	435	447	269	341	13.8	-2.8	-26.8
10–15%	Winter crops	642	543	400	417	246	312	15.4	-4.3	-26.8
Clay	Grass	639	549	416	435	261	329	14.1	-4.6	-26.1

Source: Styczen et al. Standard Parameterization of the Daisy-Model. Guide and Background. Version 1.0, Danish Hydraulic Institute, 2004. With permission.

Note: Calculated N-leaching differed up to 80%.

TABLE 8.5

Mathematical Forms Describing Soil–Moisture Retention Curves

Reference	Mathematical Form	Condition
Brooks and Corey (1964)	$S_e = \dfrac{\theta - \theta_r}{\theta_s - \theta_r} = \left(\dfrac{h_b}{h_p} \right)^\lambda$	$h_p < h_b$
Campbell (1974)	$S_r = \dfrac{\theta}{\theta_s} = \left(\dfrac{h_b}{h_p} \right)^{1/b}$	$h_p < h_b$
van Genuchten (1980)	$S_e = \dfrac{\theta - \theta_r}{\theta_s - \theta_r} = \left[\dfrac{1}{1 + \lvert \alpha\, h_p \rvert^n} \right]^m$	$h_p < 0$

Note: θ = volumetric water content; h = matrix potential; S_e = effective soil water content; S_r = relative soil water content; θ_s = volumetric soil water content at saturation; θ_r = residual soil water content; h_b = bubbling pressure, air-entry value; λ = pore size index; b = Campbell b; α = van Genuchten – α; n = van Genuchten – n; m = van Genuchten – m.

al. 1999). The former estimates Burdine–Campbell parameters based on percent of clay and sand. The latter estimates Mualem–van Genuchten parameters based on information on percentage of clay, sand, and organic carbon, and dry bulk density, and whether the horizon is the uppermost horizon. Alternatively, hydraulic properties can also be estimated outside DAISY and subsequently parameters can be fed into the model. In this case, for instance, ROSETTA (Schaap et al. 2001) can be used. ROSETTA is a computer program for estimating soil hydraulic parameters (Mualem–van Genuchten) with hierarchical pedo-transfer functions.

TABLE 8.6

Closed-Form Hydraulic Conductivity Functions Based on Soil–Moisture Retention Relationships

Retention Curve Model	Burdine Theory	Mualem Theory
Brooks and Corey (1964)	$K = K_s S_e^{(2+3\lambda)/\lambda}$	$K = K_s S_e^{(2+2.5\lambda)/\lambda}$
Campbell (1974)	$K = K_s S_r^{(2+3/b)b}$	$K = K_s S_r^{(2+2.5/b)b}$
van Genuchten (1980)	$K = K_s S_e^l \left[1 - \left(1 - S_e^{1/m} \right)^m \right]$	$K = K_s S_e^l \left[1 - \left(1 - S_e^{1/m} \right)^m \right]^2$
	$m = 1 - 2/n$	$m = 1 - 1/n$

Note: K_s = saturated hydraulic conductivity; l = tortuosity parameter.
For the other parameters, please see Table 8.5.

In cases where the only information on the soil is the mapping unit, typical parameters for the actual soil type can be used. The DAISY parameter library contains typical parameters for the most common Danish soil types. A special version of MACRO (MACRO-DB) also contains data for specific soils, not least the soils used for regulation of pesticides at European level. In APSIM, the user can create his own database.

Soil organic matter content and C/N in SOM is used in initializing the soil organic carbon turnover model and the nitrogen mineralization/immobilization model. This information is not sufficient to distribute organic matter between the individual pools. The pools can be set directly by the user; however, it is more common to use a built-in procedure that estimates the distribution between the pools based on information on the average annual amount of carbon allocated to the system during the period preceding the simulation period. This average annual amount of carbon allocation can be estimated by simulating a couple of typical crop rotations with the typical management at the site. Proper initialization of the organic pools is of utmost importance. When changes in the agricultural system takes place, the time to reach equilibrium may be hundreds of years. We are therefore typically on a sloping curve, meaning that net mineralization or immobilization of N takes place, which influences the amount of mineral N in the soil and therefore also growth and leaching. The equilibrium level of organic matter depends on the amount of C returned to the soil and the clay percentage (Figure 8.4).

Figure 8.5 shows the changes in total C content and in the SOM fraction over time in two cases. Case I corresponds to a shift from a low input system to a high input system. Case II corresponds from a high input system to a low input system. The time it takes for the model internal carbon pools to

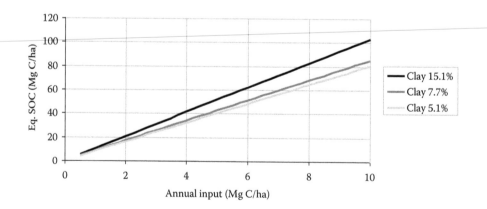

FIGURE 8.4
Equilibrium content of C in the plow-layer as a function of C input and different clay percentages, calculated with DAISY. Soil temperature and soil moisture conditions corresponds to typical Danish conditions.

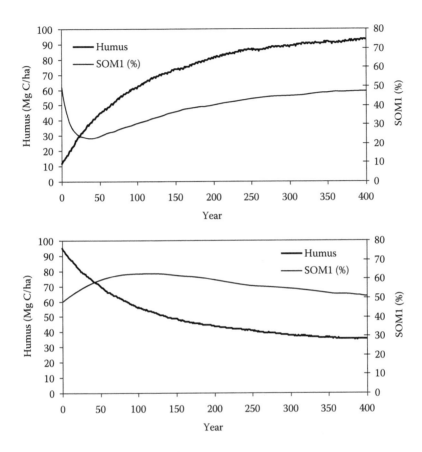

FIGURE 8.5
Changes in C content and SOM1 fraction over time. Top: shift from a low input system to a high input system. Bottom: shift from a high input system to a low input system.

reach an equilibrium corresponding to the carbon input is illustrated by the SOM1 fraction.

Soil thermal properties are estimated by pedo-tranfer functions as described by Hansen et al. (1990). Ammonium adsorption is also described by Hansen et al. (1990). Regarding solute transport, the dispersion described by the dispersivity is, by default, set to 4 cm. Annual nitrate leaching is rather insensitive to this parameter (Styczen et al. 2004).

The lower boundary characterizing the soil hydrology can be specified as: (1) a free drainage condition; (2) depth to groundwater; and (3) soil with subsurface drains. The first two are straightforward. The latter is more complicated. This condition requires the following specifications: (1) depth of tile drains; (2) distance between tile drains; (3) depth to and size of layer with low hydraulic conductivity; (4) hydraulic conductivity of this layer; and (5) hydraulic head below the layer with low hydraulic conductivity.

All other simulated variables pertaining to the soil system also need initialization. In order to make the model adjust itself to the prevailing circumstances, it is recommended to use a warm-up period of a couple of years.

8.4.3 Crop

Crop parameterization requires extensive information. Hence, in decision support studies, we normally restrict ourselves to use already parameterized crops. The available crop parameterizations can be found in the DAISY parameter library (2009; see http://code.google.com/p/daisy-model/). If the considered scenarios include crops for which a parameter set does not exist, it is common to substitute the new crops with similar parameterized crops. In this context, a similar crop is a crop with a nitrogen uptake potential and a growth pattern that resembles the selected parameterized crop. Naturally, such a substitution adds additional uncertainty to the simulations. All the other model codes also contain libraries of crop parameter, but the specific crops parameterized differ from model code to model code.

In general, the two largest entries in the field nitrogen balance are the fertilization and the crop nitrogen uptake. The former can be assumed known, whereas the latter is simulated. The crop model simulates the potential uptake, which is determined by the growth. The crop model is based on the concept of potential production, that is, it assumes optimal conditions. The "actual" production is obtained by reducing the potential production according to deficit of water and nitrogen. Other detrimental effects are not included. Hence, it is often necessary to adjust the crop production to the conditions that are being analyzed. One or a combination of the following procedures for crop production calibration is recommended (see the following subsections).

8.4.3.1 Reduction of the Photosynthesis

Gross photosynthesis is calculated by dividing the canopy into layers and applying a light response curve for a single leaf layer by layer:

$$\Delta F_i = f_w f_N f_T f_s \ \Delta L_{ai} F_m \left(1 - \exp\left(-\frac{\varepsilon}{F_m} \frac{S_{a,i}}{\Delta L_{ai}} \right) \right) \tag{8.6}$$

where ΔF_i is the gross photosynthesis for layer i for the considered crop; f_w, f_N, f_T, and f_s are stress functions adjusting for water stress, nitrogen stress, temperature stress, and senescence, respectively; ΔL_{ai} is the LAI of the considered canopy layer; F_m is a crop-specific photosynthetic rate at saturated light intensity; ε is a corresponding initial light use efficiency at low intensity; and $S_{a,i}$ is absorbed phonetically active radiation. The parameter F_m can be used for calibration. It is found in the crop parameter file under "LeafPhot."

In APSIM, the biomass increase is calculated differently, but is still dependent on water and nitrogen stress and influenced indirectly by temperatures. RZWQM uses a third method, and includes "environmental fitness factors" based on water, temperature, and nutrients. COUPmodel reacts to the same factors and salt.

8.4.3.2 Reduction of LAI

LAI is estimated as:

$$L_{ai} = S_{la} W_{leaf} \tag{8.7}$$

where L_{ai} is the LAI, S_{la} is the specific leaf area, which is assumed to be a function of DS, and W_{leaf} is the leaf weight. S_{la} can be used as a calibration parameter. It is found in the crop parameter file under "Canopy" and the name of the parameter is "SpLAI."

APSIM uses different equations and cannot be calibrated in this manner. COUPmodel contains a similar (not identical) parameter to the one used by DAISY, and in RZWQM the daily respiration as a function of photosynthate (Φ) and the biomass to leaf area conversion coef (C_{LA}) are used for calibration.

8.4.3.3 Adjusting Uptake of Nitrogen

The potential nitrogen content in the crop is estimated as:

$$N_p = \sum_x c_{p,x}(DS)W_x \tag{8.8}$$

where N_p is the potential nitrogen content of the crop; x is root, stem, leaf, or storage organ; $c_{p,x}(DS)$ is the potential nitrogen concentration at the development stage DS for the crop component x; and W_x is the corresponding weight of the component. The potential uptake rate for a time step is estimated as:

$$U_p = \frac{N_p - N_a}{\Delta t} \tag{8.9}$$

where U_p is the potential uptake rate, N_a is the actual nitrogen content of the crop, and Δt is the time step. $c_{p,x}(DS)$ can be used for calibration, and is found in the crop parameter file in the "CrpN" section. The names are "PtRootCnc," "PtStemCnc," "PtLeafCnc," "PtSOrgCnc." Each of these parameters is given as piecewise linear functions of DS.

A simpler way of adjusting the nitrogen uptake is to introduce a minimum nitrate and ammonium concentration at the root surface. Hence, the root does not act as a zero-sink any longer and the uptake will be restricted. These parameters are also found in the section "CrpN" and the names are "NO3_root_min" and "NH4_root_min."

In RZWQM, N uptake consists of a passive (with transpiration stream) and an active N uptake, and μ_1 (maximum active N uptake) is used for calibration. None of the models compared can be calibrated through a minimum concentration as used in DAISY.

8.4.4 Agricultural Management and Crop Rotation

Simulation of agricultural management is of key importance in scenario building and decision support studies. In DAISY, the management information is introduced to the model by a special input language. Introduction to this can be found in the "DAISY Tutorial" and the "DAISY Reference Manual," which can be found at the DAISY homepage. Main components of the input language are control statements and action statements.

A control statement is based on whether a certain condition is fulfilled (is true). The condition can pertain to time, time of the year, or similar. Furthermore, it can pertain to the conditions in the system: soil temperature at a certain depth, soil matrix potential at a certain depth, soil water content at a certain depth interval, mineral nitrogen content at a certain depth interval, the crop development stage, or the amount of shoot dry matter in the field. In addition, temperature sums from a certain onset can be used to regulate management.

An action statement defines a field operation, for example, sowing, harvesting, tillage (plowing, harrowing, etc.), irrigation, spraying, and fertilization. Irrigation comprises overhead, surface, and subsoil irrigation. Fertilization includes application of both mineral and organic fertilizers; and the application method includes both surface broadcasting and subsoil incorporation. Fertilizers must be characterized according to their content and the organic part must be characterized according to the turnover properties. At application, the organic part is transferred to the AOM pools of organic matter model. In case the fertilizer contains a particularly recalcitrant part, this may be transferred to the SOM2 pool of the model. The DAISY library comprises parameterizations of a number of organic fertilizers.

Based on control statements and action statements, complex-growing systems can be built. These growing systems can partly be governed by external growth conditions, and can be joined to form crop rotations. The system allows introduction of intercropping and catch crops. The system is one of the most flexible of its kind but is difficult to use because it requires that the user is able to apply the special DAISY input language. The code is built in a simple editor (e.g., TextPad).

As the growth pattern in general strongly interacts with the weather conditions, it is recommended that a scenario analysis comprises a permutation of weather and crops so every crop is grown every year.

8.5 Analysis of Simulation Results

In scenario building and decision support studies, the simulation results cannot be directly compared with observations. However, this does not mean that any results are valid. Styczen et al. (2004) recommended a procedure for a "reality" check.

Magnitude of harvested yield will often be known from agricultural statistics and can be compared with simulated yield and, hence, this comparison can be used as an indicator of the realism of the simulation. One of the entries in the nitrogen balance associated with the largest uncertainty is the nitrogen uptake by the crops. If the simulated yield is of the "right magnitude," it is an indicator that suggests that the nitrogen uptake is also of the "right magnitude."

The other entries in the nitrogen balances should also be evaluated. The nitrogen mineralization–immobilization is an entry associated with considerable uncertainty. It is suggested that one should look at the simulated development over time in soil organic nitrogen. Large changes in land use or fertilization practices may result in large values of the annual rate of change in SOM, but if this not the case this rate of change in SOM should not be too large. Experience for Danish conditions suggests that this rate of change at least should be within ± 100 kg N ha^{-1} year^{-1}.

Denitrification is subject to extreme variation and is very difficult to measure. However, very high rates are linked to wet conditions, which are enhanced by high dry bulk densities of the soil.

The entries of the water balance should be checked. Often experience on the magnitude of the annual evapotranspiration is available and can be compared to the corresponding simulation results in order to check whether the water balance is realistic. Furthermore, if the simulation study covers a large area, the simulated percolation can be compared to river runoff. One should not expect a perfect match as the simulation study is often performed in order to answer the question "What if?"

8.6 Discussion and Conclusion

The ideal model contains the relevant components at an adequate level of detail. The consequence of this statement is that the choice of model code for a particular purpose must depend on: (1) the purpose of the simulation; (2) the data availability; (3) whether the model code includes the processes, which are relevant for that particular site and purpose; and (4) whether the model generates a result that lies within acceptable limits of uncertainty. Frozen soils and their influence on hydraulic parameters are included in the

Swedish COUPmodel and the Danish DAISY model, but not in the Australian APSIM, which on the other hand includes erosion, probably reflecting what is important in different countries. Macropore inclusion is extremely important for simulation of pesticide transport to drains, but less important for N leaching. Experience with local parameterization and availability of relevant crop modules are other important points to consider when selecting a model code. The modeler and future users of the model results need to agree on objectives, data and model adequacy, and the uncertainty requirements at a very early stage of the work (see Refsgaard et al. 2005).

The fact that the user has significant influence on the result of simulations means that quality control and documentation are extremely important elements if model simulations are used for practical advice or political decisions. It is strongly recommended that local "good modeling practice" guides are developed that specify acceptable data sources, standard parameter sets for soil, crop, and manure parameterization, and set standards for initialization of the model runs through warming-up periods, specific initialization methods for particular parameters, output presentation, and documentation of input and output. The report of Styczen et al. (2004) contains, apart from the issues mentioned above, tables of yields and N content of different crops in Denmark, setting a standard for the removal of N with crops in simulations. As both of these factors have been influenced by the implemented environmental plans, values change over time, which also has to be considered in simulations. As N removal by plants is the largest "loss" in the N balance for the soil system, calibration is extremely important.

The HarmoniQua project (www.harmoniqua.org) discussed development of quality assurance (QA) guidelines for internal use or as developed in cooperation between modelers and users of model results, and although they do not deal specifically with SPAC models, the identified need for transparency and quality control at all stages of model work also applies here. Refsgaard et al. (2005) find that validation of models on independent data sets should be used to a much larger degree in order to document the prediction ability of the implemented model.

A particular issue of concern in SPAC models is heterogeneity and the difficulties in describing this heterogeneity in a suitable manner when setting up models. Issues giving rise to considerable uncertainty (in DAISY simulations, but most likely also in other process-based models) are:

1. Choice of precipitation data, which influence growth, percolation, and leaching of N.

2. Initialization of organic pools. This is an extremely difficult issue because it is seldom possible to simulate very long and realistic warm-up periods due to lack of data and the time required. Styczen et al. (2004) recommended initializing models used for administrative purposes to equilibrium with the present farming system

(assuming it has been relatively stable for "some time") to minimize the effect of earlier farming practices on simulated leaching, but it is clearly an issue for discussion. For example, if a farm moves from intensive dairy cattle production to crop production alone, leaching will be influenced by the former production method for several hundreds of years.

3. Calibration of drains. If drainage water is generated in a very uneven fashion across the field, the typical finding is that it is difficult to calibrate the model to fit production, water table, quantity of drainage water, and the period that drainage water occurs.

However, heterogeneity is a fundamental problem in model simulations and also refers to soil texture, organic matter content of the soil, macroporosity, etc. Before using model results administratively, it is necessary to try to quantify the uncertainty on the important output parameters caused by uncertainty in input data and agree on how it should be interpreted for administrative use. For leaching of nitrate, 50 mg/L nitrate or 11.3 mg NO_3^--N/L is an administrative limit in several countries, but if the uncertainty is 5 mg NO_3^--N/L, does that mean that simulation results should be below 6.3 mg/l NO_3^- -N to be acceptable? Is the result 0.11 ± 0.03 mg/L pesticide considerably worse than 0.08 ± 0.05 mg/L pesticide? And should the first fail and the second pass if tested against the general drinking water limit of 0.1 mg pesticide/L?

At present, many different model codes exist that deal with problems related to agricultural management. The major challenge in this field is probably not model development as much as the establishment of frameworks around the modeling to support good modeling practice, QA, and transparency. In addition, handling and communication of uncertainty issues to users and administrators is an issue that requires further attention if we want to make such models more acceptable outside the research community.

References

Abrahamsen, P., and S. Hansen. 2000. Daisy: An open soil–crop–atmosphere system model. *Environmental Modelling and Software* 15: 313–330.

Ahuja, L. R., K. W. Rojas, J. D. Hansen, M. J. Shaffer, and L. Ma, eds. 2000. *Root Zone Water Quality Model—Modelling Management Effects on Water Quality and Crop Production*, 372 pp. Highlands Ranch, CO: Water Resources Publications.

Allerup, P., H. Madsen, and F. Vejen. 1998. Standardværdier (1961–90) af nedbørskorrektioner. DMI. Teknisk rapport 98-10.

ANIMO. 2009. http://www.animo.wur.nl/ANIMObibliography.htm. Accessed 21 November 2010.

APSIM. 2009. http://www.apsim.info/Wiki/APSIM-Documentation.ashx. Accessed 21 November 2010.

Brooks, R. H., and A. T. Corey. 1964. Hydraulic properties of porous media. Hydrological Paper # 3. Ft. Collins, CO: Colorado State University.

Bruun, S., and L. S. Jensen. 2002. Initialisation of the soil organic matter pools of the Daisy model. *Ecological Modelling* 153 (3): 291–295.

Bruun, S., B. T. Christensen, E. M. Hansen, J. Magid, and L. S. Jensen. 2003. Calibration and validation of the soil organic matter dynamics of the Daisy model with data from the Askov long-term experiments. *Soil Biology & Biochemistry* 35 (1): 67–76.

Bruun, S., T. L. Hansen, T. H. Christensen, J. Magid, and L. S. Jensen. 2006. Application of processed organic municipal solid waste on agricultural land—a scenario analysis. *Environmental Modelling and Assessment* 11 (3): 251–265.

Burdine, N. T. 1952. Relative permeability calculations from pore-size distribution data. *Transactions of the AIME* 198: 35–42.

Boegh, E., M. Thorsen, M. B. Butts, S. Hansen, J. S. Christiansen, P. van der Keur, P. Abrahamsen, H. Soegaard, K. Schelde, A. Thomsen, C. B. Hasager, N. O. Jensen, and J. C. Refsgaard. 2004. Incorporating remote sensing data in physically based distributed agro-hydrological modelling. *Journal of Hydrology* 287: 279–299.

Boegh, E., R. N. Poulsen, M. Butts, P. Abrahamsen, E. Dellwik, S. Hansen, C. B. Hasager, J.-K. Loerup, K. Pilegaard, and H. Soegaard. 2009. Remote sensing based evapotranspiration and runoff modeling of agricultural, forest and urban flux sites in Denmark: From field to macro-scale. *Journal of Hydrology* 377: 300–316, doi:10.1016/j.jhydrol.2009.08.029.

Børgesen, C. D., J. Djurhuus, and A. Kyllingsbæk. 2001. Estimating the effect of legislation on nitrogen leaching by upscaling field simulations. *Ecological Modelling* 136: 31–48.

Campbell, G. S. 1974. A simple method for determining unsaturated conductivity from moisture retention data. *Soil Science* 117: 311–314.

Carsel, R. F., J. C. Imhoff, P. R. Hummel, J. M. Cheplick, and A. S. J. Donigan. 1998. PRZM-3, A model for predicting pesticide and nitrogen fate in the crop root and unsaturated zones. User Manual for release 3.0, National Exposure Research Laboratory, Office of Research and Development, U.S. Environmental Protectio Agency, Athens, GA, USA. (Available at (http://www.epa.gov/ceampubl/gwater/przm3/index.htm).

CENTURY. 2009. http://www.nrel.colostate.edu/projects/century5/. Accessed 21 November 2010.

Christiansen, J. S., M. Thorsen, T. Clausen, S. Hansen, and J. C. Refsgaard. 2004. Modelling of macropore flow and transport processes at catchment scale. *Journal of Hydrology* 299: 136–158.

Cosby, B. J., G. M. Hornberger, R. B. Clapp, and T. R. Ginn 1984. A statistical exploration of the relationship of soil moisture characteristics to the physical properties of soils. *Water Resources Research* 20 (6): 682–690.

Daisy. 2009. http://code.google.com/p/daisy-model/. Accessed 21 November 2010.

De Neergaard, A., H. Hauggaard-Nielsen, L. S. Jensen, and J. Magid. 2002. Decomposition of white clover (*Trifolium repens*) and ryegrass (*Lolium perenne*) components: C and N dynamics simulated with the DAISY soil organic matter submodel. *European Journal of Agronomy* 16 (1): 43–55.

Diekkrüger, B., D. Söndgerath, K. C. Kersebaum, and C. W. McVoy. 1995. Validity of agroecosystem models. A comparison of results of different models applied to the same data set. *Ecological Modelling* 81: 3–29.

Djurhuus, J., S. Hansen, K. Schelde, and O. H. Jacobsen. 1999. Modelling the mean nitrate leaching from spatial variable fields using effective parameters. *Geoderma* 87: 261–279.

Eckersten, H., P. E. Jansson, and H. Johansson. 1998. *SOILN Model: User's Manual*: version 9.2. Uppsala, Sweden: SLU.

FAO. FAO Penman–Monteith. http://www.fao.org/docrep/X0490E/x0490e06.htm.

Flanagan, D. C., and M. A. Nearing, Eds. 1995. *USDA-Water Erosion Prediction Project: Hillslope and Watershed Model Documentation*. NSERL Report No. 10. West Lafayette, IN: USDA-ARS National Soil Erosion Research Lab.

Freebairn, D. F., and G. H. Wockner. 1986. A study of soil erosion on Vertisols of the Eastern Darling Downs. *Australian Journal of Soil Research* 19: 133–146.

Gjettermann, B., M. Styczen, H. C. B. Hansen, F. P. Vinther, and S. Hansen. 2008. Challenges in modelling dissolved organic matter dynamics in agricultural soil using DAISY. *Soil Biology & Biochemistry* 40: 1506–1518, doi:10.1016/j.soilbio.2008.01.005.

Groenendijk, P., L. V. Renaud, and J. Roelsma. 2005. Prediction of nitrogen and phosphorus leaching to groundwater and surface waters. Process descriptions of the Animo4.0 model. Wageningen, Alterra-Report 983, 114 pp.

Gusman, A. J., and M. A. Mariño. 1999. Analytical modelling of nitrogen dynamics in soils and groundwater. *Journal of Irrigation and Drainage Engineering* 125 (6): 330–337.

Hansen, J. R., J. C. Refsgaard, S. Hansen, and V. Erntsen. 2007. Problems with heterogeneity in physically based agricultural catchment models. *Journal of Hydrology* 342: 1–16; doi:10.1016/j.jhydrol.2007.04.016.

Hansen, J. R., J. C. Refsgaard, V. Erntsen, S. Hansen, M. Styczen, and R. N. Poulsen. 2009. An integrated and physically based agricultural catchment modelling approach. *Hydrology Research* 40.4: 347–363, doi:10.2166/nh.2009.035.

Hansen, S. 1984. Estimation of potential and actual evapotranspiration. *Nordic Hydrology* 15: 205–212.

Hansen, S., and P. Abrahamsen. 2009. Modeling water and nitrogen uptake using a single-root concept: Exemplified by the use in the daisy model. In *Quantifying and Understanding Plant Nitrogen Uptake for System Modeling*, eds. L. Ma, L. R. Ahuja, and T. W. Bruulsema, 169–195. Boca Raton, FL: CRC Press.

Hansen, S., H. E. Jensen, N. E. Nielsen, and H. Svendsen. 1990. DAISY: Soil Plant Atmosphere System Model. NPO Report No. A 10. 272 pp. Copenhagen: The National Agency for Environmental Protection.

Hansen, S., H. E. Jensen, N. E. Nielsen, and H. Svendsen. 1991a. Simulation of nitrogen dynamics and biomass production in winter wheat using the Danish simulation model Daisy. *Fertilizer Research* 27: 245–259.

Hansen, S., H. E. Jensen, N. E. Nielsen, and H. Svendsen. 1991b. Simulation of nitrogen dynamics in the soil plant system using the Danish simulation model Daisy. In Kienitz, G., Milly, P. C. D., van Genuchten, M. Th., Rosbjerg, D., and Shuttleworth, W. J. (eds.) *Hydrological Interactions Between Atmosphere, Soil and Vegetation. IAHS Publication* 204: 185–195.

Hansen, S., H. E. Jensen, N. E. Nielsen, and H. Svendsen. 1991c. Simulation of biomass production, nitrogen uptake and nitrogen leaching by using the Daisy model. In: *Soil and Groundwater Research Report II: Nitrate in Soils, 1991: 300–309.* Final Report on Contracts EV4V-0098-NL and EV4V-00107-C. DG XII. Commission of the European Communities.

Hansen, S., M. J. Schaffer, and H. E. Jensen. 1994. Developments in modeling nitrogen transformations in soil. In *Nitrogen Fertilization and the Environment.* ed. P. Bacon, pp. 83–107. New York, NY: Marcel Dekker Publishers.

Hansen, S., M. Thorsen, E. J. Pebsma, S. Kleeschulte, and H. Svendsen. 1999. Uncertainty associated with the simulation of regional nitrate leaching. A case study. *Soil Use and Management* 15: 167–175.

Hansen, S., C. Thirup, J. C. Refsgaard, and L. S. Jensen. 2001. Modelling of nitrate leaching at different scales—application of the Daisy. In *Modeling Carbon and Nitrogen Dynamics for Soil Management*, eds. M. Shaffer, M. Liwang, and S. Hansen, pp. 511–547. Boca Raton, FL: Publishers.

Hansen, T. L., G. S. Bhander, T. H. Christensen, S. Bruun, and L. S. Jensen. 2006. Life cycle modelling of environmental impacts of application of processed organic municipal solid waste on agricultural land (EASEWASTE). *Waste Management & Research* 24 (2): 153–166.

Heidmann T., C. Tofteng, P. Abrahamsen, F. Plauborg, S. Hansen, A. Battilani, A. J. Coutinho, F. Doležal, W. Mazurczyk, J. D. R. Ruiz, J. Takác. 2008. Calibration procedure for a potato crop growth model using information from across Europe. *Ecological Modelling* 211: 209–223, doi:10.1016/j.ecolmodel.2007.09.008.

Jansson, P.-E., and L. Karlberg. 2009. Coupled heat and mass transfer model for soil–plant–atmosphere systems. Last update: 7 October, 2009. http://www.lwr.kth.se/vara%20datorprogram/CoupModel/index.htm.

Jensen, C., B. Stougaard, and N. H. Jensen. 1992. The integration of soil classification and modelling of N-balances with the DAISY model. In *Integrated Soil and Sediment Research: A Basis for Proper Protection*, eds. H. J. P. Eijsackers, and T. Hamers, pp. 512–514. The Netherlands: Kluwer Academic Publishers.

Jensen, C., B. Stougaard, and P. Olsen. 1994a. Simulation of nitrogen dynamics at three Danish locations by use of the DAISY model. *Acta Agriculturae Scandinavica, Section B, Soil and Plant Science* 44: 75–83.

Jensen, C., B. Stougaard, and H. S. Østergaard. 1994b. Simulation of the nitrogen dynamics in farm land areas in Denmark (1989–1993). *Soil Use and Management* 10: 111–118.

Jensen, C., B. Stougaard, and H. S. Østergaard. 1996. The performance of the Danish simulation model DAISY in prediction of Nmin at spring. *Fertilizer Research* 44: 79–85.

Jensen, H. E., S. Hansen, B. Stougaard, C. Jensen, K. Holst, H. B. Madsen. 1993. Using GIS-information to translate soil type patterns to agro-ecosystem management—the Daisy Model. In *Integrated Soil and Sediment Research: A Basis for Proper Protection*, eds. H. J. P. Eijsackers, T. Hamers, 401–428. The Netherlands: Kluwer Academic Publishers, Dordrecht.

Jensen, L. S., T. Mueller, S. Bruun, and S. Hansen. 2001. Application of the Daisy model for short and long-term simulation of soil carbon and nitrogen dynamics. In *Modeling Carbon and Nitrogen Dynamics for Soil Management*, eds. M. Shaffer, M. Liwang, and S. Hansen, pp. 483–509. Boca Raton, FL: Lewis Publishers.

Jensen, L. S., T. Mueller, N. E. Nielsen, S. Hansen, G. J. Crocker, P. R. Grace, J. Klir, M. Körschens, and P. R. Poulton. 1997. Simulating trends in soil organic carbon in long-term experiments using the soil-plant-atmosphere model DAISY. *Geoderma* 81: 5–28.

Johnsson, H., L. Bergstrom, P.-E. Jansson, and K. Paustian. 1987. Simulated nitrogen dynamics and losses in a layered agricultural soil. *Agriculture, Ecosystems & Environment* 18 (4): 333–356.

Jones, C. A., and J. R. Kiniry. 1986. *CERES-Maize: A Simulation Model of Maize Growth and Development.* College Station, TX: Texas A&M University Press.

Larsbo, M., and N. Jarvis. 2003. MACRO 5.0. A model of water flow and solute transport in macroporous soil. Technical description. Swedish University of Agricultural Sciences, Department of Soils Sciences, Division of Environmental Physics. Emergo 2003:6, 48 pp. ISBN: 91-576-6592-3. http://www-mv.slu.se/webfiles/bgf/MACRO/Macro50/Technical%20report%20macro%205.pdf.

Littleboy, M., D. M. Silburn, D. M. Freebairn, D. R. Woodruff, and G. L. Hammer. 1989. *PERFECT—A Computer Simulation Model of Productivity Erosion Runoff Functions to Evaluate Conservation Techniques.* Brisbane: QDPI.

Littleboy, M., D. M. Silburn, D. M. Freebairn, D. R. Woodruff, G. L. Hammer, and J. K. Leslie. 1992. Impact of soil erosion on production in cropping systems: I. Development and validation of a simulation model. *Australian Journal of Soil Research* 30: 757–774.

Makkink, G. F. 1957. Ekzameno de la formulo de Penman. *Netherland Journal of Agricultural Science* 5: 290–305.

Misra, R. K., and C. W. Rose. 1996. Application and sensitivity analysis of process-based erosion model GUEST. *European Journal of Soil Science* 47: 593–604.

Molina, J. A. E., C. E. Clapp, M. J. Shaffer, F. W. Chichester, and W. E. Larson. 1983. NCSOIL, A model of nitrogen and carbon transformations in soil: Description, calibration, and behavior. *Soil Science Society of America Journal* 47 (1).

Mualem, Y. 1976. A new model for predicting the hydraulic conductivity of unsaturated porous media. *Water Resources Research* 12: 513–522.

Muller, T., L. S. Jensen, J. Magid, and N. E. Nielsen. 1996. Temporal variation of C and N turnover in soil after oilseed rape incorporation in the incorporation in the field: Simulation with the soil–plant–atmosphere model Daisy. *Ecological Modelling* 99: 247–262.

Muller, T., J. Magid, L. S. Jensen, and N. E. Nielsen. 2003. Decomposition of plant residues of different quality in soil—DAISY model calibration and simulation based on experimental data. *Ecological Modelling* 191 (3–4): 538–544.

Müller T., K. Thorup-Kristensen, J. Magid, L. S. Jensen, and S. Hansen. 2006. Catch crops affect nitrogen dynamic in organic farming systems without livestock husbandry simulations with the DAISY model. *Ecological Modelling* 191: 538–544.

Olesen, J. E., G. H. Rubæk, T. Heidmann, S. Hansen, and C. D. Børgesen. 2004. Effect of climate change on greenhouse gas emissions from arable crop rotations. *Nutrient Cycling in Agroecosystems* 70 (2): 147–160.

Olesen, J. E., T. R. Carter, C. H. Diaz-Ambrona, S. Fronzek, T. Heidmann, T. Hickler, T. Holt, M. I. Minguez, P. Morales, J. P. Palutikof, M. Quemada, M. Ruiz-Ramos, G. H. Rubaek, F. Sau, B. Smith, and M. T. Sykes. 2007. Uncertainties in projected impacts of climate change on European agriculture and terrestrial ecosystems based on scenarios from regional climate models. *Climate Change* 81: 123–143.

Parton, W. J., D. S. Schimel, C. V. Cole, and D. S. Ojima. 1987. Analysis of factors controlling soil organic matter levels in Great Plains grasslands. *Soil Science Society of America Journal* 51: 1173–1179.

Pedersen, A., B. M. Petersen, J. Eriksen, S. Hansen, and L. S. Jensen. 2007. A model simulation analysis of soil nitrate concentrations—does soil organic matter pool structure or catch crop growth parameters matter most? *Ecological Modelling* 205: 209–220; doi:10.1016/jecolmodel.2007.02.016.

Pedersen, A., K. Thorup-Kristensen, and L. S. Jensen. 2009. Simulating nitrate retention in soils and the effect of catch crop use and rooting pattern under the climatic conditions of Northern Europe. *Soil Use and Management* 25 (3): 243–254.

Petersen, C. T., U. Jørgensen, H. Svendsen, S. Hansen, H. E. Jensen, and N. E. Nielsen. 1995. Parameter assessment for simulation of biomass production and nitrogen uptake in winter rape. *European Journal of Agronomy* 4 (1): 77–89.

Priestly, C. H. B., and R. J. Taylor. 1972. On the assessment of surface heat and evaporation using large-scale parameters. *Monthly Weather Review* 100: 81.

Refsgaard, J. C., M. Thorsen, J. Birk Jensen, S. Kleeschulte and S. Hansen. 1999. Large scale modelling of groundwater contamination from nitrogen leaching. *Journal of Hydrology* 221: 117–140.

Refsgaard, J. C., and H. J. Henriksen. 2004. Modelling guidelines—terminology and guiding principles. *Advances in Water Resources* 27: 71–82.

Refsgaard, J. C., H. J. Henriksen, W. J. Harrar, H. Scholten, and A. Kassahun. 2005. Quality assurance in model based water management—review of existing practice and outline of new approaches. *Environmental Modelling & Software* 20: 1201–1215.

Rose, C. W. 1985. Developments in soil erosion and deposition models. *Advances in Soil Science* 2: 1–63.

Rose, C. W., K. J. Coughlan, C. A. A. Ciesiolka, and B. Fentie. 1997. Program GUEST (Griffith University Erosion System Template). In *A New Soil Conservation Methodology and Application to Cropping Systems in Tropical Steeplands*, eds. K. J. Coughlan, and C. W. Rose, pp. 34. Canberra: ACIAR, ACIAR Technical Report No. 40.

Schaap, M. G., F. J. Leij, and M. T. van Genuchten. 2001. ROSETTA: A computer program for estimating soil hydraulic parameters with hierarchial pedotransfer functions. *Journal of Hydrology* 251 (3–4): 163–176.

Shaffer, M. J., A. D. Halvorson, and F. J. Peirce. 1991. Nitrate leaching and economic analysis package (NLEAP): Model description and application. In *Managing N for Groundwater Quality and Farm Profitability*. eds. R. F. Follett, et al., pp. 285–322. Madison, WI: SSSA.

Smith, P., J. U. Smith, D. S. Powlson, J. R. M. Arah, O. G. Chertov, K. Coleman, U. Franko, S. Frolking, H. K. Gunnewiek, D. S. Jenkinson, L. S. Jensen, R. H. Kelly, C. Li, J. A. E. Molina, T. Mueller, W. J. Parton, J. H. M. Thornley, and A. P. Whitmore. 1997. A comparison of the performance of nine soil organic matter models using datasets from seven long-term experiments. *Geoderma* 81 (1–2): 153–222.

Styczen, M., and B. Storm. 1993a. Modelling of N-movement on catchment scale—a tool for analysis and decision making: 1. Model Description. *Fertilizer Research* 36: 1–6.

Styczen, M., and B. Storm. 1993b. Modelling of N-movement on catchment scale—a tool for analysis and decision making: 1. A Case Study. *Fertilizer Research* 36: 7–17.

Styczen, M., S. Hansen, L. S. Jensen, H. Svendsen, P. Abrahamsen, C. D. Børgesen, C. Thirup, and H. S. Østergaard. 2004. Standard Parameterization of the Daisy-Model. Guide and Background. Version 1.0. (In Danish). Danish Hydraulic Institute, 60 pp.

SUNDIAL. 2009. http://www.rothamsted.ac.uk/aen/sundial/intro.htm. Accessed 21 November 2010.

Svendsen, H., S. Hansen, and H. E. Jensen. 1995. Simulation of crop production, water and nitrogen balances in two German agro-ecosystems using the Daisy model. *Ecological Modelling* 81: 197–212.

Thorsen, M., J. C. Refsgaard, S. Hansen, E. Pebesma, J. B. Jensen, and S. Kleeschulte. 2001. Assessment of uncertainty in simulation of nitrate leaching to aquifers at catchment scale. *Journal of Hydrology* 210: 210–227.

Van den Berg, F., and J. J. T.-I. Boesten. 1998. *Pesticide Leaching and Accumulation Model (PEST-LA), version 3.4, Description and User's Guide*, 150 pp. Wageningen, The Netherlands: Winand Staring Centre for Integrated Land, Soil and Water Research, Agricultural Research Department.

Van der Keur, P., S. Hansen, K. Schelde, and A. Thomsen. 2001. Modification of DAISY SVAT model for use of remotely sensed data. *Agricultural and Forestry Meteorology* 106: 215–231.

Van der Keur, P., J. R. Hansen, S. Hansen, and J. C. Refsgaard. 2008. Uncertainty in simulation of nitrate leaching at field scale within the Odense River basin. *Vadose Zone Journal* 7: 10–21, doi:102136/vzj/2006.0186.

Van Genuchten, M. Th. 1980. A close form equation for predicting hydraulic conductivity of unsaturated soils. *Soil Science Society of America Journal* 44: 892–898.

Veihe, A., N. H. Jensen, E. Boegh, M. W. Pedersen, and P. Frederiksen. 2006. The power of models in planning: The case of DaisyGIS and nitrate leaching. *Geografiska Annaler Series B-Human Geography* 88B (2): 215–229.

Vereecken, H., E. J. Jansen, M. J. D. Hack-ten Broeke, M. Swerts, R. Engelke, F. Fabrewiz, and S. Hansen. 1991. Comparison of simulation results of five nitrogen models using different data sets. In: Soil and Groundwater Research Report II: Nitrate in Soils, 1991: 321–338. Final Report on Contracts EV4V-0098-NL and EV4V-00107-C. DG XII. Commission of the European Communities.

WEPP. 2009. http://www.ars.usda.gov/Research/docs.htm?docid=10621. Accessed 21 November 2010.

Willigen, P. de. 1991. Nitrogen turnover in the soil-crop system; comparison of fourteen simulation models. *Fertilizer Research* 27: 141–149.

Wösten, J. H. M., A. Lilly, A. Nemes, and C. Le Bas. 1999. Development and use of a database of hydraulic properties of European soils. *Geoderma* 90: 169–185.

9

Aquaculture Systems

Bruno Díaz López

CONTENTS

9.1 Introduction..241
9.2 Ecosystem-Level Studies: Ecopath Modeling Approach 243
9.3 Impact of Aquaculture on a Coastal Ecosystem Predicted by the
   Mass-Balance Model..245
   9.3.1 Defining the System: Aranci Bay..245
   9.3.2 Application of the Ecopath Mass-Balance Model 246
      9.3.2.1 First Mass-Balance Scenario (Scenario A): State of
              the System Influenced by Aquaculture Activities......246
      9.3.2.2 Second Mass-Balance Scenario (Scenario B): State
              of the System Before the Start of Aquaculture
              Activities..247
   9.3.3 Predicted Effects of the Introduction of Finfish Culture
          Determined from the Ecopath Mass-Balance Model 248
9.4 Summary..251
Acknowledgments 253
References...253

9.1 Introduction

The worldwide expansion of marine aquaculture industries has caused growing concern regarding their environmental impact. Marine aquaculture is a significant industry that continues to grow more rapidly than all other animal food-producing sectors, with an average annual growth rate worldwide of 8.8% per year since 1970, compared with only 1.2% for capture fisheries and 2.8% for terrestrial farmed meat production systems (FAO 2007). This increase in marine aquaculture activities is thought to be in response to the increase in demand for fish, which cannot be fulfilled by traditional fisheries because of the decline in wild populations. Moreover, marine aquaculture is essentially an economic development within small- and medium-sized enterprises in areas where alternative employment is scarce.

In many cases, the ability to predict the effects of aquaculture on the marine environment is a prerequisite to establishing and expanding culture operations. Consequently, the study of aquaculture ecosystems requires consideration of biological, physical, chemical, and geological factors. According to FAO (1995), "The achievement of real marine ecosystem-based management of fisheries implies the regulation of the use of the living resources based on the understanding of the structure and dynamics of the ecosystem of which the resource is a part." This premise requires an improvement of our understanding of the structure of marine ecosystems, and the interactions between ecosystem compartments and their changes due to large-scale culturing operations (Díaz López et al. 2008).

Published literature shows that the development of the aquaculture industry has been accompanied by an increase in environmental impacts (reviewed by Pillay 1992; Fernandes et al. 2002; Cole 2002). The effects of aquaculture on the marine environment may be categorized into three types: eutrophication, sedimentation, and effects on the food web (Hargrave 2003). It has often been noted that the type of cultivated organisms, the locations of cultivation, the cultivated biomass, the quality and quantity of supplied food, and management practices are the main factors determining the extent of these effects (Beveridge 1996; Hargrave 2003; Pillay 2004; Machias et al. 2005).

Marine finfish aquaculture differs from that of shellfish farming in that bivalve culture requires minimal additions to the environment, except for the animals themselves and the infrastructures used to grow them. Their food is supplied by the environment and their wastes return nutrients and minerals to the ecosystem. Conversely, the marine finfish aquaculture, commonly practiced in cages, involves the supply of a substantial amount of nutrients with consequent impacts on the environment (Holby and Hall 1991; Hall et al. 1992). Fish production can also generate considerable amounts of effluent, such as waste feed and feces, medications, and pesticides, which can have undesirable impacts on the environment (Wu 1995; Lemarié et al. 1998; Read and Fernandes 2003). Consequently, it would be reasonable to expect effects at large spatial scales, particularly when a finfish farm is established in a coastal bay (Díaz López et al. 2008). In addition, effects on wild fish have been investigated at short spatial scales (Carss 1990, 1994; Dempster et al. 2004), indicating a considerable increase in wild fish abundance and biomass in the immediate vicinity of fish cages. There may also be undesirable effects on wild fish populations, such as genetic interactions between escaped farmed fish and wild fish (Youngson et al. 2001), disease transfer by escaped fish, or through ingestion of contaminated waste by wild fish (Heggberget et al. 1993). Additionally, aquaculture activities cause potential impacts on top predators such as modification of habitat use (Watson-Capps and Mann 2005; Díaz López et al. 2005), death, and injury through entanglement in gear (Díaz López and Shirai 2007).

Considerations of the sustainability of aquaculture include the ecological resources required to sustain the industry, that is, fish food for farmed species, and environmental capacity to assimilate waste (Read et al. 2001). However, to assess the sustainability, it is necessary to estimate the impact of aquaculture activities and to predict the change on the environment.

The impact that marine finfish aquaculture produces in the environment requires research involving field measurements as well as comprehensive modeling studies that integrate available knowledge about natural and anthropogenic parts of coastal ecosystems (Cruz-Escalona et al. 2007). Thus, any attempt to assess ecosystem-level effects of finfish aquaculture must consider the complexity of natural and human actions in coastal systems.

Most modeling studies of effects of finfish aquaculture on the environment have focused on organic matter, dissolved inorganic nutrients and oxygen, the impact of organic matter on the benthic system, and the relative potential for environmental effects of new and existing fish farm sites (i.e., Silvert 1992; Silvert and Sowles 1996; Hevia et al. 1996; Ervik et al. 1997; Findlay and Watling 1997; Black 2001; Cromey et al. 2002; Stigebrandt et al. 2004). Although these models are good tools to study the impact that finfish aquaculture has on the environment, ecosystem-level studies are needed on many areas, particularly the long-term responses of ecosystem components (phytoplankton, zooplankton, fish, marine top predators, benthos, as well as the farmed fish) to aquaculture-induced changes in system energy flow and biomass.

9.2 Ecosystem-Level Studies: Ecopath Modeling Approach

To investigate the potential effect of finfish culture on coastal ecosystems, I advocate the use of an alternative approach based on a mass-balance modeling framework. The make use of mass-balance models in aquaculture is currently at a relatively early stage of development compared to fisheries and for many other anthropogenic activities. Thus, few studies have been completed which adequately assess these potential environmental interactions of this newly developed industry (Jiang and Gibbs 2005; Díaz López et al. 2008). Ecosystem mass-balance models are very important because the complexity of ecosystems makes it difficult to gain an insight into their structure based on direct observations (Niquil et al. 1999). The major advantage of the mass-balance model is that it can be used to study the broad spectrum of ecosystem theories, including the description of trophic levels, network analysis, information theory, and thermodynamic concepts (Müller 1997). This modeling framework can also be used to investigate the possible impacts of some species and how these may affect the ecosystem as a whole (Pauly et al. 2000).

Ecopath software (Christensen and Pauly 1992; http://www.ecopath.org) was developed as a useful tool incorporating algorithms for the retrieval of the ecological, thermodynamic, and informational indices needed for network analysis (Ulanowicz 1993). Through a system of linear equations describing the mass balance for each functional component of the system, the overall ecosystem balance is obtained (Christensen and Pauly 1992; Pauly et al. 1993).

Ecopath models rely on the truism that:

1. Production by (i) = all losses by predation on (i) + nonpredation losses on (i) + export of (i).

   This applies for any group (e.g., a given fish population) and time (e.g., a year or season).

   Groups are linked through predators consuming prey, where:

2. Consumption = production + nonassimilated food + respiration.

   The basic equation that represents the balance for each trophic group, i, of the network is:

3. $$B_i \frac{P_i}{B_i} EE_i - \sum_{j=1}^{n} B_j \frac{Q_j}{B_j} DC_{ji} - EX_i = 0$$

   where DC is the diet matrix, which describes the relationships among groups whose elements DC_{ji} represent the fraction of the prey i in the average diet of the predator j; B_i is the biomass of each group; P_i/Bi is the production/biomass ratio (equal to the instantaneous rate of total mortality Z in steady-state systems) and Q_j/B_j is the consumption/biomass ratio of predator; and EE_i is the ecotrophic efficiency, which represents the part of the total production that is consumed by predators or exported; and EX_i is the export of the compartment i toward other ecosystems such as net migration and harvest by fishery (Christensen and Pauly 1993). Since the currency of the model is energy-related, the unassimilated/consumption ratio (UN/Q) is used to quantify the fraction of the food (Q_x) that is not assimilated. More details on capabilities and limitations of the Ecopath software are given by Christensen and Walters (2000).

Ecopath is a steady-state model representation for a given period of the energy flows of an ecosystem, and therefore cannot be used to simulate changes to flows with time. By contrast, this modeling framework has been used to investigate the functioning of the system and how this has changed with the introduction of intensive finfish culturing (Díaz López et al. 2008) and shellfish culturing (Jiang and Gibbs 2005).

9.3 Impact of Aquaculture on a Coastal Ecosystem Predicted by the Mass-Balance Model

The next section describes the Ecopath mass-balance model determined by Díaz López et al. (2008), refined and shaped to estimate the potential effects of finfish culture on a coastal ecosystem and, therefore, to identify the species that play a key role in the processes of ecosystems affected by aquaculture. The functioning of the Ecopath model is illustrated for two examples, and is followed by a general discussion. These examples are derived from model applications in the same area during two different periods "before" and "after" the beginning of culturing operations. The availability of published data provided the opportunity to compare these two distinct ecosystem states. Both scenarios were made in such a manner that they provide the best information on the functioning of the modeling framework.

9.3.1 Defining the System: Aranci Bay

The coastal area of 16.25 km² considered in the study was the Aranci Bay, located on the northeastern coast of Sardinia (Italy) (Figure 9.1). The selection of this area was based on the environment assessment and field studies conducted before and after the establishment of a marine finfish farm (Díaz

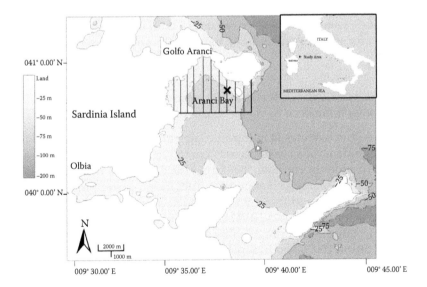

FIGURE 9.1
Map of the Aranci Bay (Sardinia, Italy), showing the area influenced by aquaculture with a line pattern. A cross indicates the location of the marine finfish farm (40°59.98′N, 9°37.09′E).

López et al. 2008). This finfish farm was set up in 1995; it covers an area of 0.04 km², which is approximately 0.25% of the Aranci Bay, and contains 850 tons of ichthyic biomass [mostly sea bass (*Dicentrarchus labrax*) and gilthead sea bream (*Sparus auratus*)].

9.3.2 Application of the Ecopath Mass-Balance Model

Definition of the functional trophic groups was based on similarities in their ecological and biological features (Pinnegar and Polunin 2004), based on their abundance and how they are affected by aquaculture. For each functional group, three out of four of the basic parameters [biomass, consumption ratio, production ratio, ecotrophic efficiency (EE)] were required to construct the Ecopath mass-balance models. All the available data for biomass, annual harvesting, and discards were converted into the same unit (t km⁻²) and expressed as wet weight. Published and unpublished sources concerning the system were used to generate input parameters; however, in some cases, it was necessary to assume from the wider literature derived for values of similar coastal systems. Furthermore, for both scenarios, the assumption was followed that the artisanal fishery effort, although low in the area, was constant.

The model was considered balanced when: (1) realistic estimates of the missing parameters of EE were calculated (EE <1); (2) gross efficiency values (GE = P/Q) for functional groups were between 0.1 and 0.35 with the exception of fast growing groups with higher values and top predators with lower values; and (3) values of R/B were consistent with the group's activities with high values for small organisms and top predators (Christensen and Walters 2004).

9.3.2.1 First Mass-Balance Scenario (Scenario A): State of the System Influenced by Aquaculture Activities

The first scenario includes 14 living and 3 detritus groups spanning the main trophic components of a coastal ecosystem influenced by finfish aquaculture (Table 9.1). In order to consider fish farm effects, the fish nourishment, harvesting, and discards, as well as outflow from the group representing farmed fish species, were introduced into the model. In addition, the mortality of some groups affected by incidental captures in the fish farm was increased according to information provided by field studies (Díaz López 2006; Díaz López and Shirai 2007), and in function of the data reported by the fish farm manager. This scenario provides a summary of current knowledge of the biomass, consumption, production, food web, and trophic structure in this area after the establishment of a marine finfish farm. Similarly, diet data for the aggregated functional trophic groups were obtained from the diet compositions reported in the model of Díaz López et al. (2008) and data gathered from published sources.

TABLE 9.1

Input Values (in Italic) and Estimates of Some Parameters in the Ecopath Mass Balance Model of the System after the Establishment of Marine Finfish Farm (Scenario A)

Group Name	TL	B (t km^{-2})	P/B (/year)	Q/B (/year)	EE	Flow to Detritus (t km^{-2} year^{-1})
Bottlenose dolphins	3.82	*6.246*	*0.331*	*3.766*	0	6.772
Cormorants	3.55	*0.554*	*0.315*	*4.225*	0	0.643
Seabirds	3.17	*0.583*	*0.251*	*1.19*	0	0.285
Cephalopods	3.17	8.889	2.34	5.3	0.961	10.227
Mugil cephalus	2.54	13.549	*0.624*	*8.587*	0.95	58.599
Mugil cephalus (juveniles)	2.54	1.866	*1.74*	*23.45*	0.781	22.585
Piscivorous fish	3.74	27.227	*0.729*	*2.88*	0.908	17.512
Zooplanktivorous fish	2.57	31.479	*1.5*	*8.86*	0.966	57.38
Farmed fish	2	*52.308*	*1.138*	*2.4*	0.656	0
Polychaetes	2	5.79	*4.8*	*11.53*	0.95	28.093
Macrobenthos detritivorous	2	8.325	*5.23*	*18*	0.837	0
Mussels	2.03	3.614	*1.8*	*6.629*	0.019	11.174
Zooplankton	2.09	6.476	*50*	*170*	0.97	560.263
Phytoplankton	1	6.57	*112.65*	–	0.95	37.006
Fish farm discards	1	*1.231*	–	–	0	0
Nourishment	1	*156.923*	–	–	0	0
Detritus	1	*631.73*	–	–	0.823	0

Note: Trophic level (TL), biomasses (*B*), production rates (*P/B*), consumption rates (*Q/B*), ecotrophic efficiency (EE), and flow to detritus used in the mass-balance model.

9.3.2.2 Second Mass-Balance Scenario (Scenario B): State of the System before the Start of Aquaculture Activities

This mass-balance model includes 11 living and 1 detritus groups spanning the main trophic components of a Mediterranean coastal ecosystem (Table 9.2). Biomasses were estimated for most trophic groups based on previous field studies, and in function of the data reported in coastal areas of similar characteristics to Aranci Bay before the aquaculture operations began. Information gathered from published sources was useful in rejecting all groups directly related with the presence of aquaculture (i.e., cormorants, farmed fish, mussels). For example, species such as farmed fish and mussels were absent in the area before the beginning of aquaculture operations. Similarly, cormorants were not present in the area because their presence was directly related with predation in the finfish farm. Likewise, diet matrix was

TABLE 9.2

Input Values (in Italic) and Estimates of Some Parameters in the Ecopath Mass Balance Model of the System before the Establishment of the Marine Finfish Farm (Scenario B)

Group Name	TL	B (t km^{-2})	P/B (/year)	Q/B (/year)	EE	Flow to Detritus (t km^{-2} year^{-1})
Bottlenose dolphins	4.04	1.679	0.875	14	0	6.172
Seabirds	3.75	0.012	0.251	1.19	0	0.006
Cephalopods	3.4	4.875	2.34	5.1	0.903	6.08
Mugil cephalus	2.54	2.887	2.824	8.087	0.967	4.939
Mugil cephalus (juvenile)	2.54	1.186	1.74	23.45	0.473	0
Piscivorous fish	4	13.02	1.56	2.88	0.902	9.496
Zooplanktivorous fish	2.72	14.95	1.5	8.86	0.938	27.885
Macrobenthos detritivorous	2	1.928	5.23	18	0.837	0
Polychaetes	2	4.90	2.67	13.36	0.909	27.403
Zooplankton	2.09	3.5	50	170	0.969	306.477
Phytoplankton	1	3.6	112.65	–	0.95	20.238
Detritus	1	356	–	–	0.824	0

Note: Trophic level (TL), biomasses (*B*), production rates (*P/B*), consumption rates (*Q/B*), ecotrophic efficiency (EE), and flow to detritus used in the mass-balance model.

obtained from the diet compositions reported in the model of Díaz López et al. (2008) and data gathered from published sources.

Where biomass estimates for this scenario were not available, the EE estimated by Ecopath for the first scenario was used as input parameter in the second scenario. In this way, Ecopath estimated biomasses under the assumption that the fraction of production used within the system is the same in the first and the second scenarios. Furthermore, although the first scenario was inherently more accurate than the second scenario, the latter represented useful estimates of the possible community states before aquaculture activities.

9.3.3 Predicted Effects of the Introduction of Finfish Culture Determined from the Ecopath Mass-Balance Model

To predict the effects of finfish aquaculture on the system, the status of the two ecosystem scenarios was compared and the system response investigated. The model as presently formulated makes it possible to predict increases in biomass groups as a result of nutrient loadings in agreement with field studies.

The structure of the ecosystem in both scenarios showed substantial differences in biomass values estimated for each group (Table 9.3). This change in all trophic groups demonstrates an increase in the biomass after the start of aquaculture activities. Biomass values were estimated by the model for zooplankton, polychaetes, macrobenthos detritivorous (amphipods and isopods), zooplanktivorous fish, piscivorous fish, cephalopods, and common gray mullets (adults and juveniles). Augment in biomass seen in benthic (i.e., polychaetes and macrobenthos detritivorous) and pelagic subsystems (i.e., fish species such as common gray mullets and zooplanktivorous) was in accordance with field studies (Dempster et al. 2004; Klaoudatos et al. 2006). Increased nutrient loading into the fish farm area resulted in greater biological activity and induced a strong coupling between the pelagic and benthic subsystems (Díaz López et al. 2008).

The calculation of "Trophic aggregations" (Ulanowicz 1995) with Ecopath, provided an accurate picture of the system and allocated the different dietary interactions to discrete trophic levels (Figures 9.2 and 9.3). This increase in biomass after the start of the aquaculture was nonmonotonic, although an increase in primary producer biomass should propagate monotonically through all trophic levels in a system based almost entirely on primary producers (Odum 1971). In cases like this (of strong coupling between the pelagic and benthic subsystems), the bottom–up control of phytoplankton

TABLE 9.3

Estimated Biomass (t km^{-2}) of Trophic Groups for the Created Scenarios

Group	Scenario B	Scenario A	Variation (%)
Bottlenose dolphin	1.679	6.246	272
Cormorant	0	0.554	
Seabirds	0.012	0.583	4 758
Cephalopods	4.875	8.889	82.3
Mugil cephalus	2.887	13.549	369
Mugil cephalus (Juveniles)	1.186	1.866	57.3
Piscivorous fish	13.023	27.227	109
Zooplanktivorous fish	14.95	31.479	110
Farmed fish	0	52.308	
Polychaetes	4.905	5.79	18
Macrobenthos detritivorous	1.928	8.325	331
Mussels	0	3.614	
Zooplankton	3.541	6.476	82.9
Phytoplankton	3.593	6.57	82.8
Discards	0	1.231	
Nourishment	0	156.923	
Detritus	356.015	631.73	77.4
Total biomass	408.594	963.36	135.7

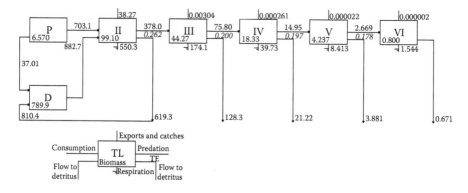

FIGURE 9.2
The complex food web for the first scenario (after the establishment of the marine finfish farm) is represented schematically in the form of the Lindeman spine, where biomasses and flows are aggregated into integer trophic levels (TL). At the first level of the chain, the primary producers (P) and detritus (D) are split for clarity.

development (i.e., nutrient loadings from fish farms) becomes less important, and the ecosystem could be more resilient to changes in external nutrient loading (Prins et al. 1998). This modeling framework confirms the important role that is played by detritus groups, and in particular those related with aquaculture (fish farm nourishment and discarded fish) in this trophic network (Díaz López et al. 2008).

Although the presence of aquaculture added two detritus groups to the Aranci Bay (nourishment and discards), the biomass stored in detritus groups in relation with the total biomass was lower after the start of

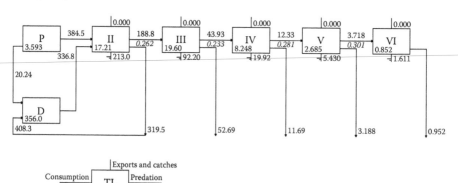

FIGURE 9.3
The complex food web for the second scenario (before the establishment of the marine finfish farm) is represented schematically in the form of the Lindeman spine, where biomasses and flows are aggregated into integer trophic levels (TL). At the first level of the chain, the primary producers (P) and detritus (D) are split for clarity.

TABLE 9.4

Summary of the Cycling Indices Estimated for the Created Scenarios

	Unit	Scenario B	Scenario A	Variance (%)
Throughput cycled (excluding detritus)	t km^{-2} year^{-1}	63.61	107.78	69.43
Predatory cycling index	% of throughput without detritus	6.57	5.42	−17.50
Throughput cycled (including detritus)	t km^{-2} year^{-1}	438.15	800.07	82.60
Finn's cycling index	% of total throughput	24.54	20.93	−14.71
Finn's mean path length	none	4.411	4.008	−9.13

aquaculture activities. This could be explained by the high concentration of species described as buffers to the eutrophication process (i.e., common gray mullets, mussels) feeding around sea cages, which may diminish the amount of organic matter that reaches the sea floor (Porter et al. 1996; Dempster et al. 2004; Lupatsch et al. 2003; Nizzoli et al. 2005; Mazzola and Sara 2001). These results support the notion that the contextual cultivation of species from different trophic levels (e.g., bivalves together with carnivorous fish) would reduce the impact that would emerge from the cultivation of only carnivorous species (Troell et al. 1999). Díaz López et al. (2008) also suggested that the role that top predators (marine mammals and seabirds) could play in the elimination of discarded fish (reducing the organic matter) implies that these species may be considered as buffers to the eutrophication process, reducing the organic matter present in the area.

The capacity of an ecosystem to entrap, withhold, and cycle nutrients increases with system "maturity" (Odum 1969), and this "maturity" has been correlated with Finn's cycling index (FCI) (Christensen 1995). The FCI, estimated through Ecopath for both trophic networks, were relatively high (Table 9.4), indicating a substantial degree of recycling before and after the start of aquaculture activities. Furthermore, the higher FCI estimated for the second scenario (before the establishment of aquaculture) confirms that recycling played an important role in the maintenance of coastal ecosystem stability. In other words, the establishment of aquaculture could reduce the recycling, and the lower the recycling level, the more slowly will the effects of perturbations be eliminated from the system.

9.4 Summary

Long-term growth of the aquaculture industry requires both ecologically sound practices and sustainable resource management. Thus, any decision related to aquaculture and environmental systems will necessarily involve management issues. In the face of the scale of aquaculture impact on coastal ecosystems, scientists have to find ways and tools to predict these effects and account for ecological interactions, especially those of a trophic nature.

The Ecopath mass-balance model provides a rigorous and relatively simple framework that is capable of accounting for the major components and trophic interactions of the ecosystem, and produce results coherent with field studies and ecological theory on ecosystem development properties. The Ecopath mass-balance model can be a valuable tool for understanding ecosystem functioning, and for designing ecosystem-scale adaptive management experiments. This modeling framework could provide new insights into the understanding of how aquaculture influences coastal ecosystems, and hence support the design of policies aimed at implementing ecosystem management principles.

The case study presented in this chapter has shown the appropriateness of the model in describing the modifications induced, at an ecosystem level, by nutrient loading into the area. Increased nutrient loading is the most obvious predicted effect from fish farms, and measures of this effect comprise the main method of regulating and controlling the size of fish farms such that the local environment is not overwhelmed. It must be highlighted that the biomass estimates from the mass balance models could serve as a guide to investigate changes in the environment attributed to the start, recovery, or cessation of farming. In addition, these estimates could provide important additional information complementary to the normal environmental assessment impact studies, and before selecting polyculture as a potential solution to some aspects of eutrophication. Finally, mass-balance models could be a useful tool, from the viewpoint of predicting how increased nutrients will affect the plankton community, particularly if there is a risk of toxic algal blooms in the area.

The accuracy of parameter estimates depends on information available to the groups. Much more refined models than this can be developed if sufficient biological data are available. Bias resulting from estimating the model parameters would have an impact on the model output. The main limitation to using this type of model is the availability of field data and the time and resources needed to obtain this information. In particular, the limited availability of parameter estimates on a seasonal or annual basis for some groups (particularly multistanza groups) reflects a need for future models. Some groups may include hundreds of species, and it is almost impossible to have an accurate parameter estimate for such groups. Data needed to estimate the model parameters (including dietary composition) are often incomplete, and estimates have to be based on studies conducted elsewhere for similar species, or rely on qualitative descriptions. These would certainly affect the model output.

Apart from the above-mentioned factors, the prediction of aquaculture's effects on the ecosystem is likely to be influenced by other factors not explicitly addressed in this model. For example, climatological variability, hydrographic conditions, bottom topography, and geography may also lead to variability in the input of new nutrients that may also ultimately influence the environmental impact of aquaculture.

Using Ecosim and Ecospace routines, in a future step, it could be possible to simulate the consequences of certain management measures, such as changes in aquaculture, on the ecosystem. Nevertheless, further research is required in order to improve input data and to support or refute the results presented in this model. Although there remains a great deal of research and development work yet to be done to improve our understanding of the environmental impacts of fish farming in marine waters, Ecopath models can help us asses these impacts and make reasonable management decisions.

Acknowledgments

The author is very grateful for the constant support that J. Andrea Bernal Shirai contributed throughout this study. Heartfelt thanks are also extended to all BDRI volunteers and interns who generously gave their time to help in the fieldwork.

References

Black, K. D. 2001. *Environmental Impacts of Aquaculture*, p. 212. Sheffield, UK: Sheffield Academic Press.

Beveridge, M. 1996. *Cage Aquaculture*. 2nd ed. Oxford: Blackwell Science Ltd, Fishing News Books.

Carss, D. N. 1990. Concentrations of wild and escaped fishes immediately adjacent to fish farm cages. *Aquaculture* 90 (1): 29–40.

Carss, D. N. 1994. Killing of piscivorous birds at Scottish fish farms, 1984–87. *Biological Conservation* 68 (2): 181–188.

Christensen, V. 1995. Ecosystem maturity—towards quantification. *Ecological Modelling* 77: 3–32.

Christensen, V., and D. Pauly. 1992. *A Guide to the Ecopath Software System* (version 2.1), 72 pp. ICLARM, Manila.

Christensen, V., and C. J. Walters. 2000. Ecopath with Ecosim: Methods, capabilities and limitations. In *Methods for Assessing the Impact of Fisheries on Marine Ecosystems of the North Atlantic. Fisheries Centre Research Reports*, eds. D. Pauly and T. J. Pitcher, 8 (2), 195: 79–105.

Christensen, V., and C. J. Walters. 2004. Ecopath with Ecosim: Methods, capabilities and limitations. *Ecological Modelling* 172: 109–139.

Cole, R. 2002. Impacts of marine farming on wild fish populations. *Final Research Report for Ministry of Fisheries Research Project ENV2000/08*, 51 pp. New Zealand: National Institute of Water and Atmospheric Research.

Cromey, C. J., T. D. Nickell, and K. D. Black. 2002. DEPOMOD—modelling the deposition and biological effects of waste solids from marine cage farms. *Aquaculture* 214: 211–239.

Cruz-Escalona, V. H., F. Arreguín-Sánchez, and M. Zetina-Rejón. 2007. Analysis of the ecosystem structure of Laguna Alvarado, western Gulf of mexico, by means of a mass balance model. *Estuarine Coastal and Shelf Science* 72: 155–167.

Dempster, T., P. Sanchez-Jerez, J. Bayle-Sempere, and M. Kingsford. 2004. Extensive aggregations of wild fish at coastal sea-cage fish farms. *Hydrobiologia* 525: 245–248.

Díaz López, B. 2006. Interactions between Mediterranean bottlenose dolphins (*Tursiops truncatus*) and gillnets off Sardinia, Italy. *ICES Journal of Marine Science* 63 (5): 775–960.

Díaz López, B., and J. A. Bernal Shirai. 2007. Bottlenose dolphin (*Tursiops truncatus*) presence and incidental capture in a marine fish farm on the north-eastern coast of Sardinia (Italy). *Journal of Marine Biology Association UK* 87: 113–117.

Díaz López, B., L. Marini, and F. Polo. 2005. The impact of a fish farm on a bottlenose dolphin population in the Mediterranean Sea. *Thalassas* 21: 53–58.

Díaz López, B., M. Bunke, and J. A. B. Shirai. 2008. Marine aquaculture off Sardinia Island (Italy): Ecosystem effects evaluated through a trophic mass-balance model. *Ecological Modelling* 212: 292–303.

Ervik, A., P. K. Hansen, J. Aure, A. Stigebrandt, P. Johannessen, and T. Jahnsen. 1997. Regulating the local environmental impact of intensive marine fish farming: I. The concept of the MOM system (Modelling–Ongrowing fish farms–Monitoring). *Aquaculture* 158: 85–94.

FAO. 1995. *Code of Conduct for Responsible Fisheries*. Rome, Italy: Food and Agriculture Organization of the United Nations.

FAO. 2007. *The State of World Fisheries and Aquaculture 2006*, Rome, Italy: FAO Fisheries and Aquaculture Department, Food and Agriculture Organization of the United Nations.

Fernandes, T. F., A. Eleftheriou, H. Ackefors, M. Eleftheriou, A. Ervik, A. Sanchez-Mata, T. Scanlon et al. 2002. *The Management of the Environmental Impacts of Aquaculture*, 88 pp. Aberdeen, UK: Scottish Executive.

Findlay, R. H., and L. Watling. 1997. Prediction of benthic impact for salmon net-pens based on the balance of benthic oxygen supply and demand. *Marine Ecology Progress Series* 155: 147–157.

Finn, J. T. 1976. Measures of ecosystem structure and function derived from analysis of flows. *Journal of Theoretical Biology* 56: 363–380.

Hargrave B. T. 2003. Far-field environmental effects of marine finfish aquaculture. A scientific review of the potential environmental effects of aquaculture in aquatic ecosystems. Volume I. *Canadian Technical Report of Fisheries and Aquatic Science* 2450: 1–35.

Hall, P. O. J., O. Holby, S. Kollberg, and M. O. Samuelsson. 1992. Chemical fluxes and mass balances in a marine fish cage farm: IV. Nitrogen. *Marine Ecology Progress Series* 89: 81–91.

Heggberget, T. G., B. O. Johnsen, K. Hindar, B. Jonsson, L. P. Hansen, N. A. Hvidsten, and A. J. Jensen. 1993. Interactions between wild and cultured Atlantic Salmon—a review of the Norwegian experience. *Fisheries Research* 18 (1–2): 123–146.

Hevia, M., H. Rosenthal, and R. J. Gowen. 1996. Modelling benthic deposition under fish cages. *Journal of Applied Ichthyology* 12: 71–74.

Holby, O., and P. O. J. Hall. 1991. Chemical fluxes and mass balances in a marine fish cage farm: II. Phosphorus. *Marine Ecology Progress Series* 70: 263–272.

Jiang, W., and M. T. Gibbs. 2005. Predicting the carrying capacity of bivalve shellfish culture using a steady, linear food web model. *Aquaculture* 244: 171–185.

Klaoudatos, S. D., D. S. Klaoudatos, J. Smith, K. Bogdanos, and E. Papageorgiou. 2006. Assessment of site specific benthic impact of floating cage farming in the eastern Hios island, Eastern Aegean Sea, Greece. *Journal of Experimental Marine Biololy and Ecology* 338: 96–111.

Lemarié, G., J. L. M. Martin, G. Dutto, and C. Garidou. 1998. Nitrogenous and phosphorous waste production in a flow-through land-based farm of European sea bass *Dicentrarchus labrax*. *Aquatic Living Resources* 11: 247–254.

Lupatsch, I., T. Katz, and D. L. Angel. 2003. Assessment of the removal efficiency of fish farm effluents by grey mullets: A nutritional approach. *Aquaculture Research* 34: 1367–1377.

Machias, A., I. Karakassis, M. Giannoulaki, K. N. Papadopoulou, C. J. Smith. and S. Somarakis. 2005. Response of demersal fish communities to the presence of fish farms. *Marine Ecology Progress Series* 288: 241–250.

Mazzola, A., and G. Sara. 2001. The effect of fish farming organic waste on food availability for bivalve molluscs (Gaeta Gulf, Central Tyrrhenian, MED): Stable carbon isotopic analysis. *Aquaculture* 92: 361–379.

Müller, F. 1997. State-of-the-art in ecosystem theory. *Ecological Modelling* 100: 135–161.

Niquil, N., J. E. Arias-González, B. Delesalle, and R. E. Ulanowicz. 1999. Characterization of the planktonic food web of Takapoto Atoll lagoon, using network analysis. *Oecologia* 118: 232–241.

Nizzoli, D., D. T. Welsh, M. Bartoli, and P. Viaroli. 2005. Impacts of mussel (*Mytilus galloprovincialis*) farming on oxygen consumption and nutrient recycling in a eutrophic coastal lagoon. *Hydrobiologia* 550: 183–198.

Odum, E. P. 1969. The strategy of ecosystem development. *Science* 104: 262–270.

Odum, E. P. 1971. *Fundamentals of Ecology*. Philadelphia, PA: W.B. Saunders.

Pauly, D., M. L. Soriano-Bartz, and M. L. D. Palomares. 1993. Improved construction, parametrization and interpretation of steady-state ecosystem models. In *Trophic Models of Aquatic Ecosystems*, eds. V. Christensen and D. Pauly, 1–13. Manila: ICLARM.

Pauly, D., V. Christensen, and C. Walters. 2000. Ecopath, Ecosim and Ecospace as tools for evaluating ecosystem impact of fisheries. *ICES Journal of Marine Science* 57: 697–706.

Pillay, T. V. R. 1992. *Aquaculture and the Environment*, 185 pp. New York, NY: John Wiley & Sons.

Pillay, T. V. R. 2004. *Aquaculture and the Environment*, 2nd ed. Oxford, UK: Blackwell Publishing.

Pinnegar, J. K., and N. V. C. Polunin. 2004. Predicting indirect effects of fishing in the Mediterranean rocky littoral communities using a dynamic simulation model. *Ecological Modelling* 172 (2–4): 249–268.

Porter, C. B., P. Krost, H. Gordin, and D. L. Angel. 1996. Preliminary assessment of grey mullet (*Mugil cephalus*) as a forager of organically enriched sediments below marine fish farms. *Israel Journal of Aquaculture-Bamidgeh* 48: 47–55.

Prins, T. C., A. C. Smaal, and R. F. Dame. 1998. A review of the feedbacks between bivalve grazing and ecosystem processes. *Aquatic Ecology* 31: 349–359.

Read, P. A., T. F. Fernandes, and K. L. Miller. 2001. The derivation of scientific guidelines for best environmental practice for the monitoring and regulation of marine aquaculture in Europe. *Journal of Applied Ichthyology* 17 (4): 146–152.

Read, P., and T. Fernandes. 2003. Management of environmental impacts of marine aquaculture in Europe. *Aquaculture* 226: 139–163.

Silvert, W. 1992. Assessing environmental impacts of finfish aquaculture in marine waters. *Aquaculture* 107: 67–79.

Silvert, W., and J. W. Sowles. 1996. Modelling environmental impacts of marine finfish aquaculture. *Journal of Applied Ichthyology* 12: 75–81.

Stigebrandt, A., J. Aure, A. Ervik, and P. K. Hansen. 2004. Regulating the local environmental impact of intensive marine fish farming: III. A model for estimation of the holding capacity in the modelling ongrowing fish farm monitoring system. *Aquaculture* 234: 239–261.

Troell, M., P. Rönnbäck, C. Halling, N. Kautsky, and A. Buschmann. 1999. Ecological engineering in aquaculture: Use of seaweeds for removing nutrients from intensive mariculture. *Journal of Applied Phycology* 11. 89–97.

Ulanowicz, R. E. 1993. Inventing the Ecoscope. In *Trophic Models of Aquatic Ecosystems*, eds. V. Christensen and D. Pauly, ix–x. Manila: ICLARM.

Ulanowicz, R. E. 1995. Ecosystem trophic foundations: *Lyndeman exonerata*. In *Complex Ecology: The Part–Whole Relation in Ecosystems*, eds. B. C. Patten and S. E. Jorgensen, 549–550. Englewood Cliffs, NJ: Prentice Hall.

Watson-Capps, J. J., and J. Mann. 2005. The effects of aquaculture on bottlenose dolphin (*Tursiops* sp.) ranging in Shark-Bay, Western Australia. *Biological Conservation* 124: 519–526.

Wu, R. S. S. 1995. The environmental impact of marine fish culture: Towards a sustainable future. *Marine Pollution Bulletin* 31 (4–12): 159–166.

Youngson, A. F., A. Dosdat, M. Saroglia, and W. C. Jordan. 2001. Genetic interactions between marine finfish species in European aquaculture and wild conspecies. *Journal of Applied Ichthyology* 17 (4): 153–162.

10

Models of Wetlands

Sven Erik Jørgensen

CONTENTS

10.1 Introduction 257
10.2 Overview of the Classical Wetland Models 258
10.3 Recent Development in Wetland Modeling 260
    10.3.1 Increase in Modeling Experience 260
    10.3.2 Recovery or Construction of Wetlands for the Control of
          Nonpoint Sources 260
    10.3.3 Contribution of Wetlands to Carbon Balance 262
    10.3.4 Influence of Expected Climate Changes on Wetlands and
          Wetland Processes Has Been Modeled 263
10.4 An Illustrative Wetland Model 264
    10.4.1 The Basic Design 264
    10.4.2 The State Variables 266
    10.4.3 Forcing Functions 268
    10.4.4 Process Equations 269
    10.4.5 Parameters 271
    10.4.6 Differential Equations 272
    10.4.7 Model Results 274
    10.4.8 A Tanzanian Case Study 274
References 275

10.1 Introduction

Wetlands are used increasingly to treat wastewater in developing countries and agricultural drainage water in developed countries. Wetlands, which are also considered important components in the landscape, contribute to the biodiversity of nature. Wetland restorations and conservations are therefore important environmental management issues. With this development in mind, it is not surprising that many wetland models have been developed to improve our understanding of wetland processes and to support environmental management.

We distinguish between surface and subsurface wetlands, and as we both construct wetlands and apply natural wetlands, the models may be classified into four groups:

1. Models of constructed subsurface wetlands
2. Models of natural subsurface wetlands
3. Models of constructed surface wetlands
4. Models of natural surface wetlands

Models have been developed for all types of wetlands: bogs, wet meadows, marshes, swamps, fens, shallow lakes, ponds, lagoons, and forested wetlands. The basic processes in the models for different wetland types are, however, the same or almost the same. The models usually include a quantitative description of decomposition of organic matter, growth of plants, determined by the available nutrients, the solar radiation and the temperature, oxidation of ammonium to nitrate, and denitrification. There is a wealth of experience supporting a proper description of these processes in the modeling literature.

In 1988, the first overview of wetland models was published in *Wetland Modelling*, which was edited by W. J. Mitsch, M. Straskraba, and S. E. Jørgensen (Elsevier, Amsterdam). The book reviewed and presented a number of models of different types of wetlands. Since then, hundreds of wetland models have been published. The next section presents an overview of the classical wetland models, and the third section reviews the characteristic features of the recently developed wetland models.

The last section illustrates wetland models by giving the details of a wetland model that was developed as a result of a Danida project in Tanzania, and further developed by the United Nations Environment Programme (UNEP) and Fleming College, Canada. The model can be used to construct subsurface wetlands or surface wetlands with a laminar flow. It has been used to construct several wetlands in Tanzania and in other African countries, and has been developed for use in the design of wastewater treatment in arctic Canada by the use of arctic tundra wetlands. The model, which is downloadable from UNEP's homepage, illustrates very clearly how the most important processes are considered in a wetland model.

10.2 Overview of the Classical Wetland Models

The classical models cover the models developed from 1975 to 2000. Table 10.1 gives an overview of models that represent the typical spectrum of models from this period, along with the corresponding reference(s). All models

TABLE 10.1

Overview of Classical Wetland Models

Wetland Type	Model Characteristics[a]	Reference
Bog	A, spatially distributed model	Alexandrov (1988)
Bog	B, carbon and nitrogen cycles	Logofet and Alexandro (1988)
Marsh	E, spatial	Costanza et al. (1988)
Forested wetland	C, productivity	Mitsch (1988)
Swamp, meadow	B, nutrient retention	Jørgensen et al. (1988)
Shallow lake	B, water quality, eutrophication	Straskraba and Mauersberger (1988)
Shallow lake	B, eutrophication	Jørgensen (1988)
Swamp	B + C, ecological networks	Patten (1988)
General	B, bacterial competition	James (1993)
General	B, nutrient retention	Mitsch and Reeder (1991)
General	B + C, wastewater treatment	Silow and Stom (1992)
Fishpond	B + D, fish production	Hagiwara and Mitsch (1994)
Lagoon	C + D, spatial distribution	Reyes (1994)
Riparian zone	A, water balance	Kluge et al. (1994)
General	B + water level	Sklar (1994)
Fen	A + B + C	Bekker (1994)
Swamp	B + C + D	Whippler and Patten (1994)
Fish pond	D, fish production, aqua culture	Jorgensen (1976)
General	B, 27 state variables, impact assessment	Peijl et al. (1994)
Swamp, meadow	B, nitrogen cyling and removal	Dørge (1991) and Jørgensen and Bendoricchio (2001)

[a] For an explanation of A, B, C, D, and E, see discussion in text.

included in this table can also be used today for environmental management including wetland conservation and restoration.

The type of wetland used for the development of the model is indicated. The model is, of course, best applied for other wetlands of the same type, but if the included processes are consistent with a considered wetland case study, there is a high probability that the model can be applied for the case study without or with minor modifications.

The processes that are included in these models can be classified into five groups:

A. Hydrodynamic processes describe the sometimes very complex water flow through the wetlands. Models of this type are often spatial models.

B. Nutrient cycling processes include decomposition of organic matter, nitrification, denitrification, decomposition of organic nitrogen and phosphorus, release and uptake of carbon dioxide, and production of methane. The models describing the cycling of nutrients can often be used to determine the water quality.

C. Growth of trees and plants, determined by solar radiation, temperature, available nutrients, and withering. Harvest of plants or trees may be included.

D. Fishery, fish growth, and fish production.

E. Land loss, succession, and biodiversity. Spatial distribution is often required for these processes.

10.3 Recent Development in Wetland Modeling

Since the year 2000, wetland modeling has developed in four directions.

10.3.1 Increase in Modeling Experience

Throughout the years, the experience we gained in modeling has accumulated to a substantial degree, which implies that today we have slightly better process formulations and increased experience in the general development of wetland models. This development is illustrated by the wetland models developed for UNEP. In the next section, one of these models, which has been applied for the design and construction of wetlands for wastewater treatment, will be discussed. It should be noted, however, that UNEP has also developed models for restoration of wetlands and for management of shallow lakes, one of which is the so-called Pamolare 2 model, which is downloadable from UNEP's homepage. It is a lake eutrophication model that considers the use of submerged vegetation and phytoplankton feeding carps for the restoration.

10.3.2 Recovery or Construction of Wetlands for the Control of Nonpoint Sources

A core question in wetland management is, where in the landscape to recover or construct wetlands for the control of nonpoint sources? This problem was solved by an EU project (Development of Ecotechnological Methods by Use of Wetlands as Buffer Zones to Protect Lakes and Rivers). The idea was to apply a point system accounting for the properties of different sites in the landscape to assist in the selection of the most appropriate areas:

- Altitude (as it would be beneficial of course to use the gravity for transportation of water to and from the recovered or constructed wetland)
- Hydraulic conductivity (as it may be relatively high for wetlands used for treatment of drainage water)

- The price of land (as it would be beneficial to use land with a low productivity potential)
- Accessibility
- Other properties of importance for specific cases

Using a model made it possible to estimate the area needed to treat the known amount of drainage water to the required level. The point system was used to select the useful wetland areas in the landscape. Thus, more detailed knowledge about the properties of the applied wetlands would be available. Consequently, it would be possible, by use of models, to test more accurately whether the selected area would meet the required standards for the treatment of drainage water. If the area was deemed insufficient, a larger area was selected by use of the point system, and the new area was afterwards tested by use of the models. The needed and the best suitable areas for solving the defined nonpoint pollution problem were thereby found step by step through an iterative method. The iterative method is described in Figure 10.1, and Figures 10.2–10.4 show how different simple models are applied for determining the result of the treatment of drainage water. Other more complex models may of course also be applied, but simple models have been used in this context to keep the measuring program and the model development at relatively low costs. Moreover, additional models—for instance, models of phosphorus removal and heavy metal removal—may also be applied.

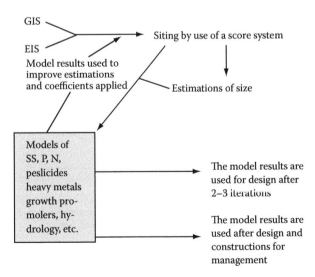

FIGURE 10.1
The iterative method used to find the needed areas and most suitable land for construction or recovery of wetlands for treatment of agricultural drainage water.

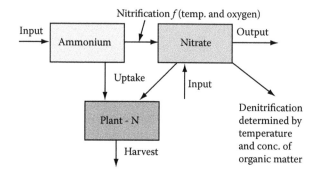

FIGURE 10.2
The nitrogen model applied to assess the nitrogen removal by the applied wetlands.

10.3.3 Contribution of Wetlands to Carbon Balance

The extent of wetlands' contribution to the global carbon balance has been widely discussed. Some researchers claim that methane production in wetlands is a major source of the greenhouse effect, whereas others underline the importance of wetlands for long-term storage of carbon and their ability to sequestrate carbon. When CO_2 sequestration and CH_4 emissions are measured and compared in the same wetland, the wetland is often determined to be a net source of radiative forcing on climate, based on the global warming potential (GWP) of about 24 for methane relative to carbon dioxide. A logical extension of this determination is that new wetlands should not be created or restored, despite the fact that ecosystem services such as flood control, water quality improvement, and habitat restoration would be provided. Methane, however, is oxidized in the atmosphere, where it has a half-life time of only about 7 years. It is therefore necessary to develop a model to consider the dynamics of the atmospheric oxidation of methane, and determine whether the positive contribution by methane emission or

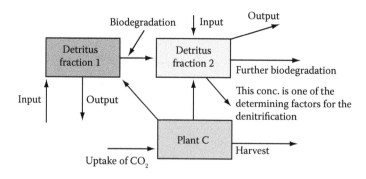

FIGURE 10.3
The organic matter model applied to assess the amount of BOD_5 removed by the wetland.

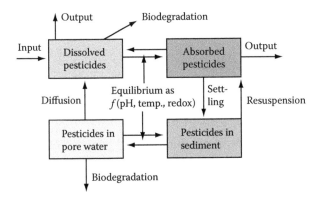

FIGURE 10.4
The pesticide model applied to assess whether the removal of pesticides by constructed or recovered wetland is sufficient.

the negative contribution by carbon sequestration in the long term is most dominant. Mitsch et al. (submitted) concluded on the basis of such a model that the world's wetlands, despite being only about 6% of the terrestrial landscape, are currently net sinks for a significant portion of the carbon released by fossil fuel combustion. They find that the 18 modeled wetlands give a net sink of carbon dioxide equivalent from 8 kg to several hundred kg per m² over a period of 100 years, in average about 50 kg of carbon dioxide equivalents per m² over a period of 100 years.

10.3.4 Influence of Expected Climate Changes on Wetlands and Wetland Processes Has Been Modeled

Holsten et al. (2009) have modeled the soil moisture dynamics in Germany (Brandenburg), and they conclude that soil moisture in nature conservation areas will decrease. Their model gives the spatial distribution of the soil moisture decline. The observations on soil moisture from 1951 to 2008 have been used to calibrate and validate the model successfully. Gaiser et al. (2009) have developed a model of carbon sequestration for various soil types. It can be used to predict the influence of temperature on carbon sequestration and thereby the future carbon sequestration of wetlands with different soil types. The growth of plants and trees in wetlands will also be influenced by changes in climate. Tatarinov and Cienciala (2009) have developed a model of the effect of climate changes on the growth of a number of European tree species. Their model can be used directly to cover the influence of climate changes on forested wetlands; however, their approach, which is simple to apply, can also be used for the influence of climate changes on plants.

There is a clear trend in the modeling literature, particularly in *Ecological Modelling*, for an increasing number of papers focusing on the effect of climate changes both on entire ecosystems and on ecological processes. This development will facilitate the possibility of modeling the effect of climate changes on specific wetlands or types of wetlands.

10.4 An Illustrative Wetland Model

The presented model was developed for the purpose of designing a subsurface wetland, based on defined removal efficiencies of organic matter (expressed in terms of BOD_5), nitrate, ammonium, organic nitrogen, and phosphorus. Thus, it is necessary to know the following: (1) the water flow; (2) the concentrations of the above-mentioned constituents in the water; and (3) the required removal efficiencies for these constituents (i.e., their concentrations in the treated water). The model contains several illustrative and fundamental process descriptions and has a very clear conceptual diagram and can furthermore be downloaded from internet. The model has therefore been used in textbooks; see Jørgensen (2009) and Jørgensen and Fath (2011).

The subsurface wetland being modeled is either a constructed wetland in which the "soil" is gravel, thereby allowing a good water flow through the wetland or a natural wetland with a lower hydraulic conductivity. The core design parameter is the volume of the wetland (denoted V). The model can also be applied to design a surface wetland, provided the flow through the wetland is laminar. The model is downloadable from UNEP's homepage. It has been developed by Danida, UNEP, and Fleming College in Canada. The presentation of the model is based on modification of the information included in the software Subwet 2.0.

10.4.1 The Basic Design

It is possible to use the model to find the volume and area, and the flow pattern that utilizes the area optimal. However, we have to distinguish between a natural and a constructed subsurface wetland. The design equations are included in the program and are also presented below. The model has a warm and a cold climate version. The difference between the two versions lies in the default parameters. The warm climate parameters have been established on the basis of nine wetlands in Tanzania, and the cold climate parameters have been determined based on observations from three wetlands in arctic Canada.

Constructed wetlands. For constructed wetlands, the hydraulic conductivity is high, implying that rainwater is insignificant compared with the water flow. For natural wetlands, however, it is necessary to select a precipitation factor to indicate how many times the water flow is increased due to rainwater (PF). For the natural wetlands, it is important to select the right value of PF, because the hydraulic conductivity is often the limiting factor for the volume/24 hours capacity.

It is possible, according to the model, to design the *constructed wetland*:

Area (AA) in m² = the width times the length of the selected wetland

Volume (VO) in m³ = AA times the selected depth of the wetland

Hydraulic loading in m³/(24 h × m²): HL = SF/(AA), where SF is the flow rate for instance in the unit m³/24 h.

The recommended horizontal flow. HF, in m/24 h is found as HF = 25 – AP*8, where AP is the % particulate matter in the water to be treated.

It is now possible to find the flow width (FW) from recommended flow (RF) = PF*SF (PF = 1.0 may have been selected for a constructed wetland, which implies of course that RF= SF).

$$FW = RF/(HF * DE)$$

If WI > or = 3 × FW, FL = WI and the number of paths, NP = LE/FW is rounded up to a whole number. FW is adjusted to LE/NP.

If 2 < WI < 3*FW, FL = LE and the number of paths, NP, = 2 and FW = WI/2.

If WI < 2 FL = LE and the number of paths, NP, = 1 and FW = WI.

The user of SubWet 2.0 is, of course, able to apply a more practical NP value than the one recommended by the model if the NP is unrealistically high.

Natural wetlands. The design of a natural wetland is based on hydraulic conductivity, which may limit the capacity of a natural wetland significantly.

Area (AA) in m² = WI × LE

Volume in m³ = WI × LE × DE

Hydraulic loading in m³/(24 h × m²): HL = SF/AA

The recommended horizontal flow: HF, in m/24 h = HC

Recommended flow should also consider the precipitation, because for natural wetland the hydraulic conductivity limits the possible flow rates. RF is therefore considered PF × SF = RF.

It is now possible to calculate the flow width as: FW = RF/(HC × S × DE).

NP is recommended for natural wetlands to be 1.0.

If WI > FW, FW is made = WI and FL = LE.

If WI < FW, FW is made equal to LE and FL = WI.

The value of the forcing function RF, the flow rate, is found above. The concentrations of the water to be treated BOD5, nitrate-N, ammonium-N, organic N, phosphorus, particulate organic matter (POM), particulate organic nitrogen (PON), and particulate organic phosphorus (POP) are all forcing functions. The concentrations after dilution with rain water is easily found from division of these concentrations by PF.

10.4.2 The State Variables

The conceptual diagram of the model is presented in Figures 10.5–10.7. Figure 10.5 illustrates the organic matter submodel, whereas Figure 10.6 illustrates the nitrogen submodel, with three groups of nitrogen compounds (organic nitrogen, ammonium, and nitrate). Figure 10.7 shows the processes and state variables of phosphorus. The model state variables include: BOD_5, nitrate (NIT), ammonium (AMM), total phosphorus (TP), and organic nitrogen (ORN) in five successive boxes, denoted A to E, a total of 25 state variables, all measured as m/L or g/m^3, as follows:

BOD_5-A, BOD_5-B, BOD_5-C, BOD_5-D. BOD_5-E (mg O_2/L)

NIT-A, NIT-B, NIT-C, NIT-D, NIT-E (mg N/L)

AMM-A, AMM-B, AMM-C, AMM-D, AMM-E (mg N/L)

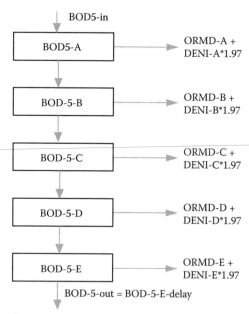

ORMD = decomposition of organic matter by oxidation
DENI* 1.97 = decomposition of organic matter by denitrification

FIGURE 10.5
The BOD_5 submodel.

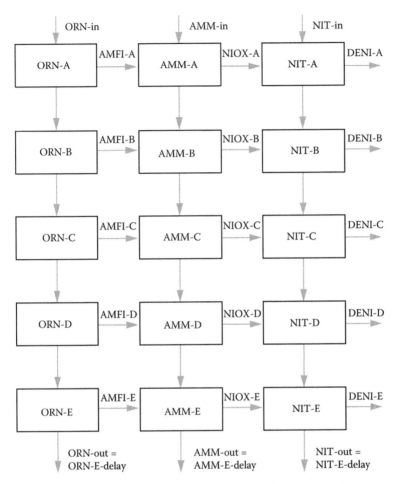

AMFI = oxidation of organic N to ammonium. NO_x = nitrification (ammonium-> nitrate. DENI = denitrification (nitrate -> dinitrogen<)

FIGURE 10.6
The nitrogen submodel, illustrating the three nitrogen compounds (organic-N, ammonium, and nitrate).

TPO-A, TPO-B, TPO-C, TPO-D, TPO-E (mg P/l)

ORN-A, ORN-B, ORN-C, ORN-D, ORN-E (mg N/L)

As noted below, the submodel variables are expressed by three letters (e.g., NIT for nitrate), followed by IN, OUT or A, B, C, D, or E, with the parameters using two letters.

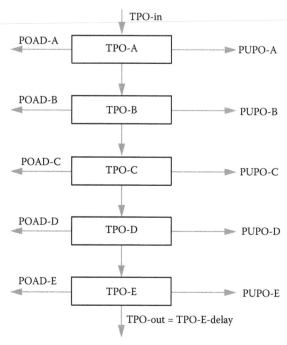

TPO = total phosphorus PUPO = plant uptake of phopshorus
POAD = adsorption of phosphorus to the gravel.

FIGURE 10.7
The phosphorus submodel.

10.4.3 Forcing Functions

The model has the following forcing functions, which the user must specify for a given model run:

Volume of wetland (m³; possible range 10–10,000,000)

Flow of water (QIN = RF, expressed as m³/24 h; possible range 1–1,000,000)

Porosity (as fraction of POR; no unit; range –1; default value 0.46)

Input concentration of BOD₅ (BOD-IN; mg O₂/L; range 0–2000)

Notice that it is the cBOD₅ which does not include nitrification. cBOD₅ is found by adding a nitrification inhibitor to the sample before the analysis.

Input concentration of ammonium (AMM-IN; mg N/L; range 0–100).

Input concentration of nitrate (NIT-IN; mg N/L; range 0–100).

Input concentration of total phosphorus (TPO-IN; mg P/L; range 0–200).

Concentration of organic nitrogen: ORN-IN mg N/L (range 0–200).

Fraction of BOD$_5$ as suspended matter (POM %; no unit; range 0–100).

POM% = (mg suspended matter/L) × (100 − ignition residue in%) × 139/cBOD5 PON% is found by replacing in this equation 139 by 7, as the average N content of organic matter is 7% and POP% is found by replacing 139 by 1, as 1% averagely of organic matter is P. It is of course also possible to determine PON% as the amount of suspended matter that is N in mg/L × 100/organic N in mg/L and POP% as the amount of suspended matter that is P in mg/L × 100/P total in mg/L.

Fraction of organic-N matter as suspended matter (PON%; no unit, range 0–100%).

Fraction of phosphorus as suspended matter (POP%; no unit, 0–100%).

Average oxygen concentration in Box A (AOX; mg/L; range 0–20).

Average oxygen concentration in Box B (BOX; mg/L; range 0–20).

Average oxygen concentration in Box C (COX; mg/L; range 0–20).

Average oxygen concentration in Box D (DOX; mg/L; range 0–20).

Average oxygen concentration in Box E (EOX; mg/L; range 0–20).

Default value for AOX, BOX, COX, DOX, EOX = 0.4 mg/L.

Average temperature (TEMP; as function of time; daily average temperature is listed for the number of days to be simulated with the model).

The length of model simulations must be indicated as number of days. The following forcing functions are calculated:

Water volume = VOL*POR

Retention time, RTT (= VOL*POR/QIN; 24 h)

Retention time per box, RTB (= RTT/5; 24 h)

Box volume, BOV (= VOL*POR/5)

Water volume = 5*BOV

10.4.4 Process Equations

A continuous transfer takes place from one state variable to another in the model simulations. This section identifies the processes that take place in the subsurface wetland. It is noted that the same processes take place in each box, although the concentrations are different in each box. Thus, the equations are repeated in the model program, with an indication of the concentrations in the five different model boxes: A, B, C, D, and E. All processes are expressed by four letters, followed by A, B, C, D, or E, corresponding to the five boxes. It is reiterated that the model expressions are the same for each box, although the applied concentrations of the modeled materials will differ for each box. Exponent is expressed by the notation (^).

The following equations are repeated in each box, with an indication of the letters of the box (see also Figures 10.1–10.3):

Ammonification = AMFI = ORN*AC* TA^(TEMP-20).

Nitrification = NIOX = AMM*NC*INOX*TN ^(TEMP-20)/(AMM + MA).

Oxidation of BOD$_5$ = ORMD = BOD$_5$*OC*INOO*TO ^(TEMP-20).

Denitrification = DENI = NIT*DC*TD ^(TEMP-20)/(NIT + MN).

INOX-A = AOX/(AOX + KO), and so on for boxes B, C, D, and E, using the notations BOX, COX, DOX, and EOX; however, KO is the same parameter for all five boxes.

INOO-A = AOX/(AOX + OO), and so on for boxes B, C, D, and E, using the notations BOX, COX, DOX, and EOX; however, OO is the same parameter for all five boxes.

Plant uptake of ammonium = PUAM = AMM*PA.

Plant uptake of nitrate = PUNI = NIT*PN.

Plant uptake of phosphorus = PUPO-A = TPO-A*PP*(1-POP) for box A, while PUPO-B = TPO-B*PP; PUPO-C = TPO-C*PP; PUPO-D = TPO-D*PP; and PUPO-E = TPO-E*PP (note that the multiplication by (1 − POP) only applied to box A).

Adsorption of phosphorus = POAD-A = TPO-A*(1-POP)*(POR) − AF* (1-POR), if POAD > 0; otherwise POAD = 0 for box A, whereas the following equation is applied for the other boxes: POAD = TPO* POR − AF*(1-POR).

The model also calculates delay values (i.e., the concentrations of the five constituents in the five boxes during one box-retention time (RTB) earlier. For example,

$$\text{AMM-A-delay} = \text{AMM-A at time } t - \text{RTB, when } t > \text{RTB;}$$
$$\text{if } t < \text{RTB, AMM-A-delay is 0.}$$

These equations are repeated for all five constituents in all five boxes, and the delay concentrations are indicated with "-delay."

It is known from particular population dynamic models that small delay values may cause bifurcations with corresponding fluctuations. To avoid these (mathematical) fluctuations, RTB is set equal to 0.7 if it is less than 0.7.

Furthermore, the simulated results are used to determine the removal efficiencies. They are found as function of time, as follows:

Efficiency of BOD$_5$ removal (%) = 100*(BOD5-in – BOD5-out)/BOD5-in

Efficiency of Nitrate removal (%) = 100*(NIT-in-NIT-out)/NIT-in

Efficiency of Ammonium removal (%) = 100*(AMM-in- AMM-out)/
AMM-in

Efficiency of Organic-N removal (%) = 100*(ORN-in – ORN-out)/ORN-in

Efficiency of Nitrogen removal (%) = 100* ((NIT-in + AMM-in + ORN-in) –
(NIT-out + AMM-out + ORN-out))/(NIT-in + AMM-in + ORN-in)

Efficiency of Phosphorus removal (%) = 100*(TPO-in – TPO-out)/TPO-in

10.4.5 Parameters

The following model parameter ranges and default values are for *warm climate conditions*:

AC = 0.05–2.0 [default value 0.5 (1/24 h)]

NC = 0.1–2.5 [default value 0.8 (1/24 h)]

OC = 0.05–2.0 [default value 0.5 (1/24 h)]

DC = 0.00–5 [default value 2.2 (1/24 h)]

TA = 1.02–1.06 [default value 1.04 (no unit)]

TN = 1.02–1.07 [default value 1.047 (no unit)]

TO = 1.02–1.06 [default value 1.04 (no unit)]

TD = 1.05–1.12 [default value 1.09 (no unit)]

KO = 0.1–2 [default value 1.3 (mg/L)]

OO = 0.1–2 [default value 1.3 (mg/L)]

MA = 0.05–2 [default value 1 (mg/L)]

MN = 0.01–1 [default value 0.1 (mg/L)]

PA = 0.00–1 [default value 0.01 (1/24 h)]

PN = 0.00–1 [default value 0.01 (1/24 h)]

PP = 0.00–1 [default value 0.003 (1/24 h)]

AF = 0–100 [default value 1.0]

The following model parameter ranges and default values are for *cold climate conditions*:

AC = 0.05–2.0 [default value 0.9 (1/24 h)]

NC = 0.1–2.5 [default value 0.9 (1/24 h)]

OC = 0.05–2.0 [default value 0.2 (1/24 h)]

DC = 0.00–5 [default value 3.5 (1/24 h)]

TA = 1.02–1.06 [default value 1.05 (no unit)]

TN = 1.02–1.09 [default value 1.07 (no unit)]

TO = 1.02–1.06 [default value 1.04 (no unit)]

TD = 1.05–1.12 [default value 1.07 (no unit)]

KO = 0.1–2 [default value 0.01 (mg/L)]

OO = 0.1–2 [default value 0.05 (mg/L)]

MA = 0.05–2 [default value 0.1 (mg/L)]

MN = 0.01–1 [default value 0.1 (mg/L)]

PA = 0.00–1 [default value 0.01 (1/24 h)]

PN = 0.00–1 [default value 0.001 (1/24 h)]

PP = 0.00–1 [default value 0.001 (1/24 h)]

AF = 0–100 [default value 0.36]

10.4.6 Differential Equations

The 25 differential equations in the model are as follows:

BOV*d BOD-5-A/dt = QIN*BOD-IN – QIN*(1- POM)*BOD-A-delay – BOV*ORMD-A-DENi-A*1.97;

BOV*d BOD-5-B/dt = QIN*(1-POM)*BOD-A-delay – BOV*ORMD-B – QIN*BOD-B-delay – DENI-B*1.97;

BOV*d BOD-5-C/dt = QIN*BOD-B-delay – BOV*ORMD-C– QIN*BOD-C-delay – DENI-C*1.97;

BOV*d BOD-5-D/dt = QIN*BOD-C-delay – BOV*ORMD-D– QIN*BOD-D-delay – DENI-D*1.97;

BOV*d BOD-5-E/dt = QIN*BOD-D-delay – BOV*ORMD-E – QIN*BOD-E-delay – DENI-E*1.97

(QIN*BOD-E-delay indicates BOD$_5$-OUT, which is eventually shown on a graph, together with measured values of the BOD$_5$-OUT, whereas BOD$_5$-A, B, C, D, and E are shown in a table as function of time);

BOV*dNIT-A/dt = QIN*NIT-IN – QIN*NIT-A-delay - BOV*DENI–A + BOV*NIOX-A – BOV*PUNI-A;

BOV*dNIT-B/dt = QIN*NIT-A-delay – BOV*DENI-B + BOV*NIOX-B – BOV*PUNI-B – QIN*NIT-B-delay;

BOV*dNIT-C/dt = QIN*NIT-B-delay – BOV*DENI-C + BOV*NIOX-C – BOV*PUNI-C – QIN*NIT-C-delay;

BOV*dNIT-D/dt = QIN*NIT-C-delay − BOV*DENI-D + BOV*NIOX-D −
BOV*PUNI-D − QIN*NIT-D-delay;

BOV*dNIT-E/dt = QIN*NIT-D-delay − BOV*DENI-E + BOV*NIOX-E −
BOV*PUNI-E − QIN*NIT-E-delay

(QIN*NIT-E-delay indicates NIT-OUT, which is eventually shown on
a graph, together with measured values of NIT-OUT, whereas
NIT-A, NIT-B, NIT-C, NIT-D, and NIT-E are all shown in a table
as function of time);

BOV*dAMM-A/dt = QIN*AMM-IN − QIN*AMM-A-delay − BOV*NIOX −
A + BOV*AMFI-A − BOV*PUAM-A;

BOV*dAMM-B/dt=QIN*AMM-A-delay+BOV*AMFI-B−BOV*NIOX-B−
BOV*PUAM-B − QIN*AMM-B-delay;

BOV*dAMM-C/dt=QIN*AMM-B-delay+BOV*AMFI-C−BOV*NIOX-C−
BOV*PUAM-C − QIN*AMM-C-delay;

BOV*dAMM-D/dt=QIN*AMM-C-delay+BOV*AMFI-D−BOV*NIOX-D−
BOV*PUAM-D − QIN*AMM-D-delay;

BOV*dAMM-E/dt = QIN*NIT-D-delay + BOV*AMFI-E − BOV*NIOX-E
− BOV*PUNI-E − QIN*AMM-E-delay

(QIN*AMM-E-delay indicates AMM-OUT, which is eventually
shown on a graph, together with measured values of AMM-OUT,
whereas AMM-A, AMM-B, AMM-C, AMM-D, and AMM-E are
all shown in a table as function of time);

BOV*dORN-A/dt = QIN*ORN-IN − QIN*(1-PON)*ORN-A-delay−
OV*AMFI-A;

BOV*dORN-B/dt = QIN*ORN-A-delay − BOV*AMFI-B − QIN*ORN-B-
delay;

BOV*dORN-C/dt = QIN*ORN-B-delay − BOV*AMFI-C − QIN*ORN-C-
delay;

BOV*dORN-D/dt = QIN*ORN-C-delay − BOV*AMFI-D − QIN*ORN-D-
delay;

BOV*dORN-E/dt = QIN*ORN-D-delay − BOV*AMFI-E − QIN*ORN-E-
delay

(IN*ORN-E-delay indicates ORN-OUT, which is eventually shown on
a graph, together with measured values of ORN-OUT, whereas
ORN-A, ORN-B, ORN-C, ORN-D, and ORN-E are all shown in a
table as function of time);

BOV*dTPO-A/dt = QIN-TPO-IN − QIN*(1-POP)*TPO-A-delay−
BOV*PUPO-A-BOV*POAD-A;

BOV*dTPO-B/dt = QIN*(1-POP)*TPO-A-delay − BOV*PUPO-B
−BOV*POAD-B − QIN*TPO-B-delay;

$BOV*dTPO\text{-}C/dt = QIN*TPO\text{-}B\text{-}delay - BOV*PUPO\text{-}C - BOV*POAD\text{-}$
$C - QIN*TPO\text{-}C\text{-}delay;$

$BOV*dTPO\text{-}D/dt = QIN*TPO\text{-}C\text{-}delay - BOV*PUPO\text{-}D - BOV*POAD\text{-}$
$D - QIN*TPO\text{-}D\text{-}delay;$

$BOV*dTPO\text{-}E/dt = QIN*TPO\text{-}D\text{-}delay - BOV*PUPO\text{-}E - BOV*POAD\text{-}$
$E - QIN*TPO\text{-}E\text{-}delay$

> (QIN*TPO-E-delay indicates TPO-OUT, which is eventually shown on a graph, together with measured values of TPO-OUT, whereas TPO-A, TPO-B, TPO-C, TPO-D, and TPO-E are all shown in a table as function of time).

10.4.7 Model Results

As mentioned above, the simulated values of BOD$_5$-out, nitrate-out (NIT-out), Ammonium-out (AMM-out), total phosphorus-out (TPO-out), and organic nitrogen-out (ORN-out) are in the software Subwet 2.0 shown in the form of both tables and as graphs. If the measured values are available, they are shown on the same graphs, to allow a direct comparison.

10.4.8 A Tanzanian Case Study

The model was applied to design a wetland for the teachers' college in Iringa, Tanzania. The wetland should treat the wastewater from 2500 students, totalling 50 m^3/24 h. Based on the design before the construction of the wetland, it was found that 625 m^2 wetland was needed to obtain a BOD$_5$ reduction from about 125 mg/L to between 25 and 30 mg/L, and a reduction of total nitrogen content from about 42 to about 10 mg/L. The recommended horizontal flow rate was found to be 9 m/24 h with 2% suspended matter in the untreated wastewater. It was furthermore found that a proper utilization of the wetland requires that four pathways be used. As the area foreseen for the wetland was 25 × 25 m, this means that the flow width should be 6.25 m. A good utilization of the wetland area requires that four pathways be used. As the area foreseen for the wetland was 25 × 25 m, this means that the flow width should be 6.25 m.

The results obtained by use of the model before the construction and the observed values are compared in Table 10.2. The forcing functions applied for the model are listed below:

Temperature: 30°C

Flow rate: 50 m^3/24 h

BOD5 untreated water: 123 mg/l

Nitrate-N untreated water: 18 mg/l

TABLE 10.2

Comparison of Model Results and Observed Values

	Model Results	Observed Results
Concentrations (treated water)		
BOD^5 (mg/L)	27	26
Nitrate-N (mg/L)	0.40	0.50
Ammonium-N (mg/L)	9.2	9.1
Organic N (mg/L)	0.45	0.50
Total N (mg/L)	10.1	10.1
Total P (mg/L)	5.2	5.0
Removal efficiencies (%)		
BOD5	79	79
Nitrate-N	98	97
Ammonium-N	23	24
Organic-N	96	96
Total N	74	74
Total P	35	38

Ammonium-N untreated wastewater: 12 mg/L

Organic nitrogen-N untreated wastewater: 12 mg/L

Total P untreated wastewater: 8 mg/L

Particulate matter 20: mg/L

As seen from the application of the model for this case study, the validation of the model is very acceptable.

References

Alexandrov, G. A. 1988. A spatially-distributed model of raised bog relief. In: *Wetland Modelling*, eds. W. J. Mitsch, M. Strakraba, and S. E. Jørgensen, 41–54. Amsterdam: Elsevier.

Bekker, S. A. 1994. Spatial and dynamic modelling: Describing the terrastrialization of fen ecosystems. In: *Global Wetlands. Old World and New*. ed. W. J. Mitsch, 555–562. Amsterdam: Elsevier.

Costanza, R., F. H. Sklar, M. L. White, and J. W. Day Jr. 1988. A dynamic spatial simulation model of land loss and marsh sucession in coastal Louisiana. In: *Wetland Modelling*, eds. W. J. Mitsch, M. Strakraba, and S. E. Jørgensen, 99–114. Amsterdam: Elsevier.

Dørge, J. 1991. Model for nitrogen cycling in freshwater wetlands. Master thesis at University of Copenhagen.

EU-Project (1996-2000). Development of ecotechnological methods by use of wetlands as buffer zones to protect lakes and rivers. 330 pp., published internal in 2000.

Hagiwara, H., and W. Mitsch. 1994. Ecosystem modelling of a multi-species integrated aquaculture pond in South China. *Ecological Modelling* 72: 41–74.

James, R. T. 1993. Sensitivity analysis of a simulation model of methane flux from the Florida Everglades. *Ecological Modelling* 68: 119–146.

Jørgensen, S. E. 1976. A model of fish growth. *Ecological Modelling* 2: 303–313.

Jørgensen, S. E., and G. Bendoricchio. 2001. *Fundamentals of Ecological Modelling*, Third edition. Amsterdam: Elsevier.

Jørgensen, S. E., C. C. Hoffmann, and W. J. Mitsch. 1988. Modeling nutrient retention by a reedswamp and wet meadow in Denmark. In: *Wetland Modelling*, eds. W. J. Mitsch, M. Strakraba, and S. E. Jørgensen, 133–152. Amsterdam: Elsevier.

Kluge, W., P. Müeller-Buschbaum, and L. Theesen. 1994. Parameter acquisition for modelling exchange processes between terrestrial and aquatic ecosystems. *Ecological Modelling* 75/76: 399–408.

Logofet, D. O., and G. A. Alexandrov. 1988. Interference between Mosses and Trees in the framework of a dynamic model of carbon and nitrogen cycling in mesotrophic bog ecosystem. In: *Wetland Modelling*, eds. W. J. Mitsch, M. Strakraba, and S. E. Jørgensen, 55–66. Amsterdam: Elsevier.

Mitsch, W. J. 1988. Productivity-hydrology-nutrient models of forested wetland. In: *Wetland Modelling*, eds. W. J. Mitsch, M. Strakraba, and S. E. Jørgensen, 115–132. Amsterdam: Elsevier.

Mitsch, W., and B. Reeder. 1991. Modelling nutrient retention of a freshwater coastal wetland: Estimating the roles of primary productivity, sedimentation, resuspension and hydrology. *Ecological Modelling* 54: 151–187.

Patten, B. C. 1988. System ecology and Okefenokee Swamp. In: *Wetland Modelling*, eds. W. J. Mitsch, M. Strakraba, and S. E. Jørgensen, 189–216. Amsterdam: Elsevier.

Peijl, Van der M. J. et al. 1994. Modelling of spatial patterns and dynamic processes in the river corridor ecosystem: Model description. Final report from FAEWE. The Netherlands.

Reyes, E., J. W. Day, and F. H. Sklar. 1994. Ecosystem models of aquatic primary production and fish migration in Laguna de Terminos, Mexico. In: *Global Wetlands. Old World and New*. ed. W. J. Mitsch, 519–536. Amsterdam: Elsevier.

Silow, E. A., and D. J. Stom. 1992. *Model Ecosystems and Models of Ecosystems in Hydrology* (In Russian). Irkutsk University Press.

Sklar, F. H., K. Gopu, T. Maxwell, and R. Costanza. 1994. Spatial explicit and implicit dynamic simulation of wetland processes. In: *Global Wetlands. Old World and New*. ed. W. J. Mitsch, 537–554. Amsterdam: Elsevier.

Straskraba, M., and P. Mauersberger. 1988. Some simulation models of water quality management of shallow lakes and reservoirs and a contribution to ecosystem theory. In: *Wetland Modelling*, eds. W. J. Mitsch, M. Strakraba, and S. E. Jørgensen, 153–176. Amsterdam: Elsevier.

Whippler, S. J., and B. C. Patten. 1994. The complex trophic structure of an aquatic bed march ecosystem in Okefenokee Swamp. In: *Global Wetlands. Old World and New*. ed. W. J. Mitsch, 593–611. Amsterdam: Elsevier.

11

Wastewater Systems

Krist V. Gernaey, Ingmar Nopens, Gürkan Sin, and Ulf Jeppsson

CONTENTS

11.1 Introduction...278
11.2 The Core Activated Sludge Model: ASM1..280
    11.2.1 Matrix Notation...281
    11.2.2 COD Fractionation in the ASM1...285
    11.2.3 Nitrogen Fractionation in the ASM1..286
    11.2.4 Activated Sludge Model No. 3...287
    11.2.5 Limitations of ASM1...289
11.3 Model Extensions Focused on N Removal...291
    11.3.1 ASM Extensions to Describe Nitrite Accumulation.......................291
    11.3.2 ASM Extensions to Describe Nitrous Gas Emissions from
              Plants...295
11.4 Model Extensions Focused on P Removal...296
11.5 Modeling Sedimentation Processes...298
    11.5.1 Gravitational Settling..298
    11.5.2 Settler Models..299
11.6 Anaerobic Digester Models..303
    11.6.1 Anaerobic Digestion Principles...303
    11.6.2 Anaerobic Digestion Models..305
    11.6.3 ADM1 Limitations...308
11.7 Current WWTP Model Development Issues...308
    11.7.1 Plant-Wide Models..309
    11.7.2 Good Modeling Practice...310
    11.7.3 Uncertainty and Sensitivity Analysis......................................311
11.8 Guidelines for Application of WWTP Models..312
References..316
Further Reading—Model Extensions Focused on N-Removal...................321
Further Reading—Bio-P Model Development.....................................322

11.1 Introduction

Wastewater systems cover a range of essential technologies to significantly reduce the impact of household and industrial wastewater on receiving water quality. Biological wastewater treatment processes are often the key to achieving wastewater purification against an acceptably low cost. One of the most widespread biological wastewater purification technologies is the activated sludge process, where the incoming wastewater is mixed with a concentrated bacterial biomass suspension (the activated sludge), which degrades the pollutants. The earliest activated sludge systems were mainly concerned with the removal of organic carbon substances from the wastewater, which could be obtained rather easily by simple process designs. However, over the years more stringent effluent standards, especially for the nutrients nitrogen (N) and phosphorus (P), has resulted in significant modifications to the design and operation of activated sludge plants to make the treatment plants suited for biological N and P removal as well.

The implementation of biological nutrient removal on wastewater treatment plants (WWTPs) coincided with a significant research effort where, for example, the processes for biological N and P removal have been studied in detail. This has resulted in an increased knowledge on those biological processes, knowledge that has also been incorporated in the activated sludge models that are used as one of the building blocks of advanced dynamic mathematical models of a WWTP. For biological phosphorus removal (bio-P), to name one example, key research contributions were published in the 1980s and the 1990s, whereas the most frequently used bio-P models that incorporated this knowledge were released from the middle of the 1990s onward (e.g., Henze et al. 1995, 1999; Van Veldhuizen et al. 1999; Brdjanovic et al. 2000).

It is indeed important here to realize the difference between an activated sludge model and a WWTP model. A standard WWTP of a reasonable size consists of a primary treatment section, a secondary treatment section, and a sludge treatment section (Figure 11.1). The primary treatment normally contains sand and grit removal, grease removal, and often also a primary clarifier. The secondary treatment consists of a set of activated sludge tanks, combined with a secondary clarifier (Figure 11.1). Depending on the specific WWTP configuration, the concentrations of dissolved oxygen and nitrate present in the different activated sludge tanks can be very different, and aerobic (oxygen present), anoxic (nitrate present, no oxygen), or anaerobic (no oxygen, no nitrate) compartments can be distinguished. Traditionally, the term WWTP model is used to indicate the combination of activated sludge model, hydraulic model, mass transfer model, and sedimentation tank model needed to describe the activated sludge part of an actual WWTP (= secondary treatment). The term "activated sludge model" specifically refers to a set of differential equations—a mechanistic model—that represent the biological (and chemical) reactions taking place in one activated sludge tank. The

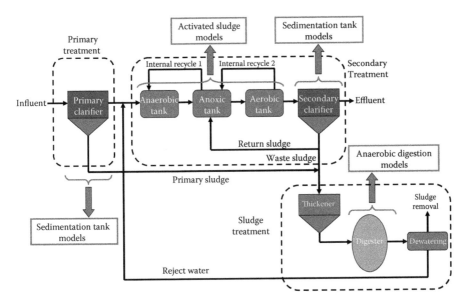

FIGURE 11.1
Schematic layout of a WWTP, in this case for a UCT plant layout. Traditionally, WWPT models were focused on the activated sludge part of the plant (secondary treatment of the wastewater). Nowadays, the scope of modeling studies extends to plant-wide models.

hydraulic model describes tank volumes, hydraulic tank behavior (e.g., perfectly mixed versus plug flow, constant versus variable volume, etc.), and the liquid flow rates in between tanks, such as return sludge flow rate and internal recycle flow rate. The sedimentation tank models are available in varying degrees of complexity (see Section 11.5). The mass transfer model typically describes the oxygen transfer from the gas phase to the liquid phase.

A recent evolution in WWTP modeling is the growing interest for so-called plant-wide models, which are models that extend beyond the activated sludge plant and also include the primary treatment and the sludge treatment line (Figure 11.1). Specifically for sludge treatment, anaerobic digestion (AD) (see Section 11.6) is often used since the AD technology allows to recover energy from the waste sludge stream. Plant-wide models are important because they allow us to study the interactions between primary, secondary, and sludge treatment, for example, by considering the effect of reject water on the dynamics of the load entering the secondary treatment.

Figure 11.1 contains the basic WWTP scheme around which this chapter has been built. In the following, an overview is provided of the most important models that are used in a WWTP modeling context. The description of the activated sludge models forms the core of the chapter, and is subdivided into three parts: first, the Activated Sludge Model No. 1 (ASM1) is introduced as the core model within the modeling of activated sludge WWTPs.

Modifications of this core model for biological N removal and for bio-P are then explained in two separate sections. The activated sludge models will be followed by an overview of sedimentation tank models and AD models. This overview of models is followed by a discussion on recent trends in WWTP modeling. The chapter ends with recommendations on the application of the WWTP models.

11.2 The Core Activated Sludge Model: ASM1

The definition of the ASM1 (Henze et al. 1987) is considered as the real starting point of WWTP modeling. Indeed, the most important contribution of the ASM1 model is that its publication has resulted in the general acceptance of WWTP modeling, first in the research community and later on also in industry. However, it should be emphasized that the model known as "ASM1" represents a summary of the "state of the art" in WWTP modeling at that time. Many of the basic concepts of ASM1 were adapted from the general activated sludge process model defined by Marais, Ekama, and coworkers in the late 1970s.

Even today, the ASM1 is the industry standard for modeling biological processes in activated sludge systems, and has been implemented—in many cases with modifications—in most of the commercial software available for modeling and simulation of WWTPs. The importance of the ASM1 is probably best illustrated by the fact that some of the basic model principles introduced in the ASM1 were retained in later activated sludge models:

- The "Matrix notation"—also known as the Petersen matrix or the Gujer matrix—is applied for the model equations (see Table 11.1).
- Chemical oxygen demand (COD) was adopted as the measure of the concentration of organic matter.
- The wide variety of organic carbon and nitrogenous compounds are subdivided into a limited number of fractions in the model, based on biodegradability and solubility considerations.

Those basic model principles will be explained below using the ASM1 as an example. Other models mentioned in this chapter will only be explained briefly, mainly focusing on how those models differ from ASM1, the core activated sludge model.

The main purpose of the ASM1, presented in matrix format in Table 11.1, was to describe the removal of organic carbon substances and nitrogen with simultaneous consumption of oxygen and nitrate as electron acceptors for municipal activated sludge WWTPs. At the same time, the ASM1 was intended to yield a good description of the sludge production in those plants. The detailed substrate flows in the ASM1 are illustrated in Figure 11.2.

11.2.1 Matrix Notation

There are four different main processes defined in ASM1 (Henze et al. 1987): (1) growth of biomass; (2) decay of biomass; (3) ammonification of organic nitrogen; (4) hydrolysis of particulate organic matter. There are two types of biomass (heterotrophs, X_{BH}, and autotrophs, X_{BA}) in the ASM1, and the model explicitly distinguishes aerobic and anoxic growth as separate processes. As a consequence, the four main processes defined in the ASM1 translate to eight processes in the final version of the ASM1 model (see Table 11.1).

The use of the matrix notation for activated sludge models (see Table 11.1 for the ASM1) allows a compact and visually appealing description of complex mathematical models (Henze et al. 1987, 2000; Noorman et al. 1991; Sin et al. 2008c). For description of a system with a number of components m and a number of processes n, the matrix formalism requires definition of the stoichiometric matrix, S, the process rates vector, ρ, and the component conversion vector, r. The component conversion vector, r, is given by:

$$r_{m\times1} = S'_{n\times m} \cdot \rho_{n\times1} \qquad (11.1)$$

Note that the subscripts in Equation 11.1 refer to the size of the different matrices and vectors. The overall component conversion vector is then coupled to a general mass balance equation around a system boundary. This finally provides a set of ordinary differential equations (ODEs) (a vector of size m) describing dynamic system behavior, which in its general form for a reactor with one inlet (influent) and one outlet (effluent) looks like:

$$\frac{d(V_L \cdot C)}{dt} = Q \cdot \left(C_{in} - C\right) - r \cdot V_L \qquad (11.2)$$

In Equation 11.2, C_{in} is a vector of concentrations in the influent stream (size m), Q is the flow rate of the inlet stream (m³/day), and V_L is the volume of the reactor (m³). For a continuous system with constant volume, Equation 11.2 can then, for example, be rewritten as:

$$\frac{dC}{dt} = \frac{Q}{V_L} \cdot \left(C_{in} - C\right) - r \qquad (11.3)$$

In Table 11.1, there are 13 system components, hence m is 13, and these are given in the first row of Table 11.1. There are eight processes considered, hence n is 8. The processes are given in the first column of Table 11.1. The stoichiometric matrix, S, has the dimension $n \times m$ and contains the stoichiometric

TABLE 11.1

Representation of the ASM1 Using the Matrix Notation

		Component (i)								
Process (j)		**1** S_I	**2** S_S	**3** X_I	**4** X_S	**5** X_{BH}	**6** X_{BA}	**7** X_P	**8** S_O	**9** S_{NO}
1	Aerobic growth of heterotrophic biomass		$-\dfrac{1}{Y_H}$			1			$-\dfrac{1-Y_H}{Y_H}$	
2	Anoxic growth of heterotrophic biomass		$-\dfrac{1}{Y_H}$			1				
3	Aerobic growth of autotrophic biomass						1		$\dfrac{4.57-Y_A}{Y_A}$	$-\dfrac{1}{Y_A}$
4	Decay of heterotrophic biomass				$1-f_p$	-1		f_p		
5	Decay of autotrophic biomass				$1-f_p$		-1	f_p		
6	Ammonification of soluble organic nitrogen									
7	Hydrolysis of slowly biodegradable substrate		1		-1					
8	Hydrolysis of organic nitrogen									

Source: Henze et al., *Activated Sludge Model No. 1.* London, UK, 1987.

coefficients for the components that are involved in each process. The **S** matrix corresponds to the gray-shaded area and is situated in the main body of Table 11.1. The process rates, ρ, are shown in the last column of Table 11.1. It is useful to note that each process may act in three different ways on the system components: neutral (zero), negative (consumption), or positive (production). Details on the meaning of the stoichiometric parameters of the ASM1 model can be found in the work of Henze et al. (1987, 2000).

The overall component conversion rates, r, can then be calculated as the sum of conversion rates of components involved in each process, for example, for heterotrophic biomass growth, this would be:

10 S_{NH}	11 S_{ND}	12 X_{ND}	13 S_{ALK}	Process Rate (ρ_j)
$-i_{XB}$			$-\dfrac{i_{XB}}{14}$	$\mu_{maxH}\dfrac{S_S}{K_S+S_S}\dfrac{S_O}{K_{OH}+S_O}X_{BH}$
$-i_{XB}$			$\dfrac{1-Y_H}{14\cdot2.86Y_H}$	$\eta_g\mu_{maxH}\dfrac{S_S}{K_S+S_S}\dfrac{K_{OH}}{K_{OH}+S_O}$
			$-\dfrac{i_{XB}}{14}$	$\dfrac{S_{NO}}{K_{NO}+S_{NO}}X_{BH}$
$-i_{XB}-\dfrac{1}{Y_A}$			$-\dfrac{2}{14\cdot Y_A}-\dfrac{i_{XB}}{14}$	$\mu_{maxA}\dfrac{S_{NH}}{K_{NH}+S_{NH}}\dfrac{S_O}{K_{OA}+S_O}X_{BA}$
		$i_{XB}-f_p i_{XB}$		$b_H X_H$
		$i_{XB}-f_p i_{XB}$		$b_A X_{BA}$
1	-1		$\dfrac{1}{14}$	$k_a S_{ND}X_{BH}$
				$k_h\dfrac{X_S/X_{BH}}{K_X+X_S/X_{BH}}\dfrac{S_O}{K_{OH}+S_O}$
				$+\eta_h\dfrac{K_{OH}}{K_{OH}+S_O}\dfrac{S_{NO}}{K_{NO}+S_{NO}}X_{BH}$
	1	-1		$\rho_7(X_{ND}/X_S)$

$$r_5 = r_{X_{BH}} = \sum_{j=1}^{n} S_{5xj}\cdot\rho_j \tag{11.4}$$

$$r_{X_{BH}} = \left(\frac{S_O}{K_{OH}+S_O}+\eta_g\frac{K_{OH}}{K_{OH}+S_O}\frac{S_{NO}}{K_{NO}+S_{NO}}\right)\mu_{maxH}\frac{S_S}{K_S+S_S}X_{BH}-b_H X_{BH} \tag{11.5}$$

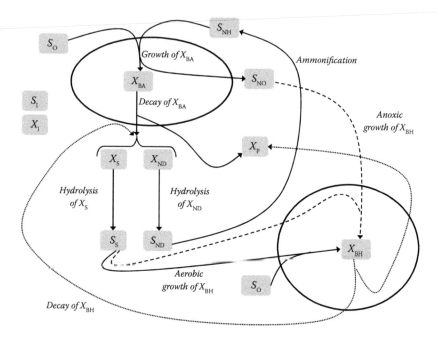

FIGURE 11.2

Substrate flows in the ASM1; for simplicity, alkalinity (S_{ALK}) is not included. The eight processes mentioned in the figure correspond to the processes in Table 11.1. S_I (soluble inert material) and X_I (particulate inert material) are assumed to be present in the influent but do not participate in any reaction. S_{NH} (ammonium nitrogen) is present in the influent or is produced via ammonification of S_{ND} (soluble organic nitrogen). S_{NH} and S_O (dissolved oxygen) are consumed during growth of autotrophic biomass (X_{BA}), and result in the production of X_{BA} and S_{NO} (nitrite + nitrate nitrogen). Decay of X_{BA} results in X_S (particulate organic matter) and X_{ND} (particulate organic nitrogen) and a fraction of nonbiodegradable organic matter (X_P). During the hydrolysis process, X_S and X_{ND} are converted to S_S (readily biodegradable substrate) and S_{ND} (soluble organic nitrogen), respectively. X_S, X_{ND}, S_S, and S_{ND} can also be present in the influent to the WWTP, and S_S is consumed during both aerobic (oxygen as electron acceptor) and anoxic (S_{NO} as electron acceptor) growth of heterotrophic biomass (X_{BH}). Similar to X_{BA}, decay of X_{BH} results in X_S and X_{ND}, as well as a fraction of X_P.

Finally, the calculated overall component conversion rates can be coupled with appropriate mass balance equations to obtain the ODE for this component. For example, for growth of heterotrophic biomass in a continuously stirred tank reactor (CSTR) with constant volume, V_L, this would be:

$$\frac{dX_{BH}}{dt} = \frac{Q}{V_L}\left(X_{BH,in} - X_{BH}\right) + r_{X_{BH}} \tag{11.6}$$

In Equation 11.6, $X_{BH,in}$ is the concentration of heterotrophic biomass in the inlet stream to the reactor, whereas X_{BH} represents the heterotrophic biomass concentration in the reactor. The ODEs for the other system components can be derived similarly. It is important here to point out that average "lumped" states are most often assumed when modeling a single activated sludge reactor. This is in many cases an acceptable simplification: it is assumed that all biomass in a reactor behaves identically thereby implicitly assuming that individual microorganisms remain in each reactor for that reactor's average hydraulic retention time (HRT).

The matrix notation is also useful since conservation principles—particularly elemental mass and charge balances for activated sludge models—are applied to each process of the system as proposed by Roels and coworkers (Roels 1980; Esener et al. 1983). This provides macroscopic information, which is useful to ensure that there are no errors in the elemental flows in the system and to derive stoichiometric relationships between process variables. The latter decreases the total number of stoichiometric parameters needed to describe the process (Esener et al. 1983; Heijnen 1999). A similar method has also been introduced successfully by Gujer and Larsen (1995). This conservation principle can be expressed in a matrix form using a composition matrix, c:

$$S_{n\times m} \cdot c'_{k\times m} = 0_{n\times k} \tag{11.7}$$

where c is the composition matrix and k stands for the total number of conservative properties in the system components. In the example (Table 11.1), COD is a conservative property. The elemental balance for N is, however, not closed because nitrogen gas is produced during anoxic growth of X_{BH}, whereas nitrogen gas is not included in the components list of the ASM1. For more details on the application of conservation principles to activated sludge models, Gujer and Larsen (1995), Henze et al. (2000), and Hauduc et al. (2010) should be consulted.

11.2.2 COD Fractionation in the ASM1

COD is selected as the most suitable variable for defining the carbon substrates in the ASM1 as it provides a link between electron equivalents in the organic substrate, the biomass and oxygen utilized. In ASM1, the COD is furthermore subdivided based on (1) solubility, (2) biodegradability, (3) biodegradation rate, and (4) viability (biomass):

1. The total COD (COD_{tot}) is divided into soluble (S) and particulate (X) components.
2. The COD is further subdivided into nonbiodegradable organic matter and biodegradable matter. The nonbiodegradable matter is

biologically inert and passes through an activated sludge system in unchanged form. The inert soluble organic matter (S_I) leaves the system in the effluent (see Figure 11.1) at the same concentration as it enters. Inert suspended organic matter in the wastewater influent (X_I) or produced via decay (X_P) becomes enmeshed in the activated sludge and is removed from the system via the sludge wastage (see Figure 11.1).

3. The biodegradable matter is divided into soluble readily biodegradable (S_S) and slowly biodegradable (X_S) substrate. It should be mentioned here that this is one of the major points of discussion related to the ASM1, because part of the slowly biodegradable matter might actually be soluble, despite the fact that X_S is defined as a particulate model component. The readily biodegradable substrate is assumed to consist of relatively simple molecules that may be taken up directly by heterotrophic organisms and used for growth of new biomass. On the contrary, the slowly biodegradable substrate consists of relatively complex molecules that require enzymatic breakdown before utilization.

4. Finally, heterotrophic biomass (X_{BH}) and autotrophic biomass (X_{BA}) are generated by growth on the readily biodegradable substrate (S_S) or by growth on ammonia nitrogen (S_{NH}). The biomass is lost via the decay process where it is converted to X_P and X_S. X_S can then be converted to S_S again via hydrolysis according to the death–regeneration concept, one of the key concepts incorporated in the ASM1.

Summarizing, the total COD of ASM1, COD_{tot}, is calculated by Equation 11.8, and illustrated in Figure 11.3.

$$COD_{tot} = S_I + S_S + X_I + X_S + X_{BH} + X_{BA} + X_P \qquad (11.8)$$

11.2.3 Nitrogen Fractionation in the ASM1

Similar to the organic matter, total nitrogen (N_{tot}) can be subdivided based on (1) solubility, (2) biodegradability, and (3) biodegradation rate:

1. N_{tot} can be subdivided into soluble (S) and particulate (X) components.

2. Nitrogen is divided into nonbiodegradable matter and biodegradable matter. The nonbiodegradable particulate organic nitrogen (X_{NI}) is associated with the nonbiodegradable particulate COD (X_I or X_P), whereas the soluble nonbiodegradable organic nitrogen (S_{NI}) is assumed to be negligible and therefore not incorporated into the model.

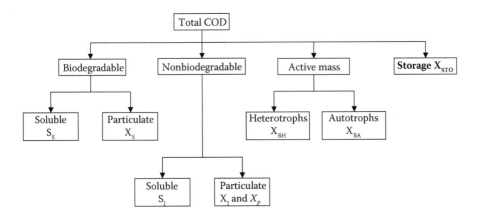

FIGURE 11.3
COD components in ASM1 and ASM3 (modified from Jeppsson, U., Modelling aspects of wastewater treatment processes, PhD thesis, IEA, Lund University, Sweden, 1996); components specifically related to ASM3 are given in bold and the ones only related to ASM1 are given in italics.

3. The biodegradable N is subdivided into ammonia nitrogen (S_{NH}), nitrate + nitrite nitrogen (S_{NO}), soluble organic nitrogen (S_{ND}), and particulate organic nitrogen (X_{ND}). The X_{ND} is hydrolyzed to S_{ND} (Figure 11.2) in parallel with the hydrolysis of the X_S (either present in the wastewater or produced via the decay process). The S_{ND} is converted to S_{NH} via ammonification. Ammonia nitrogen serves as the nitrogen source for biomass growth (the parameter i_{XB} indicates the amount of nitrogen incorporated per COD unit of biomass). Finally, the autotrophic conversion of ammonia results in nitrate nitrogen (S_{NO}), which is considered to be a single-step process in ASM1.

Summarizing, the N_{tot} balance for the components in ASM1 is defined by Equation 11.9 and further illustrated in Figure 11.4.

$$N_{tot} = S_{NH} + S_{ND} + S_{NO} + X_{ND} + X_{NI} + i_{XB} \cdot (X_{BH} + X_{BA}) + i_{XP} \cdot X_P \qquad (11.9)$$

11.2.4 Activated Sludge Model No. 3

The model concepts behind ASM1 have been altered in the Activated Sludge Model no. 3 (ASM3; Gujer et al. 1999; Henze et al. 2000), a model that also focuses on the degradation of carbon and N but allows the introduction of processes describing the storage of biopolymers under transient conditions. The major difference between the ASM1 and ASM3 models is indeed that

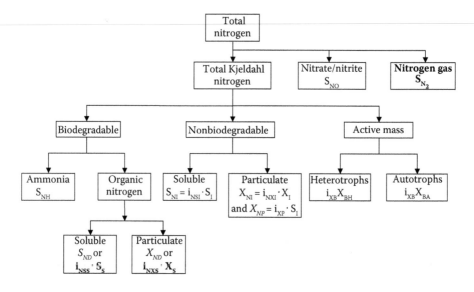

FIGURE 11.4
Nitrogen components in ASM1 (modified from Jeppsson, U., Modelling a spects of wastewater treatment processes, PhD thesis, IEA, Lund University, Sweden, 1996); components specifically related to ASM3 are given in bold and the ones only related to ASM1 in italics.

the latter recognizes the importance of storage polymers in the heterotrophic activated sludge conversions. The basic assumption of the ASM3 model is that all readily biodegradable substrate (S_S) is first taken up and stored into an internal cell component (X_{STO}) before growth. This storage principle is schematically represented in Figure 11.5. The introduction of X_{STO} implies that the heterotrophic biomass now has an internal cell structure in the model, similar to the polyphosphate accumulating organisms (PAOs) in the bio-P models (see Section 11.4). The internal component X_{STO} is subsequently used for biomass growth in the ASM3 model. Biomass growth directly on external readily biodegradable substrate (S_S) as described in the ASM1 is not considered in ASM3. A second major difference between ASM1 and ASM3 is that the circular growth–decay–growth model of the ASM1 (Figure 11.2), often called death–regeneration concept, is replaced by a growth-endogenous respiration model (Figure 11.5). The death–regeneration concept of ASM1 effectively implies that all state variables are directly influenced by a change in a parameter value. In ASM 3, the direct influence is considerably lower thus ensuring a better parameter identifiability. Koch et al. (2000), for example, used the ASM3 to describe full-scale plant data as well as data obtained from aerobic and anoxic batch experiments in the laboratory, and concluded that ASM1 and ASM3 are both capable of describing the dynamic behavior in common municipal WWTPs. Naturally, in situations where the storage

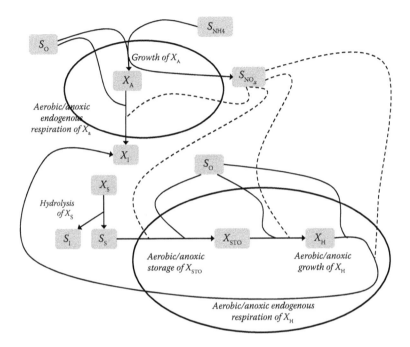

FIGURE 11.5
Carbon flows in the ASM3; for simplicity, alkalinity (S_{ALK}) consumption and nitrogen gas (S_{N2}) production are not included. Moreover, ammonium nitrogen (S_{NH4}) consumption or production is not considered either, except for the nitrification process. S_{NH4} and S_O (dissolved oxygen) are consumed during growth of X_A, and result in the production of X_A and S_{NOX} (nitrite + nitrate nitrogen). Endogenous respiration of X_A, either with S_O or S_{NOX} as electron acceptor, results in formation of particulate inert material (X_I). Slowly biodegradable particulate material (X_S) present in the influent is converted to readily biodegradable substrate (S_S) and soluble inert material (S_I) via hydrolysis. S_S is consumed during both aerobic and anoxic storage, resulting in formation of the particulate storage component X_{STO}. Heterotrophic biomass (X_H) is formed via growth on X_{STO}, both under aerobic and anoxic (=denitrification) conditions. Decay of X_H leads to formation of X_I. X_I, S_I, S_S, and S_{NH} are typically also present in the influent to the WWTP. Note that X_{STO} is also degraded through aerobic and anoxic respiration.

of readily biodegradable substrate is significant (e.g., industrial wastewater with high COD) or for WWTPs with substantial nonaerated zones, the ASM3 is expected to perform better. This is related to one of the major shortcomings of the ASM1, that is, the fact that the ASM1 does not distinguish between aerobic and anoxic decay.

11.2.5 Limitations of ASM1

The ASM1 is widely used for the modeling of activated sludge plants. However, the model has a number of important limitations (Henze et al. 1987), which have to be kept in mind when applying the ASM1:

1. The system must operate at constant temperature. A similar assumption is made in most activated sludge models. Hence, if the effect of temperature changes is to be included in the model, it is up to the model user to define suitable functions describing temperature dependency of kinetic parameters. In most cases, Arrhenius type expressions are used to describe temperature dependency of the model. Hellinga et al. (1999), for example, provide a detailed explanation of the influence of temperature on nitrification kinetics.

2. The pH is constant and near neutrality. It is known that the pH has an influence on many of the model parameters. However, only limited knowledge is available to be able to express these possible influences. Consequently, a constant pH has been assumed. The inclusion of alkalinity in the model, however, does allow for detection of pH problems.

3. No considerations have been given to changes in the nature of the organic matter within any given wastewater fractions (e.g., the readily biodegradable substrate). Therefore, the parameters in the rate expressions have been assumed to have constant values. This means that only concentration changes of the wastewater components can be handled, whereas changes in the wastewater character cannot. As a consequence, the heterotrophic biomass is homogeneous and does not undergo changes in species diversity with time, for example as a consequence of changes in wastewater composition.

4. The effects of nutrient limitations (e.g., N and P) on the cell growth have not been considered. It is, however, easy to add limitation terms in the model if needed, as illustrated by Hauduc et al. (2010).

5. The correction factors for denitrification (η_g and η_h) are fixed and constant for a given wastewater, even though it is possible that their values are depending on the system configuration.

6. The parameters for nitrification are assumed to be constant and to incorporate any inhibitory effects that wastewater constituents may have on them. As a consequence, the ASM1—and most other models—needs specific extensions to describe transient inhibitory effects.

7. The entrapment of particulate organic matter in the biomass is assumed to be instantaneous.

8. The hydrolysis of organic matter and organic nitrogen are coupled and occur simultaneously with equal rates.

9. The type of electron acceptor present does not affect the loss of biomass by decay. Siegrist et al. (1999) have shown that the type of electron acceptor does have an effect on decay. The ASM3, for example, includes electron acceptor effects on endogenous respiration (Gujer et al. 1999).

10. The type of electron acceptor does not affect the heterotrophic yield coefficient. ASM3 includes this effect (Gujer et al. 1999).

11. ASM1 is developed for simulation of treatment of municipal wastewater, and it is therefore not advised to apply the model to systems where industrial contributions dominate the characteristics of the wastewater. Modification of the ASM1 model might be necessary for description of such plants.

12. ASM1 does not include anaerobic processes. Simulations of systems with large fractions of anaerobic reactor volume may therefore lead to significant errors.

13. ASM1 cannot deal with elevated nitrite concentrations (see Section 11.3).

14. ASM1 is not designed to deal with activated sludge systems with very high load or small sludge retention time (SRT) (<1 day).

11.3 Model Extensions Focused on N Removal

An extensive recent review on this topic can be found in the work of Sin et al. (2008b).

11.3.1 ASM Extensions to Describe Nitrite Accumulation

The most frequently used activated sludge models, including the ASM models (ASM1, ASM2d, and ASM3), are developed based on the assumption that both nitrification and denitrification processes occur as single-step processes. Hence, the reaction stoichiometry (focusing on the redox reactions only) in those standard models is simply described as follows:

$$\text{Nitrification: } NH_4^+ + 2O_2 \xrightarrow{\text{autotrophs}} NO_3^- + 2H^+ + H_2O$$

$$\text{Denitrification: } NO_3^- + 6H^+ + 5e^- \xrightarrow{\text{heterotrophs}} 0.5N_2 + 3H_2O$$

In reality, however, the stoichiometry of both the nitrification and the denitrification process is far more complex than a single-step conversion. For example, the nitrification process is a two-step process that is actually carried out by two distinct groups of autotrophic organisms, ammonium oxidizing bacteria (AOB) and nitrite oxidizing bacteria (NOB). The reduction of nitrate to nitrogen gas in the denitrification process even involves four different steps. The detailed redox reactions involved in both processes are outlined in Table 11.2.

When modeling activated sludge plants, nitrite accumulation is normally insignificant in most municipal wastewater treatment systems, and hence

TABLE 11.2

Detailed Redox Reactions in the Nitrification and the Denitrification Process

Nitrification	Denitrification
$NH_4^+ + \frac{3}{2}O_2 \rightarrow NO_2^- + 2H^+ + H_2O$	$NO_3^- + 2H^+ + 2e^- \rightarrow NO_2^- + H_2O$
$NO_2^- + \frac{1}{2}O_2 \rightarrow NO_3^-$	$NO_2^- + 2H^+ + e^- \rightarrow NO + H_2O$
	$2NO + 2H^+ + 2e^- \rightarrow N_2O + H_2O$
	$N_2O + 2H^+ + 2e^- \rightarrow N_2 + H_2O$

the single-step assumption for the nitrification and denitrification processes is valid, thus avoiding the need to describe the complex reactions summarized in Table 11.2. However, in some specific situations nitrite can play a dominant role in the microbial conversions, since a number of operational factors can contribute to nitrite buildup in wastewater treatment systems: (1) unstable operation of the municipal WWTPs due to lack of sufficient oxygen, low temperature, low sludge age or the presence of inhibitory compounds; (2) operation of the plant at high temperatures; (3) side-stream processes; and (4) industrial WWTPs. In such systems, the above-mentioned single-step assumption no longer holds or is even not desired, and as a consequence nitrite needs to be considered as a model component when setting up a model for these systems. This means in practice that standard activated sludge models such as the ASM1 need to be extended with additional variables and processes.

It is also important to mention that there are situations in which nitrite buildup is purposefully induced in the system with the aim of improving process operation economics. For example, in side-stream processes such as partial nitritation and anaerobic ammonium oxidation processes (such as the Anammox process), nitrite is a key intermediate whose formation is specifically promoted by manipulation of environmental conditions (e.g., dilution rate, temperature, oxygen concentration, and pH). Hence, in order to describe such systems, single-step models cannot be used and instead one needs to have multistep nitrogen removal models.

A number of models have been proposed as extensions to ASM models to provide a more detailed description of the nitrogen removal processes. The majority of those extended models assume two-step nitrification and two-step denitrification. The redox reactions in these processes are as follows: $NH_4 \rightarrow NO_2 \rightarrow NO_3$ and $NO_3 \rightarrow NO_2 \rightarrow N_2$. These models indeed focus on nitrite as the key intermediate in both processes. Figure 11.6 provides a schematic illustration of the conceptual framework used to formulate nitrite models as extensions to the commonly used ASM family of models.

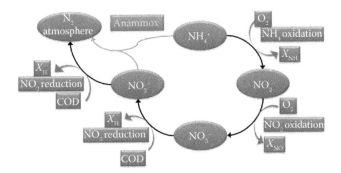

FIGURE 11.6
Conceptual framework used to formulate nitrite extensions to ASM family of models.

The common feature in the proposed extensions for modeling nitrite is that the growth rate is assumed as the rate-limiting step and used to describe the rate of other substrate conversions via stoichiometric yields—similar to the ASM1 convention. The mathematical structure of the two-step nitrification processes includes terms that describe the growth limiting and the growth inhibition effect of various compounds. For instance, the growth rate expression of the AOB as well as NOB typically includes the following terms: (1) ammonium (AOB) or nitrite (NOB) as substrate source; (2) ammonia inhibitory effect; (3) nitrous acid inhibitory effect; (4) pH effect; and (5) inorganic carbon limitation (see Table 11.3).

Similarly, the model structure of the growth rate of the anaerobic ammonium oxidation process typically includes the following terms: (1) ammonium and nitrite as rate limiting substrates limitation; (2) inhibitory effect of oxygen; (3) inhibitory/toxicity effect of nitrite as outlined in Table 11.3.

For the two-step denitrification process, the mathematical structure of the growth rate expression is not fully established due to the complex mechanism involved. Nonetheless, the following features are common in the extended nitrogen models: (1) the last two steps of the denitrification process occur faster than the former, hence the kinetic description of nitric oxide (NO) and nitrous oxide (N_2O) formation/consumption is neglected; (2) a sequential denitrification mechanism is adopted, that is, $NO_3^- \rightarrow NO_2^- \rightarrow 1/2N_2$; (3) no electron acceptor preference of the denitrifying bacteria, hence the electron acceptor is described as mixed substrate; (4) a certain fraction of heterotrophic organisms can perform denitrification and the anoxic growth rate is therefore lower compared to the aerobic growth rate of the heterotrophic organisms.

The extension of activated sludge models with nitrite also introduces a number of additional parameters that need to be provided to perform model simulations. These parameters relate to the presence of nitrite as substrate (electron donor in the case of nitrification; and electron acceptor in the case of denitrification and anaerobic ammonium oxidation). Information about

TABLE 11.3

Typical Features of the Mathematical Structure of the Growth Rate in the ASM Extensions for Modeling Nitrite

Process	Growth Rate
AOB	$\mu_{NH} X_{NH} \dfrac{S_O}{K_{O,NH}+S_O}\ \dfrac{S_{NH_3}}{K_{NH_3,NH}+S_{NH_3}}\ \dfrac{K_{INH_3,NH}}{K_{INH_3,NH}+S_{NH_3}}\ \dfrac{K_{IHNO_2,NH}}{K_{IHNO_2,NH}+S_{HNO_2}}\ \dfrac{S_{ALK}}{K_{N,ALK}+S_{ALK}}$
NOB	$\mu_{NO} X_{NO} \dfrac{S_O}{K_{O,NO}+S_O}\ \dfrac{S_{HNO_2}}{K_{HNO_2,NO}+S_{HNO_2}}\ \dfrac{K_{IHNO_2,NO}}{K_{IHNO_2,NO}+S_{HNO_2}}\ \dfrac{K_{INH_3,NO}}{K_{INH_3,NO}+S_{NH_3}}\ \dfrac{S_{ALK}}{K_{N,ALK}+S_{ALK}}$
Anaerobic ammonium oxidation	$\mu_{AN} X_{AN} \dfrac{S_{NH_4}}{K_{NH_4,AN}+S_{NH_4}}\ \dfrac{S_{NO_2}}{K_{NO_2,NO}+S_{NO_2}}\ \dfrac{K_{O,AN}}{K_{O,AN}+S_O}\ \dfrac{K_{INO_2}}{K_{INO_2,NO}+S_{NO_2}}$
Nitrate reduction	$\eta_{NO_3} \mu_H X_H \dfrac{K_{O,H}}{K_{O,H}+S_O}\ \dfrac{S_S}{K_{S,H}+S_S}\ \dfrac{S_{NO_3}}{K_{NO_3,H}+S_{NO_3}}\ \dfrac{S_{NO_3}}{S_{NO_3}+S_{NO_2}}\ \dfrac{K_{HNO_2,H}}{K_{HNO_2,H}+S_{HNO_2}}$
Nitrite reduction	$\eta_{NO_2} \mu_H X_H \dfrac{K_{O,H}}{K_{O,H}+S_O}\ \dfrac{S_S}{K_{S,H}+S_S}\ \dfrac{S_{NO_2}}{K_{NO_2,H}+S_{NO_2}}\ \dfrac{S_{NO_2}}{S_{NO_3}+S_{NO_2}}\ \dfrac{K_{HNO_2,H}}{K_{HNO_2,H}+S_{HNO_2}}$

Source: Sin et al., *Water Science and Technology*, 58(6), 1155–1171, 2008.

these parameter values does exist in the literature, but shows variations from study to study. The following recommendations can be helpful in selecting suitable parameter values when using an ASM model extended with nitrite. The common suggestion is to employ fixed (default) values for the yield and decay/maintenance parameters, while performing dedicated measurements to estimate the so-called process characteristics-dependent parameters that include affinity/inhibition constants, growth rates/reduction factors.

Note also that extending ASM models with nitrite in principle has consequences for the description of the other processes in the system, such as P uptake, endogenous respiration, hydrolysis, and storage processes. These processes will also need to be modified for a complete and consistent description of activated sludge systems including nitrite. Some of the consequences of introducing nitrite as state variable in the system are the following:

- Nitrite buildup inhibits the aerobic and anoxic P-uptake metabolism. Hence, in such a bio-P model extension, nitrite will serve as electron acceptor for the anoxic P uptake as well as an inhibitor to aerobic P uptake. Some models including such effects have been developed and partially validated using laboratory-scale tests, but comprehensive validation is still needed.

- Endogenous respiration models will need to be extended with nitrite as electron acceptor as well, since it has been shown that redox conditions affect the endogenous respiration kinetics. Some suggestions have been proposed to this end, however, a consensus on the effect of nitrite and how to describe it is still under debate.

- Similarly hydrolysis and fermentation processes—if assumed electron acceptor dependent—will have to be modified when nitrite is considered as electron acceptor in the system.

- For storage models such as ASM3, the anoxic storage metabolism will have to be extended with nitrite as electron acceptor. Some studies, albeit very few, have indeed dealt with this challenge.

11.3.2 ASM Extensions to Describe Nitrous Gas Emissions from Plants

It is important to note that other intermediates, notably NO and N_2O, have also been observed in significant amounts in wastewater treatment systems, and a number of models have been developed to describe the accumulation of these intermediates. However, thus far these intermediates have not been incorporated into the main stream–activated sludge models for a number of reasons: (1) the production of NO and N_2O does not contribute significantly to the total mass flow of nitrogen in the system [although relevant in view of greenhouse gas (GHG) issues]; and (2) in general, the models are not used to predict NO/N_2O occurrence in the plants as yet. However, due to the growing concern about climate change, there is definitely an incentive to consider

GHG emissions from the wastewater treatment works (Shiskowski 2008). Therefore, efforts to further develop N removal models particularly focusing on the production of NO/N_2O both during nitrification and denitrification processes are currently increasing.

11.4 Model Extensions Focused on P Removal

The ASM2 model (Henze et al. 1995) extends the capabilities of the ASM1 to also include the description of biological phosphorus (bio-P) removal, as well as chemical P removal via precipitation. Bio-P modeling in ASM2 is illustrated in Figure 11.7: the PAOs are modeled with cell internal structure, where all organic storage products are lumped into a single model component (X_{PHA}, storage compound in PAOs). PAOs can only grow on cell internal organic storage material. In the ASM2, it is assumed that storage is not depending on the electron acceptor conditions, but is only possible when fermentation products such as acetate are available. In practice, this assumption means that storage will usually only be observed in the anaerobic activated sludge tanks.

The ASM2 publication represented the available knowledge on bio-P processes at the time the model was published, but did not yet include all observed phenomena. For example, the ASM2d model (Henze et al. 1999) extends the ASM2 model, adding the denitrifying activity of PAOs, which should allow a better description of the dynamics of phosphate and nitrate. As already mentioned in Section 11.4, inclusion of nitrite as a model component in an activated sludge model that also describes bio-P removal means in principle that the effects of nitrite on bio-P have to be incorporated in the

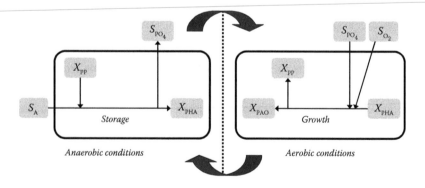

FIGURE 11.7
Substrate flows for storage and growth of PAOs in the ASM2 model. (From Henze et al., *Activated Sludge Model No. 2.* IWA Scientific and Technical Report No. 3, London, UK, 1995.)

model. The ASM3 model has been extended with a bio-P removal module as well (Ky et al. 2001; Rieger et al. 2001).

Several metabolic models were proposed as an alternative to ASM2d, where significant contributions were made by Comeau et al. (Canada), Wentzel et al. (South Africa), Mino et al. (Japan), and van Loosdrecht et al. (the Netherlands). The most widely applied metabolic bio-P model is the Technical University Delft Phosphorus (TUDP) model (van Veldhuizen et al. 1999; Brdjanovic et al. 2000; Meijer 2004), which combines the metabolic model for denitrifying and nondenitrifying bio-P of Murnleitner et al. (1997) with the ASM1 model (autotrophic and heterotrophic reactions). In contrast to ASM2/ASM2d, the TUDP model fully considers the metabolism of PAOs, modeling all organic storage components explicitly (polyhydroxy alkanoates, X_{PHA}, and glycogen, X_{GLY}), as shown in Figure 11.8. The TUDP model was validated in enriched bio-P sequencing batch reactor (SBR) laboratory systems over a range of SRT values (Smolders et al. 1995), for different anaerobic and aerobic phase lengths (Kuba et al. 1997), and for oxygen and nitrate as electron acceptor (Murnleitner et al. 1997).

The primary aim of many simulation studies of bio-P plants is to investigate the combined effect of proposed control strategies on N and P removal (Gernaey and Jørgensen 2004; Ingildsen et al. 2006). Substantial amounts of X_{PAO} should therefore be present in the activated sludge, to allow bio-P removal to take place. One of the potential control handles in a full-scale WWTP is storage of part of the activated sludge under nonaerated conditions, for example when the influent load is low. The advantages of such plant operation are: (1) the generation of fermentation products (S_A) due to hydrolysis of slowly biodegradable substrate (X_S) and fermentation, resulting in improved bio-P removal; (2) energy savings due to decreased aeration needs; (3) a lower biomass decay rate. Indeed, according to the experimental

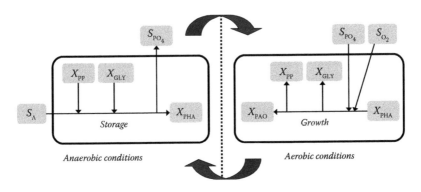

FIGURE 11.8
Substrate flows for storage and aerobic growth of PAOs in the TUDP model. (From van Veldhuizen et al., *Water Research*, 33, 3459–3468, 1999; Brdjanovic et al., *Water Research*, 34, 846–858, 2000; Meijer, PhD thesis, Delft University of Technology, the Netherlands, 2004.)

results reported by Siegrist et al. (1999), a differentiation between aerobic, anoxic, and anaerobic autotrophic biomass (X_A) decay rates seems to be justified. However, such an effect of electron acceptor conditions on decay rates is not incorporated in the ASM2d, which forms an important practical limitation of that model. Brdjanovic et al. (2000) and Rieger et al. (2001), for example, also included reduced decay rates for heterotrophs (X_H) and X_{PAO} under anoxic and anaerobic conditions in their bio-P models. Therefore, to obtain more realistic simulation results, and to promote development of sufficient amounts of X_{PAO} in the simulations, modification of the ASM2d model equations for biomass decay is proposed by Gernaey and Jørgensen (2004) to make the decay process rates electron acceptor depending, and has for example been used successfully for the modeling of a full-scale WWTP (Ingildsen et al. 2006).

In some cases, such as high pH (>7.5) and high Ca^{2+} concentrations, it can be necessary to add biologically induced P precipitation to the bio-P model (Maurer et al. 1999; Maurer and Boller 1999). Indeed, under certain conditions the bio-P reactions coincide with a natural precipitation that can account for an important P removal effect that is not related to the bio-P reactions included in the models described thus far. The formation of these precipitates, mostly consisting of calcium phosphates, is promoted by the high P concentration and increased ionic strength during the anaerobic P release of the PAOs. Model equations and components necessary to describe this precipitation process were given by Maurer and Boller (1999).

11.5 Modeling Sedimentation Processes

11.5.1 Gravitational Settling

Gravitational settling is typically used at two locations in a WWTP (see Figure 11.1): during the primary treatment (primary clarifier) and as a final step in the secondary treatment (secondary clarifier). The former has the purpose of removing excess particulate material (total suspended solids, TSS) in the influent to the biological treatment processes, whereas the latter is used to separate the purified water from the sludge. Maintaining an efficient sedimentation in the secondary clarifier is important to produce good quality effluent and maintain high concentrations of activated sludge in the system. Figure 11.9 shows an example of a circular secondary clarifier (also rectangular clarifiers exist). In all configurations, sludge enters the settler in the energy dissipating well, then flows to a flocculation zone where the formation of large flocs that settle faster is promoted. The sludge then flows toward the effluent launder or weir. In the meantime, the sludge flocs can settle out and form a sludge blanket. The performance of the settling is heavily dependent on the sludge settling properties. As the settling velocity

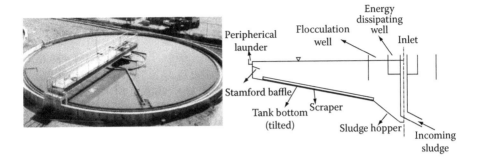

FIGURE 11.9
Illustration of a circular clarifier (left) and a side view of a cross section of a circular clarifier (right).

decreases with increasing sludge concentration, the design sludge concentration in the activated sludge tanks is limited to roughly 3–4 g/L by the settling step.

Finally, the clarified effluent leaves the treatment plant and typically contains a non-settleable fraction [or effluent suspended solids (ESS)] in the order of 5–50 mg SS/L. The sludge moves toward the sludge hopper (combination of floor slope and a sludge scraper attached to a rotating bridge; see Figure 11.9). The collected sludge has a concentration of about 8–10 g/L due to thickening and compression and is either directly disposed of or fed to the sludge treatment line (Figure 11.1).

11.5.2 Settler Models

There are several incentives for modeling of primary and secondary settling processes. For primary clarifiers, a good prediction of the load (and COD/N ratio) leaving the primary clarifier and entering the activated sludge reactors is of great importance as there is an optimum for efficient N removal. For secondary settlers, the incentives are: (1) sludge balance between biological tanks and clarifiers; (2) dynamics of sludge motion between biological tanks and settler; (3) effluent quality (ESS); (4) settler performance: sludge blanket height (SBH); (5) introduction of control systems for polymer dosage in case of bad settling; (6) sludge production (thickening); (7) settler design: placement of baffles.

Settler models differ in their spatial detail: zero-dimensional (0-D, point settler), one-dimensional (1-D, homogeneous in x- and y-directions), two-dimensional (2-D, homogeneous in y-direction), and three-dimensional (3-D, nonhomogeneous) models exist.

0-D models or point settlers are ideal separators without volume. Such a model can be regarded as an ideal flow splitter, hereby conserving mass. The only degree of freedom or parameter is the non-settleable fraction, which can in some models be an increasing function of flow rate to mimic the effect of turbulence in the settler, which will affect the ESS. The major advantage is its simplicity and it can therefore be used if none of the above incentives is part of the modeling exercise. However, the major problem of this model is that it has no residence time, which means that the typical time delays resulting from passing the mixed liquor from the activated sludge plant through the secondary settlers is absent in WWTP models that use a 0-D settler model.

1-D models discretize the 1-D partial differential equation (PDE) in Equation 11.10 using a finite difference method for the first derivative: $dX/dz = X(j) - X(j + 1)/h$.

$$\frac{\partial X}{\partial t} = \frac{\partial(D\partial X)}{\partial z^2} - \frac{\partial(V_s X)}{\partial z} - RX \tag{11.10}$$

The latter is multiplied by the sludge settling velocity, V_s, which will be discussed later. This assumes that X is uniform in the horizontal plane, and that no vertical dispersion occurs ($D = 0$). Moreover, R is often ignored assuming no biological reactions. The discretization of the 1-D model is illustrated in Figure 11.10. Ten layers have been chosen in this example, where layer six is used as feed layer where the inflow is pumped in to the clarifier. A 1-D settler model gives access to the following: effluent and underflow sludge concentrations, (discrete) SBH, and total sludge mass.

2-D and 3-D models make use of the solution of the Navier–Stokes equations for fluid hydrodynamics, including solids transport and, if required, turbulence. For a more detailed explanation of these models, the reader is referred to the specialized literature on hydrodynamics (e.g., White 1991;

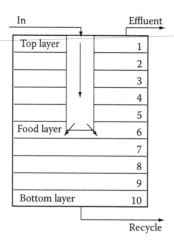

FIGURE 11.10
Illustration of the discretization of the 1-D settler model.

Hirsch 1997). A typical output of a CFD clarifier model is illustrated in Figure 11.11. It contains detailed 2-D information concerning velocity and solids concentration. One can clearly observe the sludge blanket and the concentration gradient therein.

Next to the spatial detail, settling models need an expression for the settling velocity V_s. The latter is a sludge property and is typically measured using a batch settling test, where the sludge is allowed to settle in a 1-L cylinder and the supernatant–sludge interface is recorded versus time. The linear part in this curve is used as a measure for the (hindered) settling velocity V_s. Batch settling curves change shape when the initial sludge concentration is altered. Higher concentrations result in less steep slopes, indicating slower settling. The linear slopes are then plotted against sludge concentration after which a settling velocity model is fitted to it. The most widely used settling velocity models are the models of Vesilind (1968) (Equation 11.11) and Takács et al. (1991) (Equation 11.12).

$$v_s = ke^{-nX} \tag{11.11}$$

$$v_{sj} = v_0 e^{-r_h\left(X_j - X_{min}\right)} - v_0 e^{-r_p\left(X_j - X_{min}\right)}$$
$$\text{with } 0 \le v_{sj} \le v_0' \tag{11.12}$$

Both settling velocity models are illustrated in Figure 11.12. The latter is actually an extension of the former, correcting for the unrealistically

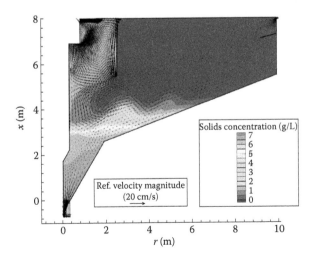

FIGURE 11.11
Illustration of a typical 2-D settler model output using CFD.

large settling velocities predicted by the Vesilind equation at small sludge concentrations.

The spatial model combined with sludge settling velocity models make up the typically used settler models and can predict the fate of particulate material in clarifiers.

At this moment, there are still a number of issues related to the modeling of sedimentation tanks where no generally accepted solution exist. There is for example no consensus on the use of reactive settler models. For the relatively simple settler models, such as the 1-D models, the sedimentation model is typically nonreactive—only sedimentation takes place, the particles that settle are considered inert—despite the fact that an important fraction of the total amount of sludge in a WWTP can be stored at the bottom of the secondary clarifier. When superimposing an activated sludge model on the settler model, resulting in a so-called reactive settler model, the limitations of some of the activated sludge models in describing the decay processes become obvious (Gernaey et al. 2006a). A general guidance here is that reactions in the secondary clarifier should be considered if a significant amount of sludge is stored in the sludge blanket.

The effect of changing operating conditions of the activated sludge plant on the settling properties of the activated sludge cannot be described with current mechanistic models. This is a major limitation since it means that the effect of the implementation of a control strategy on treatment plant perform-ance—especially secondary clarifier performance—cannot be represented properly. Recently, a risk assessment model for quantification of settling problems of microbiological origin in activated sludge systems (filamen-tous bulking, foaming, and rising sludge) has been proposed to represent the effect of activated sludge plant operation on sludge settling properties

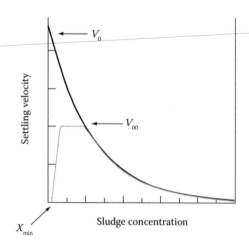

FIGURE 11.12
Illustration of the Vesilind equation (black) and the Takacs (gray) settling models.

(Comas et al. 2008), and has been used in WWTP simulation studies as well (Flores-Alsina et al. 2009). Clearly, given the limited mechanistic knowledge on those phenomena, the proposed combination of empirical and mechanistic modeling (Comas et al. 2008) is a significant improvement compared to simply ignoring plant operation effects on sludge settling properties.

Finally, for the complex 2-D and 3-D settler models, a difficult issue that still waits for a generally accepted solution is how to represent the effect of biomass concentration on the liquid viscosity.

11.6 Anaerobic Digester Models

11.6.1 Anaerobic Digestion Principles

The AD process is a well-established treatment process, primarily used for treatment of waste sludge (Figure 11.1) and high-strength organic wastes, but also for solid waste, agricultural crops, etc. The process is advantageous for several reasons but mainly due to the low biomass yields and because energy, in the form of methane, can be produced as a result of the biological conversion of organic substrates.

The AD process can decompose organic matter provided that several groups of microorganisms are present in an oxygen-free environment. In the natural environment anaerobic degradation is found in swamps, intestines of ruminants, and sediments of lakes and oceans. The temperature is an important factor for AD processes and normally they are operated in the mesophilic temperature range (30–37°C). However, there is an increased interest in operation at thermophilic conditions (50–57°C), which allows for higher reaction rates, a higher degree of degradation, and increased bacterial destruction.

The process involves several steps for degrading particulate organic matter to intermediary products that in turn can be degraded to methane and other end products. The main steps are given in Figure 11.13 and shortly outlined below:

- *Hydrolysis*—conversion of particulate and soluble polymers (e.g., protein, carbohydrates, lipids) into soluble products, such as amino acids, monosaccharides, and long chain fatty acids by enzymatic degradation. Hydrolysis is often the rate limiting reaction for AD processes where the substrate is in particular form (Vavilin et al. 1996). It is normally modeled as a first-order reaction although the true reactions are highly complex (Batstone et al. 2002).

- *Acidogenesis (i.e., fermentation)*—the products from the hydrolysis step are further degraded into several compounds, mainly acetate, propionate, butyrate, alcohols, formate, hydrogen, and carbon dioxide. The process involves a number of different types of microorganisms

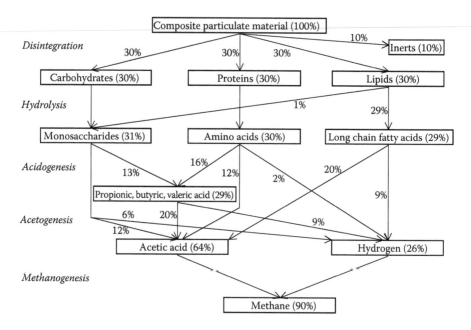

FIGURE 11.13

COD flux for a particulate composite comprised 10% inerts and 30% each of carbohydrates, proteins, and lipids (in terms of COD). Propionic acid (10%), butyric acid (12%), and valeric acid (7%) are grouped together in the figure. Reaction pathways according to the Anaerobic Digestion Model No. 1. (From Batstone et al., *Anaerobic Digestion Model No. 1*. IWA Publishing, London, UK, 2002.)

and is often the fastest step in AD of complex organic matter (Vavilin et al. 1996).

- *Acetogenesis*—acetate, hydrogen, and carbon dioxide are produced from degradation of long chain fatty acids and volatile fatty acids.
- *Methanogenesis*—methanogens (archaebacteria) obtain energy from two different pathways, either by cleaving acetate to methane and carbon dioxide (aceticlastic methanogenesis), that is,

$$CH_3COOH \rightarrow CH_4 + CO_2$$

or by converting hydrogen and carbon dioxide into methane (hydrogenotrophic methanogenesis) as

$$CO_2 + 4H_2 \rightarrow CH_4 + 2H_2O$$

The aceticlastic methanogenesis is responsible for the major part of methane production in normal AD systems (about 70%, as a rule of thumb). Minor gas quantities of nitrogen, hydrogen, ammonia, and hydrogen sulfide are also generated. Because of the slow growth rate of methanogens, this process is often rate limiting (Tchobanoglous et al. 1993).

11.6.2 Anaerobic Digestion Models

A large number of AD models exist, varying widely in terms of complexity and detail. For the majority of all AD modeling studies, the assumption of a CSTR is used (for suspended growth systems). During normal operating conditions, when the process is working properly, a number of "rules-of-thumb" can be used to estimate the steady-state methane production based on information of temperature, retention time, and loading rate (COD or VSS). However, as AD systems are complex processes that unfortunately often suffer from instability and mathematical modeling, dynamic simulations may represent the best possibility to predict their behavior. The instability is usually witnessed as a drop in the methane production rate, a drop in the pH, a rise in the volatile fatty acids concentration, causing digester failure (Lyberatos and Skiadas 1999).

AD is a multistep process involving the action of multiple microbes (see Section 11.6.1). Usually, such processes contain a "rate-limiting" step, which limits the rate of the overall process. The first attempts for modeling AD led to models describing only such a limiting step (e.g., Andrews 1971, acetogenic methanogenesis; O'Rourke 1968, conversion of fatty acids to biogas; Eastman and Ferguson 1981, hydrolysis of biodegradable suspended solids). Such models are simple and easy to use, but do not always describe very well the digester behavior, especially during transient operating conditions. Moreover, for AD processes the limiting step may depend on wastewater characteristics, hydraulic loading, temperature, etc. (Speece 1983).

Most of the early AD models are based on Monod kinetics to represent the growth rate of microorganisms (e.g., Andrews 1971; Hill and Barth 1977; Chen 1983; Hill 1983; Moletta et al. 1986), in the same way as in activated sludge models. The principle also holds for more recent AD models. However, as the number of different types of organisms that establish different reaction pathways of importance is normally higher within AD systems than in activated sludge systems the number and complexity of the differential equations increases.

Temperature represents a fundamental variable for AD processes and can affect biochemical reactions in five main ways: (1) increase the reaction rates with increasing temperature (as predicted by the Arrhenius equation); (2) decrease the reaction rates with increasing temperature above the optimum (>40°C for mesophilic and >65°C for thermophilic operation); (3) decrease yields due to increased turnover and maintenance energy with increased temperatures; (4) cause shifts in yield and reaction pathways due to changes in thermodynamic yields and microbial population; (5) increase the death rate of microorganisms due to increased lysis and maintenance.

The temperature dependence of the different groups of organisms in AD follows the Arrhenius equation up to a temperature optimum, followed by a rapid drop to zero. Often the temperature of a modeled AD process is assumed to be controlled (±3°C), in which case the biochemical parameters can be modeled without temperature dependency. In case of larger temperature fluctuation within either the mesophilic or thermophilic ranges, a double Arrhenius equation can be used (Pavlostathis and Giraldo-Gomez 1991), whereas in case of fluctuations between mesophilic and thermophilic conditions two different sets of parameters are normally used. Also, a number of physicochemical parameters depend on temperature and their variations cause impacts on the model behavior, which are generally more important than those due to changes in biochemical parameters (Batstone et al. 2002). The van't Hoff equation represents a reasonable way of describing how these parameters vary as a function of temperature (Stumm and Morgan 1996).

Because the AD system is susceptible to various types of inhibition (pH, free ammonia inhibition of aceticlastic methanogenesis, hydrogen inhibition of acetogenic bacteria, etc.), it is important to include such effects in the models. Inhibition can be modeled in numerous ways, such as noncompetitive, uncompetitive, competitive, and empirical inhibition (Batstone et al. 2002). An example of an empirical "lower pH inhibition only" function is given in Equation 11.13.

$$
I_{pH} = \begin{cases} \exp\left(-3\left(\dfrac{pH - pH_{UL}}{pH_{UL} - pH_{LL}}\right)^2\right) & : pH < pH_{UL} \\ 1 & : pH > pH_{UL} \end{cases} \tag{11.13}
$$

In Equation 11.13, index UL indicates "upper limit" and LL "lower limit." However, to dynamically model the pH of the system itself is no simple task. It requires knowledge about the acid–base system, buffer system, gas–liquid interactions (e.g., for CO_2 and NH_3), and defining an overall charge balance of the system. For this, reason alkalinity—which is easier to model—is often used as indicator variable in models to ensure "stable" pH, similar to activated sludge models. However, when total alkalinity is low the pH may become unstable and a pH model is required. Because pH significantly influences the stability and performance of AD processes, a pH model should be included.

Since the AD system is essentially a three-phase system (solids, liquid, and gas), considerable efforts are needed to model the dynamics between the gas phase and the liquid phase. Within the gas phase, it is normally assumed that the ideal gas law is obeyed but the transfer of soluble gas into the liquid phase and from the liquid into the gas phase is not at equilibrium but needs to be modeled dynamically. The most commonly used equations follow the two-film theory of Whitman (1923).

Two of the first models capable of describing all the main reaction steps in Figure 11.13 are those of Angelidaki et al. (1993, 1999) and Siegrist et al. (1993, 2002), and they have formed the basis for a significant part of the model development during the past decade. Also, the work of Vavilin et al. (1994, 1996) has been of great importance. However, all models described so far consider organic matter as a whole and do not account for the nature of the organic macromolecules in the feed composition (i.e., proteins, carbohydrates, and lipids).

In 1999, the modeling principles of AD were considered mature enough to consolidate the important features of AD modeling, as described above, and form a common platform for AD modeling and simulation: the Anaerobic Digestion Model No. 1 (ADM1; Batstone et al. 2002). The ADM1 can be considered as the core model for AD modeling. The aim of the model is to allow predictions of sufficient accuracy to be useful in process development, operation, and optimization. The open structure and common nomenclature of the model should also encourage fast and efficient development of specific add-ons as required for different purposes (see below).

The ADM1 is a structured model with disintegration, hydrolysis, acidogenesis, acetogenesis, and methanogenesis steps (Figure 11.13). Extracellular solubilization is divided into disintegration and hydrolysis, of which the first is a largely nonbiological step and converts composite particulate substrate into inerts, particulate carbohydrates, protein, and lipids. The hydrolysis step is based on enzymatic reactions converting the above into monosaccharides, amino acids, and long chain fatty acids by first-order kinetics. The disintegration step was included to simplify the use of the ADM1 for different types of substrates (primary sludge, secondary sludge, organic household waste, manure, etc.). Two separate groups of acidogens degrade monosaccharides and amino acids into mixed organic acids, hydrogen, and carbon dioxide. The organic acids are subsequently converted to acetate, hydrogen, and carbon dioxide by acetogenic groups that utilize long chain fatty acids, butyrate, valerate, and propionate. The hydrogen produced by these organisms is consumed by hydrogen-utilizing methanogenic organisms and the acetate by an acticlastic methanogenic organism group. Substrate-based Monod-type kinetics are used as the basis for all intracellular biochemical reactions. Death of biomass is represented by first-order kinetics, and dead biomass is maintained in the system as composite particulate material. Inhibition functions include pH, hydrogen, and free ammonia. The other uptake regulating functions are secondary Monod kinetics for inorganic nitrogen to prevent growth when nitrogen is limited. Mechanisms included to describe physicochemical processes are acid–base reactions to calculate concentrations of hydrogen ions, free ammonia and carbon dioxide, and nonequilibrium liquid–gas transfer. The ADM1 is considerably more complex than standard activated sludge models: If implemented as a set of ODEs, the model contains 32 dynamic state concentration variables and an additional six acid–base kinetic processes, 19 biochemical kinetic processes, and three gas–liquid transfer kinetic processes per reactor. The number of parameters included in the model is close to 100. A detailed description of all ADM1 equations, parameters, etc., can be found in the report of Rosen and Jeppsson (2006).

11.6.3 ADM1 Limitations

The ADM1 is defined for CSTR systems but the key elements of the model are largely independent of the type of application and should be more or less applicable in a number of other systems, such as solid phase digesters, plug-flow reactors, and biofilm reactors (Batstone et al. 2002). Although ADM1 represents a state-of-the-art model, a number of potentially important processes are omitted from the model, primarily due to limited available scientific knowledge. These processes include:

- Glucose alternative products (regulation of products from acidogenesis of glucose and alternatives, such as lactate and ethanol)
- Sulfate reduction and sulfide inhibition
- Nitrate, phosphate
- Weak acid and base inhibition
- Long chain fatty acids inhibition
- Acetate oxidation
- Homoacetogenesis (may compete with hydrogen utilizing methanogens to produce acetate at low temperatures)
- Solids precipitation (inorganic carbon, nitrogen, and metallic ions precipitate to solids, e.g., struvite formation)

Gradually, various add-ons to the ADM1 are presented in the scientific literature, which partly address the problems cited above.

The existing detailed AD models do not take into account the particular nature of granular sludge in systems, such as UASB and EGSB. Some suggestions of how to do this have been made, but in general it should be stated that the predictive capabilities, related to operation, failure, and possible remedies, of the models describing CSTR AD systems are significantly higher.

11.7 Current WWTP Model Development Issues

Growing computing power has a tremendous impact on WWTP modeling. In this context, several interesting developments are worth reporting and will be discussed below: increased size of simulation jobs, for example, in plant-wide models; emphasis on good modeling practice; emphasis on uncertainty and sensitivity analysis of WWTP models. Essential developments are for a large part driven by the International Water Association (IWA), through the work of the Task Group for Good Modelling Practice (GMP), the Task Group on Benchmarking of Control Strategies for WWTPs and the recently established Task Group on Design and Operations Uncertainty (DOUTGroup).

11.7.1 Plant-Wide Models

In the twentieth century, activated sludge modeling studies mainly focused on the dynamics of the secondary treatment (Figure 11.1), that is, the activated sludge tanks and the sedimentation tank, and the interactions between these units. Nowadays, plant-wide models—models where the interactions between primary, secondary, and sludge treatment train are investigated—become more common. An example is the Benchmark Simulation Model No. 2 developed by the IWA Task Group on Benchmarking of Control Strategies for WWTPs (BSM2, www.benchmarkwwtp.org) for studying and comparison of plant-wide control strategies (Jeppsson et al. 2007).

However, new modeling challenges have arisen related to plant-wide modeling, for example, when a model of the AD process is to be integrated with other models describing different processes in a WWTP. The main challenges in obtaining integrated model plants arise from the incompatibilities and different descriptions of the model components and transformations in standard unit process models such as the ASM1 or the ADM1. These include varying descriptions of organic carbonaceous substrates and organic N, as well as pH and buffer capacity in water or sludge and the different processes considered, etc. With respect to this problem, three main plant-wide modeling approaches have been proposed so far. The first approach is based on the construction of a "Supermodel" consisting of all the components and transformations needed to reproduce every process within the entire plant (Jones and Takàcs 2004). In this model, components and transformations are common to every unit process model in the WWTP and, therefore, specific transformers connecting different process models are not required. Nevertheless, the use of a unique Supermodel for any WWTP lacks the flexibility to add or remove components as well as transformations depending on the case study and model aims. Another significant drawback to this approach is the continuous increase of the model size required to progressively adapt the Supermodel to reproduce new processes. The second approach, known as the interfaces approach, is based on the construction of transformers among existing standard models. An illustrative example of transformers between the ASM1 and the ADM1 has been proposed by Nopens et al. (2009). In order to guarantee mass and charge continuity in the model interfaces, Vanrolleghem et al. (2005) proposed a general methodology (CBIM) for the interface of any two standard models. However, although the interfaces approach facilitates the construction of integrated models tailored to the case study, there are some limitations when it comes to properly transforming the model components among existing models, guaranteeing mass and charge continuity under any dynamic condition (Grau et al. 2007). Finally, Grau et al. (2007) proposed a combined approach based on the most appropriate transformations for each specific case study, called the transformation-based approach.

The complexity of models such as the ADM1 may also induce numerical problems when performing computer simulations. A system is called stiff when the range of the model time constants is large. This means that some of the system states react quickly whereas some react sluggishly. The ADM1 is a *very* stiff system with time constants ranging from fractions of a second to months. This makes the simulation of such a system challenging and in order to avoid excessively long simulation times, one needs to be somewhat creative when implementing the model. The selection of appropriate numerical solvers is also of importance. For example, by rewriting the two equations for pH and hydrogen gas in the ADM1 as implicit algebraic equations and solving those using an iterative numerical method at each integration step, the simulation speed for the ADM1 may be increased 10–100 times (Rosen et al. 2006).

11.7.2 Good Modeling Practice

Development projects such as the Benchmark Simulation Model No. 1 (BSM1) (Copp 2002) have led to a growing awareness about the fact that model implementation and exchange of models between model users is not at all straightforward (for details, see Gernaey et al. 2006b). Models are implemented into a simulation environment, and practical experience has demonstrated that it is almost impossible to obtain error-free model implementations based solely on this information, especially for the complex models mentioned above. The following error sources can be identified: (1) simplifications of the original model implementation when translating it to the publication describing the model; (2) typing errors in the paper, for example, copy and paste errors; (3) incomplete model description in the paper (intended or due to a fault); (4) scattered description of the model in the paper; (5) misinterpretation of the publication providing the model description, during the model implementation phase; (6) errors during the transformation of the equations into the modeling language; (7) general programming bugs.

The above list illustrates that many error sources are caused by the lack of standardized methods for documenting and exchanging models. Such problems were partly resolved by the recent work of the IWA Task Group for Good Modelling Practice, who have evaluated the publications describing the most popular activated sludge models and have identified the typing errors, inconsistencies, and gaps in those model publications (Hauduc et al. 2010). Moreover, Hauduc et al. (2010) also propose a systematic approach to verify models by tracking typing errors and inconsistencies in model development and software implementation. The errors found are reported in detail (Hauduc et al. 2010), and can be downloaded as a spreadsheet that provides corrected model matrices including both the calculation of all stoichiometric coefficients and an example of proper continuity checks for the models that were investigated. Note that validated implementations of some of those models (e.g., ASM1, ADM1, nonreactive 1-D Takács settler model) are also available—at least for a number of commonly used software platforms—through the work of the IWA Task Group on Benchmarking of Control Strategies for WWTPs (Jeppsson et al. 2007).

Another recent initiative worth reporting is the recent proposal of a standardized notation in WWTP modeling (Corominas et al. 2010), which should also facilitate easier communication of WWTP models and modeling studies, and can be considered as one of the main results of the recently established biyearly IWA/WEF (Water Environment Federation) Wastewater Treatment Modelling Seminar in Quebec.

11.7.3 Uncertainty and Sensitivity Analysis

Increased availability of computer power has recently resulted in a significant increase of the application of uncertainty and sensitivity analysis to WWTP models. Particularly for design purposes, the use of uncertainty analysis has been demonstrated in the studies of Rousseau et al. (2001), Bixio et al. (2002), Benedetti et al. (2008), McCormick et al. (2008), and Sin et al. (2009). Flores-Alsina et al. (2008) applied uncertainty analysis for comparing different control strategy options on a WWTP model, and demonstrated that considering uncertainty has important implications on decision making, for example, which controller is better for the WWTP in question. These studies have in common that they all demonstrated potential benefits of uncertainty analysis, for example, uncertainty analysis offers a quantitative basis to justify safety factors as well as better informed decision making thereby contributing to cost savings in engineering projects. That said, the field of uncertainty analysis of WWTP models is still in its infancy as there are a range of issues that need further research, as reflected in the recent paper on uncertainty by Belia et al. (2009). The work of the IWA Task Group on Design and Operations Uncertainty now drives the progress in this field.

Application of activated sludge models is typically based on a number of assumptions relating to wastewater composition, influent load profiles, and values of various parameters of the specified model, for example, the ASM1 (Henze et al. 2000; Sin et al. 2005, 2008a). Uncertainty about these parameters is called input uncertainty, since the model parameters are fed into the model to make a prediction. The input uncertainty may be characterized by a certain range of values reflecting the limited knowledge about the exact value of the parameter in question. For instance, the value of the aerobic heterotrophic yield in ASM1 is specified as being in the range from 0.6 to 0.7 g COD/g COD (Henze et al. 2000) rather than having an exact value.

Uncertainty is, however, not limited to input uncertainty. Indeed, there are different sources of uncertainty affecting model predictions (McKay et al. 1999; Walker et al. 2003; Refsgaard et al. 2007). Various sources of uncertainty may be grouped as (McKay et al. 1999): (1) *input (subjective) uncertainty*—reflects lack of knowledge about the model inputs as illustrated above; (2) *structural uncertainty*—relates to the mathematical form of the model (models are approximations to systems rather than an exact copy); and (3) *stochastic uncertainty*—this may be a component of the model itself (e.g., a random process to model failure events of pumps, etc.).

In general, uncertainty analysis is concerned with propagation of the various sources of uncertainty to the model output. The uncertainty analysis leads to probability distributions of model outputs, which are then used to infer the mean, variance, and quantiles of model predictions (Helton and Davis 2003). The sensitivity analysis, on the other hand, aims at identifying and quantifying the individual contributions of the uncertain inputs to the output uncertainty. Uncertainty and sensitivity analysis should preferably be performed in tandem with each other (Saltelli et al. 2008), and are both computationally very demanding, especially if global sensitivity analysis is performed.

11.8 Guidelines for Application of WWTP Models

Typical WWTP model development practice involves (1) the definition of the modeling objective; (2) data collection and quality check; (3) formulation of the mathematical models for different plant units/processes, for example, for hydraulics, biological, settling, physical, and chemical processes; (4) steady-state and dynamic model calibration; and (5) validation and scenario analysis.

In reality, there is a broad variety of process configurations employed at full-scale plants. Moreover, there are often several process-specific recycle streams within the plant (e.g., reject water, return sludge, internal recirculation), which significantly influence the plant dynamics. As a consequence, the development of a WWTP model is not a straightforward task. A successful and efficient WWTP modeling project—in terms of the cost of obtaining the required modeling quality—demands a great deal of expert and process knowledge. The latter is often defined as the ad hoc and heuristic modeling approach. For example, process engineering and modeling experience are essential for deciding which submodel to choose for describing different plant units/processes, how to fractionate influent wastewater composition, which data are needed for model calibration, which parameters should be calibrated and how, etc. In fact, attempting to develop a plant model without sufficient expert knowledge is believed to lead to nonsensical results. To this end, several guidelines were developed capitalizing on experiences resulting from numerous modeling studies performed mainly in the 1990s, primarily on municipal WWTPs, resulting in the BIOMATH protocol (Petersen et al. 2002, 2003; Vanrolleghem et al. 2003), the STOWA protocol (Hulsbeek et al. 2002), the HSG protocol (Langergraber et al. 2004), and the WERF protocol (Melcer et al., 2003). These guidelines aim to bring standardization, provide guidance to the users, and ground the modeling procedure on a systematic framework. These guidelines have a lot in common, for example, all agree on that modeling should be objective driven. However, significant differences exist too, which can be seen in a strength, weakness, opportunities,

LIST OF SYMBOLS AND ABBREVIATIONS

AD	Anaerobic digestion
ADM1	Anaerobic Digestion Model No. 1
AOB	Ammonium oxidizing bacteria
ASM1	Activated Sludge Model No. 1
ASM2	Activated Sludge Model No. 2
ASM2d	Activated Sludge Model No. 2d
ASM3	Activated Sludge Model No. 3
Bio-P	Biological phosphorus removal
BSM1	Benchmark Simulation Model No. 1
CFD	Computational fluid dynamics
COD	Chemical oxygen demand
COD_{tot}	Total chemical oxygen demand
CSTR	Continuously stirred tank reactor
EGSB	Expanded granular sludge bed
ESS	Effluent suspended solids
GHG	Greenhouse gas
GMP	Good Modeling Practice
HRT	Hydraulic retention time
N	Nitrogen
N_{tot}	Total nitrogen
NOB	Nitrite oxidizing bacteria
ODE	Ordinary differential equation
P	Phosphorus
PAO	Polyphosphate accumulating organism
PDE	Partial differential equation
SBH	Sludge blanket height
SRT	Sludge retention time
SS	Suspended solids
TSS	Total suspended solids
TUDP model	Technical University Delft Phosphorus model
UASB	Upflow anaerobic sludge blanket
WWTP	Wastewater treatment plant

TABLE 11.4

Strengths, Weaknesses, Opportunities, and Threats (SWOT) of the Different Calibration Protocols

	Strengths	Weaknesses	Opportunities	Threats
BIOMATH	– Detailed settling, hydraulic and biological characterization – Detailed influent characterization – Biomass characterization – Sensitivity analysis/parameter selection, – OED for measurement campaign design – Structured overview of protocol – Feedback loops	– Respirometric influent characterization requires model-based interpretation – OED has not been applied yet in practice but research is ongoing – OED software and specialist required – No detailed methodology for data quality check – No practical procedure for parameter calibration	– Generally applicable – Works efficiently once implemented in a simulator – Dynamic measurement campaigns can be designed and compared based on OED	– Not all modeling and simulation software have OED/Sensitivity Analysis (SA) – High degree of specialization is required for the application
STOWA	– Detailed settling and biological characterization – Process control – Time estimate for different calibration steps – Detailed data quality check – Stepwise calibration of biological process parameters – Structured overview of protocol – Feedback loops	– No detailed hydraulic characterization – BOD test gives problems (f_p) – No biomass characterization – No guidance for measurement campaign design – No detailed info on sensitivity analysis – Fixed parameter subsets for calibration of biological processes	– Easy to use – Practical experimental methods – No specialist required – Good for consultants and new modelers	– No mathematical/statistical approach for parameter selection for calibration – May not be applicable for different systems since parameter subset for calibration may change for different WWTPs

HSG	– CFD for hydraulic characterization – Biological characterization – Design of measurement campaign – A standard format for documentation – Data quality check – Structured overview of protocol	– No feedback loops in overview diagram – Provides only general guidelines – No detailed settling characterization – No particular methods for influent characterization or parameter estimation – No detailed sensitivity analysis/parameter selection	– Generally applicable – A standard format for thorough documentation/reporting of calibration studies	– Not detailed/practical enough for new practitioners – The free choice of experimental methodologies for influent/kinetic characterization may jeopardize standardization of calibration studies
WERF	– Detailed influent characterization – Detailed μ_A and b_A determination – Biomass characterization – Sensitivity analysis/parameter selection – Detailed data quality check – A tiered approach for calibration – Several examples of case studies	– No feedback loops – Settling process less emphasized – Almost no emphasis on other kinetic parameters than nitrification – No structured overview of protocol	– Based on practical experience – A tiered approach for calibration provides different calibration levels for different goals and accuracy of calibration – Good for consultants and new modelers	– Focus on μ_A determination and influent characterization – Ignoring the significance of the other compartments of the full-scale model – Laborious methods

Source: Sin et al., *Water Research,* 43, 2894–2906, 2009.

and threat analysis (SWOT) of these guidelines as presented in Table 11.4. The IWA Task Group on Good Modelling Practice is currently working on condensing existing modeling guidelines and experiences into a technical report. It is in general recommended to follow a guideline when performing a WWTP modeling study.

References

Andrews, J. F. 1971. Kinetic models of biological waste treatment processes. *Biotechnology and Bioengineering Symposium* 2: 5–33.

Angelidaki, I., L. Ellegaard, and B. K. Ahring. 1993. A mathematical model for dynamic simulation of anaerobic digestion of complex substrates: Focussing on ammonia inhibition. *Biotechnology and Bioengineering* 42: 159–166.

Angelidaki, I., L. Ellegaard, and B. K. Ahring. 1999. A comprehensive model of anaerobic bioconversion of complex substrates to biogas. *Biotechnology and Bioengineering* 63: 363–372.

Batstone, D. J., J. Keller, I. Angelidaki, S. V. Kalyuzhnyi, S. G. Pavlostathis, A. Rozzi, W. T. M. Sanders, H. Siegrist, and V. A. Vavilin. 2002. *Anaerobic Digestion Model No. 1*. IWA STR No. 13, London, UK: IWA Publishing.

Belia, E., Y. Amerlinck, L. Benedetti, B. Johnson, G. Sin, P. A. Vanrolleghem, K. V. Gernaey, S. Gillot, M. B. Neumann, L. Rieger, A. Shaw, and K. Villez. 2009. Wastewater treatment modelling: dealing with uncertainties. *Water Science and Technology* 60 (8): 1929–1941.

Benedetti, L., D. Bixio, F. Claeys, and P. A. Vanrolleghem. 2008. Tools to support a model-based methodology for emission/immission and benefit/cost/risk analysis of wastewater systems that considers uncertainty source. *Environmental Modelling and Software* 23: 1082–1091.

Bixio, D., G. Parmentier, D. Rousseau, F. Verdonck, J. Meirlaen, P. A. Vanrolleghem, and C. Thoeye. 2002. A quantitative risk analysis tool for design/simulation of wastewater treatment plants. *Water Science and Technology* 46 (4–5): 301–307.

Brdjanovic, D., M. C. M. van Loosdrecht, P. Versteeg, C. M. Hooijmans, G. J. Alaerts, and J. J. Heijnen. 2000. Modeling COD, N and P removal in a full-scale WWTP Haarlem Waarderpolder. *Water Research* 34: 846–858.

Chen, Y. R. 1983. Kinetic analysis of anaerobic digestion of pig manure and its design implications. *Agricultural Wastes* 8: 65–81.

Comas, J., I. Rodriguez-Roda, M. Poch, K. V. Gernaey, C. Rosen, and U. Jeppsson. 2008. Risk assessment modelling of microbiology-related solids separation problems in activated sludge systems. *Environmental Modelling and Software* 23: 1250–1261.

Copp, J. B., ed. 2002. *The COST Simulation Benchmark—Description and Simulator Manual*. ISBN 92-894-1658-0, Luxembourg: Office for Official Publications of the European Communities.

Corominas, L. L., L. Rieger, I. Takács, G. Ekama, H. Hauduc, P. A. Vanrolleghem, Oehmen, A., K. V. Gernaey, M. C. M. van Loosdrecht, and Y. Comeau. 2010. New framework for standardized notation in wastewater treatment modelling. *Water Science and Technology* 61 (4): 841–857.

Eastman, J. A., and J. F. Ferguson. 1981. Solubilization of particulate organic carbon during the acid phase of anaerobic digestion. *Journal Water Pollution Control Federation* 53: 352–366.

Esener, A. A., J. Roels, and N. W. F. Kossen. 1983. Theory and applications of unstructured growth models: Kinetic and energetic aspects. *Biotechnology and Bioengineering* 25: 2803–2841.

Flores-Alsina, X., J. Comas, I. Rodriguez-Roda, K. V. Gernaey, and C. Rosen. 2009. Including the effects of filamentous bulking sludge during the simulation of wastewater treatment plants using a risk assessment model. *Water Research* 43: 4527–4538.

Flores-Alsina, X., I. Rodríguez-Roda, G. Sin, and K. V. Gernaey. 2008. Multi-criteria evaluation of wastewater treatment plant control strategies under uncertainty. *Water Research* 42: 4485–4497.

Gernaey, K. V., and S. B. Jørgensen. 2004. Benchmarking combined biological phosphorus and nitrogen removal wastewater treatment processes. *Control Engineering Practice* 12: 357–373.

Gernaey, K. V., U. Jeppsson, D. J. Batstone, and P. Ingildsen. 2006a. Impact of reactive settler models on simulated WWTP performance. *Water Science and Technology* 53 (1): 159–167.

Gernaey, K. V., C. Rosen, D. J. Batstone, and J. Alex. 2006b. Efficient modelling necessitates standards for model documentation and exchange. *Water Science and Technology* 53 (1): 277–285.

Grau, P., M. de Gracia, and P. A. Vanrolleghem. 2007. A new plant-wide modelling methodology for WWTPs. *Water Research* 41: 4357–4373.

Gujer, W., M. Henze, T. Mino, and M. C. M. van Loosdrecht. 1999. Activated Sludge Model No. 3. *Water Science and Technology* 39 (1): 183–193.

Gujer, W., and T. A. Larsen. 1995. The implementation of biokinetics and conservation principles in ASIM. *Water Science and Technology* 31 (2): 257–266.

Hauduc, H., L. Rieger, I. Takács, A. Héduit, P. A. Vanrolleghem, and S. Gillot. 2010. A systematic approach for model verification: Application on seven published Activated Sludge Models. *Water Science and Technology* 61 (4): 825–839.

Heijnen, J. J. 1999. Bioenergetics of microbial growth. In *Encyclopedia of Bioprocess Technology: Fermentation, Biocatalysis and Bioseparation*, eds. M. C. Flickinger, and S. W. Drew, 267–291. John Wiley & Sons Inc.

Hellinga, C., M. C. M. van Loosdrecht, and J. J. Heijnen. 1999. Model based design of a novel process for nitrogen removal from concentrated flow. *Mathematical and Computer Modelling of Dynamical Systems* 5: 351–371.

Helton, J. C., and F. J. Davis. 2003. Latin hypercube sampling and the propagation of uncertainty in analyses of complex systems. *Reliability Engineering and System Safety* 81: 23–69.

Henze, M., C. P. L. Grady, Jr., W. Gujer, G. v. R. Marais, and T. Matsuo. 1987. *Activated Sludge Model No. 1.* IAWQ Scientific and Technical Report No. 1, London, UK.

Henze, M., W. Gujer, T. Mino, T. Matsuo, M. C. Wentzel, and G. v. R. Marais. 1995. *Activated Sludge Model No. 2.* IWA Scientific and Technical Report No. 3, London, UK.

Henze, M., W. Gujer, T. Mino, T. Matsuo, M. C. Wentzel, G. v. R. Marais, and M. C. M. van Loosdrecht. 1999. Activated Sludge Model No. 2d, ASM2D. *Water Science and Technology* 39 (1): 165–182.

Henze, M., W. Gujer, T. Mino, M. C. M., van Loosdrecht. 2000. *Activated Sludge Models ASM1, ASM2, ASM2d and ASM3*. IWA Scientific and Technical Report No. 9, London, UK: IWA Publishing.

Hill, D. T. 1983. Simplified Monod kinetics of methane fermentation of animal wastes. *Agricultural Wastes* 5: 1–16.

Hill, D. T., and C. L. Barth. 1977. A dynamic model for simulation of animal waste digestion. *Journal Water Pollution Control Federation* 49 (10): 2129–2143.

Hirsch, C. 1997. *Numerical Computation of Internal and External Flows. Vol. 1, Fundamentals of Numerical Discretization*, 515 pp. West Sussex, UK: John Wiley & Sons.

Hulsbeek, J. J. W., J. Kruit, P. J. Roeleveld, and M. C. M. van Loosdrecht. 2002. A practical protocol for dynamic modelling of activated sludge systems. *Water Science and Technology* 45 (6): 127–136.

Ingildsen, P., C. Rosen, K. V. Gernaey, M. K. Nielsen, and B. N. Jakobsen. 2006. Modelling and control strategy testing of biological and chemical phosphorus removal at Avedøre WWTP. *Water Science and Technology* 53 (4–5): 105–113.

Jeppsson, U. 1996. Modelling aspects of wastewater treatment processes. PhD thesis, Sweden: IEA, Lund University.

Jeppsson, U., M. N. Pons, I. Nopens, J. Alex, J. Copp, K. V. Gernaey, C. Rosen, J.-P. Steyer, and P. A. Vanrolleghem. 2007. Benchmark Simulation Model No. 2—General protocol and exploratory case studies. *Water Science and Technology* 56 (8): 67–78.

Jones, R. M., and I. Takàcs. 2004. Importance of anaerobic digestion modelling on predicting the overall performance of wastewater treatment plants. In: *Proceedings 10th World Congress on Anaerobic Digestion*, Montréal, Canada, Aug. 29–Sept. 2, 2004.

Koch, G., M. Kuhni, W. Gujer, and H. Siegrist. 2000. Calibration and validation of activated sludge model no. 3 for Swiss municipal wastewater. *Water Research* 34: 3580–3590.

Kuba, T., M. C. M. van Loosdrecht, E. Murnleitner, and J. J. Heijnen. 1997. Kinetics and stoichiometry in the biological phosphorus removal process with short cycle times. *Water Research* 31: 918–928.

Ky, R. C., Y. Comeau, M. Perrier, and I. Takács. 2001. Modelling biological phosphorus removal from a cheese factory effluent by an SBR. *Water Science and Technology* 43 (3): 257–264.

Langergraber, G., L. Rieger, S. Winkler, J. Alex, J. Wiese, C. Owerdieck, M. Ahnert, J. Simon, and M. Maurer. 2004. A guideline for simulation studies of wastewater treatment plants. *Water Science and Technology* 50 (7): 131–138.

Lyberatos, G., and I. V. Skiadas. 1999. Modelling of anaerobic digestion—a review. *Global Nest: The International Journal* 1 (2): 63–76.

Maurer, M., D. Abramovich, H. Siegrist, and W. Gujer. 1999. Kinetics of biologically induced phosphorus precipitation in waste-water treatment. *Water Research* 33: 484–493.

Maurer, M., and M. Boller. 1999. Modelling of phosphorus precipitation in waste-water treatment plants with enhanced biological phosphorus removal. *Water Science and Technology* 39 (1): 147–163.

McCormick, J., B. Johnson, and P. A. Vanrolleghem. 2008. Risk-based wastewater treatment plant design: Applying Monte Carlo simulation to process evaluation. In: *Proceedings 1st IWA/WEF Wastewater Treatment Modelling Seminar, June 1–3 2008*: Mont Sainte-Anne, Quebec, Canada.

McKay, M. D., J. D. Morrison, and S. C. Upton. 1999. Evaluating prediction uncertainty in simulation models. *Computer Physics Communications* 117: 44–51.

Meijer, S. C. F. 2004 Theoretical and practical aspects of modelling activated sludge processes. PhD thesis, Delft University of Technology, the Netherlands.

Melcer, H., P. L. Dold, R. M. Jones, C. M. Bye, I. Takacs, H. D. Stensel, A. W. Wilson, P. Sun, and S. Bury. 2003. Methods for wastewater characterisation in activated sludge modeling, VA, USA: Water Environment Research Foundation (WERF), Alexandria.

Moletta, R., D. Verrier, and G. Albagnac. 1986. Dynamic modeling of anaerobic digestion. *Water Research* 20: 427–434.

Murnleitner, E., T. Kuba, M. C. M. van Loosdrecht, and J. J. Heijnen. 1997. An integrated metabolic model for the aerobic and denitrifying biological phosphorus removal. *Biotechnology and Bioengineering* 54: 434–450.

Noorman, H. J., J. J. Heijnen, and K. Ch. A. M. Luyben. 1991. Linear relations in microbial reaction systems: A general overview of their origin, form and use. *Biotechnology and Bioengineering* 38: 603–618.

Nopens, I., D. Batstone, J. B. Copp, U. Jeppsson, E. I. P. Volcke, J. Alex, and P. A. Vanrolleghem. 2009. A practical ASM/ADM model interface for enhanced dynamic plantwide simulation. *Water Research* 43: 1913–1923.

O'Rourke, J. T. 1968. Kinetics of anaerobic treatment at reduced temperatures. PhD thesis, Stanford, California, USA: Stanford University.

Pavlostathis, S. G., and E. Giraldo-Gomez. 1991. Kinetics of anaerobic treatment: a critical review. *Critical Reviews in Environmental Control*, 21: 411–490.

Petersen, B., K. Gernaey, M. Henze, and P. A. Vanrolleghem. 2002. Evaluation of an ASM1 model calibration procedure on a municipal–industrial wastewater treatment plant. *Journal of Hydroinformatics* 4: 15–38.

Petersen, B., K. Gernaey, M. Henze, and P. A. Vanrolleghem. 2003. Calibration of activated sludge models: a critical review of experimental designs. In *Biotechnology for the Environment: Wastewater Treatment and Modeling, Waste Gas Handling*, eds. S. N. Agathos, W. Reineke, 101–186. Dordrecht, the Netherlands: Kluwer Academic Publishers.

Refsgaard, J. C., J. P. van der Sluijs, A. L. Højberg, and P. A. Vanrolleghem. 2007. Uncertainty in the environmental modelling process—a framework and guidance. *Environmental Modelling and Software* 22: 1543–1556.

Rieger, L., G. Koch, M. Kühni, W. Gujer, and H. Siegrist. 2001. The EAWAG bio-P module for Activated Sludge Model No. 3. *Water Research*, 35: 3887–3903.

Roels, J. A. 1980. Application of macroscopic principles to microbial metabolism. *Biotechnology and Bioengineering* 22: 2457–2514.

Rosen, C., and U. Jeppsson. 2006. Aspects on ADM1 Implementation within the BSM2 Framework. Tech. report LUTEDX/(TEIE-7224)/1-35/(2006). Lund, Sweden: Department of Industrial Electrical Engineering and Automation, Lund University.

Rosen, C., D. Vrecko, K. V. Gernaey, M. N. Pons, and U. Jeppsson. 2006. Implementing ADM1 for plant-wide benchmark simulations in Matlab/Simulink. *Water Science and Technology* 54 (4): 11–20.

Rousseau, D., F. Verdonck, O. Moerman, R. Carrette, C. Thoeye, J. Meirlaen, and P. A. Vanrolleghem. 2001. Development of a risk assessment based technique for design/retrofitting of WWTPs. *Water Science and Technology* 43 (7): 287–294.

Saltelli, A., M. Ratto, T. Andres, F. Campolongo, J. Cariboni, D. Gatelli, M. Saisana, and S. Tarantola. 2008. *Global Sensitivity Analysis: The Primer*, West Sussex, England: John Wiley & Sons.

Shiskowski, D. M. 2008. Change—more than just the climate. *Water Environment Research* 80: 99–100.

Siegrist, H., I. Brunner, G. Koch, Phan Linh Con, and Le Van Chieu. 1999. Reduction of biomass decay rate under anoxic and anaerobic conditions. *Water Science and Technology* 39 (1): 129–137.

Siegrist, H., D. Renggli, and W. Gujer. 1993. Mathematical modelling of anaerobic mesophilic sewage sludge treatment. *Water Science and Technology* 27 (2): 25–36.

Siegrist, H., D. Vogt, J. L. Garcia-Heras, and W. Gujer. 2002. Mathematical model for meso and thermophilic anaerobic digestion. *Environmental Science and Technology* 36: 1113–1123.

Sin, G., D. J. W. De Pauw, S. Weijers, and P. A. Vanrolleghem. 2008a. An efficient approach to automate the manual trial and error calibration of activated sludge models. *Biotechnology and Bioengineering* 100: 516–528.

Sin, G., K. V. Gernaey, M. B. Neumann, M. C. M. van Loosdrecht, and W. Gujer. 2009. Uncertainty analysis in WWTP model applications: a critical discussion using an example from design. *Water Research* 43: 2894–2906.

Sin, G., D. Kaelin, M. J. Kampschreur, I. Takács, B. Wett, K. V. Gernaey, L. Rieger, H. Siegrist, and M. C. M. van Loosdrecht. 2008b. Modelling nitrite in wastewater treatment systems: A discussion of different modelling concepts. *Water Science and Technology* 58 (6): 1155–1171.

Sin, G., S. W. H. Van Hulle, D. J. W. De Pauw, A. van Griensven, and P. A. Vanrolleghem. 2005. A critical comparison of systematic calibration protocols for activated sludge models: a SWOT analysis. *Water Research* 39: 2459–2474.

Sin, G., P. Ödman, N. Petersen, A. Eliasson Lantz, and K. V. Gernaey. 2008c. Matrix notation for efficient development of first-principles models within PAT applications: integrated modeling of antibiotic production with *Streptomyces coelicolor*. *Biotechnology and Bioengineering* 101: 153–171.

Smolders, G. J. F., J. M. Klop, M. C. M. van Loosdrecht, and J. J. Heijnen. 1995. Metabolic model of the biological phosphorus removal process: I. Effect of the sludge retention time. *Biotechnology and Bioengineering* 48: 222–233.

Speece, R. 1983. Anaerobic biotechnology for industrial wastewater treatment. *Environmental Science and Technology* 17: 416–426.

Stumm, W., and J. J. Morgan. 1996. *Aquatic Chemistry: Chemical Equilibria and Rates in Natural Waters*, New York, NY: John Wiley & Sons.

Takács, I., G. G. Patry, and D. Nolasco. 1991. A dynamic model of the clarification-thickening process. *Water Research* 25: 1263–1271.

Tchobanoglous, G., H. Theisen, and S. Vigil. 1993. *Integrated Solid Waste Management: Engineering Principles and Management Issues*, New York, NY: McGraw-Hill.

Vanrolleghem, P. A., G. Insel, B. Petersen, G. Sin, D. De Pauw, I. Nopens, S. Weijers, and K. Gernaey. 2003. A comprehensive model calibration procedure for activated sludge models. In: *Proceedings: WEFTEC 2003 October 11–15, 2003: 76th Annual Technical Exhibition and Conference*. Los Angeles, CA, USA (on CD-ROM).

Vanrolleghem, P. A., C. Rosen, U. Zaher, J. Copp, L. Benedetti, E. Ayesa, and U. Jeppsson. 2005. Continuity based interfacing of models for wastewater systems described by Peterson matrices. *Water Science and Technology* 52 (1–2): 493–500.

Van Veldhuizen, H. M., M. C. M. van Loosdrecht, and J. J. Heijnen. 1999. Modelling biological phosphorus and nitrogen removal in a full scale activated sludge process. *Water Research* 33: 3459–3468.

Vavilin, V. A., S. V. Rytov, and L. Y. Lokshina. 1996. A description of hydrolysis kinetics in anaerobic degradation of particulate organic matter. *Bioresource Technology* 56: 229–237.

Vavilin, V. A., V. B. Vasiliev, A. V. Ponomarev, and S. V. Rytow. 1994. Simulation model methane as a tool for effective biogas production during anaerobic conversion of complex organic matter. *Bioresource Technology* 48: 1–8.

Vesilind, P. A. 1968. Theoretical considerations: design of prototype thickeners from batch settling tests. *Water and Sewage Works* 115: 302–307.

Walker, W. E., P. Harremoes, J. Rotmans, J. P. Van der Sluijs, M. B. A. Van Asselt, P. Janssen, and M. P. Krayer von Krauss. 2003 Defining uncertainty a conceptual basis for uncertainty management in model-based decision support. *Integrated Assessment* 4: 5–17.

White, F. M. 1991. *Viscous Fluid Flow*, 614 pp. Singapore: McGraw-Hill.

Whitman, W. G. 1923. The two-film theory of gas absorption. *Chemical and Metallurgical Engineering* 29: 146–148.

Further Reading—Model Extensions Focused on N Removal

Almeida, J. S., M. A. M. Reis, and M. J. T. Carrondo. 1995. Competition between nitrate and nitrite reduction in denitrification by *Pseudomonas fluorescens*. *Biotechnology and Bioengineering* 46: 474–484.

Anthonisen, A. C., R. C. Loehr, T. B. S. Prakasam, and E. G. Srinath. 1976. Inhibition of nitrification by ammonia and nitrous acid. *Journal of the Water Pollution Control Federation* 48: 835–852.

Hao, X., J. J. Heijnen, and M. C. M. van Loosdrecht. 2002. Model-based evaluation of temperature and inflow variations on a partial nitrification—ANAMMOX biofilm process. *Water Research* 36: 4839–4849.

Hellinga, C., M. C. M. van Loosdrecht, and J. J. Heijnen. 1999. Model based design of a novel process for nitrogen removal from concentrated flow. *Mathematical and Computer Modelling of Dynamical Systems* 5: 351–371.

Hiatt, W. C., and C. P. L. Grady Jr. 2008. An updated process model for carbon oxidation, nitrification and denitrification. *Water Environment Research* 80: 2145–2156.

Kaelin, D., R. Manser, L. Rieger, J. Eugster, K. Rottermann, and H. Siegrist. 2009. Extension of ASM3 for two-step nitrification and denitrification and its calibration and validation with batch tests and pilot scale data. *Water Research* 43: 1680–1692.

Kampschreur M. J., H. Temmink, R. Kleerebezem, M. S. M. Jetten, and M. C. M. van Loosdrecht. 2009. Nitrous oxide emission during water treatment. *Water Research* 43: 4093–4103.

Knowles R. 1982. Denitrification. *Microbiology Reviews* 46: 43–70.

von Schulthess, R., and W. Gujer. 1996. Release of nitrous oxide (N2O) from denitrifying activated sludge: Verification and application of a mathematical model. *Water Research* 30: 521–530.

Further Reading—Bio-P Model Development

Barker, P. S., and P. L. Dold. 1997. General model for biological nutrient removal activated sludge systems: model presentation. *Water Environment Research* 69: 969–984.

Comeau, Y., K. J. Hall, R. E. W. Hancock, and W. K. Oldham. 1986. Biochemical model for enhanced biological phosphorus removal. *Water Research* 20: 1511–1521.

Hu, Z., M. C. Wentzel, and G. A. Ekama. 2007. A general kinetic model for biological nutrient removal activated sludge systems: model development. *Biotechnology and Bioengineering* 98: 1242–1258.

Hu, Z., M. C. Wentzel, and G. A. Ekama. 2007. A general kinetic model for biological nutrient removal activated sludge systems: model evaluation. *Biotechnology and Bioengineering* 98: 1259–1275.

Mino, T., W. T. Liu, F. Kurisu, and T. Matsuo. 1995. Modeling glycogen storage and denitrification capability of microorganisms in enhanced biological phosphate removal processes. *Water Science and Technology* 31 (2): 25–34.

Mino, T., D. C. San Pedro, and T. Matsuo. 1995. Estimation of the rate of slowly biodegradable COD (SBCOD) hydrolysis under anaerobic, anoxic and aerobic conditions by experiments using starch as model substrate. *Water Science and Technology* 31 (2): 95–103.

Mino, T., H. Satoh, and T. Matsuo. 1994. Metabolism of different bacterial populations in enhanced biological phosphate removal processes. *Water Science and Technology* 29 (7): 67–70.

Wentzel, M. C., P. L. Dold, G. A. Ekama, and G. v. R. Marais. 1985. Kinetics of biological phosphorus release. *Water Science and Technology* 17 (11/12): 57–71.

Wentzel, M. C., G. A. Ekama, and G. v. R. Marais. 1992. Process and modelling of nitrification denitrification biological excess phosphorus removal systems—a review. *Water Science and Technology* 25 (6): 59–82.

Section II

Models of Environmental Problems

12

Eutrophication Models

Sven Erik Jørgensen

CONTENTS

12.1 Introduction...325
12.2 Eutrophication Models: An Overview...329
12.3 An Illustrative Eutrophication Model..329
12.4 Application of SDMs to Improve the Results
     of Eutrophication Models 341
12.5 Future Development of Eutrophication Models...................................344
References..345

12.1 Introduction

Many aquatic ecosystems suffer from eutrophication: lakes, reservoirs, estuaries, lagoons, fjords, and bays. Eutrophication models have been developed and applied for these ecosystems. In particular, many eutrophication models have been applied for environmental management, particularly for lakes and reservoirs.

The word *eutrophy* is generally taken to mean "nutrient rich." Nauman introduced in 1919 the concepts of oligotrophy and eutrophy, distinguishing between oligotrophic lakes containing little planktonic algae and eutrophic lakes containing much phytoplankton. The eutrophication of aquatic ecosystems all over the world has increased rapidly during the past decade due to increased urbanization and consequently increased discharge of nutrient per capita. The production of fertilizers has grown exponentially in this century, and the concentration of phosphorus in many lakes reflects this growth.

The word eutrophication is used increasingly in the sense of artificial addition of nutrients, mainly nitrogen and phosphorus, to waters. Eutrophication is generally considered to be undesirable, although this is not always true. The green color of eutrophied lakes makes swimming and boating more unsafe due to the increased turbidity, and from an aesthetic point of view the chlorophyll concentration should not exceed 100 mg m^{-3}. However, the most critical effect from an ecological point of view is the reduced oxygen content of the hypolimnion, caused by the decomposition of dead algae, particularly in the fall. Eutrophic aquatic ecosystems sometimes show a high oxygen concentration at the surface during summer, but low oxygen concentrations in the hypolimnion, which may be lethal to fish.

Water in plant tissue constitutes 75–90% of the total wet weight. It means that, except for oxygen and hydrogen, the composition on dry weight basis would be 4–10 times higher. For phytoplankton, the amounts of carbon, nitrogen, and phosphorus on dry weight basis are approximately 40–60%, 6–8%, and 0.75–1.0%, respectively. Phosphorus is considered the major cause of eutrophication in lakes, as it was formerly the growth-limiting factor for algae in the majority of lakes. Its use, however, has increased tremendously during the past decades. Nitrogen is a limiting factor in a number of East African lakes as a result of the nitrogen depletion of soils by intensive erosion in the past. Nitrogen is furthermore often a limiting factor in coastal ecosystems. Today, however, nitrogen may become limiting to growth in lakes as a result of the tremendous increase in the phosphorus concentration caused by discharge of wastewater, which contains relatively more phosphorus than nitrogen. Although algae use 5–10 times more nitrogen than phosphorus, wastewater generally contains only three times as much nitrogen as phosphorus. In lakes, a considerable amount of nitrogen is lost by denitrification (nitrate $\rightarrow$ N_2). In environmental management, however, the key question is *not which element is the limiting factor, but which element can be more easily be controlled as a limiting factor.*

The growth of phytoplankton is the key process of eutrophication, and it is therefore important to understand the interacting processes that regulate growth. Primary production has been measured in great detail in a number of aquatic ecosystems. This process represents the synthesis of organic matter, and can be summarized as follows:

$$\text{Light} + 6CO_2 + 6H_2O \rightarrow C_6H_{12}O_6 + 6\,O_2 \tag{12.1}$$

The composition of phytoplankton is not constant. The composition of phytoplankton and plants in general reflects to a certain extent the concentration of the water. If, for example, the phosphorus concentration is high, the phytoplankton will take up relatively more phosphorus—this is called luxury uptake. Phytoplankton consists mainly of carbon, oxygen, hydrogen, nitrogen, and phosphorus; without these elements, no algal growth can take place. This has led to the concept of the limiting nutrient, which has been developed by Liebig as the law of the minimum. However, the concept has been considerably misused due to oversimplification. First of all, growth might be limited by more than one nutrient. The composition, as mentioned above, is not constant, but varies with the composition of the environment. Furthermore, growth is not at its maximum rate until the nutrients are used, and is then stopped, but the growth rate slows down as the nutrients become depleted.

The sequences of events leading to *eutrophication* has often been described as follows: Oligotrophic waters will have a ratio of N:P greater than or equal to 10, which means that phosphorus is less abundant than nitrogen for the needs of phytoplankton. If sewage is discharged into the aquatic ecosystem the ratio will decrease, since the N:P ratio for municipal wastewater is 3:1, and consequently nitrogen will be less abundant than phosphorus relative to the needs of phytoplankton. In this situation, however, the best remedy for the excessive growth of algae is not necessarily the removal of nitrogen from the sewage, because the mass balance might then show that nitrogen-fixing algae will give an uncontrollable input of nitrogen into the system. It is particularly the case if the aquatic ecosystem is a lake or reservoir. It is necessary to set up mass balances for each of the nutrients, and these will often reveal that the input of nitrogen from nitrogen-fixing blue green algae, precipitation, and tributaries is contributing too much to the mass balance for the removal of nitrogen from the sewage to have any effect. On the other hand, the mass balance may reveal that most of the phosphorus input (often more than 95%) comes from sewage, which means that it is better management to remove phosphorus from the sewage than nitrogen. Thus, it is not important which nutrient is most limiting, but which nutrient can more easily be made to limit algal growth.

The next section will present a short overview of eutrophication models. The overview is far from being complete, as several hundred eutrophication models have been published during the past four decades. The overview, however, gives an expression of the wide spectrum of complexity for the available models. It implies that a model is available for almost any combination of aquatic ecosystem and databases. Although the latest developments in eutrophication models have already been covered in Chapter 2, a few of the recent models are included in Table 12.1.

The third section of this chapter presents a eutrophication model of medium to high complexity. It has been chosen to represent a well-established eutrophication model that has been applied on 25 different case studies, including lakes, reservoirs, estuaries, fjords, and lagoons. Furthermore, the model illustrates the use of independent nutrient cycles and the generality of a more complex model.

The fourth section reviews the application of structurally dynamic modeling (SDM) approach for the development of better eutrophication models. In this section, a few illustrative cases are described. Basic information about SDM is given in Section 2.7, where also an example is presented.

The last section discusses the expected future development of eutrophication models, including how the well-established model presented in Section 12.3 and how the structurally dynamic eutrophication models could be further improved by use of the latest obtained results in eutrophication modeling.

TABLE 12.1

Various Eutrophication Models

Model Name	Number of st. var. per Layer or Segment	Nutrients	Segments	Dimension (D) or Layer (L)	CS or NC[a]	C and/ or V[b]	Number of Case Studies
Vollenweider	1	P (N)	1	1L	CS	C + V	many
Imboden	2	P	1	2L, 1D	CS	C + V	3
O'Melia	2	P	1	1D	CS	C	1
Larsen	3	P	1	1L	CS	C	1
Lorenzen	2	P	1	1L	CS	C + V	1
Thomann 1	8	P, N, C	1	2L	CS	C + V	1
Thomann 2	10	P, N, C	1	2L	CS	C	1
Thomann 3	15	P, N, C	67	2L	CS	–	1
Chen&Orlob	15	P, N, C	sev.	2L	CS	C	min. 2
Patten	33	P, N, C	1	1L	CS	C	1
Di Toro	7	P, N	7	1L	CS	C + V	1
Biermann	14	P, N, Si	1	1L	NC	C	1
Canale	25	P, N, Si	1	2L	CS	C	1
Jørgensen[c]	17–20	P, N, C	1	1–2L	NC	C + V	25
Cleaner	40	P, N, C, Si	sev.	sev.	CS	C	many
Nyholm, Lavsoe	7	P, N	1–3	1–2L	NC	C + V	25
Aster/Melodia	10	P, N, Si	1	2L	CS	C + V	1
Baikal	>16	P, N	10	3L	CS	C + V	1
Chemsee	>14	P, N, C, Si	1	profile	CS	C + V	many
Minlake	9	P, N	1	1L	CS	C + V	>10
Salmo	17	P, N	1	2L	CS	C + V	16
Pamolare[d]	17–20	P, N, Si	1	2L	NC	C + V	many?
Zao et al. (2008)	16	P, N, Si, C	1	1–2L	NC	C + V	2
Zhang et al. (2008)	18	P, N, C	1	2D model	CS	C + V	1

[a] CS, constant stoichiometry; NC, independent nutrient cycle.
[b] C, calibrated; V, validated.
[c] The model is denoted the Glumsø Model and modification of the model includes Si.
[d] Downloadable from www.unep.or.jp/ietc/pamolare.

12.2 Eutrophication Models: An Overview

Several eutrophication models with a wide spectrum of complexity have been developed. As for other models, the right complexity of the model is dependent on the available data and the ecosystem. Table 12.1 reviews various eutrophication models.

The table lists the characteristic features of the models, the number of case studies to which a particular model has been applied (of course with some modification from case study to case study, as a general model is nonexisting and as site-specific properties should be reflected in the selected modification, unless the model is very simple), and whether the model has been calibrated and validated.

It is of course not possible to cover all of the more complex models in detail. Therefore, one model of this type was selected and presented in more detail in Section 12.3. The results obtained by the use of a complex eutrophication model demonstrate what can be achieved with respect to prognosis validation, provided that sufficient effort is expended to obtain good data and good ecological background knowledge about the modeled ecosystem.

12.3 An Illustrative Eutrophication Model

The model presented in this section was chosen to illustrate the typical results that can be obtained by using a biogeochemical eutrophication model. The advantages of using this example are rooted in the following:

1. The same model has been applied in a total of 25 cases after modifications. It means that the model is able to illustrate the generality of the modeling approach.
2. The model has been used to set up prognoses in a few cases, and in two cases it was possible to validate the prognoses.
3. The model has been applied after modifications at both temperate and tropical climate.
4. The model has inspired the development of structurally dynamic eutrophication models.
5. The model has, of course, been calibrated and validated, and in few cases the calibration and validation were based on a database of particularly good quality. Very frequent measurements have been applied in this context (see Jørgensen 2009; Jørgensen and Bendoricchio 2001).

6. The model is relatively complex, with 17–20 state variables and has in a few cases been applied on lakes with a simple hydrology, which means that mainly the expression of the ecological processes was challenged by the calibration and validation of the eutrophication model.

7. The eutrophication model has been applied on different aquatic ecosystems: lakes, reservoirs, lagoons, estuaries, and fjords.

8. The model is based on an independent cycling of phosphorus, nitrogen, and carbon (denoted NC in Table 12.1), but has in a few cases also been compared with a model based on constant stoichiometry (denoted CS in Table 12.1). These results are therefore able to assess the difference between the two approaches, NC and CS.

9. It should finally also be mentioned that the author has, together with coworkers, developed this model, which has the advantage that the author knows all the details about this model. The model is very representative for what can be obtained by the use of a eutrophication

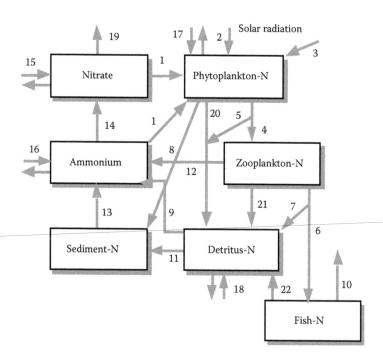

FIGURE 12.1
The conceptual diagram of a nitrogen cycle in an aquatic ecosystem. The processes that connect the state variables and forcing functions are: (1) uptake of nitrate and ammonium by algae; (2) photosynthesis; (3) nitrogen fixation; (4) grazing with loss of undigested matter; (5)–(7) predation and loss of undigested matter; (8) settling of algae; (9) mineralization; (10) fishery; (11) settling of detritus; (12) excretion of ammonium from zooplankton; (13) release of nitrogen from the sediment; (14) nitrification; (15)–(18) inputs/outputs; (19) denitrification; (20)–(22) mortality of phytoplankton, zooplankton, and fish, respectively.

model, and it has therefore been applied as typical example in text-books, see Jørgensen (2009) and Jørgensen and Fath (2011).

The conceptual diagram shown in Figure 12.1 is used for the nitrogen cycle of the selected typical eutrophication model. The corresponding phosphorus cycle is shown Figure 12.2. The model is called the Glumsø Model, because it was applied for the first time on Glumsø Lake. The state variables and processes were selected based on a mass balance in each of the 25 cases where the model was used. It is always beneficial to use a mass or energy balance to choose the most important components, forcing functions, and state variables, to be included in the model. Let us illustrate the needed mass balance considerations by use of an example. Let us anticipate that it is an open question, whether birds should be included in a eutrophication model as a state variable. Birds may contribute considerably to the inputs of nutrients by their droppings. If the nutrients—nitrogen and phosphorus—coming from the birds' droppings are insignificant compared with the amounts of nutrients

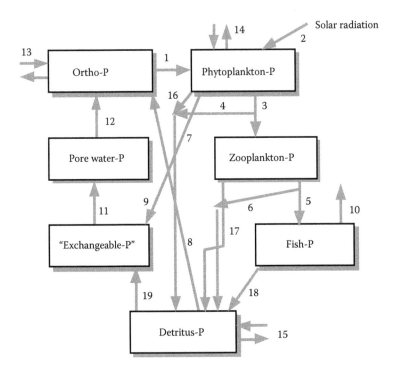

FIGURE 12.2
The phosphorus cycle. The processes are: (1) uptake of phosphorus by algae; (2) photosynthesis; (3) grazing with loss of undigested matter; (4)–(5) predation with loss of undigested material; (6), (7), (9) settling of phytoplankton; (8) mineralization; (10) fishery; (11) mineralization of phosphorous organic compounds in the sediment; (12) diffusion of pore water P; (13)–(15) inputs/outputs; (16)–(18) mortalities; (19) settling of detritus.

coming from drainage water, precipitation, and wastewater, inclusion of birds both as a state variable and as a contributing component (forcing function) in the model is only an unnecessary complication that only would contribute to the uncertainty. There are, however, a few cases where birds may contribute as much as 25% or at least more than 5% of the total inputs of nutrients. In such cases, it is of course important to include birds as a model component, or at least as an important forcing function. A mass balance is needed to uncover the main sources of the problem—in the example, the inputs of nutrients. In none of the 25 cases has it been necessary to include birds as state variables. The conceptual diagrams in Figures 12.1 and 12.2 present the basic model that was applied for Lake Glumsø and a few others of the 25 cases.

The so-called Michaelis–Menten's equation is applied in the constant stoichiometry version for the growth, gr = grmax NS/(kn + NS), or gr = grmax PS/(kp + PS), depending on which nutrient is limiting, N or P. grmax, kn, and kp are parameters. If both nutrients are limiting in different periods of the year, the formulation is

$$gr = grmax * min(NS/(kn + NS), PS/(kp + PS)) \qquad (12.2)$$

The product or average of several limiting factors has also been proposed and applied.

For the influence of temperature, there are two possible formulations:

$$K_t{}^\wedge(TEMP - 20) \text{ (the so-called Arrhenius equation)} \qquad (12.3)$$

or

$$\exp(A * (TEMP\text{-}OPT)/(TEMP\text{-}MAXTEMP)) \qquad (12.4)$$

where K_t is a parameter, which in most cases is between 1.04 and 1.06 (in average, 1.05). OPT is the optimum temperature for phytoplankton growth and MAXTEMP is the maximum temperature. OPT, MAXTEMP, and A are all parameters that are different for different phytoplankton species. In most cases, Equation 12.4 gives a more correct description of temperature dependence than Equation 12.3, but it also contains more parameters that have to be determined. In the Lake Glumsø case, intensive measurements were used to assess that Equation 12.4 would facilitate the calibration and also improve the validation. It is of course possible to include a limitation by silica and/or iron and other possible nutrients in a similar manner to that shown above for nitrogen and phosphorus.

The description of phytoplankton growth by equations through the CS approach is a simplification, because phytoplankton growth is in reality a two-step process, as illustrated in Figure 12.3, and applied in the NC approach. The first step is uptake of nutrients and the second step is growth

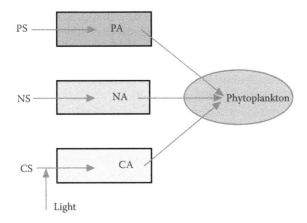

FIGURE 12.3
The two-step model of phytoplankton growth. The first step is uptake of nutrients PS, NS, and CS, followed by a growth of phytoplankton dependent on the nutrient concentrations in the phytoplankton cells, PA, NA, and CA. Uptake of carbon is dependent on light, whereas uptake of phosphorus and nitrogen can take place even in darkness. This is a more physiologically correct description of phytoplankton growth than that portrayed when using the constant stoichiometry (CS) approach.

of phytoplankton (increase in biomass). The more correct description can be formulated mathematically by the following equations:

$$\text{Uptake rate P} = dPA/dt = PA * \text{maxupp} * (PS/(kp + PS)) *$$
$$((PAMAX - PA)/(PAMAX - PAMIN))$$

$$\text{Parallel for uptake rate N} \tag{12.5}$$

$$\text{Uptake rate C} = dCA/dt = CA * \text{maxupc} * (CS/kc + CS) * ((CAMAX - CA)/$$
$$(CAMAX\ CAMIN)) * ((L/KL + L) - RESP$$

if $L < L_1$, L is used; if $L_2 > L > L_1$, L_1 is used; if $L > L_2$, $L_1 + L_2 - L$ is used for L
$$\tag{12.6}$$

PA, NA, and CA are the state variables that denote the amount of phosphorus, nitrogen, and carbon, respectively, in the form of phytoplankton expressed as mg P, N, or C per liter of water. Note that the unit is mg in 1 L of water. Maxupp, maxupn, macups, kp, kn, kc, PAMAX, PAMIN, NAMAX, NAMIN, CAMAX, CAMIN, KL, L_1, and L_2 are all parameters. PAMAX, PAMIN, NAMAX, NAMIN, CAMAX, and CAMIN are, however, known fairly well. They are the phytoplankton concentration times respectively 0.025, 0.005, 0.12, 0.05, 0.6, and 0.4 with good approximations. It is of course more difficult to calibrate the two-step growth equations than the CS approach because of the higher number of parameters in the NC equations, although the approximate knowledge that is available to the six parameters (PAMAX, PAMIN, NAMAX, NAMIN, CAMAX, CAMIN) facilitates the calibration slightly. Notice that the uptake of phosphorus, nitrogen, and carbon is according to the equations dependent on both the concentrations of nutrients in the water and concentrations of nutrients in the cells.

The closer the nutrient concentrations in phytoplankton are to the minimum, the faster is the uptake. On the other hand, when a nutrient concentration has reached the maximum value, the uptake stops. Carbon uptake, in contrast to the uptake of phosphorus and nitrogen, is also dependent on light and, by a Michaelis–Menten's expression, includes light prohibition. At a light intensity $L > L_2$, growth decreases in an approximately linear manner according to the equation for uptake of carbon. At light intensity L_1, the saturation level is reached. A graph showing growth as a function of light intensity is shown in Figure 12.4. RESP denotes respiration. Only carbon, of course, is involved in respiration.

The growth process is quantified by the following equation:

Growth = grmax * phytoplankton * (min((PA – PAMIN)/(PAMAX – PAMIN)), ((NA –NAMIN)/(NAMAX – NAMIN)), ((CA – CAMIN)/ (CAMAX – CAMIN)))   (12.7)

where grmax is a parameter in line with the corresponding parameter in Equation 12.1.

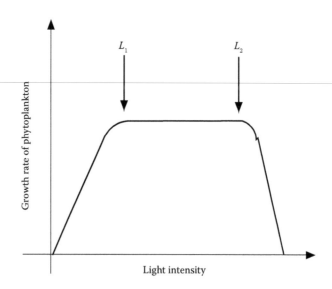

FIGURE 12.4
Phytoplankton growth plotted versus light intensity. At light intensity L_1, saturation is reached. At light intensity $>L_2$, photo prohibition takes place.

The phytoplankton growth model has four state variables: PA, NA, CA, and phytoplankton, which are all assumed to be expressed in mg/l. As the minimum and maximum values are presumed to be a parameter times the phytoplankton concentration, they are also expressed in mg/l. This is in accordance with the discussion in Chapter 1 (about verification): always check the units.

The two-step description is more difficult to calibrate and validate and use in general, which of course raises the question: "When should the two-step description be applied instead of the easier applicable constant stoichiometric approach?" As pointed out by Jørgensen (1976), Jørgensen et al. (1978), and Jørgensen (1986a, 1986b, 1986c), it is sometimes impossible to calibrate the one-step growth of phytoplankton, because the reaction rates at a high nutrient concentration are too fast. It has been clearly concluded, in at least four case studies, where it was possible to compare the two approaches. Di Toro and Conolly (1980) have revealed that the need for two-step description increases with the shallowness and the eutrophication of the aquatic ecosystem (see Figure 12.5). Two-step description is recommended for shallow and very eutrophied aquatic ecosystems, whereas it is hardly needed for deep mesotrophic or oligotrophic aquatic ecosystems.

The exchange of nutrients between sediment and water may be a very important process, because the amount of nutrients stored in the sediment is frequently several times larger than the amount of nutrients in the water phase, particularly for shallow lakes. In the model presented here, which has a medium to high complexity, it is therefore natural to question the submodel used for the nutrient exchange between sediment and water.

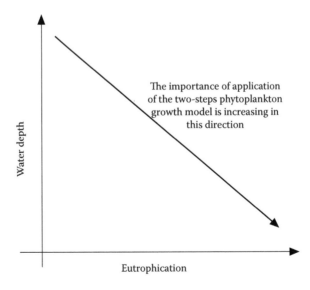

FIGURE 12.5
The need for two-step description of phytoplankton growth increases with nutrient concentration and decreases with depth.

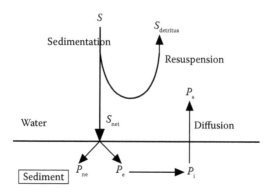

FIGURE 12.6
The sediment–water exchange submodel applied in the Glumsø eutrophication model. A part
of the settled material is resuspended, a part net settled material is exchangeable, and a part is
nonexchangeable. The exchangeable part is decomposed (it is mainly organic) and forms dissolved phosphorus in the pore water (for nitrogen, ammonium is formed). The phosphorus (or
for nitrogen the ammonium) is by a diffusion process returned to the water phase.

It has been found that the submodel (shown in Figure 12.6) gives a proper
description, which has been applied in the Glumsø Model. Sedimentation, S,
corresponds to processes (9) and (19) in Figure 12.2. A part of the sediment
is resuspended, particularly in shallow lakes. A part of the settled material
is exchangeable and a part is nonexchangeable. The exchangeable part is
decomposed (it is mainly organic material) and form dissolved phosphorus
in the pore water or interstitial water. A similar submodel can be applied for
the nitrogen exchange between water and sediment. The dissolved nitrogen
in the pore water is ammonium. The parameter k (ratio between exchangeable and total nutrient sediment) can be determined by using a phosphorus
and nitrogen sediment profile. In particular, k is different for phosphorus
and nitrogen, but even the active layer may be different for the two nutrients. Figure 12.7 shows such a profile taken. The active layer, a_1, is in this
case 5 cm = LUL. k is $B/(A + B) = 4.5/6.0 = 0.75$. Three-fourths or 75% of the
top layer of phosphorus in the sediment is able to return to the water phase,
whereas 1/4 or 25% is bound very firmly chemically and cannot return to the
water phase. Usually, k is higher for nitrogen than for phosphorus, because
both ammonium and nitrate form soluble inorganic compounds, whereas
there are several insoluble inorganic phosphorus compounds particularly
with calcium and iron. As noted previously, it is crucial for the prognoses of
recovery of aquatic ecosystems after a significant eutrophication to account
for the part of the nutrient that is not returned to the water phase. The results
of the model presented in this section have in three lake case studies been
compared with the results based on the assumption that all nutrients in the
active layer of the sediment would be exchangeable. The results are significantly different. According to the model, it takes a significantly longer time
for the lake to recover if all the nutrients would be exchangeable.

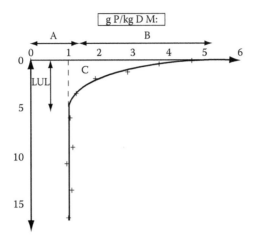

FIGURE 12.7
A phosphorus sediment profile for Lake Esrom in Denmark. As shown in the active layer, al (denoted LUL) is 5 cm and the ratio of exchangeable to total phosphorus in the sediment is $k = 4.5/6.0 = 0.75$. In particular, k is different for phosphorus and nitrogen, whereas the active layer, al, is often the same as or approximately the same for phosphorus and nitrogen.

The grazing of phytoplankton by zooplankton can be expressed as a Michaelis–Menten's equation with modifications. When the phytoplankton concentration is very low, no grazing is observed, because the zooplankton can only graze a certain volume per day. A threshold concentration is therefore introduced to account for this situation.

On the other hand, if the phytoplankton concentration is very high, zooplankton will grow too fast in accordance with the Michaelis–Menten's expression, because it is no longer the amount of food that is limiting the zooplankton growth, but the carrying capacity of the aquatic ecosystem, which is considered to be the concentration of zooplankton that can be maintained in the aquatic ecosystem due to the predation by fish and the available shelter for protection of eggs and offsprings (nesting space). A regulation of the growth in accordance with the concentration of zooplankton in relation to the carrying capacity can be applied to account for this growth limitation.

According to the presented considerations, the following equation can be proposed for a proper description of grazing:

$$\text{Grazing} = \text{Growth rate of zooplankton} =$$
$$\text{zoo} * \text{maxgraz} * ((\text{Phyt} - \text{trc})/(\text{kph} + \text{Phyt})) * (1 - \text{zoo}/\text{CC}) \qquad (12.8)$$

where zoo denotes zooplankton, Trc is the concentration below which there is no grazing, and CC is the carrying capacity that is dependent on the environmental conditions (available nesting space).

A Michaelis–Menten equation was chosen for nitrification, based on general experience. For the decomposition of detritus, a first-order reaction was selected, due to a fast regulation of a microbiological population to the available organic matter. The model was also used experimentally. More groups of zooplankton were introduced to the model—feeding on the herbivorous zooplankton, feeding on detritus and omnivorous—but this did not change the model results significantly. Therefore, the model based on the conceptual diagrams shown in Figures 12.1 and 12.2 was selected to be used for Glumsø Lake to set up prognoses for the removal of phosphorus from the discharged wastewater. Frequent measurements during the spring and summer blooms were applied in two cases for a clearly improved determination of the most crucial parameters. For details about the application of frequent measurements to obtain a better parameter estimation before the calibration, see Jørgensen and Bendoricchio (2001), Jørgensen et al. (2004), Jørgensen et al. (2009), and Jørgensen (2009).

In the case of Lake Glumsø, a good performance of the calibration was obtained, and the validation gave acceptable results (see below).

Standard deviation between modeled and observed values:

Phytoplankton concentration (average of 15 comparisons)	10%
Peak phytoplankton concentration (1 determination)	15%
Production (g C/(m² 24 h)) (average of 15 comparisons)	8%
Production peak value (g C/(m² 24 h)) (1 determination)	5%
Zooplankton peak value (1 determination)	0%
Zooplankton concentration (average of 15 comparisons)	27%
Transparency (average of 15 comparisons)	15%
All 17–20 state variables (average of 225 comparisons)	16%

The validation of the total assessment of the eutrophication of Lake Glumsø by the model can be found by dividing the standard deviation for one comparison of the modeled and observed values considering all state variables—it is 16%—divided by $\sqrt{225}$, as the number of comparisons is 225. It yields a standard deviation for the total assessment of the eutrophication of slightly more than 1%.

As the validation was acceptable, the model was applied to set up a prognosis for the production, phytoplankton concentration, and transparency, if the phosphorus input to the lake was reduced by 90%, which was easy to achieve (e.g., by a well-controlled chemical precipitation). Before the reduction of the phosphorus input, the lake was very eutrophic, which can be seen by the following typical observations:

Total P (g/m³)	1.1
Phytoplankton concentration peak value (mg chl a/m³)	850

| Production (g C/(m² year)) | 1050 |
| Minimum transparency at spring bloom (m) | 0.18 |

Fortunately, the water residence time of Glumsø Lake is only 6 months, which implied that it was possible to properly validate the prognosis after a few years. A comparison of the prognosis and the actual observations after the 90% reduction of phosphorus input is shown in Table 12.2. The indicated standard deviations in the table are for the prognosis based on the validation results shown above and for the measurement based on a general determination of the standard deviations for measurement on 10% relatively.

The prognosis validation is fully acceptable (no significant difference determined by a t-test) except for the daily production (g C/(m² 24 h)) at spring bloom in the second year. In the beginning of the second year, the phytoplankton species shifted from *Scenedesmus* (green algae) to various species of diatoms.

The shift from green algae to diatoms implied that the well-determined parameters for phytoplankton were no longer valid, which may explain the discrepancy between the prognosis and the observations, particularly in the second year after reduction of phosphorus loading. The shift, however, has clearly demonstrated the need for an SDM approach. SDM was introduced and discussed in Section 2.7 to explain the ecosystem behavior of lakes—that lakes in certain nutrient ranges have two possible structures. The use of SDMs for development of eutrophication models will be presented in more detail in the next section. See also Jørgensen et al. (2004) and Jørgensen (2009).

TABLE 12.2

Validation of the Prognosis for Glumsø Lake

Comparison of	Time (years)	Prognosis Value	St. Dev.	Measurements Value	St. Dev.
Min. transparency	1	0.20 m	0.03	0.20 m	0.02
Min. transparency	2	0.30 m	0.05	0.25 m	0.025
Min. transparency	3	0.45 m	0.07	0.50 n	0.05
Max. production	2	6.0 g C/24 h m²	0.03	11.0 g C/24 h m²	1.1
Max. production	3	5.0 g C/24 h m²	0.03	6.2 g C/24 h m²	0.6
Max. chl. a.	1	750 mg/m³	112	800 mg/m³	80
Max. chl. a.	2	520 mg/m³	78	550 mg/m³	55
Max. chl. a.	3	320 mg/m³	48	380 mg/m³	38
Annual production	2	720 g C/year m²	15[a]	750 g C/year m³	19[a]
Annual production	3	650 g C/year m³	13[a]	670 g C/year m³	17[a]

[a] Standard deviation is used 8% for the prognosis divided by $\sqrt{15}$ and 10% divided by $\sqrt{15}$ for the measurements, because the determination of the annual production is based on 15 measurements and 15 prognosis values.

TABLE 12.3

Revision of the Glumsø Eutrophication Model by Applications for Other Case Studies

Case Studies	Revisions Made
Lake Victoria, Lake Kyoga, Lake Edward, Africa	Thermocline, other food chain, boxes
Fure Lake, Denmark	Nitrogen fixation, SDM, restoration
Ringkøbing Fjord, Denmark	Nitrogen fixation, boxes
Esrom Lake, Denmark	Silica included as nutrient
Lyngby Lake, Denmark	Recalibration only
Four Lakes in Zealand, Denmark	Resuspension
Lake Gyrstinge, Denmark	Level fluctuations, sediment exposed to air
Roskilde Fjord, Denmark	Complex hydrodynamics, SDM
The Lagoon of Venice, Italy	*Ulva* and *Zostera* included, SDM
Stadsgraven and the internal lakes, Copenhagen	interconnected basins
Søbygaard Lake, Veng Lake, Denmark	SDM
Lake Annone, Italy	SDM and two connected basins
Lake Mogan, Turkey	SDM, only P cycle
Biomanipulation, general	SDM
Bergunda Lake, Sweden	Mixture of data from different years
Broia Reservoir, Brazil	Two different types of zooplankton
Mondego Estuary	Macroalgae, SDM to compare ecological indicators

The Glumsø model has been applied with modifications in 24 other case studies. These modifications are shown in Table 12.3. As shown, in some case studies an SDM modification of the model has been developed because major changes in the species composition took place. As the model is relatively complex, it is not surprising that it was necessary to revise the model from case to case. A simpler model could be applied more generally. Most of the case studies have included calibration and validation (denoted levels 5–6). Three case studies have given occasion to validate prognoses (denoted level 7),

TABLE 12.4

Levels of Development for the 25 Case Studies Where the Glumsø Model Has Been Applied with Modifications

Level	Number of Case Studies
1: Conceptual diagram only	0
2: Verification made	1
3: Verification and sensitivity analysis	1
4: Calibration is performed either by use of very frequent measurements or use of annual measurements	3
5: Two calibrations are made	5
6: Calibration and validation	12
7: Validation of previously published prognoses	3

whereas in two case studies this move was stopped before the calibration because of lack of data of sufficient quality. Table 12.4 gives an overview of the number of case studies at different development levels.

The following levels of indication were applied (only levels 6 and 7 are fully acceptable):

Level 1: Conceptual diagram developed

Level 2: Verification carried out

Level 3: Sensitivity analysis carried out

Level 4: One calibration by either very frequent or annual measurements carried out

Level 5: Two calibrations are made—in most cases, by very frequent and annual measurements

Level 6: Two calibrations and validation carried out

Level 7: Validation of the prognoses was possible

12.4 Application of SDMs to Improve the Results of Eutrophication Models

The parameters are constantly varied in SDMs to account for adaptations and shifts in species composition. The changes in parameters are based on either expert knowledge or optimization of a so-called goal function that can describe the fitness to the changed conditions. Eco-exergy has been the most applied goal function (used in 21 case studies) in the development of SDMs. This new approach seemingly overcomes some of the weaknesses encountered when using traditional biogeochemical models (Zhang et al. 2010):

1. Fixed and rigid parameter sets are used in these models, although we know that the properties and the compositions of the species change according to the prevailing conditions of the ecosystem.

2. Calibration is often difficult, because we have to deal with a number of uncertain parameters simultaneously and test them within a wide range of possible values. In SDMs, a few parameters are determined by the goal function and therefore the number of parameters to be calibrated is reduced. In addition, current eco-exergy calibrations can be used to limit the ranges of possible values because the values resulting in significant decreases of eco-exergy can often be excluded.

The development of a structural dynamic model using eco-exergy as the goal function is described in Figure 12.8. If the optimization of eco-exergy

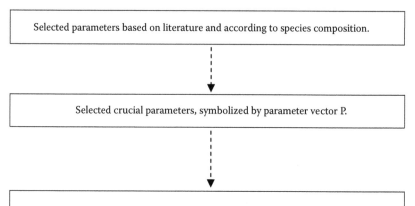

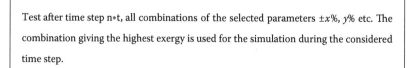

FIGURE 12.8
The procedure applied for the development of SDMs.

involves many parameters and combinations of variations in those parameters, the time required to select the best combinations of parameters that can maximize eco-exergy may be enormous, especially if a high level of selection of parameter changes is employed. For instance, for n number of parameters, 3^n number of combinations are needed to be tested for each time step even for only three levels ($\pm x$ %) of present values. Alternatively, size functions are introduced in the calibration phase when developing eutrophication models by relating all the biotic parameters of phytoplankton and zooplankton to their corresponding sizes. As a result, estimated parameters are reduced to two (sizes of phytoplankton and zooplankton) (Zhang et al. 2003a, 2003b).

Zhang et al. (2003a) presented an improved calibration using eco-exergy as the goal function combined with allometric principle in the study of Lake Mogan. The results with SDM approach in calibration were compared to the ones from a trial-and-error method for the state variables of phytoplankton, zooplankton, submerged plant, and total phosphorus. The SDM calibration shows a better fit between the simulations and the observations for phytoplankton and total phosphorus in comparison with other approaches, and for the other two state variables they are about the same.

Zhang et al. (2004) also applied the SDM approach to modeling of Lake Glumsø. The SDM gives considerable time saving in the calibration in comparison to the trial-and-error method. In addition, it yields almost equally good results in calibration, validation, and prognosis as the model that was developed specifically for Lake Glumsø, which used a trial-and-error calibration. In this case study, they compared four models that were applied to the case of Lake Glumsø. The first model was Glumsø-78, a specifically developed model for this particular lake with a trial-and-error calibration. The model was later further developed using the SDM approach with eco-exergy as the goal function and combined with size functions into SDM-Glumsø. The comparison of these two models shows that SDM-Glumsø, which requires less time for calibration, gives good results of simulating state variables compared to Glumsø-78. The study also compared the results of two other models—the two-layer PAMOLARE model and the SDM-PAMOLARE model—which were applied to the case study of Lake Glumsø. The two-layer PAMOLARE model was developed as a general model with a trial-and-error calibration, whereas the SDM-PAMOLARE model is also a two-layer general model but with structurally dynamic approach through optimizing eco-exergy and using size-related relationships as well as implementing an automatic calibration procedure for several physical parameters. Results of the comparison also show that the SDM decreases the time spent on calibration and yields a significant improvement in terms of simulation results.

The importance of using the SDM approach—which can account for the changes in ecosystem properties, structure, and species composition, and provides an improved result when the model is used for predictions and prognoses—has been illustrated in the case study of modeling Lake Glumsø (Jørgensen 1986c). Using eco-exergy as a goal function in this case gives a significant improvement of prognosis of eutrophication, which has also been validated by later observations. The occurring shift between phytoplankton species compositions is also successfully predicted. More recently, the SDM approach was also used for setting up prognosis scenarios for a lake restoration project taking place in Lake Furesø in Denmark, using the SDM version of Pamolare (downloadable from www.unep.or.jp/ietc/pamolare). The SDM approach was seen as a significant move, because it was expected that as a result of the restoration adaptation and shifts to other better-fitted species might take place in the lake and water quality would be improved as well. An SDM-PAMOLARE model was developed for this case study, which eventually yielded acceptable calibration and validation results (Gurkan et al. 2006). Next, the model was applied for testing the possible effects of bio-manipulation and aeration processes on the restoration of the lake, which has been implemented since 2003. The result indicates that it is possible to achieve significant improvements in water quality and ecological conditions in the lake, as has been earlier expected from the restoration project.

The recent development in this approach has also been applied for exploring biological patterns relating to the system response when disturbance factors are imposed. Nielsen (1992a, 1992b) applied a zero-dimensional model to describe the structural dynamics of the macrophyte societies in the Danish estuary, Roskilde Fjord. The results show that the simple model, through optimizing eco-exergy, is able to simulate the competition and shift between eelgrass (*Zostera marina*) and sea lettuce (*Ulva lactuca*) and demonstrate spatial coexistence in time. Similarly, Coffaro et al. (1997) used an SDM based on eco-exergy as the goal function to estimate the spatial variability of primary producers in the Lagoon of Venice. The results show that SDM with eco-exergy maximization can simulate the spatial variability of the community of primary producers and competition between macroalgae and seagrasses.

Jørgensen and Bernardi (1998) successfully explained the structural changes between a zooplankton/carnivorous fish controlled state and a phytoplankton/planktivorous fish controlled state in a lake when the total phosphorus concentrations were in the range of 60–125 µg L^{-1}. A shift between a clear water state dominated by the rooted submerged plants and a turbid water state dominated by phytoplankton in Lake Mogan was also predicted using this approach when total phosphorus was between 100 and 250 µg L^{-1} (Zhang et al. 2003b). The result is also consistent with the observations in other studies (Scheffer et al. 2001). In another example, SDM was used for examining the effect of fish biomanipualtion by addition of silver carp to suppress algal blooms in Lake Taihu (Zhang 2004). Silver carp (*Hypophthalmichthys molitrix*) that feeds on both phytoplankton and zooplankton through size selection may yield controversial results when used for biomanipulation. The approach may be expanded to use of variable parameters that may express the change in ecosystems influenced by external environmental conditions.

12.5 Future Development of Eutrophication Models

In the coming years, developments in computer resources and techniques will only increase and facilitate the development of more 3D eutrophication models. However, it should not be forgotten that databases are often limiting the selection of model complexity. A number of different ecological networks and different processes will inevitably be tested in our efforts to improve eutrophication models, which implies that we will continuously gain more experience in the development of eutrophication models in a growing spectrum of aquatic ecosystems.

For modelers, the challenges of developing future SDMs are still in determining the proper model structure and the mechanisms involved as well as the final parameterizations. It is expected that, as computer resources and techniques develop and better algorithms are applied in calibration, more state variables and parameters may be examined and included in the development of SDMs. More prognosis validations of SDMs (as for e.g., Lake Furesø) based on different case studies should be tested for use in environmental management and to enhance our understanding of ecosystem properties so we can gain sufficient experience in this type of eutrophication models. Suitable constraints for both calibration and validation phases should be tested during the process of optimizing eco-exergy. It is highly probable that SDMs will be used increasingly in the coming years because this model type captures an important feature of ecosystems: that biological components can change their properties by adaptation and even be replaced by other biological components better fitted to the *steadily changing properties.*

It may also be important to consider different ecological networks by the development of SDMs since it is known that a shift in the ecological network is also a sign of possible change in the ecological structure.

References

Coffaro, G., M. Bocci, and G. Bendoricchio. 1997. Application of structural dynamic approach to estimate space variability of primary producers in shallow marine water. *Ecological Modelling* 102: 97–114.

Di Toro, D. M., and J. F. Conolly Jr. 1980. Mathematical models of water quality in large lakes: Part II. Lake Erie. U.S. Environmental Protection Agency, Ecological Research Series, EPA-600/3-80-065.

Gurkan, Z., J. Zhang, and S. E. Jørgensen. 2006. Development of a structurally dynamic model for forecasting the effects of restoration of lakes. *Ecological Modelling* 197: 89–103.

Jørgensen, S. E., and G. Bendoricchio. 2001. *Fundamentals of Ecological Modelling*, 3rd ed. 628 pp. Amsterdam: Elsevier.

Jørgensen, S. E. 1976. A eutrophication model for a lake. *Ecological Modelling* 2: 47–166.

Jørgensen, S. E. 1986a. Examination of the generality of eutrophication models. *Ecological Modelling* 32: 251–266.

Jørgensen, S. E. 1986b. Validation of "Prognosis based upon a eutrophication model." *Ecological Modelling* 32: 165–182.

Jørgensen, S. E. 1986c. Structural dynamic model. *Ecological Modelling* 31: 1–9.

Jørgensen, S. E. 2009. *Ecological Modelling, An Introduction,* 192 pp. Southhampton: WIT.

Jørgensen, S. E., H. F. Mejer, and M. Friis. 1978. Examination of a lake model. *Ecological Modelling* 4: 253–278.

Jørgensen, S. E., and R. de Bernardi. 1998. The use of structural dynamic models to explain successes and failures of biomanipulation. *Hydrobiologia* 359: 1–12.

Jørgensen, S. E., H. Löffler, W. Rast, and M. Straskraba. 2004. *Lake and Reservoir Management.* Elsevier, Amsterdam, 503 pp.

Jørgensen, S. E., T. Chon, and F. Recknagel. 2009. *Handbook of Ecological Modelling and Informatics.* WIT, Southampton, 432 pp.

Nielsen, S. N. 1992a. Application of maximum energy in structural dynamic models. PhD thesis, National Environmental Research Institute, Denmark.

Nielsen, S. N. 1992b. Strategies for structural–dynamic modelling. *Ecological Modelling* 63: 91–100.

Scheffer, M., S. Carpenter, J. A. Foley, C. Folke, and B. Walker. 2001. Ecology of shallow lakes. *Nature* 413: 591–596.

Zhang, H., D. A. Culver, and L. Boegman. 2008. A two dimensional ecological model of Lake Erie: Application to estimate dreissenid impacts on large lake plankton population. *Ecological Modelling* 214: 219–240.

Zhang, J. 2004. A structurally dynamic approach to ecological and environmental models. PhD thesis, Copenhagen University.

Zhang, J., S. E. Jørgensen, C. O. Tan, and M. Beklioglu. 2003a. A structurally dynamic modelling—Lake Mogan, Turkey as a case study. *Ecological Modelling* 164: 103–120.

Zhang, J., S. E. Jørgensen, C. O. Tan, and M. Beklioglu. 2003b. Hysteresis in vegetation shift–Lake Mogan prognoses. *Ecological Modelling* 164: 227–238.

Zhang, J., S. E. Jørgensen, and H. Mahler. 2004. Examination of structurally dynamic eutrophication models. *Ecological Modelling* 173: 213–333.

Zhang, J., Z. Gurkan, and S. E. Jørgensen. 2010. Application of eco-energy for assessment of ecosystem health and development of structurally dynamic models. *Ecological Modelling* 221: 693–702.

Zhao, J., M. Ramin, V. Cheng, and G. B. Arhonditsis. 2008. Plankton community patterns across a trophic gradient: The role of zooplankton functional groups. *Ecological Modelling* 213: 417–436.

13

Modeling Oxygen Depletion in a Lagunar System: The Ria de Aveiro

José Fortes Lopes

CONTENTS

13.1 Introduction 347
13.2 The Study Area 351
13.3 Description of the Model 354
    13.3.1 The Water Quality Model 355
    13.3.2 The Eutrophication Model 360
                13.3.2.1 Oxygen Processes in the Water Column 366
                13.3.2.2 Oxygen Production 367
                13.3.2.3 Oxygen Consumption 367
                13.3.2.4 Oxygen Reaeration 369
13.4 Results 370
13.5 Discussion and Conclusions 380
References 384

13.1 Introduction

During the past decades, most of the estuarine and coastal waters have changed from balanced and productive ecosystems to ones experiencing sudden trophic changes, biogeochemical alterations, and a deterioration in habitat quality. Nuisance and sometimes harmful phytoplankton blooms, accompanied by oxygen depletion, toxicity, fish death, and shellfish mortality, are becoming more common (Pinckney et al. 2001; Paerl et al. 1998; Roelke and Buyukates 2001; Smayda 1990). These changes are associated with the so-called eutrophication processes and are, in general, dependent on the length and extent of nutrient over enrichment of these areas. The assessment of eutrophication in coastal systems, and particularly in estuaries, is highly complex and the symptoms are diverse. Although there is an association between pressure and state, the relationship between them is strongly influenced by estuarine geomorphology and hydrodynamics: estuaries subject to similar nutrient-related pressure often exhibit totally different eutrophication symptoms, and in some cases no symptoms at all

(Ferreira et al. 2005). Factors such as water residence time (Ketchum 1954; Lucas 1997; Lucas et al. 1999a; Tett et al. 2003), tidal range (Alvera-Azcarate et al. 2003), and turbidity (May et al. 2003) play a major role in determining the nature and magnitude of symptom expression. Biological interactions, particularly those attributed to grazing (Cloern 1982, 1987, 1996; Gallegos 1989; Lucas et al. 1999b), may provide top-down control of eutrophication symptoms. These may occur in similar types of estuaries, due to natural variability, but also due to human activities such as shellfish aquaculture (Nunes et al. 2003). In eutrophic ecosystems, phytoplankton carbon loading, especially following bloom events, can produce large-scale bottom water anoxia (Pinckney et al. 1997, 1998). Dissolved oxygen is one of the most important indicators of the state of coastal water ecosystems, including estuaries and lagoons. Hypoxia is operationally defined as a situation corresponding to dissolved oxygen concentration below 2 mg/l, as bottom trawls fail to capture demersal fish, shrimp, or crabs below that level (Renaud 1986). Prolonged exposure to low dissolved oxygen levels (5–6 mg/l) may not directly kill an organism, but will increase its susceptibility to other environmental stresses. Exposure to <30% saturation (2 mg/l oxygen) for 1 to 4 days may kill most of the biota in a system (Gower 1980; Radwan et al. 2003). In general, oxygen may be added to or removed from the water, by various physical, chemical, or biological reactions. Oxygen in the aquatic environment is produced by photosynthesis of algae and plants and consumed by respiration of plants, animals, and bacteria, biological oxygen demand, and sediment oxygen demand. It is reaerated through interchange with the atmosphere (Radwan et al. 2003). The oxygen consumption from the degradation of organic material is normally measured as BOD, a process combining several oxygen consumption processes into one variable, quantifying the oxidizable matter through biochemical processes, that is, the amount of oxygen required for the stabilization of organic matter. When the oxygen concentration falls below the saturation level, the deficit is offset by the transfer of gas from the atmosphere, through the surface, whereas when the oxygen concentration is greater than the saturation level, the supersaturation is reduced by the transfer from the water column to the air. Such interactions are driven by the partial pressure gradient in the gas phase and the concentration gradient in the liquid phase. The oxygen transfer in natural waters depends on the physical properties of the water: internal mixing and turbulence due to the velocity gradients and fluctuations, wind mixing, and water temperature. Although oxygen concentrations fluctuate under natural conditions, severe oxygen depletion usually results from human activities. Introducing large quantities of biodegradable organic materials into surface waters can rapidly consume the available oxygen. Pollution involving organic wastes also provides a continuous supply of food for bacteria, which accelerates bacterial activity and bacterial population growth. In polluted waters, the consumption of oxygen in the bacterial decomposition of organic materials can rapidly outpace oxygen replenishment from the atmosphere and photosynthesis performed by algae and aquatic plants resulting in a net decline in oxygen concentrations in the water. On the other hand, high level of pollutants dissolved in waters by killing algae, aquatic weeds, or fish, provides additional organic matter to the oxygen-consuming bacteria, contributing indirectly to lowering the oxygen concentrations in the water. Nitrogen and organic matter inputs from riverine, groundwater, and atmospheric sources intensify hypoxia/anoxia potentials by providing significant amounts of diverse forms of N that support phytoplankton growth, biomass, and blooms (Paerl et al. 1998). The current understanding about the causes of hypoxia in estuaries and coastal waters is that riverine nutrient loading, principally as nitrate-N, increased in the last half of the twentieth century, and stimulated in situ production of organic material which, when it sinks to the bottom layer, consumes oxygen faster than it can be replaced by vertical mixing through a stratified water column (Turner et al. 2005). Shallow, well-mixed estuaries are less susceptible to anoxic situations because wave action and circulation patterns and exchanges with the atmosphere can easily supply the deeper waters with oxygen. As estuaries have shorter residence times than, for example, lakes, the estuarine water tends to be nitrogen-limited because the fixed nitrogen is rapidly swept away (Allanson and Winter 1999; Allanson 2001). Long residence times, low flushing rates, and periodic vertical stratification promote hypoxic/anoxic conditions that can persist for weeks and cover large areas (Paerl et al. 1998; Pinckney et al. 2001). Hypoxia in estuaries can be periodic because of wind events (Yin et al. 2004; Stanley and Nixon 1992), spring-neap tidal cycles (Diaz et al. 1992; Montani et al. 1998; Uncles et al. 1998), and other events such as rainfall (Lapointe and Matzie 1996; Paerl et al. 1998) and storms (Wilson et al. 1991; Glasgow and Burkholder, 2000). The rain favors the development of hypoxia (Portnoy 1991; Lapointe and Matzie 1996), but winds destroy the hypoxia due to vertical mixing, a result of the storm (Paerl et al. 1998; Stanley and Nixon 1992).

Ecological models, which include eutrophication and water quality models, are important tools for the assessment of the changes in coastal ecosystems status resulting from changes in natural or anthropogenic forcings. Ecological models are essential tools for better knowledge of the coastal waters ecosystems and prediction of their evolution as well as for their effective management. Water quality models deal with the basic aspects of water quality of coastal areas influenced by human activities, namely, the oxygen depletion as a result of nutrients and organic matter loadings. They are the most commonly used approaches to tackle the various processes involved that lead to the rapid degradation of the ecosystems (Bonnet and Wessen 2001). The need for water quality management tools has arisen as a result of increased eutrophication of lakes and coastal ecosystems throughout the world. Nowadays, most of those models are focused on the dissolved oxygen balance in the estuary (Baird and Whitelaw 1992). Eutrophication models describe, in general, nutrient cycling, phytoplankton and zooplankton growth, growth and distribution of rooted vegetation, and macroalgae. In addition, they allow us to focus on phytoplankton-related processes: the primary production and the oxygen budget within the water column. Bonnet and Wessen (2001) and Hang et al. (2009) reviewed a number of ecological models used in the past to simulate natural ecosystems. Several authors— Fransz and Verhagen (1985), Franks et al. (1986), Frost (1987), Fasham et al. (1990), Taylor et al. (1993), Sharples and Tett (1994), McCreary et al. (1996), Bonnet and Wessen (2001), and Drago et al. (2001)—have used those models to describe eutrophication in rivers, lakes, and reservoirs. On the other hand, ecohydrodynamic models have been developed during the past decades for well-known coastal and estuarine ecosystems; examples include a model that was developed to evaluate the effectiveness of nutrient reduction on bay eutrophication in the Baltic Sea (Fennel 1995), and another model whose aim was to study the eutrophic processes in the Chesapeake Bay (Neumann 2000; Neumann et al. 2002; Fennel and Neumann 2004) using the widely known CE-QUALICM eutrophication model (Cerco and Cole 1993, 1995; Cole and Cloern 1987; Cole et al. 1992).

The objective of this study is to apply a system of ecological models in the study of the influence of physical and the biogeochemical factors in the oxygen conditions in Ria de Aveiro lagoon. Two model systems were applied: (1) a water quality model (Mike3-WQ; Mike3 2005b) integrating the main biochemical processes associated to the phytoplankton photosynthesis and respiration, oxidation due to the biochemical oxygen demand (BOD), nitrification, sediment oxygen demand, phytoplankton, and bacterial respiration, and able to deal with the basic aspects of water quality associated to oxygen depletion and the transforming processes of the main biochemical compounds, namely, BOD and ammonia levels resulting from the organic matter loadings; (2) a eutrophication model (Mike3-EU; Mike3 2005c), which includes the basic aspects of the lower trophic status of the water column, and is able to describe the nutrient cycling, including, organic matter (detritus), organic and inorganic nutrients, the phytoplankton and the zooplankton growth, the primary production, and the dissolved oxygen.

The results show that the lagoon dissolved oxygen (DO) distribution is characterized by high DO values for the main lagoon areas (in general, greater than 7 mg O_2/l) even in a situation of organic matter contamination at the rivers' boundaries, evidencing the influence of the exchange with the ocean high oxygenated waters as well as the high productivity inside the lagoon. Maximum oxygen production by phytoplankton occurs, in general, at the central and the northernmost areas, along the S. Jacinto channel and the Ovar channel, reflecting conditions of high phytoplankton growth and productivity of those areas. Hypoxic situations occur over limited time and spatial scale, and are presented at the far ends of the lagoon, characterized by low flushing rates and high residence time, and low residual currents remain, in general, confined in those areas, even though their influence may reach the central areas of the lagoon, as the tidal transport and mixing are the dominant processes. This leads to the conclusion that in the present situation, eutrophication processes in the Ria de Aveiro are unlikely to occur, even in the worst-case scenarios of high organic contamination at the river boundaries.

13.2 The Study Area

Ria de Aveiro (Figure 13.1) is a lagoon situated on the northwest Atlantic coast of Portugal (40°38′N, 8°45′W), 45 km long and 10 km large, covering an area of 66–83 km², respectively, at low and high tide (Dias et al. 2000). It is

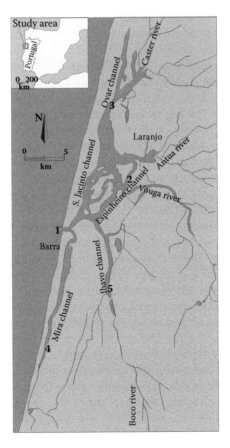

FIGURE 13.1
Study area: the Ria de Aveiro lagoon, with the location of the main lagoon stations.

characterized by several branching channels, the most important of which are S. Jacinto, Ovar, Espinheiro, Mira, and Ilhavo, which are directly connected to the lagoon mouth and the ocean by a single tidal channel. The central areas of the lagoon is constituted by a myriad of other small channels dominated by shallow and tidal flat areas, contributing to a strong damping of the currents and an increase of the phase delay of the tidal wave (Dias 2001). The bathymetry (Figure 13.1) shows that Ria de Aveiro is a very shallow lagoon (average depth of 1 m). The deepest areas of the lagoon are confined to the inlet channel and to small areas close to the lagoon mouth, at the western boundary of the lagoon. In those areas, the depth may be of the order or greater than 10 m, whereas elsewhere, at the inner parts of the lagoon, the depth barely reaches values beyond 3 m. The hydrological characterization of the lagoon (Dias et al. 1999, 2000; Dias 2001) reveals that, excluding the areas close to the lagoon mouth, the salinity and the temperature vertical profiles have very well mixed structures. Tides, which are semidiurnal, are the main forcing of the circulation in the Ria de Aveiro lagoon. The tidal range varies between 3.2 and 0.6 m, and the estimated freshwater input is very small (about 1.8×10^6 m^3 during a tidal cycle) when compared with the mean tidal prism at the mouth (about 70×10^6 m^3). In general, fluvial input to the lagoon occurs at small steady flow alternated by intense pulses due to the sudden increase of the river's tributaries flows. Two main rivers, the Antuã river (5 m^3 s^{-1} average flow) and the Vouga river (50 m^3 s^{-1}), which are located at the eastern side of the lagoon, account for almost 90% of the freshwater input into the lagoon. Other small rivers situated at the far end of the lagoon, which include Caster (situated at the far end of the Ovar channel), may contribute substantially to the salinity patterns of the lagoon during the rainy period. Therefore, in normal situation, as the lagoon is dominated by tides, the freshwater flows only affects the restricted areas, close to the river mouth, along the main channels (Dias et al. 1999; Dias 2001). The wind regime is characterized by northwestern winds during summer and a more variable wind regime during winter, with strong fluctuations in direction and intensity (Fiúza et al. 1982). The annual average precipitation in the Aveiro region is about 913 mm and 75% of the precipitation occurs during the first semester of the hydrological year (October to March), whereas 25% occurs in summer. The highest precipitation occurring during winter is due to the presence of cold and maritime polar air masses, which originates low pressures over the Atlantic coast. During winter and spring, the runoff usually ranges from 20 to 60 m^3 s^{-1} but periodic high flow events (>100 m^3 s^{-1}) can cause extensive flooding of the land bordering the inner areas of the lagoon. During summer, the displacement of the Azores anticyclone to highest latitudes creates a blocking situation affecting the arriving of the precipitation through the north (Manso et al. 1996; Vaz et al. 2005). The base flow is typically reduced to 3–4 m^3 s^{-1} (Silva 1994), and the evaporation exceeds the freshwater input, resulting in higher salinity values (>36 psu) when compared to the local sea salinity (Silva et al. 2002).

Salinity is an important physical variable for the understanding of the dynamic inside the lagoon, which is dominated by tides (Dias et al. 2001). During the dry season, which generally coincides with summer, the river's flow is very low and high salinity values (30–34), which are typical for the lagoon mouth, are observed inside the lagoon and even at its far ends, such as during flood tides. These locations are close to the river mouths and generally present extremely low values of salinity, between 0 and 5, during the wet season (autumn and spring).

From the biogeochemical point of view, Ria de Aveiro lagoon may be considered as a mesotrophic shallow estuarine system (Almeida et al. 2005). It is characterized by a rich biodiversity as well as by an increasing pressure posed by anthropogenic activities near its margins, that is, building and land occupation as well as agricultural and industrial activities. This has resulted in a significant change of the lagoon morphology, and in a constant input of a large volume of anthropogenic nutrients as well as of contaminant loads, with the consequent negative impact in the water circulation, as well as in the water quality of the lagoon. The lagoon is subjected to considerable inputs of industrial and domestic discharges, which occur mainly at its periphery, and is currently receiving a mean annual influxes of total nitrogen (N) and total phosphorus (P) of 6118 and 779 t year^{-1}, respectively, from its influent rivers (Silva et al. 2002). The assessment of the lagoon ecological state is given by the dissolved oxygen distribution. In general, DO (dissolved oxygen concentration) inside the lagoon varies between 8 and 12 mg O_2/l. Given the salinity and the temperature distribution inside the lagoon, the last values correspond to near oxygen saturation condition (Almeida et al. 2005). The far end of the lagoon, that is, at the Ovar channel and near the Vouga river mouth, may show extreme DO values, the highest values (12 mg O_2/l or greater) as well as the lowest values (below 5 mg O_2/l). The first case occurs in extreme situations corresponding to conjugation of high nutrient input (nitrate and nitrites (NO_3 + NO_2) concentrations at the river boundaries greater than 1.0 mg l^{-1}) and light availability. These conditions, which in general occur in late winter or early summer and autumn, reflect the situation of CHL-*a* concentrations (of the order of or greater than 50 µg/l); these values are particularly high when compared to typical ones observed for other locations of the lagoon (in general, below 7 µg/l). The water quality of Ria de Aveiro results, therefore, from a complex balance between pollutant emissions from natural and anthropogenic origin and its water auto-depuration capacity. The main anthropogenic sources of pollution are domestic and industrial discharge flows, as well as contaminated water infiltration from agricultural and cattle raising activities. The areas more affected by agricultural and industrial activities are as follows: Laranjo (near the Antuã's river mouth); the far end of the Espinheiro channel (near the Vouga's rivers mouth); the far end of the Ovar channel and Vista Alegre in the Ílhavo channel. Except for the period starting from January 1989 to December 1990, BOD$_5$ is rarely reported in surveys conducted in Ria de Aveiro. The highest

BOD_5 concentrations were observed near the Vouga River and at the far end of the Ovar channel, where the maximum values of 10 mg O_2/l were found in the summer of 1989, whereas elsewhere in the lagoon values lower than 2 mg O_2/l were generally observed (Alcântara et al. 1992, 1994).

13.3 Description of the Model

The physical model solves the hydrodynamic and the transport equations, and includes the main physical processes occurring in the water column as well as the exchanges in the air–water interface between the atmosphere and the water system. The hydrodynamic model, the Hydrostatic Version of Mike3 FLOW MODEL (Mike3 2005a), solves the continuity and the momentum equations. It is based on the solution of the three-dimensional (3-D) incompressible Reynolds averaged Navier–Stokes equations, subjected to the assumptions of Boussinesq and hydrostatic pressure; therefore, vertical accelerations are assumed to be negligible. The transport-diffusion equation is coupled to the hydrodynamic model and it is used to solve for the temperature, T, salinity, S, and any scalar quantity (biogeochemical variables), as well as k, the turbulent kinetic energy per unit mass, and ε, the dissipation of the turbulent kinetic energy. It integrates radiative exchanges between the atmosphere and the water system, including solar and nonsolar radiation (Gill 1982). The hydrodynamic and transport models are described by Dias and Lopes (2006), Lopes and Silva (2006), and Lopes et al. (2008). The model domain covers a rectangular area corresponding to 266×654 points, with a grid size of 60×60 m. It makes use of the so-called alternating direction implicit (ADI) technique to integrate the mass and the momentum equations in the space–time domain. The equation matrices, which result for each direction and each individual grid, are resolved by a Double Sweep (DS) algorithm. It results in a third-order convective matricial equation that is solved for each direction and each grid line by a double sweep algorithm (Leendertse and Gritton 1971; Abbott et al. 1981; Casuli 1999). The time step for the hydrodynamic and the transport model has been chosen to be small (2 s) in order to assure stable solutions for the transport equation. The transport model makes use of the ULTIMATE-QUICKEST scheme (Leonard 1979) for the integration of the advection–diffusion equation in the space–time domain, extended to the third dimension. The scheme ensures mass conservation through the control volume formulation of the transport terms. The scheme is simple and computationally fast and is, thus, well suited for coupling to the 3-D hydrodynamic model in order to simulate density-driven flows. It results in a very efficient, stable, and third-order accurate explicit algorithm (Vested et al. 1992). The validation of the 3-D hydrodynamic and transport models was performed using the same data set and periods as those used to validate the 2-D models, such as those described by Dias and Lopes (2006), Lopes and Silva (2006), and Lopes et al. (2008), in order to obtain similar results as the authors. The same strategy and methodologies were therefore adopted. The hydrodynamic model is set up by spinning up the lagoon from rest. The sea surface elevation is specified at the western open boundary at each time step. The lateral boundary conditions correspond to no-slip velocity and insulation for temperature and salinity on the side walls. Initial conditions and boundaries conditions for the transport models were obtained from Almeida et al. (2005). During the simulation period, while the biogeochemical variables were kept constant at the river boundaries, the salinity and the temperature were allowed to vary at each time step, as a year time series were defined based on typical seasonality and historical data for these variables. Concerning the ocean open boundary, the von Newmann boundary conditions were imposed. The atmospheric parameters needed to compute the latent and sensible heat fluxes, as well as the radiation fluxes (relative humidity, air temperature, wind speed, and cloudiness) were specified at each time step.

Ecological models have been applied to the study area (Lopes et al. 2005; Lopes and Silva 2006; Lopes et al. 2008) and are revised in the following subsection:

- The Water Quality model (Mike3-WQ; Mike3 2005b), which deals with the basic aspects of water quality influenced by human activities—for example, oxygen depletion and the transforming processes of the main biochemical compounds, that is, BOD and ammonia levels resulting from the organic matter loadings.
- The Eutrophication model (Mike3-EU; Mike3 2005c), a biogeochemical model that deals with the basic aspects of the lower trophic status of the water column.

The models are coupled to the 3-D hydrodynamic model described above, through the transport equation, and require the definition of the concentrations at model boundaries, flow and concentrations from pollution sources, water temperature, and irradiance.

13.3.1 The Water Quality Model

The Water Quality model solves the system of differential equations that describes the chemical and biological state of the coastal waters. The variables of the model are the following: dissolved BOD (BOD_d), suspended BOD (BOD_s), sedimented BOD (BOD_b), ammonia (NH_4), nitrite (NO_2), nitrate (NO_3), phosphates (PO_4) (not considered in this study), dissolved oxygen (DO), chlorophyll-*a* (CHL-*a*). Table 13.1 presents the parameters representing the main biochemical processes for the water quality model. The key biochemical

TABLE 13.1

Main Parameters for the Water Quality Model

State Variables	Symbol	Unit
Phytoplankton carbon	PC	g C/m³
Phytoplankton nitrogen	PN	g N/m³
Phytoplankton phosphorus	PP	g P/m³
Chlorophyll-*a*	CH	g/m³
Zooplankton	ZC	g C/m³
Detritus carbon	DC	g C/m³
Detritus nitrogen	DN	g N/m³
Detritus phosphorus	DP	g P/m³
Ammonium	NH	g N/m³
Nitrate	N_3	g N/m³
Inorganic phosphorus	IP	g P/m³
Dissolved oxygen	DO	g/m³
Benthic vegetation carbon	BC	g C/m³

processes are phytoplankton photosynthesis and respiration, oxidation due to BOD, nitrification, sediment oxygen demand, phytoplankton, and bacterial respiration (Ambrose et al. 1993; Zheng 2004). Photosynthetic carbon fixation, which leads to oxygen production by photosynthesis of algae and plants, is one of the most important sources of DO in the water column. It is proportional to phytoplankton density and growth rate. On the other hand, dissolved oxygen in the water column is consumed by the processes associated with oxygen demand and oxidation of organic matter, phytoplankton and bacteria respiration, nitrification, sediment oxygen demand, etc. Generally, the importance of sediment processes affecting DO is magnified in shallower waters (Stow et al. 2005; Eldridge and Morse 2008). The reaeration through interchange with the atmosphere is an important source or sink of DO, depending on the partial pressure gradient in the gas phase and the concentration gradient in the liquid phase (Radwan et al. 2003). Physical processes that affect the DO within the water column include current advection and turbulent mixing. The oxygen transfer in natural waters depends on internal mixing and turbulence due to velocity gradients and fluctuations, temperature, and wind mixing.

The processes above described are included in the dissolved oxygen budget and are described by the following first-order differential equation (Malmgren-Hansen et al. 1984)

$$\frac{dDO}{dt} = k_2 * (C_s - DO) + FBOD + FNI + FRE + FPHO + FSED \quad (13.1)$$

The first term represents the reaeration process that is described as occurring at the air–water interface and follows the first-order kinetic law, where C_s is the saturation concentration for the dissolved oxygen:

$$C_S = 14.652 - 0.0841 * S + [-0.41022 + 0.00256 * S + A(S,T) * T] * T \quad (13.2)$$

with

$$A(S,T) = 0.007991 - 0.0000374 * S - 0.000077774 * T$$

where T and S denote temperature and salinity, respectively. The reaeration kinetic constant k_2 is influenced by the water temperature, the river flow characteristics (flow velocity, water depth, and slope of river), and the meteorological conditions (air temperature and wind).

The biochemical degradation of organic matter consumes dissolved oxygen and it is represented by the term FBOD, which includes dissolved, suspended, and sedimented organic matter.

The term FNI represents the mineralization process, which consumes dissolved oxygen in order to produce ammonia from organic matter.

Respiration and photosynthesis involve exchanges of dissolved oxygen with the water column, and are represented by the term FRE and FPHO, respectively.

Finally, the term FSED represents the consumption of oxygen by sediments. These terms are represented by the following empirical relations:

$$FBOD = -k_{d3} * BOD_d * \theta_{d3}^{T-20} - k_{s3} * BOD_s * \theta_{s3}^{T-20} - k_{b3} * BOD_b * \theta_{b3}^{T-20}$$

$$FNI = -Y_1 * k_4 * NH_3 * \theta_4^{T-20} - Y_2 * k_5 * NO_2 * \theta_5^{T-20}$$

$$FRE = -R_1 * \theta_1^{T-20} * F(N,P) - R_2 * \theta_2^{T-20} \quad (13.3)$$

$$FPHO = PP * F(N,P)$$

$$FSED = -SOD * \frac{DO}{HS\_SOD\text{-}DO}$$

where k_{d3}, k_{s3}, and k_{b3} are the degradation constants for BOD_d, BOD_s, and BOD_b, respectively, at 20°C; k_4 is the nitrification rate at 20°C; k_5 is the specific rate for conversion of nitrite to nitrate at 20°C; Y_1 and Y_2 are yield factors describing the amount of oxygen used in the nitrification; R_1 and R_2 are the autotrophic and heterotrophic respiration (mg O_2 l^{-1} day^{-1}), respectively; θ is the Arrhenius temperature; T is the temperature; PP is the actual productivity (mg O_2 l^{-1} day^{-1}); and $F(N,P)$ is a limitation function by the potential nutrient limitation on photosynthesis (N and P are related to the nitrogenous and the phosphorous compounds, respectively).

The oxygen consumption by the degradation of the organic material is normally measured as the BOD. In this study, only the BOD_5 was considered. The BOD may be separated into oxygen consumption by degradation of materials and that used in the nitrification of ammonia (Kaspar 1982; Malmgren-Hansen and Hanne 1992; Henriksen et al. 1981; Henriksen and Kemp 1988; Rysgaard et al. 1993). Organic matter can be found in the dissolved phase, BOD_d, the suspended phase, BOD_s, or at the bottom, BOD_b. Each phase of the BOD can be described by the following first-order differential equations.

- Dissolved *BOD*:

$$\frac{dBOD_d}{dt} = -k_{d3} * BOD_d * \theta_{d3}^{T-20}$$ (13.4)

Particulate BOD, including the interaction between the water column and the bottom sediments, is represented by the last two terms of the following equations:

$$\frac{dBOD_s}{dt} = -k_{s3} * BOD_s * \theta_{s3}^{T-20} + S_1 * BOD_d / H - S_2 * BOD_s / H$$

$$\frac{dBOD_b}{dt} - k_{b3} * BOD_b * \theta_{b3}^{T-20} - S_1 * BOD_d / H + S_2 * BOD_s / H$$ (13.5)

where S_1 is the resuspension rate (m day^{-1}), S_2 is the sedimentation rate (m day^{-1}), and H is the water depth.

The nitrogenous stages of the BOD test include conversion of organic nitrogen into ammonia. Ammonia is subsequently oxidized into nitrite, which will then be oxidized into nitrate.

The ammonium/ammonia process includes the following balance: BOD_d decay, nitrification, uptake by plants, uptake by bacteria, and heterotrophic respiration. The differential equation describing the ammonium/ammonia reaction is:

$$\frac{dNH_3}{dt} = BOD\_decay + Nitr + UP + UB + HR$$ (13.6)

where BOD_decay represents the temperature-dependent BOD decay, Nitr is the nitrification process, UP and UB are the uptake by plants and bacteria, respectively, and HR is the heterotrophic respiration. The different terms in 13.6 are parameterized as:

$$BOD\_decay = +Y_b * k_{b3} * BOD_b * \theta_{b3}^{(T-20)} + Y_d * k_{d3} * BOD_d * \theta_{d3}^{(T-20)}$$

$$+ Y_s * k_{s3} * BOD_s * \theta_{s3}^{(T-20)}$$

$$Nitr = -k_4 * NH_3 * \theta_4^{(T-20)}$$

$$UP = UN_p * \left(P - R_1 * \theta_1^{(T-20)}\right) * F(N,P) \tag{13.7}$$

$$UB = UN_b * \frac{NH_3}{NH_3 + HS\_NH_3} * \left[k_{b3} * BOD_b * \theta_{b3}^{(T-20)}\right.$$

$$\left. + k_{d3} * BOD_d * \theta_{d3}^{(T-20)} + k_{s3} * BOD_s * \theta_{s3}^{(T-20)}\right]$$

$$HR = UN_p * R_2 * \theta_2^{(T-20)}$$

where UN_p is the ammonia uptake by plants (mg N/mg O_2); UN_b is the ammonia uptake by bacteria (mg N/mg BOD); Y_b, Y_d, Y_s (mg NH_4-N/mg BOD) are the nitrogen content in sedimented organic matter, dissolved organic matter, and suspended organic matter, respectively; $F(N,P)$ is the nutrient limitation on the photosynthesis; and HS_NH_3 is the half-saturation concentration for N uptake by bacteria (mg N/l); and k_4 is the nitrification rate. During nighttime, the ammonium uptake by plants is assumed to be constant.

Nitrite and nitrate processes include nitrification and denitrification. The reactions that influence nitrite and nitrate concentration are given respectively, by:

$$\frac{dNO_2}{dt} = k_4 * NO_3 * \theta_4^{(T-20)} * \frac{DO}{DO + HS\_Nitr} - k_5 * NO_2 * \theta_5^{(T-20)} \tag{13.8}$$

$$\frac{dNO_3}{dt} = k_5 * NO_2 * \theta_5^{(T-20)} - k_6 * NO_3 * \theta_5^{(T-20)} \tag{13.9}$$

where k_6 is the denitrification rate and θ_6 is the Arrhenius temperature coefficient. The nitrification terms are the same as those described in the oxygen differential equation. In the model, the decay rates are expressed in days^{-1}.

The differential equation describing the chlorophyll-*a* (CHL) evolution is given by:

$$\frac{dCHL}{dt} = K_{10} * K_{11} * \left(PP - R_1 * \theta_1^{(T-20)} \right) * F(N,P) - K_8 * CHL - K_9 * CHL / H \quad (13.10)$$

where k_8 is the death rate of chlorophyll-*a*, k_9 is the settling rate of chlorophyll-*a*, k_{10} is the chlorophyll-*a*/carbon ratio, and k_{11} is the carbon to oxygen ratio at primary production. The actual productivity has been parameterized as a function of the daily photosynthesis activity:

$$PP = P_{max} \cos(2\pi t/a) \quad (13.11)$$

where P_{max} (mg O_2 l^{-1} day^{-1}) is the maximum productivity at noon, *a* the relative daylength, and *t* is the time from sunrise to sunset.

13.3.2 The Eutrophication Model

The ecological model describes the nutrient cycling, the phytoplankton and the zooplankton growth, and the growth and the distribution of rooted vegetation

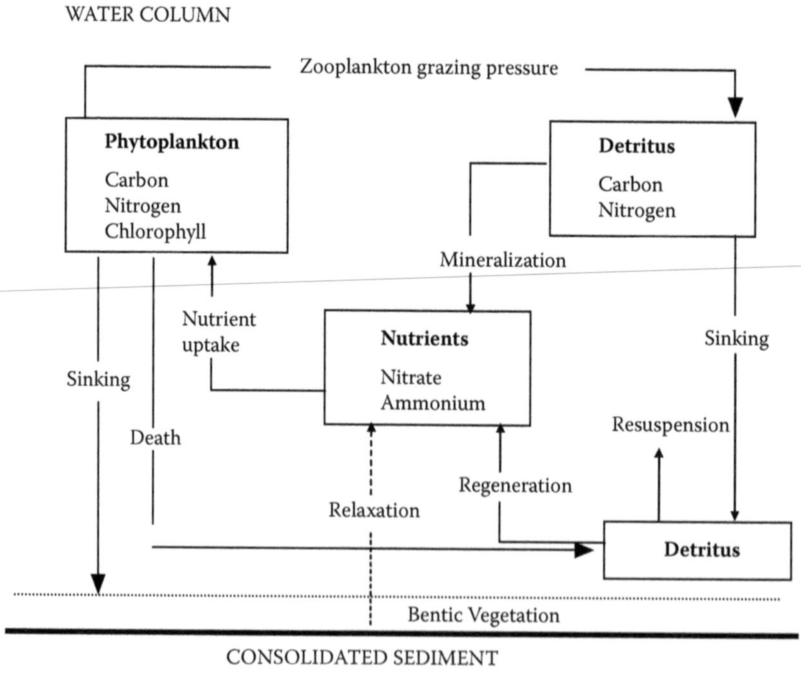

FIGURE 13.2.
Conceptual diagram for the ecological model.

and macroalgae. In addition, it simulates the concentrations of phytoplankton, chlorophyll-*a*, zooplankton, organic matter (detritus), organic and inorganic nutrients, dissolved oxygen, and the area-based biomass of benthic vegetation over time, as well as derived variables such as the primary production, the total nitrogen and phosphorus concentrations, the sediment oxygen demand, and the Secchi disk depth.

The eutrophication module, which is described by Lessin and Raudsepp (2006) and Erichsen and Rasch (2001), consists of 13 state variables represented by four functional groups (phytoplankton, zooplankton, benthic vegetation, and detritus), nutrients, and oxygen (Table 13.1). Figure 13.2 summarizes the conceptual diagram of the ecological model. The model is based on nitrogen and phosphorus but for the Ria de Aveiro lagoon nitrogen is generally the only limiting nutrient (Almeida et al. 2005). Table 13.2 presents the parameters representing the main biochemical processes for the eutrophication model.

The mathematical formulations of the biological and chemical processes and transformations for each state variable are described by a first-order ordinary differential equation. Phytoplankton is represented by one functional group and three state variables PX representing: phytoplankton carbon (PC), phytoplankton nitrogen (PN), and phytoplankton phosphorus (PP). The time evolution of each state variable X is related to its mass balance within the water column:

$$\frac{\partial X}{\partial t} = \text{PRODUCTION} - \text{RESPIRATION} - \text{GRAZING}$$
$$- \text{SEDIMENTATION} - \text{DEATH}$$

(13.12)

Once the advection–dispersion terms and the sources and sinks terms of the transport equation are evaluated, the model makes an explicit time integration of 13.12, then calculates the concentrations to the next time step, following a Runge–Kutta 4 integration method. Although each variable is characterized by specific processes, the mass balance is similar to 13.12. For instance, concerning CH (chlorophyll-*a*), the following processes are considered: production, death, and sedimentation; for ZC, representing important grazers such as copepods and various microzooplankters, the following are considered: production, respiration, and death. Detritus are defined in the model as particles of dead organic material in the water, and its pool receives the dead primary producers, dead zooplankton grazers, and unassimilated material left over after grazing. Sedimentation and mineralization are the only processes draining the detritus pools. Instead, the mass balance for detritus (DC, DN, and DP) includes regeneration, mineralization, and sedimentation processes. The inorganic nitrogen is represented by two state variables: oxidized forms (sum of nitrate and nitrite) and reduced forms (sum of ammonia, NH, and urea). The main balance for the inorganic nitrogen and phosphorus includes the inputs from mineralization and the uptake.

TABLE 13.2

Main Parameters for the Eutrophication Model

Parameters	Values
Phytoplankton Parameters	
Phytoplankton C growth rate	4.2 day^{-1}
Settling rate	0.1 day^{-1}
Settling velocity	0.2 m day^{-1}
Coagulation induced settling at high phytoplankton concentration	0.02 mg^{-2} day^{-1}
First-order death rate	0.07 day^{-1}
Temperature growth rate coefficient	1.07
Coefficient for min. chlorophyll-*a* production	0.4
Coefficient for max. chlorophyll-*a* production	1.1
Min. intracellular concentration of nitrogen	0.07 g N (g C)$^{-1}$
Max. intracellular concentration of nitrogen	0.17 g N (g C)$^{-1}$
Min. intracellular concentration of phosphorus	0.003 g P (g C)$^{-1}$
Max. intracellular concentration of phosphorus	0.02 g P (g C)$^{-1}$
Shape factor for sigmoid for nitrate uptake function	0.01
Half-saturation constant for NH$_4$ uptake	0.05 mg N l^{-1}
Half-saturation constant for NO$_3$ uptake	0.1 mg N l^{-1}
Half-saturation constant for P uptake	0.2 mg P l^{-1}
Maximum N uptake	0.3 g N (g C)$^{-1}$
P uptake under limiting conditions	0.05 mg P l^{-1}
Fraction of nutrients release at phytoplankton death	0.1
Light saturation intensity at 20°C	200 µmol m^{-2} s^{-1}
Temperature dependency for light saturation intensity	1.04
Algae respiration rate	0.06 day^{-1}
Coefficient for nitrate dark uptake by phytoplankton	0.6
N release coefficient for N mineralization	1
Zooplankton Parameters	
Maximum grazing rate	1 day^{-1}
Death rate (first order, second order)	0.05 day^{-1}, 0.002 day^{-1}
Grazing threshold	0.02 mg l^{-1}
Half-saturation concentration for grazing PC	0.5 mg l^{-1}
Temperature dependency for maximum grazing rate	1.1
Temperature dependency for C mineralization	1.14
N:C ratio	0.17 g N/g C
P:C ratio	0.03 g P/g C
Respiration due to specific dynamic	0.2
Basal metabolism	0.04 day^{-1}
Assimilation factor	0.28
Detritus c mineralization rate	0.05 day^{-1}
Detritus c settling rate < depth 2 m	0.05 day^{-1}

(continued)

TABLE 13.2 (Continued)

Main Parameters for the Eutrophication Model

Parameters	Values
Detritus c settling velocity > depth 2 m	0.1 m day^{-1}
N release coefficient for N mineralization	1
N release coefficient for N mineralization	1
Macroalgae Parameters	
Sloughing rate at 20°	0.01 day^{-1}
Production rate at 20°	0.25 day^{-1}
Temperature dependency sloughing rate	1.07
Temperature dependency production rate	1.05
N:C ratios	0.137 g N/g C
P:C ratios	0.016 g P/g C
Half-saturation concentration for N uptake	0.5 mg l^{-1}
Half-saturation concentration for P uptake	0.01 mg l^{-1}
Light saturation intensity at 20°C	180 µEinstein m^{-2} s^{-1}
Temperature dependency for light saturation	1.04
Macroalgae Parameters	
Proportional factor for sediment respiration	1
Proportional factor for N release from sediment	1
Proportional factor for P release from sediment	1
Temperature dependency for N release from sediment	1.1
Temperature dependency for P release from sediment	1.1
N release from anoxic conditions	0.05 g N m^{-2} day^{-1}
P release from anoxic conditions	0.01 g P m^{-2} day^{-1}
Temperature factor for sediment respiration	1.07
Diffusion between coefficient for NO$_3$ and NH$_4$ between sediment and water	0.002 m^2 day^{-1}
Max. denitrification in sediment at 20°C	0.2 mg N l^{-1} day^{-1}
First constant for DO penetration into sediment	0.00124 m
Second constant for DO penetration into sediment	0.0000403 m^4/g O$_2$
Third constant for DO penetration into sediment	0.0000473 m^3 day/g O$_2$
Half-saturation conc. DO, for NH$_4$ or NO$_3$ release from sediment	3
Specific nitrification rate pelagic water, 20°C	0.05 day^{-1}
Theta value, nitrification	1.04
Half-saturation conc. NO$_3$, for denitrification	0.2 mg N l^{-1}
Half-saturation conc., for DC denitrification	0.2 mg C l^{-1}
Half-saturation conc., for DO denitrification	0.5 mg O$_2$ l^{-1}
g O$_2$ used to oxidize 1 g NH$_4$ to 1 g NO$_3$	4.3 g O$_2$/g NH$_4$-N

The primary source of ammonia is the input from respiration and mineralization processes, whereas the sinks include uptake by primary producers (phytoplankton and benthic vegetation, and nitrification from nitrate). The mineralization of NH is expressed as a fraction of the sedimentation of organic matter. Under anoxic conditions, the release of nutrients is not only a result of recently sedimented material, but also a zero-order function, where large amounts of nutrients buried in the sediment will be released. Mineralization of organic matter is the main input of inorganic phosphorus (IP), and corresponds to the sum of mineralization of detritus, zooplankton, and phytoplankton phosphorus as well as the release from the sediment, the last one being only relevant for the bottom layer. The benthic vegetation, BC, is assumed to be rooted and/or attached to stones, etc. Fixed ratios of nitrogen to carbon and phosphorus to carbon are assumed. The mass balance for the benthic vegetation includes production and losses.

Hereafter, the attention will be focused on the PC- and DO-related processes. The mass balances for PN and PP are similar to the mass balance of PC, and are therefore omitted here. The net production of PC depends on light intensity, nutrient availability, and ambient temperature. The gross production is computed with a multiplicative approach considering the maximal rate of production (μ), the influence of light ($F(I)$), temperature ($F(T)$), and the internal concentrations of nitrogen and phosphorus $F(N,P)$).

The production of phytoplankton depends on light intensity, availability of nutrients, and ambient temperature. The phytoplankton gross production is computed with a multiplicative approach considering the maximal rate of production (μ), the influence of the light intensity I, ($F(I)$), the temperature ($F(T)$), and the internal concentrations of nitrogen and phosphorus $F(N,P)$):

$$\text{PRODUCTION} = \mu * F(I) * F(T) * F(N,P) * PC \qquad (13.13)$$

The gross production–irradiance parameterization is written according to the Jassby and Platt (1976) relation between photosynthesis and light without considering light inhibition:

$$F(I) = 1 - \exp\left(\alpha - \frac{\text{IPAR}}{\mu PC}\right) \qquad \text{IPAR} = I_S \frac{\exp(\eta dz) - 1}{\eta dz} \qquad (13.14)$$

where α is the initial slope in production–irradiance relationship (g C/g C/(μE/m^2/s), I_S is the surface irradiance or irradiance at the surface of any layer, and $\eta = \eta_0 + \eta_1 \text{CHL} + \eta_2 \text{DC}$, where η_0 is the background extinction constant (m^{-1}), η_1 is the chlorophyll-a–specific absorption constant (mg CHL-a)$^{-1}$ m^{-1}, and η_2 is the detritus specific absorption constant (mg CHL-a)$^{-1}$ m^{-1}.

Phytoplankton at low temperatures maintains higher concentrations of photosynthetic pigments, enzymes, and carbon (Steemann and Jørgensen

1968), enabling more efficient use of light. The influence of temperature on phytoplankton production is parameterized by an Arrhenius function $F(T) = \theta_G^{(T-20)}$, where θ_G is the temperature coefficient for growth.

The nutrient-dependent growth limitation $F(N,P)$ is calculated from the relative saturation of the internal N and P pools following Droop (1973, 1975), Nyholm (1977), Mommaerts (1978), Tett et al. (1986), and Lancelot and Rousseau (1987):

$$F(N,P) = \frac{2}{\dfrac{1}{myn} + \dfrac{1}{myp}} \qquad (13.15)$$

where

$$myn = \frac{PN/PC - pnmi}{pnma - pnmi} \qquad myp = \frac{(PN/PC - ppmi)(k_c + ppma - ppmi)}{(ppma - ppmi)(k_c + PP/PC - ppmi)}$$

pnmi and pnma are the minimum and the maximum internal nitrogen content in algae (g N/g C), respectively; ppmi and ppma are the minimum and the maximum phosphorus content in algae (g P/g C), respectively, and k_c is the half-saturation constant for phosphorus in phytoplankton (g P/g C).

Loss due to respiration is represented by a basal metabolic expenditure to maintain life processes, and is considered a constant fraction of biomass PC, with a dependence on temperature represented by the Arrhenius function:

$$\text{RESPIRATION} = \text{rphc} * \theta_G^{T-20} * PC \qquad (13.16)$$

where rphc is the fraction of biomass (day^{-1}).

Loss of phytoplankton due to grazing by zooplankton depends on zooplankton carbon, and is regulated by zooplankton grazing function modified by a temperature function. The grazing rate of phytoplankton carbon by zooplankton is represented by a saturation equation relating food concentration to grazing rate, a threshold food concentration below which no grazing takes place (Kiørboe and Nielsen 1994), the Arrhenius temperature function, and a function of dissolved oxygen suppression grazing at low DO (Roman et al., 1993):

$$\text{GRAZING} = \text{MGR} * F_5(T) * F(DO) \frac{PC - PCT}{PC - PCT - HSG} * ZC \qquad (13.17)$$

where MGR is the maximum grazing rate constant at 20°C (day^{-1}), PCT is the threshold food concentration (mg C/l), HSG is the half-saturation concentration of PC for carbon uptake for zooplankton (mg C/l), and $F_5(T)$ is the temperature Arrhenius dependence function for the zooplankton grazing and $F(DO) = \dfrac{DO^2}{DO^2 + MDO}$, with MDO the oxygen concentration indicating depressed grazing rates due to oxygen depletion

Nutrient-replete phytoplankton is able to adjust its buoyancy and hence, to minimize sinking rate. Under conditions of nutrient stress, with the internal nutrient pools at lower levels, sinking rates increase (Smayda 1970). At very high concentrations of PC, coagulation rate between cells increases (Jackson 1990) resulting in higher sedimentation rates.

The sedimentation rate of PC is described by a first-order equation depending on internal nutrient status and a second-order equation representing coagulation:

$$ \text{SEDIMENTATION} = \frac{1}{2}\,ksp1\left(\frac{pnma}{\dfrac{PN}{PC}} + \frac{ppma}{\dfrac{PP}{PC}} \right) * PC + kps2 * PC^2 \qquad (13.18) $$

where ksp1 is a sedimentation rate parameter (day^{-1}), ksp2 is the sedimentation rate constant ((mg/l)$^{-1}$ day^{-1}), pnma is the maximum internal nitrogen content in algae (g N/g C), and ppma is the maximum phosphorus content in alga.

Natural mortality of phytoplankton, or autolysis, has been shown as a significant phenomenon in the marine ecosystem (Jassby and Goldman 1974), and this decay of blooms is partly mineralized in the water column (Lancelot et al. 1987). The mortality of phytoplankton is described by a first-order equation, with a dependence on temperature represented by the Arrhenius function:

$$ \text{DEATH} = deac * F(T) * PC \qquad (13.19) $$

where deac is the death rate at 20° (day^{-1}).

The mass balances for PN and PP are similar to the mass balance of PC as described by Lessin and Raudsepp (2006).

13.3.2.1 Oxygen Processes in the Water Column

Dissolved oxygen mass balance plays a crucial role in the water column oxygen budget. It includes the photosynthesis, the respiration, and the mineralization processes. The oxygen balance (DO) includes the production of oxygen by the primary producers and the benthic vegetation, the oxygen consumption, and the exchange of oxygen between water and air, that is, reaeration. The oxygen consumption is due to mineralization of organic matter in water and sediment, oxidation of ammonia (nitrification), respiration of zooplankton and phytoplankton, and mineralization of the part of the phytoplankton, which is mineralized immediately without entering the detritus pool. The oxygen balance (DO) follows a similar equation as 13.12, and includes the production of oxygen by primary producers and benthic vegetation, the consumption of oxygen, and the exchange of oxygen between water and air.

13.3.2.2 Oxygen Production

Oxygen is produced during the phototrophic production by phytoplankton and benthic vegetation. Once the PRODUCTION term of the PC mass balance (13.12) is computed, it is included in the balance.

The production by benthic vegetation, PRBV, which is assumed to be rooted and/or attached to stones and considering fixed ratios of nitrogen to carbon and phosphorus to carbon, is:

$$PRBV = \mu_b * RD * F'(1) * F'(T) * F'(N,P) * BV \qquad (13.20)$$

where μ_b is the net specific growth rate at 20°C; RD is the relative daylength; and $F'(I)$, $F'(T)$, and $F'(N,P)$ are the production–irradiance, the temperature Arrhenius dependence production, and the nutrient-dependent growth limitation functions for the benthic vegetation, respectively; and BV is the concentration of the benthic vegetation.

13.3.2.3 Oxygen Consumption

Respiration and photosynthesis involve exchanges of dissolved oxygen within the water column and includes the autotrophic and the heterotrophic respirations. The oxygen consumption by respiration of phytoplankton is considered a constant fraction of its biomass and depends on temperature:

$$RSPC = \mu_r F_3(T) * PC \qquad (13.21)$$

where μ_r is the constant fraction of biomass (day^{-1}) and $F_3(T)$ is the temperature Arrhenius dependence function for the phytoplankton respiration.

The oxygen consumption by respiration of zooplankton is considered a constant fraction of its biomass and depends on temperature:

$$RSZC = \mu_{rz0}GZRP + \mu_{rz}F_4(T) * ZC \qquad (13.22)$$

where μ_{rz} is a proportionality constant for specific dynamic action (day^{-1}), μ_{rz} is a constant fraction of biomass (day^{-1}), and $F_4(T)$ is the temperature Arrhenius dependence function for the zooplankton respiration.

The grazing rate of phytoplankton carbon by zooplankton is represented by a saturation equation relating food concentration to grazing rate, a threshold food concentration below which no grazing takes place (Kiørboe at al. 1985; Kiørboe and Nielsen 1994), as well as by a temperature function and a function of dissolved oxygen suppression grazing at low DO (Roman et al. 1993):

$$\text{GRZP} = \text{MGR} * F_5(T) * F(\text{DO}) \frac{\text{PC} - \text{PCT}}{\text{PC} - \text{PCT} - \text{HSG}} * \text{ZC} \tag{13.23}$$

where MGR is the maximum grazing rate constant at 20°C (day^{-1}), PCT is the threshold food concentration (mg C/l), HSG is the half-saturation concentration of PC for carbon uptake for zooplankton (mg C/l), and $F_5(T)$ is the temperature Arrhenius dependence function for the zooplankton grazing, and $F(\text{DO}) = \dfrac{\text{DO}^2}{\text{DO}^2 + \text{MDO}}$ with MDO, the oxygen concentration indicating depressed grazing rates due to oxygen depletion.

Oxygen consumption in the mineralization of dead phytoplankton resulting from the natural mortality of phytoplankton, or autolysis, which is a significant phenomenon in the marine ecosystem (Jassby and Goldman 1974; Lancelot et al. 1987) is computed as:

$$\text{MINPC} = \text{VM} * \text{DEAC} * F_1(T) * \text{PC} \tag{13.24}$$

where VM is the fraction of dead phytoplankton undergoing immediate mineralization and DEAC is the death rate at 20°C (day^{-1}).

The oxygen consumption involving the oxygen demand by the nitrification process deals with transformation of ammonia into nitrate and nitrite into nitrate:

$$\text{ODNI} = {}^*Y_1 * k_4 * \text{NH}_3 * \theta_4^{T-20} - Y_2 * k_5 * \text{NO}_2 * \theta_5^{T-20} \tag{13.25}$$

where k_4 is the nitrification rate at 20°C, k_5 is the specific rate for conversion of nitrite into nitrate at 20°C, and Y_1 and Y_2 are yield factors describing the amount of oxygen used in the nitrification, that is, the mass of O_2 used to oxidize 1 g NH_4 into 1 g NO_3.

The sediment oxygen demand is related to carbon mineralization in the sediment, which again is related to the sedimentation of organic matter (detritus and phytoplankton):

$$\text{SOD} = K_{\text{sod}} * F_6(T) * F(\text{DO}) * (\text{SEPC} + \text{SEDC}) \tag{13.26}$$

where k_{sod} is a proportionality factor and $F_6(T)$ is the temperature Arrhenius dependence function for the sediment oxygen demand.

SEDC is the sedimentation of detritus and is modeled similarly to the phytoplankton sedimentation:

$$SEDC = \frac{U_d * DC}{DZ} \qquad (13.27)$$

where U_d is the sedimentation rate parameter for detritus (m day^{-1}).

SEPC is the sedimentation rate of PC which depends on internal nutrient status and on coagulation:

$$SEPC = \frac{1}{2} ksp1 \left(\frac{pnma}{\frac{PN}{PC}} + \frac{ppma}{\frac{PP}{PC}} \right) * PC + kps2 * PC^2 \qquad (13.28)$$

where ksp1 is a sedimentation rate parameter (day^{-1}), ksp2 is the sedimentation rate constant (mg/l)$^{-1}$ day^{-1}, pnma is the maximum internal nitrogen content in algae (g N/g C), and ppma is the maximum phosphorus content in alga.

Depending on the oxidation state of inorganic nitrogen that are assimilated, the mass ratio Vo = O_2:C (g O_2/g C) varies between 2.67 and 3.47 g O_2 per g C produced. Therefore, the production and the consumption rates obtained in the preceding section are multiplied by Vo for the mass balance of oxygen.

13.3.2.4 Oxygen Reaeration

The reaeration term in the oxygen balance is a source of oxygen to the water column, and is calculated from the oxygen saturation concentration:

$$Rear = (K_2(O_{sat,air} - DO))/\Delta z \qquad (13.29)$$

where $O_{sat,air}$ is the oxygen saturation concentration (g m^{-3}):

$$O_{sat,air} = 14.652 - 0.0841 * S + T * \{0.00256 * S - 0.41022 + $$
$$T * (0.007991 - 0.0000374 * S - 0.000077774 * T)\} \qquad (13.30)$$

where T is the water temperature (°C) and S denotes salinity. The reaeration rate K_2 (day^{-1}) is influenced by the flow characteristics (flow velocity, water depth, and slope of the bed), and is calculated from horizontal current velocity and wind speed (at 10 m) using the empirical model:

$$K_2 = 3.93 * (v_{sp})^{0.5}/(\Delta z)^{1.5} + (2.07 + 0.215 * (w_{sp})^{1.7}) * 24/100 \qquad (13.31)$$

where v_{sp} is the horizontal current velocity, w_{sp} is the wind speed at 10 m above water surface, and Δz is the depth of layer.

13.4 Results

Ria de Aveiro is a temperate lagoon that may be subjected to episodic hypoxia situation because of the presence of high oxygen demand waters at the far end areas, generally occurring during dry seasons. The application of numerical models to study the water quality in the lagoon has provided some information about the role of physical and biogeochemical processes in the DO distribution in the lagoon (Lopes et al. 2005, 2008; Lopes and Silva 2006). It was shown that in general DO values remain quite high in the main lagoon areas because of two factors: oxygen productivity and water circulation. Hypoxic conditions only occur over a limited time and spatial scale, as under the influence of tidal transport, rich oxygenated waters coming from the ocean and the central areas are dispersed and diluted at the far end areas. The lagoon is a tidal dominated system well connected to the ocean, which contributes to a constant oxygen renewal of its water (Dias et al. 2003; Dias and Lopes 2006). Lopes and Silva (2006) and Lopes et al. (2008) have characterized the lagoon in two main areas. (1) One is the central area under the influence of the tide, where DO values are relatively high, typically of the order or greater than 7 mg O_2/l. These areas are characterized by strong currents, along the S. Jacinto and the Espinheiro channels, particularly close to the mouth, where intensity may reach values as high as 2 m s^{-1} (Figure 13.3), which progressively decrease toward the inner areas of the channels. (2) The far end areas, which are mainly intertidal very shallow areas (Figure 13.1), where currents may reach low values (0.1 m s^{-1}; Figure 13.3) and the residence time is 2 weeks (Dias et al. 2001, 2003). These areas, which are dominated by the biogeochemical processes, may show extreme DO values. Under very favorable situations of high productivity inside the lagoon, DO can reach very high values, close to 12 mg O_2/l, corresponding in general to oxygen oversaturation conditions (Almeida et al. 2005). On the other hand, in situations

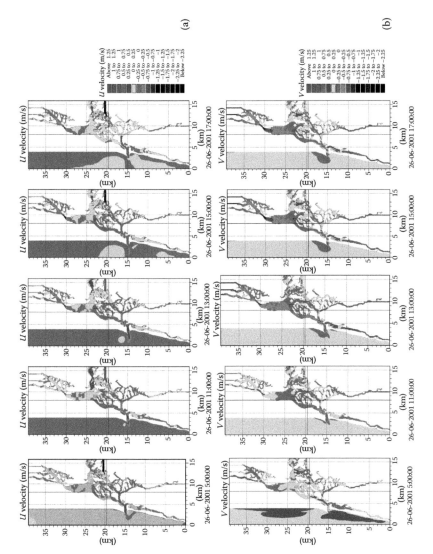

FIGURE 13.3
Maps of currents field: (a) *U* component; (b) *V* component.

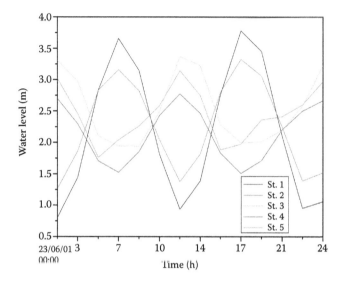

FIGURE 13.4
Time series of water level for the main lagoon stations for 23-06-01.

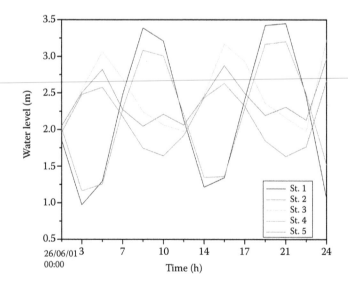

FIGURE 13.5
Time series of water level for the main lagoon stations for 26-06-01.

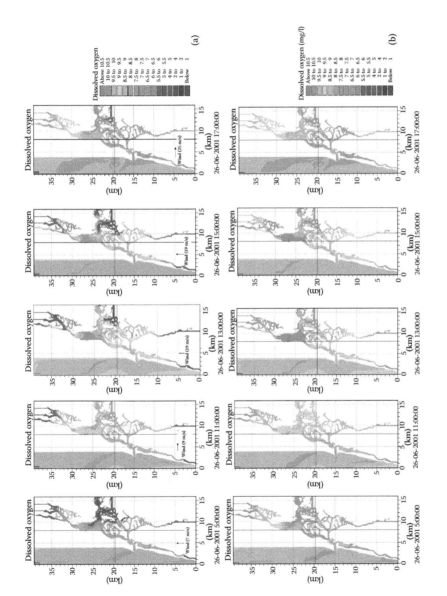

FIGURE 13.6
DO distributions for: (a) high BOD input; (b) low BOD input, at the far ends' boundaries.

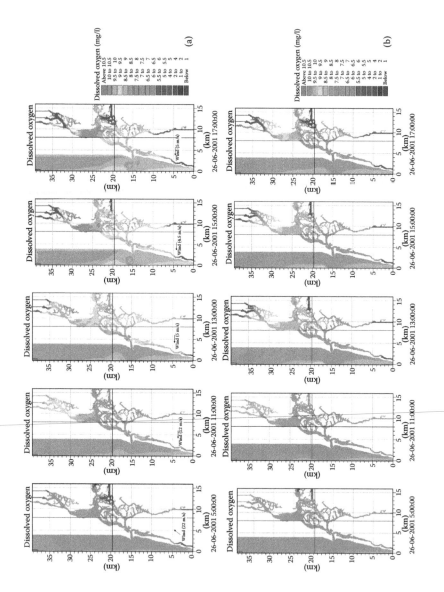

FIGURE 13.7

DO distributions for: (a) reference situation; (b) wind intensity twice the value of the reference situation.

of high oxygen demand and consumption, associated to degradable organic matter contamination, DO values can be very low (below 5 mg O_2/l).

The models described in the previous sections were applied to investigate the interplay between the physical forcing and the biogeochemical conditions and their impact on DO distribution. In all simulations, it was considered that an abnormal situation of high summer river flow occurred in the year 2001, which was considered a very wet year (Almeida et al. 2005). The model was forced, using data from the Aveiro meteorological station for the referred year. In order to better analyze the influence of the tidal phase, the water level time series for the main lagoon stations are presented in Figures 13.4 and 13.5.

Figure 13.6a and b shows the DO distributions in two extreme situations corresponding to high and low BOD values, respectively, at the river boundaries. In Figure 13.4, it can be observed that the first maps (5:00:00; 23-06-2001) correspond to a rising tide at both the lagoon mouth (station 1) and near the Vouga river (station 2), whereas the other stations show a falling tide. The wind direction was variable during that day: it was northward during early morning, then turned eastward and again northward, and finally ended in the eastward direction. It can be observed that, despite the presence of hypoxic waters at the far ends of the lagoon, the DO values remained quite high in the main lagoon areas, when compared to the open sea values (9 mg O_2/l). Typically, the main areas of the lagoon show relatively high DO values, whereas the far ends show the lowest levels (below 5 mg O_2/l), reflecting the situation of high BOD values (Figure 13.6a). In early morning (5:00 A.M.), the central areas of the lagoon and the open ocean show DO values below 8.5 mg/l, whereas in the far ends, where the tide is ebbing, extremely low values are observed (below 5.0 mg/l). In the situation of low BOD values the DO distribution pattern (Figure 13.6b) is similar to the previous one, but no DO values below 8.0 mg/l are observed. DO values progressively increase during the day until early afternoon, at which time maximum values (of the order of 10 mg/l) are observed in areas situated at the northernmost part of the S. Jacinto channel and extending to the Ovar channel. In early evening (3:00 P.M.) the far ends, which are again experiencing an ebbing situation, show low DO values. After reaching the middle of the lagoon (5:00 P.M.), the low oxygenated water mixes and dilutes with higher DO waters of the central areas of the lagoon. Frontal structures, separating high and low oxygenated waters, are set up between the central and the far end areas, and propagate after the tidal cycle. During flooding tide, the currents are strong enough to push the low oxygenated waters toward their origin, again replenishing the lagoon with high oxygenated waters. During ebbing tide, low DO waters are transported from the far ends and disperses throughout the main areas and even reach the lagoon mouth. The reversing tide brings very low oxygenated water from the rivers boundaries into the central area, mixing and diluting the lagoon central waters. These results confirm the influence of the tidal processes in the DO transport and distribution within the lagoon. As the flooding tide at the lagoon mouth brings oxygenated waters (8.5 mg/l) to

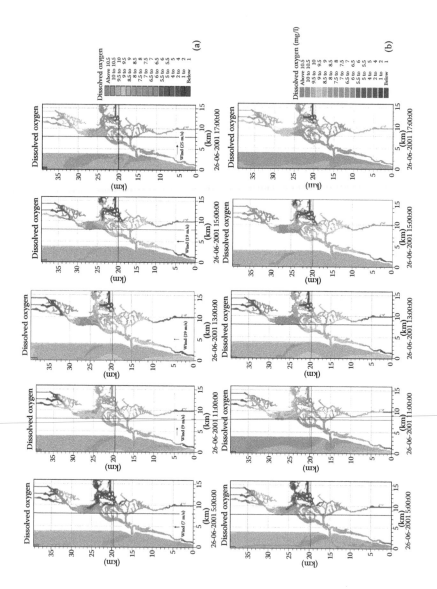

FIGURE 13.8
DO distributions for: (a) reference situation; (b) SOD increased by 10%.

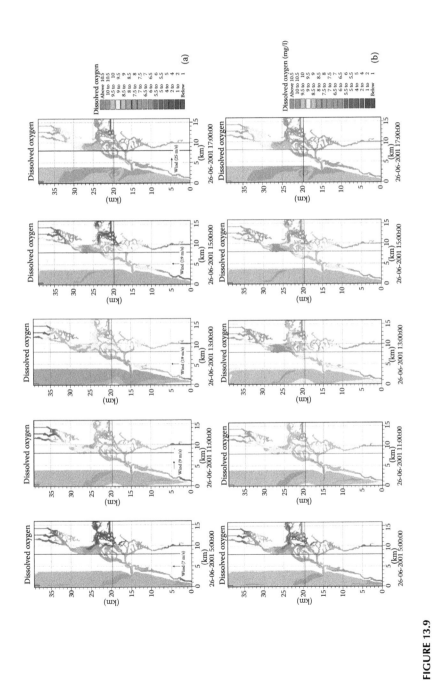

FIGURE 13.9
DO: (a) reference situation; (b) high organic detritus concentration at the river boundaries and high phytoplankton grow rate simulated with the eutrophication model.

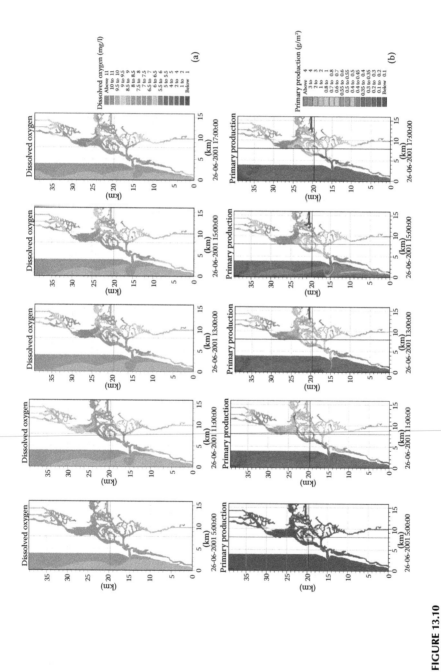

FIGURE 13.10

(a) CHL-*a* distributions; (b) primary production distribution simulated with the eutrophication model.

the central areas, mixing with and reoxygenating the local waters, the ocean waters can be seen as a major source of oxygen reservoir to the lagoon.

Other factors, besides those related to tidal processes, influence the DO variability inside the lagoon. Meteorological conditions such as air temperature, wind speed, and direction also influence the oxygen budget within the water column. Wind condition has a direct influence on the currents' intensity and direction, through the wind stress as well as through the reaeration process at the air–water interface. During the study period (June), wind intensity was null or very weak, even though it had significantly increased of intensity during the second half of the month to values in the order of 10 m/s. To emphasize the influence of wind direction on DO transport, it was decided to choose a situation in which wind direction was almost stationary during the day. Figure 13.7b presents the DO distributions for a typical summer situation, in which wind intensity was increased twice the value relative to a reference situation (Figure 13.7a), with the wind direction remaining unchanged. In Figure 13.5, it can be concluded that the first map (5:00:00; 26-06-2001) corresponds to a falling tide at both the lagoon mouth (station 1) as well as at Vouga (station 2), whereas the other stations show a rising tide. In the early morning, the wind direction was northeastward, then turned westward and maintained this direction during the day. Therefore, wind direction is in phase with the tidal currents, during the ebbing at the Vouga river, and the fronts between low and high oxygenated waters are pushed far westward. The intensification of wind intensity has therefore enhanced the exchange between the very low oxygenated water from the eastern river boundaries and the high oxygenated water from the central area of the lagoon. The increase of the turbulent mixing within the water column has therefore induced an overall decrease in DO values inside the lagoon, whereas the maximum DO located at the central areas of the lagoon has completely disappeared.

Figure 13.8b presents the DO distributions for the sediment oxygen demand (SOD) increased by 10% compared to the reference situation (Figure 13.8a). No significant differences are observed between these two simulations, which led us to conclude that a small increase in the sediment oxygen demand does not lead to net changes in the oxygen budget or a decrease in DO values within the water column, due to the leading effect of tidal transport and oxygen production, thereby emphasizing the importance of the exchanges with the ocean.

Figure 13.9a presents the DO distributions, which are simulated with the eutrophication model, for the same situation described in Figure 13.6a. In order to simulate a situation of organic contamination, high organic detritus concentration was imposed at the river boundaries and high phytoplankton growth rate, twice the value found in the reference situation (Figure 13.9b). The results show a very clear pattern of DO distribution, with the concentration steadily rising from noon to late evening, reaching values as high as 11 mg/l. The central areas of the lagoon—those situated between the northernmost part of the S. Jacinto and the Ovar channel—show a pattern corresponding to maximum DO values (11 mg/l), contrasting with those of the remaining areas, which was well below 9 mg/l. The high DO values simulated in those areas reflect the high phytoplankton productivity, with relatively high values for the maximum oxygen production (4 g/(m² day)) (Figure 13.10b). On the other hand, the low DO values simulated for the far end areas reflect the high oxygen demand and consumption by the organic detritus.

13.5 Discussion and Conclusions

It is generally accepted that one of the main criteria for the assessment of the water quality of a given aquatic system is its oxygen content, as hypoxia or anoxia conditions reflect potential chemical or biological contaminations. Oxygen is involved in the main processes occurring in the water column: photosynthesis, organic matter degradation, and bacterial nitrification. Ecological models, which may include both water quality and eutrophication models, are important tools for the assessment of estuarine ecosystem status, including the role played by main interactions and forcings. Several models have been developed and applied to natural ecosystems to describe eutrophication in rivers, lakes, reservoirs, and coastal waters (Bonnet and Wessen 2001; Hang et al. 2009). Dissolved oxygen in coastal systems has been studied using both eutrophication and water quality models. Malmgren-Hansen et al. (1984) modeled oxygen depletion in Limfjord, Denmark. They showed that the bottom oxygen concentration was sensitive to two main factors: the vertical mixing and the biological processes producing and consuming oxygen. They also demonstrated that the number of periods with oxygen depletion will increase exponentially with the rates of the oxygen consuming processes. Carlsson et al. (1999), using a model based on both statistical regressions and dynamic interactions, predicted seasonal changes of dissolved oxygen in the Baltic Sea archipelago areas. They showed that the model can be used in practical water management to assess the sensitivity of coastal areas to increased nutrient load, and could account for the seasonal change in dissolved oxygen, simply from the information obtained from bathymetric maps, and could predict the total oxygen consumption, that is, benthic pelagic community respiration and mean values for the oxygen production for a given period. In particular, they showed that seasonal variation in dissolved oxygen could be predicted from water turnover times, organic load, and default seasonal patterns for particulate organic carbon and temperature. Plus et al. (2003) have developed a 3-D model coupling physical and biological processes for the Thau lagoon (France) in order to assess the relationships between macrophytes and the oxygen and nitrogen cycles. Their results show that during winter the oxygen concentrations remain rather high in the central areas of the lagoon, because of the phytoplankton production and water mixing by the wind, but in summer the oxygen concentrations are lower due to several factors leading to oxygen depletion: faster mineralization processes, lower solubility in warm waters, and accumulation of dead organic material coupled with higher oxygen consumption due to high temperatures. They showed that the introduction in the model of macrophyte populations and of the mineralization of organic matter change significantly the dissolved oxygen distribution in the Thau lagoon. They found that the far end areas of the lagoon have potential risks of oxygen depletion and anoxic crisis, whereas the central and the southern areas of the lagoon may exhibit higher oxygen concentrations because of the influence of marine oxygenated waters. Their results confirm that the oxygen concentrations not only depend on the in situ biogeochemical processes, but also on the hydrodynamics of the lagoon. Tuchkovenko and Lonin (2003) applied a mathematical model of eutrophication/oxygen based on phytoplankton, bacteria, detritus, dissolved organic matter, phosphate, ammonium, nitrite, nitrate, and dissolved oxygen processes for the Cartagena Bay (Colombia) with the main goal of designing strategies to improve the water quality of the bay. They showed that the oxygen deficit on the bottom layer of the internal bay will disappear by reducing the nutrient discharge from the industrial sources by 80% of the current level, and that both nutrient reduction and fresh water flows would reduce the oxygen deficit in the bottom layer. Zheng et al. (2004) developed a coupled 3-D physical and water quality model for the Satilla River Estuary, Georgia. The physical model was a modified ECOM-si, a 3-D estuarine and coastal ocean model, developed originally by Blumberg (1993), with inclusion of the flooding/draining process over the intertidal salt marsh. The water quality model was a modified 3-D model based on *WASP5*, developed originally by Ambrose et al. (1993), with the inclusion of the nutrient fluxes from the bottom sediment layer. It constitutes a complex of four interacting systems: dissolved oxygen, carbonaceous biochemical oxygen demand (CBOD), nitrogen cycle (including ammonium nitrogen, nitrate, and nitrite nitrogen), phosphorus cycle (including orthophosphorus or inorganic phosphorus, organic nitrogen, and organic phosphorus), and phytoplankton dynamics. Their results showed that the intertidal salt marsh has high sediment oxygen demand and acts as a major consumer of DO and is, therefore, the major factor responsible for the low DO values. They also emphasized the importance of the heterotrophic bacterial respiration process in DO distribution inside the estuary. Lianyuan et al. (2004) used the same models employed by Zheng and Zhang (2004) to study the Satilla river estuary water quality. Their results showed a DO distribution with a high-spatial variability, with high concentration values (7 mg O_2 l^{-1}) in the inner shelf and mouth of the estuary and lower values (below 4 mg O_2 l^{-1}) in the upstream end of the estuary. They pointed out that DO balance is dominantly controlled by the SOD, reflecting the nature of an estuarine–salt marsh ecosystem, with low anthropogenic influence. Liu et al. (2005) applied a water quality model in the separation waterway of the Yulin offshore industrial park in Taiwan. The model was based on the principle of conservation of eight interlinked water quality state variables: DO, chlorophyll-*a*, CBOD, organic nitrogen, ammonia nitrogen, nitrite–nitrate nitrogen, organic phosphorus, and inorganic phosphorus. They demonstrated the influence of the increase of the tidal flushing, at the seaward end of the domain, in the increase of the dissolved oxygen concentration in the seaward direction. On the other hand, they showed that increasing the waterway width, results in a better water quality in the separation waterway, reaching typical values for the coastal water quality standards of Taiwan (average DO = 5 mg/l, average CBOD = 3.6 mg/l). Babu et al. (2006) used the MIKE21 model to study water quality around an offshore outfall off Kochi (India), assuming a high BOD effluent discharge through an outfall located at a distance of 6.8 km from the shore at a depth of 10 m. Several scenarios with different discharge rates and different BOD loads were carried out to estimate the assimilation capacity of the waters off Kochi, assuming a high BOD load (50 mg l^{-1}). They estimated the BOD assimilative capacity of the waters off Kochi, emphasizing the advantage of this area for the disposal of treated effluent, and showed that for high discharge rates a significant BOD reduction was observed at the outfall after 48 h, whereas high BOD values were confined to an area of 8 km^2 around the outfall. Hang et al. (2009) applied the MIKE3 (2005b) ecosystem model to study the dynamics of phytoplankton dissolved oxygen and other water quality constituents in response to the variation in boundary and weather conditions in the Ariake Sea coastal sea, a semiclosed sea in the west coast of Kyushu (Japan). Their purposes were to identify and quantify how the physical, biological, and chemical processes, and their interactions, control the spatial distribution of water quality constituents in the sea. Their results showed that the dissolved oxygen concentrations are typically lower at night and early morning, reflecting the absence of photosynthesis to balance plant and animal respiration, and that the inner part of the study area is sometimes in anoxic or hypoxic conditions since it is close to the point sources (river discharges) of pollutant loads. They also showed that the variability of temperature, winds, and daily light levels influence the DO concentration in the inner part, with the lowest bottom DO level observed in September (about 5.5 mg/l) and the highest surface DO level in February (about 10.0 mg/l). More recently, Eldridge and Roelk (2010) coupled two available models—a plankton food web model involving competing phytoplankton groups (Roelke et al. 1999; Roelke 2000) and a multielement geochemical model (Morse and Eldridge 2007; Eldridge and Morse 2008)—to areas of increasing salinity moving westward away from the Mississippi River point of discharge at the Gulf of Mexico. The phytoplankton groups were allowed to compete for nitrogen, phosphorus, light, and zooplankton grazing. The model predicts temporal and spatial variations in hypoxia, showing that a 40% decrease in nutrient loading would substantially reduce the duration and area extent of the Gulf of Mexico hypoxia under present-day inflows.

They also showed that river discharge is the dominant factor in stimulating the hypoxia state of the lagoon and that a relatively small increase in river discharge would offset a large reduction in river nutrient concentration.

The results presented in the previous section concerning the Ria de Aveiro lagoon have emphasized the influence of the physical and the biogeochemical factors in the water quality of the lagoon, that is, its DO distribution. Except for the remote areas, the main lagoon areas show relatively high DO values (in general, greater than 7 mg O_2/l, even in a situation of organic matter contamination at the rivers boundaries). Since these concentrations are very close to those observed at the open sea, these results emphasize the role of the lagoon hydrodynamics in the DO distribution. The hypoxic situations presented at the far ends of the lagoon remain confined in those areas, even though their influence may reach the central areas of the lagoon, due to tidal transport and mixing, where DO may show values lower than 7 mg O_2/l. In general, the far end areas are characterized by low flushing rates, low residence time and low residual currents (Dias et al. 2003; Lopes and Dias 2007), and therefore, in some particular situation when the oxygen budget is negative—that is, when its consumption by the phytoplankton respiration, nitrification, organic matter degradation, and sediment oxygen exceeds its production by the phytoplankton and advection by the tidal flow—it can generate hypoxic situations. Therefore, hypoxic situations may occur over a limited time and spatial scale at the far end of the lagoon, and persist for some time. When this situation occurs, hypoxic water may be transported to the lagoon central areas, leading to a decrease in the oxygen content of the water column, through mixing and dilution with high oxygenated water. On the other hand, the oxygen production by phytoplankton, with maximum values observed at the central and the northernmost areas, along the S. Jacinto channel and the Ovar channel, reflects the conditions of high phytoplankton growth and productivity of those areas. Therefore, it can be concluded that the lagoon may be divided into two areas: (1) the main area where the dissolved oxygen concentrations are relatively high, because of the influence of the local production and of the dynamic processes induced by tides, which allow constant interaction between the ocean and the lagoon and constant renewal of the dissolved oxygen; (2) and the far end areas, under the influence of weak tidal currents, where the biogeochemical processes may become dominant and the dissolved oxygen concentrations may reach very low values, which can occur in situations corresponding to the occurrence of episodic organic contamination. Nevertheless, in the same area, the dissolved oxygen concentrations may reach high levels if a favorable situation (available light and high nutrient loads) occurs, thereby increasing the phytoplankton productivity (Lopes et al. 2005; Lopes and Silva 2006). Meteorological conditions, such as air temperature, wind speed, and direction, influence the oxygen budget within the water column. Wind condition has a direct influence on current intensity and direction, through wind stress as well as the reaeration process at the air–water interface. Wind intensity and direction influence the tidal currents and therefore the position of the fronts between low and high oxygenated waters, as well as the turbulent mixing within the water column, thereby allowing a greater exchange in water with different DO values, that is, central waters and far end waters. An intensification of wind intensity or a change in its direction may enhance the exchange between the very low oxygenated water from the western river boundaries and the high oxygenated water from the central area of the lagoon.

In a scenario where the geomorphodynamical and the hydrodynamical conditions of the Ria de Aveiro remain unchanged, a setup of a permanent oxygen depletion situation in most of the lagoon areas is unlikely to occur, even in the worst-case scenario of high organic contamination at the river boundaries. The good connection between Ria de Aveiro lagoon and the sea, and the circulation pattern inside the lagoon as well as the primary productivity are, therefore, responsible for the relatively high level of oxygen water content in the water column, throughout its main channels. Changes in the river flows result in an increase of the residual circulation toward the lagoon mouth (Dias et al. 2001; Dias and Lopes, 2007), increasing the export of low oxygenated waters toward the lagoon central areas during episodes of organic contamination, but may not be enough to dramatically impact the water quality and the DO distribution due to lagoon circulation patterns. Changes in the lagoon morphodynamics resulting from the anthropogenic activities inside the lagoon and near its border, infilling the lagoon mouth or its most important channels, may induce changes in the hydrodynamic regime, contributing to an increase of lagoon flushing and residence times, and therefore reflecting negatively on the DO transport and distribution inside the lagoon. In addition, if these changes are followed by an increase in organic and inorganic load, the negative impact on its water quality will be greater.

References

Abbott, M. B., A. D. McCowan, and I. R. Warren. 1981. Numerical modelling of free-surface flows that are two-dimensional in plan. In *Transport Models for Inland and Coastal Waters*, ed. H.B. Fischer, 362–407. New York: Academic Press.

Alcântara, F., M. G. Pereira, M. A. Almeida, and M. A. Cunha. 1992. Qualidade microbiológica da água da Ria de Aveiro, 1991. Polaveiro Repport. Departamento de Biologia, Universidade de Aveiro, Portugal, p. 90.

Alcântara, F., M. G. Pereira, M. A. Almeida, and M. A. Cunha. 1994. Qualidade Microbiológica da Água da Ria de Aveiro, 1989–1993. Polaveiro Repport. Departamento de Biologia, Universidade de Aveiro, Portugal, p. 85.

Allanson, B. R. 2001. Some factors governing the water quality of microtidal estuaries in South Africa. *Water SA* 27 (3): 373–383.

Allanson, B. R., and P. Winter. 1999. Chemistry. In *Estuaries of South Africa*, ed. B. R. Allanson, and D. Baird, 53–90. Cambridge: Cambridge Univ. Press.

Ambrose, Jr., R. B., T. A. Wool, and J. L. Martin. 1993. *The Water Quality Analysis Simulation Program, WASP5, Part A: Model Documentation*. U.S. Environmental Protection Agency, Athens, GA, p. 202.

Almeida, M. A., M. A. Cunha, and F. Alcântara. 2005. Relationship of bacterio-plankton production with primary production and respiration in a shallow estuarine system (Ria de Aveiro, NW Portugal). *Microbiological Research* 160, 315–328.

Babu, M. T., D. V. Kesava, and P. Vethamony. 2006. BOD–DO modeling and water qual-ity analysis of a waste water outfall off Kochi, west coast of India. *Environment International* 32: 165–173.

Baird, J. I., and K. Whitelaw. 1992. Water quality aspect of estuary modelling. In *Water Quality Modelling*. Ashgate, ed. Falconer, R. A., 119–125. Burlington, VT.

Billen, G. and J. Garnier. 1997. The Physon river plume: coastal eutrophication in response to changes in land use and water management in watershed. *Aquatic Microbial Ecology* 18: 301–312.

Blumberg, A. F. 1993. *A Primer of ECOM-si*. Technical Report. HydroQual, Inc., Mahwah, NJ.

Bonnet, M. P. and K. Wessen. 2001. ELMO, a 3-D water quality model for nutrients and chlorophyll: first application on a lacustrine ecosystem. *Ecological Modelling* 14: 19–33.

Alvera-Azcarate, A., J. G. Ferreira, and J. P. Nunes. 2003. Modelling eutrophication in mesotidal and macrotidal estuaries. The role of intertidal seaweeds. *Estuarine and Coastal Shelf Science* 57 (4): 715–724.

Carlsson, L., J. Persson, and L. Hakanson. 1999. A management model to predict sea-sonal variability in oxygen concentration and oxygen consumption in thermally stratified coastal waters. *Ecological Modelling* 119: 117–134.

Casuli, V. 1999. A semi-implicit finite difference method for nonhydrostatic, free-surface flows. *International Journal for Numerical Methods in Fluids* 30: 425–440.

Cerco, C., and T. Cole. 1993. Three dimensional euthrophication model of Chesapeake Bay. *Journal of Environmental Engineering* 119: 1006–1025.

Cerco, C. F. and T. Cole. 1995. *User's Guide to the CE-QUAL-ICM Three-Dimensional Eutrophication Model*, Release Version 1.0, Technical Report EL-95-15. US Army Engineer Waterways Experiment Station, Vicksburg, MS.

Cloern, J. E. 1982. Does the benthos control phytoplankton biomass in South San Francisco Bay? *Marine Ecology Progress Series* 9: 191–202.

Cloern, J. E. 1996. Phytoplankton bloom dynamics in coastal ecosystems: a review with some general lessons from sustained investigation of San Francisco Bay, California. *Reviews of Geophysics* 34 (2): 127–168.

Cloern, J. E. 1987. Turbidity as a control on phytoplankton biomass and productivity in estuaries. *Continental Shelf Research* 7: 1367–1381.

Cole, B. E., and J. E. Cloern. 1987. An empirical model for estimating phytoplankton productivity in estuaries. *Marine Ecology Progress Series* 36: 299–305.

Cole, J. J., N. F. Caraco, and B. L. Peierls. 1992. Can phytoplankton maintain a positive carbon balance in a turbid, freshwater, tidal estuary? *Limnology and Oceanography* 37: 1608–1617.

Dias, J. M. 2001. Contribution to the study of the Ria de Aveiro hydrodynamics. PhD thesis, Universidade de Aveiro, Portugal, 288 pp.

Dias, J. M., J. F. Lopes, and I. Dekeyser. 1999. Hydrological characterisation of Ria de Aveiro lagoon, Portugal, in early summer. *Oceanologica Acta* 22: 473–485.

Dias, J. M., J. F. Lopes, and I. Dekeyser. 2000. Tidal propagation in Ria de Aveiro lagoon, Portugal. *Physics and Chemistry of the Earth B* 25: 369–374.

Dias, J. M., J. F. Lopes, and I. Dekeyser. 2001. Lagrangian transport of particles in Ria de Aveiro lagoon, Portugal. *Physics and Chemistry of the Earth* B 26: 729–734.

Dias, J. M., J. F. Lopes, and I. Dekeyser. 2003. A numerical system to study the transport properties in the Ria de Aveiro lagoon. *Ocean Dynamics* 53: 220–231.

Dias, J. M., and J. F. Lopes. 2006. Implementation and evaluation of hydrodynamic, salt and heat transport models: the case of Ria de Aveiro Lagoon (Portugal). *Environmental Modelling and Software* 21: 1–15.

Diaz R. J., R. J. Neubauer, L. C. Schaffner, L. Pihl, and S. P. Baden. 1992. Continuous monitoring of dissolved oxygen in an estuary experiencing periodic hypoxia and the effect of hypoxia on macrobenthos and fish. *Science of the Total Environment* (Supplement 1992); 1055–1068.

Drago, M., B. Cescon, and L. Iovenitti. 2001. A three-dimensional numerical model for eutrophication and pollutant transport. *Ecological Modelling* 145: 17–34.

Droop, M. R. 1973. Some thoughts on nutrient limitation in algae. *Journal of Phycology* 9: 264–272.

Droop, M. R. 1975. The nutrient status of algal cells in batch cultures. *Journal of the Marine Biological Association of the United Kingdom* 55: 541–555.

Eldridge, P. M., and J. W. Morse. 2008. Origins and temporal scales of hypoxia on the Louisiana shelf: importance of benthic and sub-pycnocline water metabolism. *Marine Chemistry* 108: 159–171.

Eldridge, P. M., and D. L. Roelke. 2010. Origins and scales of hypoxia on the Louisiana shelf: Importance of seasonal plankton dynamics and river nutrients and discharge. *Ecological Modelling* 221: 1028–1042.

Erichsen, A. C., and P. S. Rasch. 2001. Two- and three-dimensional model system predicting the water quality of tomorrow. In *Proc. of the Seventh International Conference on Estuarine and Coastal Modeling*, ed. M. L. Spaulding, 165–184. Virginia, USA: American Society of Civil Engineers.

Fasham, M. J. R., H. W. Ducklow, and S. M. McKelvie. 1990. A nitrogen-based model of plankton dynamics in the oceanic mixed layer. *Journal of Marine Research* 48: 591–639.

Fennel, W. 1995. Model of the yearly cycle of nutrients and plankton in the Baltic Sea. *Journal of Marine Systems* 6: 313–329.

Fennel, W., and T. Neumann. 2004. *Introduction to the Modeling of Marine Ecosystems.* Elsevier Oceanographic Series, vol. 72. Elsevier, Amsterdam, 297 pp.

Ferreira, J. G., W. J. Wolff, T. C. Simas, and S. B. Bricker. 2005. Does biodiversity of estuarine phytoplankton depend on hydrology? *Ecological Modelling* 187: 513–523.

Fiúza, A. F. G., M. E. Macedo, and M. R. Guerreiro. 1982. Climatological space and time variation of the Portuguese coastal upwelling. *Oceanologica Acta* 5 (1): 31–40.

Franks, P. J. S., J. S. Wroblewski, and G. R. Flierl. 1986. Behavior of a simple plankton model with food-level acclimation by herbivores. *Marine Biology* 91: 121–129.

Fransz, H. G., and J. H. G. Verhagen. 1985. Modelling research on the production cycle of phytoplankton in the Southern Bight of the North Sea in relation to river-borne nutrient loads. *Netherlands Journal of Sea Research* 19: 241–250.

Frost, B. W. 1987. Grazing control of phytoplankton stock in the open subarctic Pacific Ocean: a model assessing the role of mesozooplankton, particularly the large calanoid copepods, *Neocalanus* spp. *Marine Ecology Progress Series* 39: 49–68.

Gallegos, C. L. 1989. Microzooplankton grazing on phytoplankton in the Rhode River, Maryland: nonlinear feeding kinetics. *Marine Ecology Progress Series* 57: 23–33.

Gill, A. E. 1982. *Atmosphere–Ocean Dynamics*. International Geophysics Series, 30. New York, NY: Academic Press.

Glasgow, Jr., H. B. and J. M. Burkholder. 2000. Water quality trend and management implication from a five-year study of eutrophic estuary. *Ecological Application* 10: 1024–1046.

Jackson, G. A. 1990. A model of the formation of marine algal flocs by physical coagulation processes. *Deep-Sea Research* 37: 1197–1211.

Jassby, A. D. and C. R. Goldman. 1974. Loss rates from a lake phytoplankton community. *Limnology and Oceanography* 19: 618–627.

Jassby, A. D. and T. Platt. 1976. Mathematical formulation of the relationship between photosynthesis and light for phytoplankton. *Limnology and Oceanography* 21: 540–547.

Gower, A. M. 1980. *Water Quality in Catchment Ecosystems*. New York, NY: John Wiley & Sons, 335 pp.

Hang N. T. M., C. D. Nguyen, A. Hirokyuki, Y. Hirokyuki, and K. Kenichi. 2009. Applications of a new ecosystem model to study the dynamics of phytoplankton and nutrients in the Ariake Sea, west coast of Kyushu, Japan. *Journal of Marine Systems* 75: 1–16.

Henriksen, K., J. I. Hansen, and T. H. Blackburn. 1981. Rates of nitrification, distribution of nitrifying bacteria, and nitrate fluxes in different types of sediments from Danish waters. *Marine Biology* 61: 299–304.

Henriksen, K., and W. Kemp. 1988. Nitrification in estuarine and coastal marine sediments. In *Nitrogen Cycling in Coastal Marine Environments*, ed. T. H. Blackburn, and J. Sorensen, 201–249. New York, NY: John Wiley & Sons, Inc.

Kaspar, H. F. 1982. Denitrification in marine sediment: measurement of capacity and estimate of *in situ* rate. *Applied and Environmental Microbiology* 43: 522–527.

Ketchum, B. H. 1954. Relation between circulation and planktonic populations in estuaries. *Ecology* 35: 191–200.

Kiørboe, T., F. Møhlenberg, and H. U. Riisgård. 1985. In situ feeding rates of planktonic copepods: a comparison of four methods. *Journal of Experimental Marine Biology and Ecology* 88: 67–81.

Kiørboe, T., and T. G. Nielsen. 1994. Regulation of zooplankton biomass and production in a temperate, coastal ecosystem. I: Copepods. *Limnology and Oceanography* 39: 493–507.

Lancelot, C., and V. Rousseau. 1987. ICES intercalibration exercise on the 14C 33 method for estimating phytoplankton primary production. Phase 2: experiments conducted on board of RV DANA. Preliminary report, 35 pp.

Lancelot, C., G. Billen, A. Sourina, T. Weisse, F. Colijn, M. J. W. Veldhuis, A. Davies, and P. Wassman. 1987. Phaeocystis blooms and nutrient enrichment in the continental coastal zones of the North Sea. *Ambio* 16: 38–46.

Lapointe, B. E., and W. R. Matzie. 1996. Effect of storm water nutrient discharges on eutrophication processes in nearshore water of the Florida Key. *Estuaries* 19 (2B): 422–435.

Leendertse, J. J., and E. C. Gritton. 1971. A *Water-Quality Simulation Model for Well-Fixed Estuaries and Coastal Seas: Vol. 2: Computation Procedures*. Rand Corporation, R-708-NYC. New York, NY: The New York City Rand Institute.

Leonard, A. 1979. A stable and accurate convective modeling procedure based on quadratic upstream interpolation. *Computer Methods in Appl Mechanics and Engineering* 19: 59–98.

Lessin, G., and U. Raudsepp. 2006. Water quality assessment using integrated modelling and monitoring in Narva Bay, Gulf of Finland. *Environmental Modeling and Assessment* 11: 315–332.

Liu, W. C., J. T. Kuob, and A. Y. Kuoc. 2005 Modelling hydrodynamics and water quality in the separation waterway of the Yulin offshore industrial park, Taiwan. *Environmental Modelling & Software* 20: 309–328.

Lopes, J. F., J. M. Dias, A. C. Cardoso, and C. I. Silva. 2005. The water quality of the Ria de Aveiro lagoon, Portugal: from the observations to the implementation of a numerical model. *Marine Environmental Research* 60: 594–628.

Lopes, J. F., and C. Silva. 2006. Temporal and spatial distribution of dissolved. *Ecological Modelling* 197: 67–88.

Lopes, J. F., and J. M. Dias. 2007. Residual circulation and sediments transport in Ria de Aveiro lagoon, Portugal. *Journal of Marine Systems* 68: 507–528.

Lopes, J. F., C. Silva, and A. C. Cardoso. 2008. Validation of a water quality model for the Ria de Aveiro lagoon, Portugal. *Environmental Modelling & Software* 23: 479–494.

Lucas, L. V. 1997. A numerical investigation of coupled hydrodynamics and phytoplankton dynamics in shallow estuaries. PhD thesis, Stanford University.

Lucas, L. V., J. R. Koseff, S. G. Monismith, J. E. Cloern, and J. K. Thompson. 1999a. Processes governing phytoplankton blooms in estuaries: I. The role of horizontal transport. *Marine Ecology Progress Series* 187: 1–15.

Lucas, L. V., J. R. Koseff, S. G. Monismith, J. E. Cloern, and J. K. Thompson. 1999b. Processes governing phytoplankton blooms in estuaries: II. The role of horizontal transport. *Marine Ecology Progress Series* 187: 17–30.

Malmgren-Hansen, A. and K. B. Hanne. 1992. Modelling river water quality and impact from sewers overflows. In *Water Quality Modelling*. Ashgate, ed. R. A. Falconer, 69–80. Burlington, VT.

Malmgren-Hansen, A., P. Mortensen, and B. Møller. 1984. Modelling of oxygen depletion in coastal waters. *Water Science and Technology* 17: 967–978.

Manso, M. D., J. M. Dias, and F. Macedo. 1996. Aplicación de un modelo de cadenas de Markov de primero y Segundo orden, *Avances en Geofísica y Geodesia* 1 (1): 151–157.

May, C. L., J. R. Koseff, L. V. Lucas, J. E. Cloern, and D. H. Schoellhamer. 2003. Effects of spatial and temporal variability of turbidity on phytoplankton blooms. *Marine Ecology Progress Series* 254: 111–128.

McCreary, J. P., K. E. Kohler, R. R. Hood, and D. B. Olson. 1996. A four-component ecosystem model of biological activity in the Arabian Sea. *Progress in Oceanography* 37: 117–165.

Mike3. 2005a. *Flow Model, Hydrodynamic and Transport Module*. DHI Water & Environment Denmark, p. 42.

Mike3. 2005b. *Water Quality Model, A Scientific Description*. DHI Water & Environment. Denmark, p. 36.

Mike3. 2005c. *Eutrophication Model, A Scientific Description*. DHI Water & Environment. Denmark, p. 36.

Mommaerts, J. P. 1978. Systeembenadering van en gesloten mariene milieu, met de nadruk op de rol van het fytoplankton. Doctoral thesis, Vrije Universiteit Brussel, p. 335.

Montani, S., P. Magni, M. Shimamoto, N. Abe, and K. Okutani. 1998. The effect of a tidal cycle on the dynamic of nutrients in a tidal estuary in the Seto Inland Sea. *Japan Journal of Oceanography* 54: 65–76.

Morse, J. W., and P. M. Eldridge. 2007. A non-steady-state diagenetic model for changes in sediment biogeochemistry in response to seasonally hypoxic/anoxia conditions in the "dead zone" of the Louisiana Shelf. *Marine Chemistry* 106: 239–255.

Neumann, T. 2000. Towards a 3-D ecosystem model of the Baltic Sea. *Journal of Marine Systems* 25: 405–419.

Neumann, T., W. Fennel, and C. Kremp. 2002. Experimental simulations with an ecosystem model of the Baltic Sea: a nutrient load reduction experiment. *Global Biogeochemical Cycles* 16: 7_1–7_19002E.

Nyholm, N. 1976. A mathematical model for the growth of phytoplankton. Int. Symp. on Experimental Use of Algal Cultures in Limnology, Sandefjord, Norway, Oct. 26–28, 1976.

Nunes, J. P., J. G. Ferreira, F. Gazeau, J. Lencart-Silva, X. L. Zhang, M. Y. Zhu, and J. G. Fang. 2003. A model for sustainable management of shellfish polyculture in coastal bays. *Aquaculture* 219 (1–4): 257–277.

Paerl, H. W., J. L. Pinckney, J. M. Fear, and B. L. Peierls. 1998. Ecosystem responses to internal and watershed organic matter loading: consequences for hypoxia in the eutrophying Neuse River estuary, North Carolina, U.S.A. *Marine Ecology Progress Series* 166: 17–25.

Pinckney, J. L., H. W. Paerl, T. Patricia, and L. R. Tammi. 2001. The role of nutrient loading and eutrophication in estuarine ecology. *Environmental Health Perspectives* 109 (5): 699–701.

Pinckney, J. L., D. F. Millie, B. T. Vinyard, and H. W. Paerl. 1997. Environmental controls of phytoplankton bloom dynamics in the Neuse River estuary, North Carolina. *Canadian Journal of Fisheries and Aquatic Sciences* 54: 2491–2501.

Pinckney, J. L., H. W. Paerl, M. B. Harrington, and K. E. Howe. 1998. Annual cycles of phytoplankton community structure and bloom dynamics in the Neuse River Estuary, North Carolina. *Marine Biology* 131: 371–381.

Pinckney, J. L., H. W. Paerl, T. Patricia, and L. R. Tammi. 2001. The role of nutrient loading and eutrophication in estuarine ecology. *Environmental Health Perspectives* 109 (5): 699–701.

Plus, M., A. Chapelle, P. Lazure, I. Auby, G. Levasseur, M. Verlaque, T. Belsher, J.-M. Deslous-Paoli, J.-M. Zaldivar, and C. N. Murray. 2003. Modelling of oxygen and nitrogen cycling as a function of macrophyte community in the Thau lagoon. *Continental Shelf Research* 23: 1877–1898.

Portnoy, J. W. 1991. Summer oxygen depletion in a diked New England estuary. *Estuaries* 14 (2): 122–129.

Radwan, M., P. Willems, A. El-Sadek, and J. Berlamont. 2003. Modelling of dissolved oxygen and biochemical oxygen demand in river water using a detailed and a simplified model. *International Journal of River Basin Management* 1 (2): 97–103.

Renaud, M. 1986. Hypoxia in Louisiana coastal waters during 1983: implications for fisheries. *Fishery Bulletin* 84: 19–26.

Roelke, D. L. 2000. Copepod food-quality threshold as a mechanism influencing phytoplankton succession and accumulation of biomass, and secondary productivity: a modeling study with management implications. *Ecological Modeling* 134: 245–274.

Roelke, D. L. and Y. Buyukates. 2001. The diversity of harmful algal bloom-triggering mechanisms and the complexity of bloom initiation. *Human and Ecological Risk Assessment* 7: 1347–1362.

Roelke, D. L., P. M. Eldridge, and L. A. Cifuentes. 1999. A model of phytoplankton competition for limiting and nonlimiting nutrients: Implications for development of estuarine and nearshore management schemes. *Estuaries* 22: 92–104.

Roman, M., A. Gauzerns, W. Rhinehart, and J. White. 1993. Effects of low oxygen waters on Chesapeake Bay zooplankton. *Limnology Oceanography* 38: 1603–1614.

Rysgaard, S., N. Risgaard-Petersen, L. P. Nielsen, and N. P. Revsbech. 1993. Nitrification and denitrification in lake and estuarine sediments measured by the n dilution technique and isotope pairing. *Applied and Environmental Microbiology* 59 (7): 2093–2098.

Sharples, J., and P. Tett. 1994. Modelling the effect of physical variability on the mid-water chlorophyll maximum. *Journal of Marine Research* 52: 219–238.

Silva, J. F. 1994. Circulação da Água na Ria de Aveiro: Contribuição para o Estudo da Qualidade da Água. Tese de Doutoramento, Universidade de Aveiro.

Silva, J. F., R. W. Duck, T. S. Hopkins, and M. Rodrigues. 2002. Evaluation of the nutrient inputs to a coastal lagoon: the case of the Ria de Aveiro, Portugal. *Hydrobiologia* 475–476: 379–385.

Smayda, T. J. 1970. The suspension and sinking of phytoplankton in the sea. *Oceanography and Marine Biology, An Annual Review* 8: 357–414.

Smayda, T. J. 1990. Novel and nuisance phytoplankton blooms in the sea: evidence for a global epidemic. In *Toxic Marine Phytoplankton*, ed. Granéli, E., B. Sunderström, L. Edler, and D. M. Anderson, 20–40. New York: Elsevier.

Stanley, D. W., and S. W. Nixon. 1992. Stratification and bottom-water hypoxia in the Pamlico River Estuary. *Estuaries* 15: 270–281.

Steemann, N. E., and E. G. Jørgensen. 1968. The adaptation of plankton algae: III. With special consideration of the importance in nature. *Physiologia Plantarum* 21: 647–654.

Stow, C. A., S. S. Qian, and J. K. Craig. 2005. Declining threshold for hypoxia in the Gulf of Mexico. *Environmental Science and Technology* 39: 716–723.

Taylor, A. H., D. S. Harbour, and R. P. Harris. 1993. Seasonal succession in the pelagic ecosystem of the North Atlantic and the utilization of nitrogen. *Journal of Plankton Research* 15: 875–891.

Tett, P., A. Edwards, and K. Jones. 1986. A model for the growth of shelf-sea phytoplankton in summer. *Estuarine, Coastal and Shelf Science* 23: 641–672.

Tett, P., L. Gilpin, H. Svendsen, C. P. Erlandsson, U. Larsson, S. Kratzer, E. Fouilland, et al. 2003. Eutrophication and some European waters of restricted exchange. *Continental Shelf Research* 23: 1635–1671.

Tuchkovenko, Y. S., and S. A. Lonin. 2003. Mathematical model of the oxygen regime of Cartagena Bay. *Ecological Modelling* 165: 91–106.

Turner, R. E., N. N. Rabalais, E. M. Swenson, M. Kasprzak, and T. Romaire. 2005. Summer hypoxia in the northern Gulf of Mexico and its prediction from 1978 to 1995. *Marine Environmental Research* 59: 65–77.

Uncles, R. J., I. Joint, and J. A. Stephens. 1998. Transport and retention of suspended particulate matter and bacteria in the Humber-Ouse Estuary, United Kingdom, and their relationship to hypoxia and anoxia. *Estuaries* 21 (4A): 597–612.

Vaz, N., J. M. Dias, and I. Martins. 2005. Dynamics of a temperate fluvial estuary in early winter. *Global NEST Journal* 7 (3): 237–243.

Vested, H. J., P. Justesen, and L. Ekebjærg. 1992. Advection–dispersion modelling in three dimensions. *Applied Mathematical Modelling* 16: 506–519.

Wilson, J. G., M. Brennan, and B. Brennan. 1991. Horizontal and vertical gradients in sediment nutrients on mud flats in the Shannon Estuary, Ireland. Marine and Estuarine Gradients, 1993. *Netherlands Journal of Aquatic Ecology* 27 (2–4): 173–180.

Yin, K., L. Zhifeng, and K. Zhiyuan. 2004. Temporal and spatial distribution of dissolved oxygen in the Pearl River Estuary and adjacent coastal waters. *Continental Shelf Research* 24: 1935–1948.

Zheng, L., C. Chen, and F. Y. Zhang. 2004. Development of water quality model in the SatillaRiver Estuary, Georgia. *Ecological Modelling* 178: 457–482.

14

Models of Pollution by Heavy Metals

Sven Erik Jørgensen

CONTENTS

14.1 Introduction and Standard Processes for Heavy Metal Models.........393
14.2 Overview of Heavy Metal Models by a Short Presentation of
Typical Examples 397
14.3 An Illustrative Heavy Metal Model: Contamination of Plants by
Cadmium 404
14.4 A Structurally Dynamic Heavy Metal Model 408
References..412

14.1 Introduction and Standard Processes for Heavy Metal Models

Ecotoxicological models of distribution and effects of heavy metals in eco-systems have been applied extensively in environmental management the past 3–4 decades. Because of the characteristic properties of ecotoxicological models, they are often considered a special model type (see, e.g., Jørgensen et al. 2009), although most ecotoxicological models are either biogeochemical or population dynamic models. The heavy metal models integrate the different species of metal compounds and the processes that determine the distribution of heavy metals in the environment. As for ecotoxicological models, in general, the core questions leading to the conceptual diagram are:

1. Which form of the heavy metal is most harmful?
2. Which processes determine the concentration of this most harmful form of the heavy metal?
3. Which forms and processes should be included in the conceptual diagram as a consequence of the answers to questions 1 and 2?

The most important processes of heavy metals in the environment can be quantified by equations that are generally applied in ecological models.

The adsorption equilibrium between solution (usually water) and solid (usually soil or suspended matter) can, in most cases, be expressed by one of the following two equations (see, e.g., Jørgensen et al. 2000):

Freundlich's adsorption isotherm:

$$A = bC^d \tag{14.1}$$

Langmuir's adsorption isotherm:

$$A = bC/(C + k) \tag{14.2}$$

where A is the concentration of the heavy metal in the solid expressed, for instance, in mg/kg and C is the concentration of the solution expressed in mg/m³. The terms b, d, and k are constants. Figures 14.1 and 14.2 illustrate the equations as graphs. Figure 14.1 is a log, log diagram because Freundlich's adsorption isotherm yields a straight line with the slope d in a log, log diagram: log A = log b + d log C. Langmuir's adsorption isotherm, which is the same expression as the Michaelis–Menten's equation used in Chapter 19, expresses the growth of plants as a function of the nutrient concentrations.

Heavy metals may precipitate in aquatic systems (often as carbonates or hydroxides) and the settling of the precipitated matter can be described via a first-order equation:

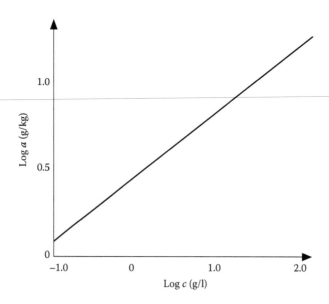

FIGURE 14.1
Freundlich's adsorption isotherm is a straight line in a log–log diagram.

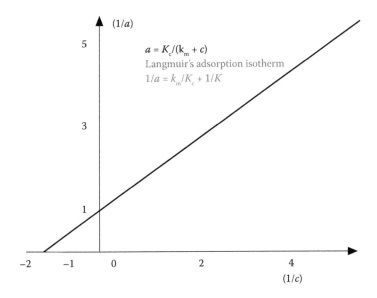

FIGURE 14.2
The plot of $1/a$ versus $1/c$ yields for Langmuir's adsorption isotherm a straight line, which can be used to find constants K and k_m.

$$dM/dt = sM \qquad (14.3)$$

where M is the concentration of the heavy metal as suspended matter and s is a constant (parameter).

Heavy metals can easily form complex compounds with a wide spectrum of different ligands: chloride, other halogenides, amino acids, hydroxides, carbonate, and humic acid. Complex compounds are usually less toxic than free metal ions and may therefore be crucial in determining how much of the heavy metal is in the form of free ions and in the form of various complex compounds. For instance, for cadmium, the chloride complexes are much less toxic than the free cadmium ion Cd^{2+}. In the marine environment, the chloride complexes are a significant part of the total concentration of cadmium, which implies that cadmium is less toxic in the marine environment than in freshwater aquatic systems. The equilibrium expressions for the formation of complexes are used to find the ratio of the concentration of free ions to the total concentration of the heavy metal. The ratio can be used to obtain the toxicity of the total cadmium concentration.

Plants take up heavy metals, and the rate is close to the growth rate of plants. The uptake rate, UP, is therefore expressed as:

$$UP = u * C * PG(time) \qquad (14.4)$$

where u is a parameter, C is the concentration of the heavy metal in the water (mg/m^3), and PG(time) is the plant growth expressed as g or kg biomass dry weight as function of time. The uptake rate parameter, u, varies for different plants and for different forms of heavy metals. The parameter u is determined by the uptake of water in relation to plant growth and the affinity of the heavy metal to the roots of the plants. Usually, the uptake rate is slower for complex compounds than for free ions. The amount of heavy metal in plants as a function of time is found via Equation 14.4 for the most important forms of the heavy metal. The concentration of heavy metals in the plants is obtained as the amount of heavy metals in the plants as a function of time divided by PG(time). This implies that the increase in heavy metal concentration in plants becomes a first-order reaction: $dC/dt = uC$.

For animals, there are two possible pathways for the uptake of heavy metals from the environment: with the food and with the air or water. For a fish that is living in water polluted by heavy metals, we get the following equations for the description of heavy metal contamination in the fish:

$$dW/dt = aW^b - rW^d \text{ b} \approx 0.67; d \approx 0.75 \tag{14.5}$$

where W is the weight of the fish as a function of time, and a, b, r, and d are parameters (constants).

$$\text{Uptake from water} = BCF * C_W * dW/dt \tag{14.6}$$

where BCF is the biological concentration factor, which is dependent on the heavy metal and the forms of heavy metals. BCF is dependent on the size of the fish according to allometric principles (see, e.g., Jørgensen and Bendoricchio 2001 and Jørgensen 2009a). C_w is the concentration of the heavy metal in the water.

$$\text{Uptake from food} = aW^b * C_w * \text{eff} \tag{14.7}$$

where C_f is the concentration of the heavy metal in the food and eff is the efficiency with which the heavy metal is taken up. eff is usually 0.05–0.1, provided that the heavy metal is in inorganic form. For organic metal compounds, for instance, methyl mercury, eff is much higher and may reach as high as 0.9.

$$\text{Excretion} = \text{exc} * C_{org} \tag{14.8}$$

where C_{org} is the concentration of the heavy metal in the fish and exc is a parameter that is dependent on the size of the fish according to allometric principles.

$$dTx/dt = BCF * C_w * dW/dt + aW^b * C_f * \text{eff} - \text{exc} * C_{org} \tag{14.9}$$

where Tx is the total amount of heavy metal in the fish.

$$C_{org} = Tx/W = f \text{ (time)} \qquad (14.10)$$

The ratio concentration in soil water divided by the concentration in soil depends on the composition of the soil. It has been found (Jørgensen 1976, 1993), that the ratio is decreased by (1) decreasing the pH, (2) increasing the clay content of the soil, (3) increasing the humus content of the soil, and (4) increasing the ion exchange capacity of the soil. For instance, for cadmium, the ratio (R) is obtained using the following equation:

$$R = 0.00001(80.01 - 6.14 * \text{pH} - 0.2603 * \text{CL} - 0.5189 * \text{HU} - 0.93 * \text{CEC}) \qquad (14.11)$$

where CL is the clay fraction content, HU is the humus fraction content, and CEC is the ion exchange capacity in eqv. per kg soil.

Equations 14.1–14.11 cover a wide range of processes that are included in most heavy metal models.

The next section gives an overview of a wide range of heavy metal models. Several hundreds of heavy metal models have been published so far. Because it is not possible to include them all in this report, the overview provides a representative impression of the many different types of heavy metal models. The characteristic features of a wide spectrum of heavy metal models are presented. The third section presents in detail a model focusing on the cadmium contamination of plants. The model represents a typical heavy metal model, and illustrates how the process equations are applied in a modeling context. The last section shows that structurally dynamic models (SDMs) have important applications in heavy metal modeling. The model focuses on how zooplankton adapt to contamination by copper ions via a change in size.

14.2 Overview of Heavy Metal Models by a Short Presentation of Typical Examples

All ecotoxicological models can be classified in accordance with the focus of the model. The model may focus on a global or regional contamination, or it may cover an ecosystem, an ecological network (as a trophic chain, for instance), a population, or an organism.

Very few heavy metal models report on results involving global or regional contamination, but a mercury model looking into global cycling and global distribution of the contamination has been developed (Vandal et al. 1993). Orlob et al. (1980) give an example on modeling of the contamination of an ecosystem, namely, contamination of a marine ecosystem by copper. Thomann et al. (1974) and Thomann (1984) have modeled the biomagnification of cadmium through the food chain. Early on, Seip (1978) presented a metal model focusing on zinc in a population of algae. Lam and Simons (1976) give a typical example of modeling the concentration and effect of one tropic level—in this case, lead. Fagerstrøm and Aasell (1973) provided another example, that of methyl mercury on fish. The models of these six classes are developed in a straightforward manner, building on the experience gained through the development of population dynamic models or eutrophication models. Other considerations are the basis for heavy metal models in the examples given below. The considerations often concentrate on how to develop a simple model while maintaining the focus of the model.

Figure 14.3 shows a copper model that has a different focus. Copper ions are toxic to phytoplankton and bacteria, and the main impact of copper contamination is therefore the sublethal effects exerted by free copper ions on these organisms. Thus, it would be sufficient to model, as shown in Figure 14.3, the copper concentrations on various possible forms in the considered ecosystem in order to determine the ion copper concentration, which can easily be translated into an effect.

Mogensen and Jørgensen (1979)—see also Jørgensen and Bendorichio (2001)—illustrate another approach, which is interesting because

1. The model is simple and can even be solved analytically.
2. Time is replaced by distance as the independent variable.
3. The model prognosis has been validated.

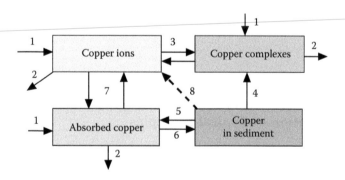

FIGURE 14.3
The model considers for copper forms the following: free ions, copper complexes (with chloride, amino acids, humic acid, etc.), copper in the sediment, and copper adsorbed to suspended matter. The copper ion concentration can be translated into an effect on phytoplankton and bacteria. This idea is also used in the structurally dynamic model presented in the last section of this chapter.

A tanning plant has, for decades, been discharging wastewater with a high concentration of chromium(III) into Faaborg fjord, Denmark. A model was developed to uncover the consequences of this contamination. An overall analysis showed the following important processes:

1. Settling of the precipitated chromium(III) hydroxide and other insoluble chromium compounds.
2. Diffusion of the chromium, mainly as suspended matter, throughout the fjord caused mainly by tides. This implies that an eddy diffusion coefficient has to be found. Advection may also play a role, but because of the relatively small amount of tanning wastewater, the importance of advection is probably less than the diffusion caused by the tide.
3. Bioaccumulation from sediment to benthic fauna, particularly to the blue mussels.

Items (1) and (2) can be combined into one submodel, whereas process (3) requires a separate submodel. Both submodels could be solved analytically and did not require the use of a computer. The model is described in more detail by Jørgensen and Bendoricchio (2001) and Jørgensen (2009a), but the most important equations and the prognosis validation are given below since the model is very illustrative as to what can be achieved by the use of even simpler models.

A mathematical description of advection and diffusion processes is combined with a description of settling in accordance with processes (1) and (2) mentioned above: The distribution model is based on the following simple chromium(III) transport equation (see, e.g., other similar cases in the work of Weber 1972):

$$\partial C/\partial t = D * \partial^2 C/\partial X^2 - Q * \partial C/\partial X - K * (C - C_o)/h \qquad (14.12)$$

where C is the concentration of total chromium in water (mg/l), C_o is the solubility of chromium(III) in seawater at pH = 8.1 (mg/l), Q is the inflow to the fjord = outflow by advection (m³/24 h), D is the eddy diffusion coefficient considering the tide (m²/24 h), X is the distance from the discharge point (m), K is the settling rate (m/24 h), and h is the mean depth (m).

For a tidal fjord such as Faaborg Fjord, the advection Q is—as already indicated—insignificant compared with the tide, and it is possible with good approximations to set Q to 0. The tanning plant has discharged an almost constant amount of chromium(III) during the past two decades. We can therefore consider the stationary situation:

$$\partial C/\partial t = 0 \qquad (14.13)$$

Accordingly, Equation 14.12 takes the form:

$$D * \partial^2 C/\partial X^2 = K * (C - C_o)/h \qquad (14.14)$$

This second-order differential equation has an analytical solution. The total discharge of chromium in g per 24 h, denoted CR, is known. This information is used together with F = cross-sectional area (m²) to state the boundary conditions. The following expression is obtained as an analytical solution (see any mathematical textbook or handbook):

$$C - C_o = (CR/F) * \sqrt{(h/D * K)} \exp\left[-\sqrt{(K/h * D)} * X\right] = IK \qquad (14.15)$$

F is known only approximately in this equation because of the nonuniform geometry of the fjord. The total annual discharge of chromium is 22,400 kg. Both the consumption of chromium by the tanning factory and the analytical determinations of the wastewater discharged by the factory confirm this number. h, the mean depth, is about 8 m on average. IK is an integration constant.

Equation 6.4 may be transformed to:

$$Y = K * (C - C_o) = (Cu/F) * \sqrt{(h * K/D)} * \exp\left[-\sqrt{(K/h * D)} * X\right] + K * IK \qquad (14.16)$$

As shown, $Y = K * (C - C_o)$, i.e., the amount of chromium (g) settled per 24 h and per m². The equation gives Y as a function of X. Y, however, is known from the sediment analysis at the 10 stations. A typical chromium profile for a sediment core is shown in Figure 14.4.

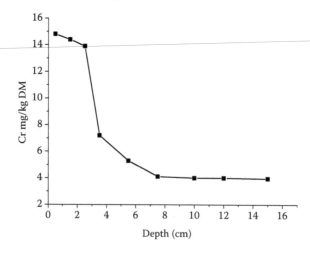

FIGURE 14.4
Typical chromium profile for a sediment core.

As we know that the increase in chromium concentration took place about 25 years before the model was built, it is possible to find the sediment rate in mm or cm per year: 75 mm/25 years = 3 mm/year. As we furthermore know the concentration of chromium in the sediment, we can calculate the amount of chromium settled per year, or 24 h, and per m², and this is Y. The Y values found for the 10 stations using this method are plotted versus X in Figure 14.5. It is easily seen that the decrease in settled chromium as a function of the distance from the discharge point is close to an exponential decay.

Note that it has been possible by the transformations to express chromium concentration as a function of X = distance from the discharge point, because the time variation is of minor importance and was eliminated by Equation 14.13. Furthermore, the second-order differential equation could easily be solved.

The second submodel focuses on the chromium contamination of the benthic fauna. It may be shown (Jørgensen 1979) that, under steady-state conditions, the relation between the concentration of a contaminant in the nth level of the food chain and the corresponding concentration in the $(n - 1)$th link, which is the food for the nth level, can be expressed using the following equation:

$$C_n = (\mu(n) * C_{n-1} * Y(n))/(MY(n) * F(n) - RESP(n) + EXC(n)) = K_c * C_{n-1} \qquad (14.17)$$

where $\mu(n)$ is the maximum growth rate for nth level of the food chain (1/day), C_n is the chromium concentration in the nth level of the food chain (mg/kg), C_{n-1} is the chromium concentration in the $(n - 1)$th level of the food chain (mg/kg), $Y(n)$ is the utility factor of chromium in the food for the nth level of the food, $F(n)$ is the utility factor of the food in the nth level of the food

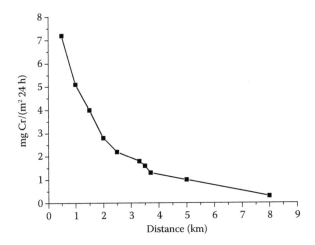

FIGURE 14.5
Settled amount of chromium as $f(X = $ distance from discharge point).

chain, RESP(n) is the respiration rate of the nth level of the food chain (1/day), EXC(n) is the excretion rate of chromium for the nth link of the food chain (1/day), and K_c is the accumulation coefficient for the transfer of chromium from the (n − 1)th level to the nth level of the food chain.

For blue mussels present in Faaborg Fjord, these parameter values can be found in the literature (see, e.g., Jørgensen et al. 2000). The mussels, *Mytilus edulis* (blue mussels), that were found in almost all the stations, were used to find the accumulation from the sediment to benthic animals The following parameters are valid for *Mytilus edulis* in the literature:

$\mu(n) = 0.03$ 1/day

$Y(n) = 0.07$

$F(n) = 0.67$

$RESP(n) = 0.001$ 1/day

$EXC(n) = 0.04$ 1/day

The use of these values implies that K_c = 0.036 for *M. edulis*. In other words, the concentration of chromium in *M. edulis* should be expected to be 0.036 times the concentration in the sediment. As the concentration of organic matter that is the food for *M. edulis* is in the order of 0.4 ppt, the concentration of chromium in *M. edulis* is about 90 times = 0.036/0.0004 higher than in the organic matter in the sediment. The major part of the sediment is sand, which is indigestible and contains very little chromium.

The distribution model has been used to assess the total allowable discharge of chromium (kg/year), if the chromium concentration in the sediment should be reduced to 70 mg per kg dry matter in the most polluted areas (stations 1 and 2, which are about 0.5 km from the discharge point). It was found that the total discharge of chromium should be reduced to a maximum 2000 kg or less per year to achieve a reduction of about 92%. Consequently, the environmental authorities required the tanning plant to reduce its chromium discharge to ≤2000 kg/year. The tanning plant has complied with the standards since 1982.

A few samples of sediment (6) and mussels (10) were taken in 1988–1989 to validate the model prognosis. The analytical results of the sediment and the mussels are presented in Table 14.1. Settled chromium in mg/m² day was calculated on the basis of the previous determined sedimentation rate (see above). The prognosis validation was fully acceptable as the deviation between prognosis and observed average values for chromium in mussels is approximately 12%. Based on the successful validation of the prognosis, it can be concluded that the model—although simple and consists only of two equations that are solved analytically—has been able to capture the determining processes. For other metals, it may be necessary to include more processes, because several metals are able to form organic metal compounds in contrast to chromium. For instance, for mercury, which is able to react

TABLE 14.1

Validation of the Prognosis

Item	Observed Value	Range mg per kg Dry Matter	Prognosis
Cr in sediment	65	57–81	70
Cr in mussels	2.2	1.4–4.5	2.5
mg Cr/m² day	0.59	0.44–0.83	0.67

with the organic matter of the sediment and form volatile methyl–mercury compounds, it has been found necessary to include the reactions between organic matter and mercury to be able to develop a workable model.

The model gave a surprisingly good result of the prognosis validation considering that the model in reality has only two equations. It is, however, not always the case for heavy metal models, as they often can react with organic matter, which of course implies that more state variables are also needed when dealing with these forms of heavy metals. Moreover, as noted in Section 14.1, the heavy metals can form complexes (see also Figure 14.1), which means that it is necessary to include the complexes as state variables. Figure 14.6 shows a conceptual diagram of a mercury model for Mex Bay, Egypt, which includes more state variables than the chromium model.

Gustafsson (2009) has developed a model of metal sorption in soils that is able to give a better quantitative description of the exchange between metals in soil water and metals in soil than Equation 14.11. The model considers adsorption, ion exchange, precipitation, complex formation, and the influence

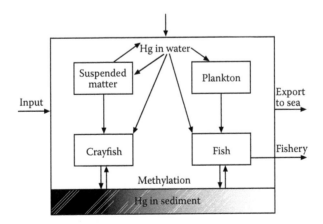

FIGURE 14.6

Mercury model of Mex Bay, Egypt. The model has the following state variables: suspended matter, inorganic mercury in the water, organic mercury in the water, mercury adsorbed to the suspended matter, mercury in plankton, mercury in the crayfish, mercury in the carnivorous fish (tuna fish), and mercury in the sediment.

of pH on these processes. A more detailed model of uptake of heavy metals by plants than the one presented in Section 14.3 could be developed by using Gustafsson's model for the exchange between soil water and soil for metals. Nyholm et al. (1984) have applied a similar approach, although their model included fewer processes.

Kim et al. (2008) have developed a very complex mercury model that focuses on the bioaccumulation of methyl mercury. It considers both carbon flow and methyl mercury flow and has 25 state variables. The model is used to consider the effect on filter feeders and on sediment resuspension, and the role of the methylation process in the sediment. The model is based on a comprehensive discussion of the many mercury processes included in the model, and can therefore be recommended as a reference for other mercury models.

The overview of heavy metal models presented in this section has shown that the models are not very different from general biogeochemical models or population dynamic models. The processes in heavy metal models are described similarly to what is applied in biogeochemical models. It is, however, beneficial to have a good knowledge of the many possible processes in which heavy metal ions can participate. It is often necessary, in addition to a description of bioaccumulation through the food chain and adsorption processes, to consider the many possibilities for formation of complexes that may change the toxicity, the solubility, and the uptake and excretion processes. Mercury has a particularly complicated chemistry because it is able to form methyl mercury that is released from sediment and is able to bioaccumulate.

14.3 An Illustrative Heavy Metal Model: Contamination of Plants by Cadmium

Plants are able to take up toxic substances dissolved in the soil water. This makes it possible to remove toxic substances by plants from contaminated soil—an ecological engineering method that belongs to a group of methods called phytoremediation. It is, on the other hand, a problem in that edible crops would be contaminated by toxic substances, taken up from contaminated soil. For more details on this ecological engineering method, see Jørgensen (2009b).

Sludge from wastewater treatment and inorganic fertilizers contain small concentrations of various toxic organic compounds and heavy metals. There are standards for the concentrations of a few toxic organic compounds and for heavy metal concentrations that can be accepted for sludge applied as soil conditioner in agriculture. The standards reflect the uptake of toxic compounds by plants and the toxicity of the heavy metal.

For heavy metals, the distribution between soil and soil water determines how much the plants can take up of the total amount of heavy metals. The distribution is determined by the composition of soil: How much clay and humus does the soil contain? And what is the pH and the ion exchange capacity of the soil, CEC? (See Equation 14.11 and Jørgensen 1976, 1993.) The uptake of cadmium by plants is dependent on the protein content of the plants (in percentage of dry matter).

Figure 14.7 shows a conceptual STELLA diagram for a model focusing on the contamination of plants by cadmium. The model is presented in more detail, including its applicability for other heavy metals than cadmium, by Jørgensen (1976, 1993). The sources for cadmium contamination are the addition of cadmium to soil due to use of fertilizers and/or sludge as soil conditioner and the atmospheric fallout mainly by dry deposition. Notice that the model has two state variables to cover cadmium in soil. It was initially

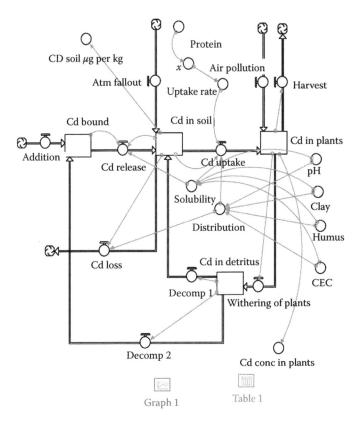

FIGURE 14.7
STELLA diagram of a cadmium model. The model relates the cadmium in fertilizers and soil conditioners added to the soil and the dry deposition of cadmium on plants and soil with the cadmium concentration in the plants as a function of time.

TABLE 14.2

Model Equations (STELLA Format)

Cd_bound(t) = Cd_bound(t – dt) + (decomp_2 + Addition – Cd_release) * dt
INIT Cd_bound = 0.19
INFLOWS:
decomp_2 = Cd_in_detritus*0.012
Addition = PULSE(0.14,1,960)
OUTFLOWS:
Cd_release =
IF(Cd_in_soil<solubility)THEN(0.000012*Cd_bound)ELSE(Cd_bound*0.0000012)

Cd_in_detritus(t) = Cd_in_detritus(t - dt) + (Withering_of_plants - decomp_1 - decomp_2) * dt
INIT Cd_in_detritus = 0.27
INFLOWS:
Withering_of_plants = PULSE(0.6*Cd_in_plants,180,360)+PULSE(0.6*Cd_in_plants,181,360)
OUTFLOWS:
decomp_1 = IF(TIME>180)THEN(0.005*Cd_in_detritus)ELSE(0.0001*Cd_in_detritus)
decomp_2 = Cd_in_detritus*0.012

Cd_in_plants(t) = Cd_in_plants(t - dt) + (Cduptake + air_pollution - Withering_of_plants - harvest) * dt
INIT Cd_in_plants = 0.00002
INFLOWS:
Cduptake = Cd_in_soil*distribution*uptake_rate
air_pollution = 0.0000028+STEP(-0.0000028,180)+STEP(0.0000028,360)+STEP(-0.0000028,540)+ STEP(0.0000028,720)+STEP(-0.0000028,900)
OUTFLOWS:
Withering_of_plants = PULSE(0.6*Cd_in_plants,180,360)+PULSE(0.6*Cd_in_plants,181,360)
harvest = PULSE(0.4*Cd_in_plants,180,360)+PULSE(0.4*Cd_in_plants,181,360)

Cd_in_soil(t) = Cd_in_soil(t - dt) + (Cd_release + decomp_1 + atm_fall_out - Cduptake - Cd_loss) * dt
INIT Cd_in_soil = 0.08
INFLOWS:
Cd_release = IF(Cd_in_soil<solubility)THEN(0.000012*Cd_bound) ELSE(Cd_bound*0.0000012)
decomp_1 = IF(TIME>180)THEN(0.005*Cd_in_detritus)ELSE(0.0001*Cd_in_detritus)
atm_fall_out = 0.0000014
OUTFLOWS:
Cduptake = Cd_in_soil*distribution*uptake_rate
Cd_loss = 0.1*Cd_in_soil*distribution
Cd_conc_in_plants = 5000*Cd_in_plants/14
Cd_soil_microg_per_kg = Cd_in_soil*1000
CEC = 8
clay = 4
distribution = 0.00001*(800-6.14*pH-0.26*clay-0.52*humus-0.93*CEC)
humus = 2
pH = 7
protein = 47
solubility = 10^(6.273-1.505*pH+0.00212*humus+0.002414*CEC-0.00122*clay)*112.4*350
uptake_rate = x+STEP(-x,180)+STEP(x,360)+STEP(-x,540)+STEP(x,720)+STEP(-x,900)
x = 0.00216*(-0.377+0.4544*protein)

attempted to have only one state variable to cover cadmium in soil, but it was almost impossible to calibrate the model. As it is known that heavy metals are bound in the soil that is either very solid or less solid, it was tested whether two fractions of cadmium in soil could improve the model calibration and as it was the case, the model shown in Figure 14.7 was selected. In the conceptual diagram, the two fractions are denoted Cd bound (solid bound cadmium) and Cd in soil (less solid bound cadmium).

The equations of the model are shown in Table 14.2, and Figure 14.8 gives the validation results of the model. The parameters and the calibration were based on pot experiments, and the validation was based on large-scale experiments adding cadmium containing fertilizers and sludge to the soil at time = 0. The cadmium content in plants was determined at harvest in the first, second, and third year, and these results were used for the validation of the model. The following concentrations were found (average of 20 randomly collected maize plants):

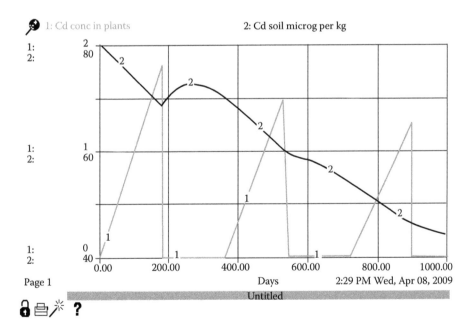

FIGURE 14.8
The graph shows the cadmium concentration in plants and soil (µg/kg dry matter). The harvest takes place at days 180, 540, and 900. The observed cadmium concentrations were found during three harvests (first, second, third) to be 1.7, 1.1, and 0.8 µg/kg dry matter, respectively. Cadmium content in soil is reduced over the simulation period from about 80 µg/kg dry matter in soil to about 45 µg/kg dry matter in soil. The figure is from Jørgensen (*Ecological Modelling*, 2, 59–68, 1976; with permission).

1. Harvest 1.7 µg/kg dry matter
2. Harvest 1.1 µg/kg dry matter
3. Harvest 0.8 µg/kg dry matter

These values can be compared with prognosis by the model: 1.7 µg/kg dry matter, 1.6 µg/kg dry matter, and 1.2 µg/kg dry matter, respectively (see Figure 14.8). When one considers the fact that the model is relatively simple—with only four state variables—the result of the validation seems acceptable, and also because in ecotoxicological models, a high deviation—in this case, about 50% at the second and third harvest—can be accepted. The predicted values are higher than the observed ones, which means that the model is at least not increasing the risk for cadmium contamination by crops harvested in areas that have received sludge as soil conditioner in addition to the cadmium contamination of the usually applied fertilizers and the general background air pollution in the countryside. The concentration of cadmium is, of course, decreasing by subsequent harvests (in the second and third year), because no additional sludge is added to the soil after time 0, but a part of the cadmium is removed by harvest each year and a minor part is also lost to groundwater (indicated in the conceptual diagram as Cd loss). Moreover, the amount of cadmium in detritus will slow down the cycling of cadmium and thereby reduce the amount of easily available cadmium. The cadmium in soil is reduced over the simulation period from about 80 µg/kg dry matter in soil to about 45 µg/kg dry matter in soil (see Figure 14.8).

14.4 A Structurally Dynamic Heavy Metal Model

The model presented below has been reported by Jørgensen (2009c) and illustrates the use of SDM in developing heavy metal models. Basic information about the development of SDMs has been presented in Section 2.7 and Section 19.4 illustrates the use of SDM for development and improvement of eutrophication models.

The conceptual diagram used to illustrate the use of SDMs for development of heavy metal models is shown in Figure 14.9, using the modeling software STELLA. Copper is an algaecide that causes an increase in the mortality of phytoplankton (Kallqvist and Meadows 1978) and a decrease in phosphorus uptake and photosynthesis (see Hedtke 1984). Copper also reduces the carbon assimilation of bacteria (Havens 1994). In the literature, changes in the following three parameters in the model have been reported: growth rate of phytoplankton, mortality of phytoplankton, and mineralization rate of detritus with increased copper concentration (Kallquist and Meadows 1978; Hedtke 1984; Havens 1994; Xu et al. 1999). As a result, the zooplankton is

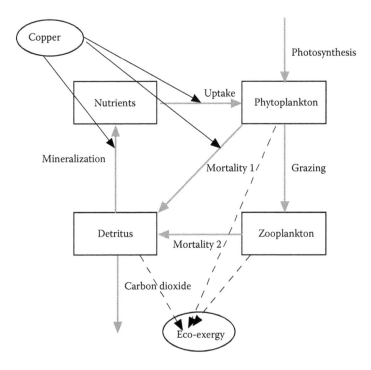

FIGURE 14.9
Conceptual diagram of a heavy metal model focusing on the influence of copper on the photosynthetic rate, phytoplankton mortality rate, and the mineralization rate. The boxes are the state variables, the thick gray arrows symbolize processes, and the thin black arrows indicate the influence of copper on the processes and the calculation of eco-exergy from the state variables. Due to the change in these three rates, it is an advantage for the zooplankton and the entire ecosystem to decrease their size. The model is therefore made structurally dynamic by allowing zooplankton to change its size, and thereby the specific grazing rate and the specific mortality rate according to the allometric principles. The size yielding the highest eco-exergy is currently found.

reduced in size (Xu et al. 1999), which according to the allometric principles means an increased specific grazing rate and specific mortality rate (Xu et al. 1999). It has been observed that the size of zooplankton in a closed system (e.g., a pond) is reduced to less than half the size at a copper concentration of 140 mg/m³ compared with a copper concentration of less than 10 mg/m³ (Xu et al. 1999). In accordance with allometric principles (Peters 1983), this would more than double the grazing rate and the mortality rate.

The model shown in Figure 14.9 was made structurally dynamic by varying the size of zooplankton and using an allometric equation to determine the corresponding specific grazing rate and specific mortality rate. The equation expresses that the two specific rates are inversely proportional to the linear size (Peters 1983). Four different copper concentrations, varying from

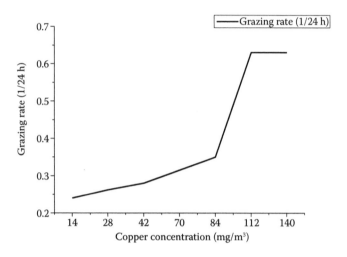

FIGURE 14.10
The grazing rate yielding the highest eco-exergy is shown at different copper concentrations. The grazing rate is increasing more and more rapidly as the copper concentration is increased but at a certain level, it is not possible to increase the eco-exergy further by changing the zooplankton parameters, because the amount of phytoplankton is becoming the limiting factor for zooplankton growth.

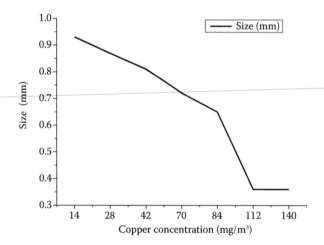

FIGURE 14.11
The zooplankton size that yields the highest eco-exergy is plotted versus the copper concentration. The size is decreasing more and more rapidly as the copper concentration is increasing but at a certain level, it is not possible to increase the eco-exergy further by changing the zooplankton size, because the amount of phytoplankton is becoming the limiting factor for zooplankton growth.

10 to 140 mg/m³, are found by the model in which zooplankton size yields the highest eco-exergy. In accordance with the presented SDM approach, it is expected that the size yielding the highest eco-exergy would be selected (see Section 12.4). The results of the model runs are shown in Figures 14.10–14.12. The specific grazing rate, the size yielding the highest eco-exergy, and the eco-exergy are plotted versus the copper concentration in these three figures. As expected and illustrated in the last figure, the eco-exergy is even at the zooplankton size yielding the highest eco-exergy, decreasing as copper concentration increases due to the toxic effect on phytoplankton and bacteria.

The selected size at 140 mg/m³ is—also according to the literature—less than half, about 40% of the size at 10 mg/m³. The eco-exergy decreases from 198 kJ/l at 10 mg/m³ to 8 kJ/l at 140 mg m³. The toxic effect of copper, in other words, leads to an eco-exergy reduction to about 4% of the original eco-exergy level, which can be considered a very significant toxic effect. The decrease in eco-exergy may be considered an indirect measure of toxicity and contamination by copper, as eco-exergy measures within a system the amount of energy that can perform work, i.e., the work capacity. If the zooplankton had been unable to adapt to the toxic effect by changing its size and thereby the parameters, the reduction in eco-exergy would have been more pronounced even at low copper concentrations. It is therefore important for model results that the model be made structurally dynamic and thereby able to account for the change of parameters when the copper concentration is changed.

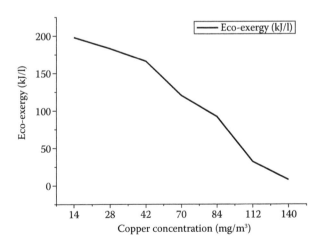

FIGURE 14.12
The highest eco-exergy obtained when varying the zooplankton size is plotted versus the copper concentration. The eco-exergy is decreasing almost linearly with increasing copper concentration. The discrepancy from approximately a linear plot may be due to model uncertainty and discontinuous change in copper concentration and the zooplankton size.

References

Fagerstrøm, T., and B. Aasell. 1973. Methyl mercury accumulation in an aquatic food chain. A model and implications for research planning. *Ambio* 2: 164–171.

Gustafsson, J. P. 2009. Modelling metal sorption in soils. In *Modelling of Pollutants in Complex Environmental Systems*, vol. I, ed. G. Hanrahan, 145–176. Hertfordshire, U.K.: ILM Publications.

Havens, K. E. 1994. Structural and functional responses of a freshwater community to acute copper stress. *Environmental Pollution* 86: 259–266.

Hedtke, S. F. 1984. Structure and function of copper-stressed aquatic microcosms. *Aquatic Toxicology* 5: 227–244.

Jørgensen, S. E. 1976. Model of heavy metal contamination of crops and ground water. *Ecological Modelling* 2: 59–68.

Jørgensen, S. E. 1993. Removal of heavy metals from compost and soil by ecotechnological methods. *Ecological Engineering* 2: 89–100.

Jørgensen, S. E. 2009a. *Ecological Modelling, An Introduction.* 192 pp. Southhampton: WIT.

Jørgensen, S. E. 2009b. *Applications in Ecological Engineering.* 380 pp. Amsterdam: Elsevier.

Jørgensen, S. E. 2009c. *The Application of Structurally Dynamic Models in Ecology and Ecotoxicology.* In *Ecotoxicology Modeling*, ed. J. Devillers, 77–394. New York: Springer.

Jørgensen, L. A., S. E. Jørgensen, and S. N. Nielsen. 2000. *Ecotox*, CD. Amsterdam: Elsevier.

Jørgensen, S. E., and G. Bendoricchio. 2001. *Fundamentals of Ecological Modelling*, 3rd edn, 530 pp. Amsterdam: Elsevier.

Jørgensen, S. E., T. Chon, and F. Recknagel. 2009. *Handbook of Ecological Modelling and Informatics.* 432 pp. Southampton: WIT.

Kim, E., R. P. Mason, and C. M. Bergeron. 2008. A modeling study on methylmercury bioaccumulation and its controlling factors. *Ecological Modelling* 218: 267–289.

Kallqvist, T., and B. S. Meadows. 1978. The toxic effects of copper on algae and rotifer from a soda lake. *Water Research* 12: 771–775.

Lam, D. C. L., and T. J. Simons. 1976. Computer model for toxicant spills in Lake Ontario. In *Metals Transfer and Ecological Mass Balances, Environmental Biochemistry*, vol. 2, ed. J. O. Nriago, 537–549. Ann Arbor, MI: Ann Arbor Science.

Mogensen, B., and Jørgensen, S. E. 1979. Modelling the distribution of chromium in a Danish fjord. In *Proceedings of the 1st International Conference on State of the Art in Ecological Modelling*, Copenhagen, 1978. ed. S. E. Jørgensen, 367–377. Copenhagen: International Society of Ecological Modelling.

Nyholm, N., T. K. Nielsen, and K. Pedersen. 1984. Modelling heavy metals transpon in an arctic fjord system polluted from mine tailings. *Ecological Modelling* 22: 285–324.

Orlob, G. T., D. Hrovat, and F. Harrison. 1980. Mathematical model for simulation of the fate of copper in a marine environment. *American Chemical Society, Advances in Chemistry Series* 189: 195–212.

Peters, R. H. 1983. *The Ecological Implications of Body Size.* 329 pp. Cambridge: Cambridge University Press.

Seip, K. L. 1978. Mathematical model for uptake of heavy metals in benthic algae. *Ecological Modelling* 6: 183–198.

Thomann, R. V. 1984. Physico-chemical and ecological modeling the fate toxic substances in natural water system. *Ecological Modelling* 22: 145–170.

Thomann, R. V., D. Szumski, D. M. DiToro, and D. J. O'Connor. 1974. A food chain model of cadmium in western Lake Erie. *Water Research* 8: 841–851.

Vandal, G. M., et al. 1993. Variation in mercury deposition to Antarctica over the past 34,000 years. *Nature* 62: 621–623.

Weber, W. J. 1972. *Physicochemical Processes for Water Quality Control.* 642 pp. New York: Wiley-Interscience.

Xu, F.-L., S. E. Jørgensen, and S. Tao. 1999. Ecological Indicators for assessing the freshwater ecosystem health. *Ecological Modelling* 116: 77–106.

15

Models of Pharmaceuticals in the Environment

Sven Erik Jørgensen and Bent Halling Sørensen

CONTENTS

15.1 Introduction.. 415
15.2 Environmental Risk Assessment... 416
15.3 Characteristics and Structure of Ecotoxicological Models, in
      Comparison with Other Models Applied in Ecological and
      Environmental Management 424
15.4 Overview of Pharmaceutical Models.. 427
15.5 An Illustrative Example of a Pharmaceutical Model........................... 431
      15.5.1 The Equations... 432
References.. 437

15.1 Introduction

Ecosystems are open systems, and it is therefore inevitable that sooner or later all the chemicals we are using today will find their way into the environment. Consequently, this raises the question: What harmful effects can the chemicals used in pharmaceuticals cause in nature, and what harm can their accumulation impose on humans, either indirectly or directly? The modern industrialized society uses about 100,000 chemicals, many of which have been detected in nature in concentrations that may cause harm. More than 3000 compounds out of a total of 100,000 are pharmaceutical compounds. They are in principle not different from other organic compounds with respect to the potential harm they pose to nature and human beings, but because of their medical value, it is necessary to strike a balance between their positive and negative effects. These compounds are used as pharmaceutical components because they have a biological effect, and therefore it should not be surprising that several pharmaceuticals become harmful and have an adverse effect on nature and humans at certain concentrations. An obvious question in this context would be, "Is it sensible and feasible to assess potential adverse effects in ecosystems and their organisms?" Because of constant release in most cases, a chronic—rather than an acute—effect will probably be observed.

In accordance with the registration process, information and data about fate and effects are necessary for all new chemicals in industrialized countries. This implies that an environmental risk assessment is a prerequisite for registration. Although registration of pharmaceuticals has been implemented for decades, it was only in the past few years that information about their environmental impacts has become a requirement. A tiered approach is applied, and the necessary information depends on the amounts produced. For new substances applied as veterinary medicine, environmental data have been required in the European Union (EU) since 1998, and the risk assessment methodology has been proposed to be applied in this context. It requires that an environmental concentration be calculated for various compartments in the environment: water, soil, air, specific organisms, groundwater, and so on—the so-called predicted environmental concentration. This is hardly possible without the use of environmental models.

The next section gives a short introduction to environmental risk assessment in order to lay down the basis for an environmental model development. Section 15.3 presents the characteristic features of ecotoxicological models compared with other ecological models. Relatively few models of pharmaceuticals in the environment have been developed so far. Nevertheless, Section 15.4 gives a brief overview of the spectrum of existing models, including a short discussion of the selection of model type. The last section presents a detailed model to illustrate how this model type is developed in contrast to other models.

15.2 Environmental Risk Assessment

Treatment of industrial wastewater, solid waste, and smoke is very expensive. Consequently, many industries attempt to adopt changes so as to make their products and production methods more environmentally friendly in order to reduce treatment costs. Industries, therefore, need to know how much various chemicals, components, and processes are polluting the environment. Or expressed differently, what is the environmental risk of using a specific material or chemical compared with other alternatives? If industries can reduce their pollution just by switching to another chemical or process, they will consider doing so to reduce their environmental costs or improve their green image. An assessment of the environmental risk associated with the use of a specific chemical and a specific process gives industries the possibility of making the right selection of materials, chemicals, and processes to the benefit for the economy of the enterprise and the quality of the environment. Similarly, society needs to know the environmental risks of all chemicals applied so as to phase out the most environmentally threatening chemicals and set standards for the use of all other chemicals.

Modern abatement of pollution, therefore, includes environmental risk assessment (ERA), which may be defined as the process of assigning magnitudes and probabilities to the adverse effects of human activities. The process involves identifying hazards such as the release of toxic chemicals to the environment by quantifying the relationship between an activity associated with an emission to the environment and its effects. The entire ecological hierarchy is considered in this context—this includes the effects on the cellular (biochemical) level, the organism level, the population level, the ecosystem level, and the entire ecosphere.

The application of ERA is rooted in the recognition that:

1. *The elimination cost of all environmental effects is impossibly high.*

2. Practical environmental management decisions must always be made on the basis of incomplete information.

We use about 100,000 chemicals in such amounts that they might threaten the environment, but we know only less than 1% of what we need to know to be able to make a proper and complete ERA of all these chemicals.

ERA is in the same family as environmental impact assessment (EIA), which attempts to assess the impact of a human activity. EIA is predictive, comparative, and concerned with all possible effects on the environment, including secondary and tertiary (indirect) effects, whereas ERA attempts to assess the probability of a given (defined) adverse effect as a result of a considered human activity.

Both ERA and EIA use models to find the expected environmental concentration (EEC), which is translated into impacts for EIA and to risks of specific effects for ERA.

Legislation and regulation of domestic and industrial chemicals with respect to the protection of the environment have been implemented in Europe and North America for decades. Both regions distinguish, as indicated above, between existing chemicals and introduction of new substances. For existing chemicals, the EU requires a risk assessment to humans and environment according to a priority setting. An informal priority setting (IPS) is used for selecting chemicals among the 100,000 listed in the European Inventory of Existing Commercial Chemical Substances. The purpose of IPS is to select chemicals for detailed risk assessment from among the EEC high production volume compounds, that is, >1000 t/year (about 2500 chemicals). Data necessary for the IPS and an initial hazard assessment are called Hedset, and cover issues such as environmental exposure, environmental effects, exposure to humans, and human health effects.

Uncertainty plays an important role in risk assessment. Risk is the probability that a specified harmful effect will occur or, in the case of a graded effect, the relationship between the magnitude of the effect and its probability of occurrence.

Risk assessment has emphasized risks to human health and has, to a certain extent, ignored ecological effects. However, some chemicals that have no or only little risk to human health cause severe effects on ecosystems such as aquatic organisms. Examples are chlorine, ammonia, and certain pesticides. An up-to-date risk assessment comprises considerations of the entire ecological hierarchy, which is the ecologist's worldview in terms of levels of organization. Organisms interact directly with the environment, and it is organisms that are exposed to toxic chemicals.

Uncertainty is addressed using an assessment (safety) factor from 10 to 1000. The choice of assessment factor depends on the quantity and quality of toxicity data (see Table 15.1). The assessment or safety factor is used in step 3 of the ERA procedure presented below. Other relationships aside from the uncertainties originating from randomness, errors, and lack of knowledge may be considered when the assessment factors are selected, for instance, cost–benefit. This implies that the assessment factors for drugs and pesticides may be given a lower value because of their possible benefits.

Lack of knowledge results in undefined uncertainty that cannot be described or quantified. It is a result of practical constraints on our ability to accurately describe, count, measure, or quantify everything that pertains to a risk estimate. Clear examples are the inability to test all toxicological responses of all species exposed to a pollutant, and the simplifications needed in the model used to predict the EEC.

The most important feature distinguishing risk assessment from impact assessment is the emphasis in risk assessment on characterizing and quantifying uncertainty. Therefore, it is of particular interest in risk assessment to analyze and estimate the analyzable uncertainties. They are natural stochasticity, parameter errors, and model errors. Statistical methods may

TABLE 15.1

Selection of Assessment Factors to Derive PNEC (See also Step 3 of the Procedure Presented in Text)

Data Quantity and Quality	Assessment Factor
At least one short-term LC_{50} from each of the three trophic levels of the base set (fish, zooplankton, and algae)	1000
One long-term NOEC (non-observed effect concentration, either for fish or *Daphnia*)	100
Two long-term NOECs from species representing two trophic levels	50
Long-term NOECs from at least three species (normally fish, *Daphnia*, and algae) representing three trophic levels	10
Field data or model ecosystems	Case by case

provide direct estimates of uncertainties. They are widely used in model development.

The use of statistics to quantify uncertainty is complicated in practice by the need to consider errors in both dependent and independent variables, and to combine errors when multiple extrapolations should be made. The Monte Carlo analysis is often used in the development of models to overcome these difficulties (see, e.g., Bartell et al. 1992).

Model errors include inappropriate selection or aggregation of variables, incorrect functional forms, and incorrect boundaries. The uncertainty associated with model errors is usually assessed by field measurements utilized for calibration and validation of the model. The modeling uncertainty for ecotoxicological models is, in principle, not different from what was already discussed in the other chapters.

Chemical risk assessment may be divided into nine steps (see, e.g., Jørgensen and Bendoricchio 2001, which is the source of this description of ERA), which are shown in Figure 15.1. The nine steps correspond to questions that the risk assessment attempts to answer to quantify the risk associated with the use of a chemical. The nine steps are presented in detail below, with reference to Figure 15.1.

Step 1. Which hazards are associated with the application of the chemical? This involves gathering data on the types of hazards—possible environmental damage and human health effects. The health effects include congenital, neurological, mutagenic, endocrine disruption (the so-called estrogen), and carcinogenic effects. It may also include characterization of the behavior of the chemical within the body (interactions with organs, cells, or genetic material). What is the possible environmental damage including lethal effects and sublethal effects on growth and reproduction of various populations?

As an attempt to quantify the potential danger posed by chemicals, a variety of toxicity tests has been devised. Some of the recommended tests involve experiments with subsets of natural systems (e.g., microcosms) or with entire ecosystems. The majority of testing new chemicals for possible effects has, however, been confined to studies in the laboratory on a limited number of test species. Results from these laboratory assays provide useful information for quantification of the relative toxicity of different chemicals. They are used to forecast effects in natural systems, although their justification has been seriously questioned (Cairns et al. 1987).

Step 2. What is the relation between dose and responses of the type defined in step 1? It implies knowledge of non-effect concentration (NEC), LD_x (the dose which is lethal to $x\%$ of the organisms considered), LC_y (the concentration which is lethal to $y\%$ of the organisms considered), and EC_z values (the concentration giving the indicated effect to $z\%$ of the considered organisms), where x, y, and z express a probability of harm. The answer can be found by laboratory examination or we may use estimation methods. Based on these answers, a most probable level of no effect (NEL) is assessed. Data needed for steps 1 and 2 can be obtained directly from scientific libraries, but

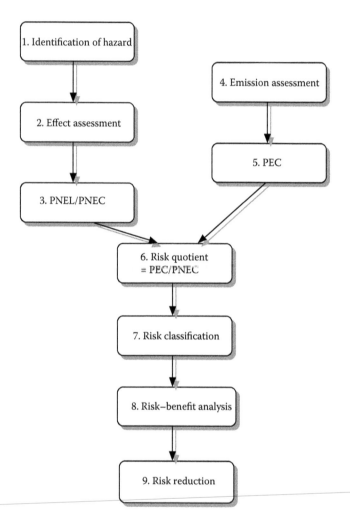

FIGURE 15.1

Nine-step procedure to assess the risk of chemical compounds. Steps 1–3 require extensive use of ecotoxicological handbooks and ecotoxicological estimation methods to assess the toxicological properties of the chemical compounds considered, whereas step 5 requires the selection of a proper ecotoxicological model.

are increasingly found via online data searches in bibliographic and factual databases. Data gaps should be filled with estimated data. It is very difficult to obtain complete knowledge about the effect of a chemical on all levels from cells to ecosystems. Some effects are associated with very small concentrations, such as the estrogen effect. It is, therefore, far from sufficient to know NEC, LD_x, LC_y, and EC_z values.

Step 3. Which uncertainty (safety) factors reflect the amount of uncertainty that must be taken into account when experimental laboratory data or empirical estimation methods are extrapolated to real situations? Usually, safety factors of 10–1000 are used. The choice is discussed above and will usually be in accordance with Table 8.1. If good knowledge about the chemical is available, then a safety factor of 10 may be applied. If, on the other hand, it is estimated that the available information has a very high uncertainty, then a safety factor of 10,000 may be recommended. Most frequently, safety factors of 50–100 are applied. NEL times the safety factor is named the predicted non-effect level (PNEL). The complexity of ERA is often simplified by deriving the predicted no effect concentration, PNEC, for different environmental components (water, soil, air, biotas, and sediment).

Step 4. What are the sources and quantities of emissions? The answer requires thorough knowledge of the production and use of the chemical compounds considered, including an assessment of how much of the chemical is wasted in the environment by production and use. The chemical may also be a waste product, which makes it very difficult to determine the amounts involved. For instance, the very toxic dioxins are waste products from incineration of organic waste.

Step 5. What is (are) the actual exposure concentration(s)? The answer to this question is named the predicted environmental concentration (PEC). Exposure can be assessed by measuring environmental concentrations. It may also be predicted by a model when the emissions are known. The use of models is necessary in most cases either because we are considering a new chemical, or because the assessment of environmental concentrations requires a very large number of measurements to determine the variations in concentrations in time and space. Furthermore, it provides an additional certainty to compare model results with measurements, which implies that it is always recommended to both develop a model and make at least a few measurements of concentrations in the ecosystem components when and where it is expected that the highest concentration will occur. Most models will demand an input of parameters, describing the properties of the chemicals and the organisms, which also will require extensive application of handbooks and a wide range of estimation methods. The development of an environmental, ecotoxicological model requires extensive knowledge of the physical–chemical–biological properties of the chemical compound(s) considered. The selection of a proper model is discussed in this chapter and in Chapter 2.

Step 6. What is the PEC/PNEC ratio? This ratio is often called the risk quotient. It should not be considered an absolute assessment of risk, but rather a relative ranking of risks. The ratio is usually found for a wide range of ecosystems such as aquatic ecosystems, terrestrial ecosystems, and groundwater.

Steps 1–6, shown in Figure 15.2, agree with the data shown in Figure 15.1 and the information given above.

Step 7. How will you classify the risk? Risk valuation is made to decide on risk reductions (step 9). Two risk levels are defined: (1) the upper limit, that is, the maximum permissible level (MPL) and (2) the lower limit, that is,

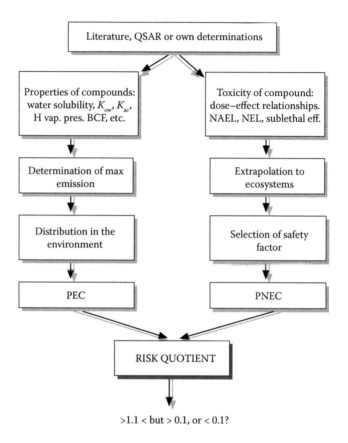

FIGURE 15.2
Steps 1–6 are shown in more detail for practical applications. The result of these steps also leads to assessment of the risk quotient.

the negligible level (NL). It may also be defined as a percentage of MPL, for instance, 1% or 10% of MPL.

The two risk limits create three zones: a black, unacceptable, high-risk zone > MPL; a gray, medium-risk level; and a white, low-risk level < NL. The risk of chemicals in the gray and black zones must be reduced. If the risk of the chemicals in the black zone cannot be reduced sufficiently, then it should be considered to phase out the use of these chemicals.

Step 8. What is the relation between risk and benefit? This analysis involves examination of socioeconomic, political, and technical factors, which are beyond the scope of this volume. The cost–benefit analysis is difficult because the costs and benefits are often of a different order and dimension.

Step 9. How can the risk be reduced to an acceptable level? The answer to this question requires an extensive technical, economic, and legislative investigation. Assessment of alternatives is often an important aspect in risk reduction.

Steps 1, 2, 3, and 5 require knowledge of the properties of the focal chemical compounds, which again implies an extensive literature search and/or selection of the best feasible estimation procedure. In addition to "Beilstein," it can be recommended to have at hand the following very useful handbooks of environmental properties of chemicals and methods for estimation of these properties in case literature values are not available:

S. E. Jørgensen, S. N. Nielsen, and L. A. Jørgensen, 1991. *Handbook of Ecological Parameters and Ecotoxicology*, Elsevier, Amsterdam. Year 2000 published as a CD called Ecotox. It contains three times the amount of parameter in the 1991 book edition. See also Chapter 2 for further details about Ecotox.

P. H. Howard et al., 1991. *Handbook of Environmental Degradation Rates.* Lewis Publishers, Chelsea, MI.

K. Verschueren, *Handbook of Environmental Data on Organic Chemicals.* Van Nostrand Reinhold, New York, NY. Several editions have been published, the latest in 2007.

D. Mackay, W. Y. Shiu, and K. C. Ma. *Illustrated Handbook of Physical–Chemical Properties and Environmental Fate for Organic Chemicals.* Lewis Publishers, Chelsea, MI.

Volume I. *Mono-aromatic Hydrocarbons. Chloro-benzenes and PCBs.* 1991.

Volume II. *Polynuclear Aromatic Hydrocarbons, Polychlorinated Dioxins, and Dibenzofurans.* 1992.

Volume III. *Volatile Organic Chemicals.* 1992.

S. E. Jørgensen, H. Mahler, and B. Sørensen, 1997. *Handbook of Estimation Methods in Environmental Chemistry and Ecotoxicology.* Lewis Publishers, Chelsea, MI.

Steps 1–3 are sometimes denoted as effect assessment or effect analysis and steps 4–5 exposure assessment or effect analysis. Steps 1–6 may be called risk identification (Figure 15.2), whereas ERA encompasses all the nine steps presented in Figure 15.1. In particular, step 9 is very demanding, as several possible steps in reduction of the risk should be considered, including treatment methods, cleaner technology, and substitutes to replace the examined chemical.

In North America, Japan, and EU, medicinal products are considered in a similar manner to other chemical products as there is, in principle, no difference between a medicinal product and other chemical products. At present, technical directives for human medicinal products do not, in the EU, include any reference to ecotoxicology and the assessment of their potential risk. However, a detailed technical draft guideline issued in 1994 indicates that the approach applicable for veterinary medicine would also apply to human medicinal products. Presumably, ERA will be applied to all medicinal products in the near future when sufficient experience with veterinary medicinal products has been achieved. Veterinary medicinal products, on the other hand, are released in larger amounts to the environment as manure, for instance, which in spite of its possible content of veterinary medicine, is utilized as fertilizer on agricultural fields.

It is also possible to perform an ERA where the human population is in focus. It uses the non-adverse effect level (NAEL) and non-observed adverse effect level (NOAEL) to replace the predicted non-effect concentration and the PEC is replaced by the tolerable daily intake (TDI).

This type of ERA is of particular interest for veterinary medicine, which may contaminate food products for human consumption. For instance, the use of antibiotics in pig feed has attracted a lot of attention, as they may be found as residue in pig meat or may contaminate the environment through the application of manure as natural fertilizer.

Selection of a proper ecotoxicological model, which is a first initial step in the development of an environmental exposure model, will be discussed in more detail in the next section.

15.3 Characteristics and Structure of Ecotoxicological Models, in Comparison with Other Models Applied in Ecological and Environmental Management

Toxic substance models are most often biogeochemical models because they attempt to describe the mass flows of the considered toxic substances, although there are effect models of the population dynamics, which include the influence of toxic substances on birth rate and/or mortality, and therefore should be considered toxic substance models. Fugacity models are also applied, as will be mentioned in the next section.

Toxic substance models differ from other ecological models in that:

1. The need for parameters to cover all possible toxic substance models is great, and general estimation methods are therefore widely used.

2. The safety margin, assessment factors, should be high, for instance, expressed as the ratio between the predicted concentration and the concentration that gives undesired effects; see also Table 15.1

3. They require possible inclusion of an effect component, which relates the output concentration to its effect. It is easy to include an effect component in the model; it is, however, often a problem to find a well-examined relationship to base it on.

4. Because of items (1) and (2), simple models are needed. Our limited knowledge of process details, parameters, sublethal effects, and antagonistic and synergistic effects is limited.

These characteristic properties of ecotoxicological models imply that the normal procedure for the development of an ecological model is slightly changed. The following six points can be recommended as an applicable procedure:

1. Obtain the best possible knowledge about the possible processes of the toxic substances in the ecosystem.
2. Attempt to obtain parameters from the literature and/or from experiments (in situ or in the laboratory).
3. Estimate all parameters using available methods (i.e., QSAR methods, use of allometric principles, and use of mass balance equations; see, e.g., Jørgensen 2009).
4. Compare the results from (2) and (3) and attempt to explain discrepancies. Physical–chemical parameters are often determined with a standard deviation of 20–30% relatively, whereas biological parameters will often have a standard deviation of 50% and toxicological parameters even more than 100%.
5. Estimate which processes and state variables it would be feasible and relevant to include in the model. When in doubt at this stage, it is better to include too many processes and state variables rather than too few.
6. Use a sensitivity analysis to evaluate the significance of the individual processes, parameters, and state variables. This may often lead to further simplification or need for laboratory determinations of important parameters.

To summarize, ecotoxicological models differ from ecological models in general by:

1. Being most often more simple (due to the high assessment factor and our limited knowledge about processes and parameters)
2. Requiring more parameters (to cover all possible chemicals of ecotoxicological interest, including the more than 3000 pharmaceutical compounds)
3. A wider use of parameter estimation methods (because of our limited knowledge of many parameters that are needed to cover the entire spectrum of chemical compounds)
4. A possible inclusion of an effect component

Ecotoxicological models may be divided into five classes according to their focus. The five classes, which also illustrate the possibilities of simplification by including only the important processes, are:

A. Models that focus on nonliving components in the environment, for instance, soil, fresh water, groundwater, etc.

B. Models that focus on an organism or species

C. Models that focus on a population

D. Models that focus on an ecosystem (site-specific models)

E. Models that focus on regional or global levels, or a specific model is applied more generally

It is often beneficial to develop a model for a specific ecosystem based on good data, and then apply this very well developed model more generally for pollution of other ecosystems with other characteristics but still with most processes described by the same equations. It is, in most cases, easy to see how the characteristics of the ecosystem, for instance, quality of soil, slope, groundwater level, dominant plants, and so on, can be considered in the use of a more general model from one ecosystem to another ecosystem. The model presented in Section 15.5 to illustrate the details of a model of pharmaceuticals in the environment is a model developed for a specific case, namely, pollution by two antibiotics on basis of a comprehensive measuring program. Since it is not possible to cover the costs of such a comprehensive analytical program in all cases where these antibiotics are applied or where antibiotics are applied generally, it is assumed that the model—because of its transparency—can be applied generally by changing the parameters describing the physical–chemical properties of the antibiotics and the site-specific parameters and forcing functions.

The model can also be classified into two groups based on whether the effect of the pharmaceutical is included (indicated as EF) or not included (NE). The latter class is sometimes called a fat model to underline the fact that it gives "only" the fate of the chemicals in the environment.

Finally, we can divide ecotoxicological models including models of pharmaceuticals by the type of model applied. The following model types (see overview of model types in the Introduction section of Jørgensen 2009) have been applied for development of ecotoxicological models:

I. Biogeochemical models

II. Population dynamic models

III. Fugacity models, also denoted equilibrium criterion models (EQC)

IV. Steady-state models

V. Use of artificial neural networks (ANN)

VI. Individual-based models (IBM)

VII. Structurally dynamic models (SDM)

If we use all three classifications simultaneously, we obtain $5 \times 2 \times 7 = 70$ different classes. As models of pharmaceuticals in the environment have only found limited application, only a few of these 70 classes have been applied to assess the distribution and effect of pharmaceuticals in the environment. In the next section, a short overview of pharmaceutical models is given, and it will be indicated to which class the models belong.

15.4 Overview of Pharmaceutical Models

Ecosystems are open systems, which implies that all the chemicals we are using will sooner or later reach the environment. A model attempts to relate this "leakage" to the environment with the corresponding concentrations in various parts of the environment: soil, water, porewater, groundwater, sediment, air, specific species, our food, entire populations, and so on. The possible pathways for veterinary medicine are illustrated in Figure 15.3. A model should therefore contain a quantitative description of the veterinary medicine used and how much is transferred (lost) to the environment, at which rate it is exchanged among the compartments of the environment, and at which rate it is decomposed. As always, in modeling, we need to estimate which of the possible pathways we need to include in the model and are negligible.

The decomposition of pharmaceuticals is frequently an important process. In soil, a significant part of the chemical is often adsorbed firmly to the soil and is thus not decomposed bacteriologically. It can be expressed by the following decomposition equation (Krogh et al. 2008):

$$C(t) = C_0 \exp(-k/a(1 - \exp(-at))) \qquad (15.1)$$

which is denoted the availability first-order reaction. The decomposition can also in some cases be described by two first-order reactions, corresponding to a different decomposition rate for the adsorbed (or bound by other mechanisms) and the soluble fraction.

Fugacity models have been developed for pharmaceuticals in the environment both for ecosystems and for regions, both with and without effect. They cover, therefore, III D and III E models. Fugacity models can be developed on four dynamic levels. Level 1 represents the environmental situation, where a fixed quantity is discharged to the environment under steady-state and equilibrium conditions. If advection and chemical reactions are included

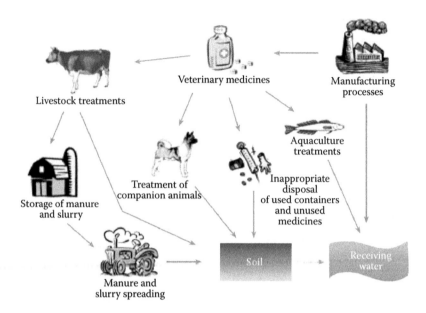

FIGURE 15.3
Pathways of veterinary medicines into the environment. The processes shown should be considered in models of veterinary medicine. (From Boxall et al., *Environ. Sci. Technol.*, 38 (19), 368A–375A, 2004. With permission.)

in the one or more phases, but equilibrium is still valid, we have level 2. Level 3 presumes steady state but no equilibrium between the phases. The transfer rates are proportional to the equilibrium difference between two differences. Level 4 is a dynamic version of level 3, where all the concentrations and emissions are changed over time. Levels 1–3 have been applied for pharmaceutical models, both site-specific and regional. Six main compartment are considered in most fugacity models: air, water, soil, sediment, suspended matter, and biota (e.g., fish). Further details about fugacity models can be found in the work of Mackay (1991) and Jørgensen and Bendoricchio (2001), and Jørgensen and Fath (2011). The following pharmaceuticals have been modeled by fugacity models: cyclophosphamide, diazepam, ivermectin (see Kümmerer 2001). Mackay et al. (2001) have developed a regional level 2 model for testosterone and progesterone, as illustrated in Figure 15.4. The distribution of these two hormones are roughly 3 times in water and 1 times in soil and sediment. Khan and Ongerth (2004) have developed a fugacity model describing the distribution of pharmaceuticals in a sewage treatment plant, using the following questions: How much is decomposed? Which concentration is found in the treated sewage water? And which concentration has the sludge? All the fugacity models mentioned here are NE models, but

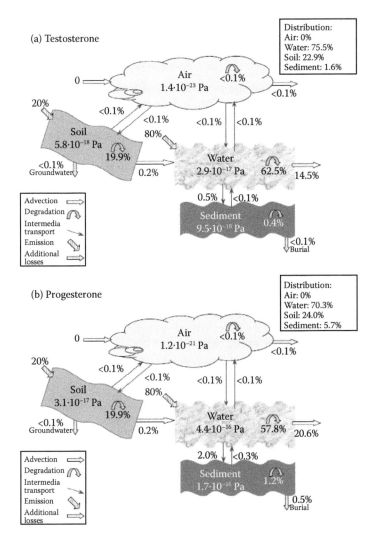

FIGURE 15.4
Fugacity model of the distribution of hormones in the environment. (From Kümmerer, K., *Pharmaceutical in the Environment*, Springer, Heidelberg, 2001. With permission.)

could easily be expanded to also include an effect by translating the found concentration into a relevant effect.

Montforts et al. (1999) and Spaepen et al. (1997) have developed an exposure model for oxytetracycline on grassland and how it contaminates the groundwater. The model is biogeochemical and a class I E model, but could easily be expanded to include the I F class. The model has several features in common with the model presented in detail in the next section. It is, furthermore, using principles from models of groundwater leaching of pesticides (see Linden and Boesten 1989).

Another biogeochemical model was developed by Jørgensen et al. (1998) for olaquindox and tylosine (see Figure 15.5). The model was supported by determinations in the laboratory of important parameters. The model is not calibrated on the basis of data from ecosystems, but the parameters have been determined with sufficient accuracy in the laboratory to allow a more general application of the model for the contamination of the two growth promoters in the environment.

The model has seven state variables: the growth promoters in soil, in soil water in the root zone, in deeper soil, in pore water in contact with the solid below the root zone, in crops, in surface water, and in groundwater. The model was used to find the highest concentrations of the two growth promoters in crops (at harvest), fish, surface water, and groundwater. It was found that the highest concentrations in these compartments were attained after about 20–30 days. The highest concentrations were in fish and surface water (several ppb), and in crops 30–70 ppb at harvest, whereas the concentrations in groundwater were very low. The model also showed that the biodegradability of the growth promoters is the most important and sensitive parameter, which should be determined with the highest possible accuracy to reduce the uncertainty of the model results.

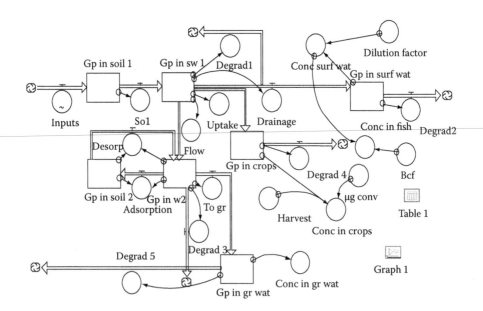

FIGURE 15.5
STELLA diagram of a model developed for environmental risk assessment of growth promoters. (Reprinted from Jørgensen et al., *Ecol. Modell.*, 107, 63–72. Copyright 1998, with permission from Elsevier.)

TABLE 15.2

Models of Pharmaceuticals in the Environment, an Overview

Components	Reference	Model Class
Cyclophosphamide, diazepam, ivermectin	Kümmerer (2001)	III, NE, D + E
Hormones	Mackay (2001)	III, NE, D + E
Pharmaceuticals in a sewage plant	Khan and Ongerth (2004)	III, NE, D
Oxytetracycline	Montforts et al. (1999)	I, EE, D
Oxytracycline	Spaepen et al. (1997)	I, EE, D
Growth promoters	Jørgensen et al. (1998)	I, E + NE, D
Tylosine, CTC	See next section	I, NE, D

The models of pharmaceuticals in the environment are summarized in Table 15.2. The focal components, the reference, and the model class are also given in the table.

The overview of the applied pharmaceutical models has shown that either fugacity models or biogeochemical models are developed. Fugacity models have a significantly higher standard deviation than biogeochemical models. However, development of a general applicable biogeochemical model requires, in most cases, that good data are available for one case study and that the most crucial parameters are determined for the pharmaceutical in the laboratory as a supplement to the in situ measurements that are used for calibration and validation of the examined case study. The forcing functions and the properties of the site (e.g., the soil properties) must be determined in each individual case. If another chemical (pharmaceutical) is applied in the next case study, it is of course necessary to determine the ecotoxicological properties of the chemical by laboratory examination to obtain the needed information on compound-specific parameters. If these guidelines are followed, it is possible by using a biogeochemical model to achieve reasonably accurate results of the prognoses. The next section shows a typical result, where the discrepancy between measured and modeled concentrations of an antibiotic typically varies from 5% to 15% (see Figures 15.7–15.8).

15.5 An Illustrative Example of a Pharmaceutical Model

The model focuses on contamination of an agricultural field by antibiotics. The model has been calibrated and validated for two antibiotics, tylosine and CTC, but it could in principle be applied for any toxic substance. The parameters, surprisingly, are not dependent on the toxic substance considered in the model. Such ecotoxicological properties as water solubility, K_{ow}, adsorption isotherms, and biodegradability are, of course, substance-dependent.

The model has seven state variables:

- Toxic substance in soil water, tsw (µg/L)
- Toxic substance in soil, tss (µg/kg)
- Metabolites of the toxic substance in soil water, msw (µg/L)
- Metabolites of the toxic substance in soil, mss (µg/kg)
- Water in the root zone, wrz (L/m³)
- Toxic substance concentration in plant, tpl (µg/kg)

The model has five forcing functions (note that capital letters are used for the symbols):

- Input of toxic substance to the soil water as a function of time, ISW (µg/L 24 h)
- Input of the metabolites to the soil water as a function of time, IMW (µg/kg 24 h)
- Precipitation, PRE (mm/24 h)
- Radiation, RAD (MJ/m² 24 h)
- Temperature, TEM (°C)

15.5.1 The Equations

The conceptual diagram is shown in Figure 15.6.

The number on the diagram and the abbreviations are used in the equations for the processes (note that the processes are indicated with three letters and the parameters are indicated with two letters):

1. ads = tsw * ac − tss
2. dsp = tss/ac − tsw
3. dcp = dr * tsw * $1.05^{(TEM-20)}$
4. adm = msw * ad − mss
5. dsm = mss/ad − msw
6. dcm = dm * mss * $1.05^{(TEM-20)}$
7. drt = tsw * dra
8. drm = msw * dra
9. eva = eps + epp
10. dra = if wrz >= wf * 400 then (0.1*(wrz − 400 * wf) else (sl * (400*wf − wrz)/(wf − wp)
11. epp = (fa * la * 0.931 * dl * RAD)/((dl + 66.7) * 2.47)

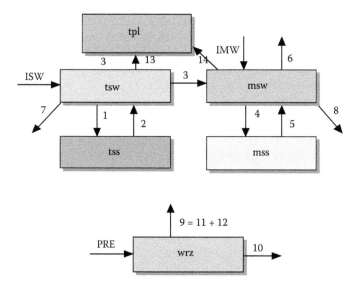

FIGURE 15.6
Conceptual diagram of the model. The processes are indicated with numbers; see below the equations of the processes, covering adsorption, desorption, decomposition, drainage with the drainage water, evaporation, and uptake by plants of the toxic substance and uptake by plants of the metabolites. Three of the five forcing functions: precipitation (PRE), input of the toxic substance (ISW), and input of metabolites of the toxic substances (IMW), are directly shown in the conceptual diagram.

12. eps = 0.7 ∗ dl ∗ RAD ∗ fa ∗ exp(−0.4 ∗ la)/((dl + 66.7)∗2.47)

13. put = tpl ∗ epg ∗ RAD ∗ tsw/(rz ∗ 4.1 ∗ (tsw + msw)

14. pum = tpl ∗ eg ∗ RAD ∗ msw/(rz ∗ 4.1 ∗ (tsw + msw)

The equations have several help—parameters that are calculated from the basic parameters or forcing functions:

dl = 40 + 3 ∗ TEM + 0.17 ∗ TEM²

fa = if wrz> = wf ∗ 400 then 1; else ((400 ∗ wf − wrz)/(400wf)

la = lm/(1+exp(1 − 0.01 ∗ TIME) (TIME is the time in days, which is used as the time unit). Note that it should be possible to select between the application of calculations for la by the shown equations or to give information about la as a function of time. Note that the time applicable for the shown equation is the growing season and if there are no plants, it is a matter of setting lm = 0.

wf = 0.46 ∗ cl + 0.305 ∗ si + 0.25 ∗ oc

wp = 0.33 ∗ cl + 0.12 ∗ si + 1.6 ∗ oc

wxs = 0.69 ∗ cl + 0.55 ∗ si + 4.28 ∗ oc

The basic parameters are shown in Table 15.3, together with the possible ranges and the recommended default values, which are taken from the application of the model on tylosine. These parameter values have been determined by calibration and validation of the model using observed data.

The differential equations (as coded in C++):

$$tsw = isw + tsw - ads + dsp - dcp - drt$$
$$tss = iss + tss - dsp + ads$$
$$msw = msw - dcp - drm - adm + dsm$$
$$mss = mss + adm - dsm$$
$$wrz = wrz + PRE - eva - dra$$

The toxic substance concentration in the plants, tpl, is found by the use of a partition coefficient The following equation is valid:

$$tpl = (tsw + msw) * (pw + 1.22 * pl * K_{ow}^b) * (dp/dw)$$

TABLE 15.3

Basic Parameters

Parameter	Symbol Used	Unit	Range	Default Value
Slope	sl	–	0.00–0.1	0
Clay	cl	–	0.001–0.2	0.113
Silt	si	–	0.001–0.2	0.107
Organic C	c	–	0.000–0.2	0.016
Adsorption coefficient	ac	kg/L	0–10,000	128
Adsorption coefficient for metabolites	ad	kg/L	0–10,000	128
Decomposition rate	dr	1/24 h	0–100	0.35
Decomposition rate for metabolites	dm	1/24 h	0–100	0.20
Water cont. in plants	pw	kg/kg	0–1.00	0.3 (plant-dependent)
Lipid cont. in plants	pl	kg/kg	0–100	0.1 (plant-dependent)
Octanol–water coefficient	K_{ow}	–	0–10^6	1000[a]
Eff. plant growth	eg	kg/MJ	0.001–0.1	0.025
Density plants	dpl	kg/L	0.9–1.5	1.000
Density water	dwa	kg/L	0.95–1.05	0.998 (at 20°C)
Exponent	b	–	0.5–1.0	0.80 (plant-dependent)
Root zone depth	rz	m	0.2–1.0	0.5

[a] Strongly dependent on the toxic substance. For estimation methods, see help file "Parameter estimation methods."

Usually, it is not possible to use observed tsw and tss separately and msw and mss separately. Therefore, two additional variables are introduced:

tts = tsw + tss and
tms = msw + mss

A software has been developed for the presented model. The software is using the following screen images:

1. The Model (GEM). The user should indicate the following information on the second screen image page: (a) the name of the toxic substance, (b) the initial value for the five state variables, (c) the parameter values.

   For the case of tylosine, the name of the toxic substance is tylosine and the following initial values are applied tsw = 0.3, tss = 29.4, msw = 0.3, mss = 29.4, and wrz = 5, and the parameter values are the default values in Table 15.3.

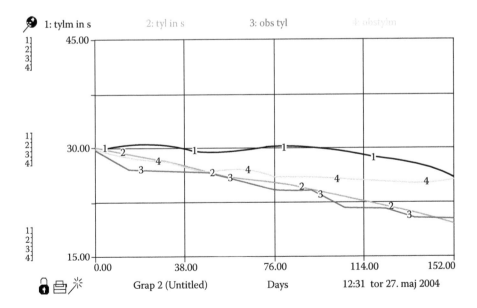

FIGURE 15.7
The model has been applied to follow tylosine in soil (parameters from Table 15.3 are valid). Tylosine in soil, modeled (tyl in s), can be compared with the observed values of tylosine in soil (obs. Tyl). The metabolites of tylosine in soil modeled (tylm in s) are shown together with the observed values (obstylm).

2. The third screen image is devoted to the forcing functions and the observations as a function of TIME. It means that the table should contain the following columns:

TIME (the days with observations), PRE (precipitation (mm/24)), TEM (temperature), RAD (radiation), ISW (input of the toxic substance to the soil water), IMW (input of the metabolites to the soil), tts-observed, tms-observed, and tpl-observed.

The fourth screen image focuses on the simulation. The integration step is selected as the number of integration cycles per time unit (days). On a graph, the simulation result of the six state variables, including tpl, +tts, and tms, can be shown. For the last two variables, the observed values can—if it has been selected—also be shown on the graphs of tts and tms. The simulated values are all in red and the observed values of tts and tms are in blue.

A table of the simulation results for the six state variables, including tpl, plus tts, and tms, is shown on the fifth screen image. All concentrations are expressed in µg/kg or liter.

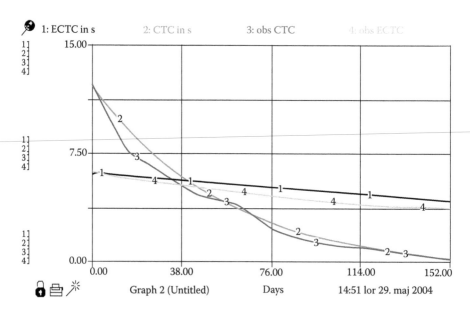

FIGURE 15.8
Modeled values of CTC in soil (CTC in s) are compared with the observed values (obs CTC). A similar comparison is made for metabolites of CTC: modeled (ECTC) and observed (obs ECTC).

The presented model has been applied for tylosine and CTC. Manure applied on agricultural fields may contain the two antibiotics, and the concentrations of the antibiotics and their metabolites as a function of time were observed in two cases. The observed values were compared with the modeled results, and the comparisons are shown in Figures 15.7 and 15.8 for tylosine and CTC, respectively. The figures show a fully acceptable accordance between modeled and observed values. The concentrations in plants were not observed, and it is therefore not possible to validate the modeled concentrations of the two antibiotics in plants. The CTC case applied the parameters corresponding to the characteristic properties of this antibiotic and to the soil where the manure with CTC was applied. Moreover, the forcing functions were determined for this case study and applied in the model for CTC.

References

Bartell, S. M., R. H. Gardner, and R. V. O'Neill. 1992. *Ecological Risk Estimation*. Boca Raton, FL: Lewis Publishers.

Boxall, A. B. A., C. J. Sinclair, K. Fenner, D. Kolpin, and S. J. Maud. 2004. When synthetic chemicals degrade in the environment. *Environmental Science & Technology* 38 (19): 368A–375A.

Cairns, Jr., J., K. L. Dickson, and A. W. Maki. 1987. *Estimating Hazards of Chemicals to Aquatic Life*. Philadelphia, PA: STP. 675, American Society for Testing and Materials.

Jørgensen, S. E. 2009. *Ecological Modelling: An Introduction*, 190 pp. WIT, Southampton.

Jørgensen, S. E., H. C. Lützhøft, and B. Halling Sørensen. 1998. Development of a model for environmental risk assessment of growth promoters. *Ecological Modelling* 107: 63–72.

Jørgensen, S. E., and G. Bendoricchio. 2001. *Fundamentals of Ecological Modelling*, 3rd ed., 628 pp. Amsterdam: Elsevier.

Jørgensen, S. E., and B. Fath. 2011. *Fundamentals of Ecological Modelling*, 4th ed., 400 pp. Amsterdam: Application in Environmental Management and Research. Elsevier.

Khan, S. J., and J. E. Ongerth. 2004. Modelling of pharmaceutical residues in Australian sewage by quantities of use and fugacity calculations. *Chemosphere* 54 (3): 355–367.

Krogh, K. A., T. Soeborg, B. Brodin, and B. Halling-Sorensen. 2008. Sorption and mobility of ivermectin in different soils. *Journal of Environmental Quality* 37 (6): 2202–2211.

Kümmerer, K. 2001. *Pharmaceutical in the Environment*, 265 pp. Heidelberg: Springer.

Linden, A. van der, and J. J. Boesten. 1989. Calculations of leakage and accumulation of substances. RIVM- report # 7288000003 (in Dutch).

Mackay, D. 1991. *Multimedia Environmental Models. The Fugacity Approach*, 258 pp. Chelsea, MI: Lewis Publishers.

Mackay, D., L. S. McCarty, and M. MacLeod. 2001. On the validity of classifying chemicals for persistence, bioaccumulation, toxicity, and potential for longrange transport. *Environmental Toxicology and Chemistry* 20 (7): 1491–1498.

Montforts, M. et al. 1999. The exposure assessment for veterinary medicinal products. *Science of the Total Environment* 225: 119–133.

Spaepen, D. A. et al. 1997. A uniform procedure to estimate the predicted environmental concentration of the residues of veterinary medicines in soil. *Environmental Toxicology and Chemistry* 16: 1977–1982.

16

Models of Global Warming and the Impacts of Climate Changes on Ecosystems

Sven Erik Jørgensen

CONTENTS

16.1 Introduction 439
16.2 Brief Overview of Global Warming Models 440
16.3 A Global Average Model Presented by Use of the Software
     STELLA 447
16.4 Models of Global Warming Consequences for Nature 454
References 455

16.1 Introduction

Hundreds of models have been developed to describe and predict global warming. Several mathematical models have been constructed to describe the features of the climate and predict how these features may be changed because of the changes in atmospheric carbon dioxide. No models offers a complete and exact description or prediction, but appreciation of the strengths, weaknesses, and assumptions of the different models will lead to a deeper understanding of the carbon dioxide problem and to what extent we can rely on model scenarios.

We can distinguish between global warming models—those that consider the Earth as one global system, for which an annual average temperature is predicted [let us denote these models Global Average Models (GAM)], and models that are able to make predictions of the seasonal climate changes (temperature, precipitation, and humidity) regionally. The first type of models are of course less complex, although they may entail many important feedback mechanisms. These are often developed by researchers with a physical–chemical–ecological background. The latter type is often based on meteorological models and will often require huge computer power. They are most often developed by meteorological researchers at meteorological institutes. Let us therefore call these models Meteorological Models of the Green House Effect (MGHE).

A short overview of global warming models will be presented in the next section to give the reader an idea about the spectrum of available models—from the simplest to the most complex. MGHEs are included in this overview, but only very few characteristics of these models will be mentioned as they are very complex. Presentation of one MGHE in all details could hardly be made in this handbook, even if we would devote all the pages to such a model.

The first models of the GAM type were developed in the 1970s. The author of this chapter has already developed such a model in 1976 (see Jørgensen and Mejer 1976a, 1976b), which gave predictions that are very much in accordance with the global warming model results reported today. It contains many interacting processes with feedbacks. A new version of this model is presented in Section 16.3. It contains even a few more, interacting processes, but nevertheless gives values that are close to the same results. Before the presentation of the updated version, which also in one of the versions applied today includes the greenhouse effect of methane, the main results from the early model in 1976 will be presented. It will furthermore be pointed out which new interacting processes have been included in the new versions, and why it was found necessary to include them. The model is of course only one out of many, but it illustrates the typical considerations involved in GAMs.

The last section of this chapter presents the most characteristic properties of several recently developed models of impact of climate changes on ecosystems. They represent a new challenge for ecological modeling that several modelers have met. The core question is: How can we model the response of an ecosystem to increasing temperature and changed precipitation pattern? Which processes are mostly touched by these changes, and how will the entire ecosystem react to these changes? Can we make predictions of the reactions on the ecosystem level?

16.2 Brief Overview of Global Warming Models

Water vapor and carbon dioxide are the strongest absorbers of thermal infrared radiation in the atmosphere, but methane and nitrogen dioxide are also absorbers of infrared radiation. Water vapor concentration varies according to temperature and weather, whereas carbon dioxide is almost uniformly distributed in the atmosphere to an altitude of about 100 km. The carbon dioxide concentration has increased from about 280 molecules of carbon dioxide in year 1900 to about 390 per million molecules of air in the atmosphere today, representing an increase of 39% in 110 years. A good average scenario predicts a concentration of 600 ppm (volume basis or as number of molecules relatively to molecules of air) by the end of this century.

Methane and nitrogen oxide and dioxide are emitted from wetlands including rice fields (see, e.g., Ramanathan, Lian, and Cess 1979). They are several times better absorbers of infrared radiation than carbon dioxide, but fortunately emission of these greenhouse gases is much smaller and they have a much shorter half-life in the atmosphere.

The idea behind GAMs is the Stefan–Boltzmann equation:

$$F = \sigma T^4 \tag{16.1}$$

where σ is the Stefan–Boltzmann constant. We receive on average a solar radiation corresponding to 342 W/m², of which 30% or 102 W/m² is reflected—or expressed differently, the albedo is 35%. This means that, of the total, 240 W/m² is absorbed by Earth and its atmosphere. If the temperature remains constant, the same amount of energy should be transferred to outer space by infrared radiation. F is therefore 240 W/m² and the corresponding temperature is 255 K, which is the temperature at 6 km above Earth's surface. The global emission of thermal infrared radiation can be visualized as equivalent to the radiation that would be emitted by a black shell at the temperature 255 K. The temperature at Earth's surface corresponding to the temperature at 6 km above the surface, however, is 287.5 K. By differentiation of Equation 16.1, we obtain:

$$dF/dT = 4\sigma T^3 \tag{16.2}$$

which implies that:

$$\Delta T = \Delta F/\lambda \tag{16.3}$$

where

$$\lambda = 4\sigma T^3 = \text{approximately 4 W/(m}^2\text{ K)} \tag{16.4}$$

We can calculate the temperature increase if we know ΔF, which can be determined if we know the trapping of infrared radiation by various constituents of the atmosphere. We know the relationship between the concentration of the actual constituents—ozone, carbon dioxide, water, clouds, methane, and nitrogen oxides—and their ability to absorb infrared radiation.

This implies that we have the means to model global warming by following these modeling steps:

1. Model the carbon dioxide in the atmosphere, determined by the processes of global carbon cycle, including carbon dioxide exchange between the atmosphere and the hydrosphere, photosynthesis, respiration of organisms including microbiological respiration as a result of the decomposition of detritus. A good submodel for the

uptake of carbon dioxide by the hydrosphere is compulsory because the uptake by oceans is more than 50% of the entire carbon dioxide emission. It is necessary to consider in the model many oceanic layers, because the diffusion of carbon dioxide from the surface to the deep sea requires a very long time—several years.

2. Model the irradiative effects of increased carbon dioxide. The physical basis for this part of the model is known.
3. Model the influence of the changes in climate parameters.

Back in the 1970s, a number of models of climate changes as a consequence of emissions of greenhouse gases, mainly carbon dioxide, have already been published (see, e.g., Manabe and Wetherald 1975). The different models vary by the number of details, feedbacks, and processes that are included, but the influence of one process or feedback mechanism more or less when the model already contains many details is, of course, minor. Therefore, the different models yielded different results, although the differences between most model results are not very significant when they are based on the same assumption about the following factors (forcing functions):

1. Population growth
2. Economic development
3. Energy use
4. Efficiency of energy use
5. Use of various energy technologies

This means that the results are robust, and it is feasible to base political decisions on the models. It is widely accepted that our decisions on these five factors are determining the development of the climate, and that the model results give approximately accurate predictions corresponding to these decisions.

The models used today by the Intergovernmental Panel on Climate Change (IPCC) are of course improved versions of the previous global models, but GAMs have not been improved significantly, whereas MGHEs have been radically improved, because we have today the computing power needed for these types of models. The results of GAMs adopted by the IPCC and the state of the art of GAMs today are shown in Figure 16.1. As seen in the figure, the different models do not produce very significant results, whereas the differences are more pronounced for scenarios based on different developments of the five factors cited above. The figure shows the results from three climate model scenarios used by the IPCC, with predictions starting in 1990 and ending in the year 2100. In all three, the global population rate rises during the first half of the century, then declines, which seems to be a realistic scenario. The following different developments are foreseen:

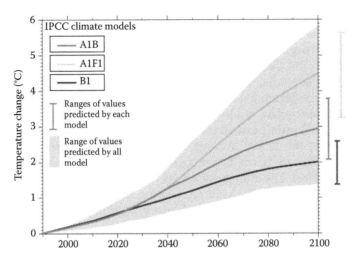

Adapted from IPCC, Third Assessment Report on Climate Change, 2001.
Global temperature increases predicted by three different IPCC climate models. With permission

FIGURE 16.1
Results of several GAMs for three different scenarios (for details of the scenarios, see text). The prognoses of the temperature increase are not very different for the different models (the range is about 1°C for B1, 1.5°C for A1B, and 2°C for A1F1), whereas the different scenarios give from 1.5°C to 5.5°C in temperature increase.

- The A1B model assumes rapid economic growth and increased equity—the reduction of regional differences in per-person income. New and more efficient technologies are introduced, without relying heavily on a single energy source.

- The A1F1 model is the same as A1B, but assumes the continued use of fossil fuel-intensive technologies.

- In the B1 model, the world moves rapidly from a producer–consumer economy toward a service and information economy. There is a reduction in the use of raw materials, and an emphasis on clean and efficient technologies and improved equity.

Other models have been developed, each based on a different set of assumptions.

Google presents a very rich overview of the global model literature including a review of the differences among the many models and the criticism that have been published about the various models. See, in this context, "Limitations of integrated assessment models of climate change," by Ackerman (2009). Further details about GAMs are therefore not given in this handbook. The model presented in the next section represents the GAMs in the sense that the model results are very similar to those of other GAMs, and that the number of details, processes, and feedbacks is not very different from most other GAMs. It contains some details that cannot be found in most other GAMs (e.g., the carbon cycle is interacting with the nitrogen and phosphorus cycle), but there also are some processes in other GAMs that are not included in the model. This, of course, does not preclude a discussion regarding which model among the wide spectrum of models that are offered would be most appropriate in a given context.

Climate change within particular regions is usually of much more practical interest than the global average temperature change. Hence, three-dimensional models are needed to recognize the detailed consequences of the greenhouse effect. The climate system is characterized by redistribution of thermal energy from locations of net heating to locations of net cooling. The redistribution is both vertical and horizontal. It implies that complete models for climate changes must consider properly the horizontal transport processes, at least in a summarized form. The complexity of MGHEs is presented in Figure 16.2, where the schematic structure of an atmospheric general circulation model is shown.

The results of using MGHEs compared with GAMs are, however, very important. MGHEs predict very roughly that the increase in temperature in the high latitudes (mainly the polar regions) is higher than that in low latitudes (the tropical regions). They forecast faster melting of ice in the polar regions, thereby changing the albedo more than that expected by GAMs (see Figure 16.3). This implies that the increase in the annual average temperature

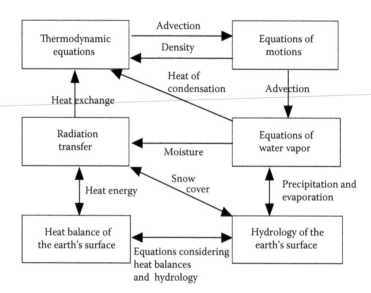

FIGURE 16.2
Schematic structure of an MGHE. The six boxes show six models that are linked as indicated by arrows.

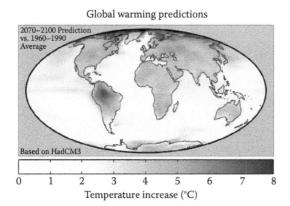

Global warming predictions

Temperature increase (°C)

FIGURE 16.3
Distribution of global temperature changes predicted by MGHEs applied by IPCC.

is slightly too low. Figure 16.3 shows the global distribution of the regional temperature changes—a result of the MGHEs applied by IPCC.

The temperature changes also imply that the water vapor in the atmosphere is changed. It is possible via MGHEs to make prognoses for the regional changes in precipitation, which of course is a source of enormous practical interest.

Earlier, we have mentioned that in global warming models, we integrate the carbon cycle and the climate changes, mainly the temperature changes. By using MGHEs, we are able to obtain the regional climatic changes. The models that we have mentioned can give the consequences of climate changes attributed to political decisions about our energy consumption, which is determined by the amount of energy we are consuming (per capita in average), the growth of the population, the efficiency with which we are using the energy, and the form of energy applied (fossil fuel, renewable energy, nuclear energy, etc.). See also the five factors listed above. However, to have a comprehensive overview of all the consequences of these political decisions, we need to know the consequences of changed climate. Below, we list six of the most important of these consequences:

1. Melting of the polar ice, reducing the albedo, and thereby increasing the temperature
2. Melting of the permafrost, reducing the albedo, and increasing the methane emission, but also increasing the photosynthesis of the tundra
3. Rising of the sea due to ice and snow melting, whereby the land areas and its photosynthesis will be reduced
4. Formation of deserts due to reduced precipitation in certain regions. It would imply reduced photosynthesis, but increased albedo

5. Flooding, which implies reduced photosynthesis on the flooded land

6. Increased number of hurricanes of level 5 and above

Figure 16.4 shows how models of these six consequences could be linked to global warming models (either GAM or MGHE), consisting of a description of the carbon cycle and the climate changes due to an unbalanced carbon cycle, and with the five factors listed above as determining forcing functions. The figure also indicates the five additional models covering the consequences (1)–(5), to what extent they cause changes—give feedbacks—to the carbon cycle model, and the climate change model. We need the entire complex of models to be able to give recommendations on the best possible political decision. The GAM, which is presented in detail in the next section, contains the feedbacks from mechanisms (1), (2), and (3). It would not be too difficult to also include (5), at least semiquantitatively, but (4) would require a proper MGHE, because desert formation is a consequence of regional climate changes, and (6) is probably not yet possible to include. It would probably require a more comprehensive knowledge about the formation of hurricanes than we have today, and it would be impossible to predict where future hurricanes would destroy towns and ecosystems.

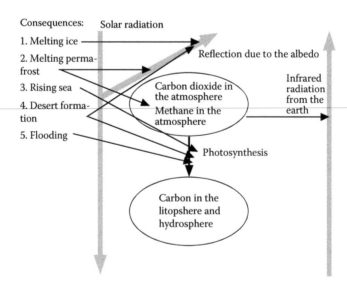

FIGURE 16.4
Five consequences of global warming. They all give changes in global C-cycle models and global climate change models. Therefore, it is necessary to develop models of these consequences and integrate them with the global C-cycle and climate change models.

16.3 A Global Average Model Presented by Use of the Software STELLA

A global warming model that can predict the increase in global average temperature is presented in this section to illustrate an ecological modeling approach for this central environmental problem. The model is developed by using STELLA, and the diagram is shown in Figure 16.5. It is characterized by including partially the nitrogen and phosphorus cycle, as these two elements are nutrients that influence photosynthesis. The model equations using the STELLA format are shown Table 16.1. The model is a further development of

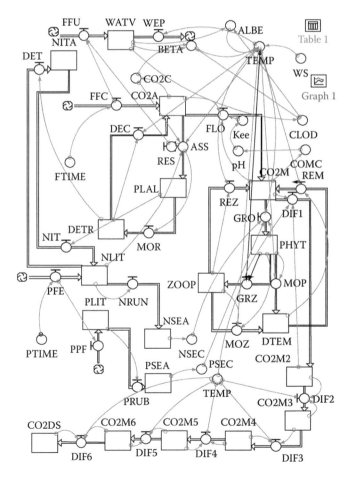

FIGURE 16.5
The global warming model presented in Section 16.3.

TABLE 16.1

The Global Warming Model: Equations in the STELLA Format

CO2A(t) = CO2A(t - dt) + (FFC + DEC + RES - ASS - FLO) * dt
INIT CO2A = 775

INFLOWS:
FFC = 6*FTIME*1.02^TIME
DEC = DETR*66*1.05^(TEMP-15.2)/1500
RES = 0.5*ASS
OUTFLOWS:
ASS = (120*13*CO2C)/(7*(CO2C+300))*(NLIT/865)*(1.05^(TEMP-15.2))*(1-0.00004*(TEMP-15.71))
FLO = 0.15*CO2M*(((CO2A-775)/CO2A)/(Keel))
CO2DS(t) = CO2DS(t - dt) + (DIF6) * dt
INIT CO2DS = 92094

INFLOWS:
DIF6 = 3987*((CO2M6/125-CO2DS/2800)/(2800))*(TEMP/15.2)^0.5
CO2M(t) = CO2M(t - dt) + (FLO + REZ + REM - GRO - DIF1) * dt
INIT CO2M = 2466.8

INFLOWS:
FLO = 0.12*CO2M*(((CO2A-775)/CO2A)/(Keel))
REZ = ZOOP*10*1.05^(TEMP-15.2)
REM = 0.02*DTEM*1.05^(TEMP-15.2)
OUTFLOWS:
GRO = PHYT*(1-PHYT/20)*1.05^(TEMP-15.2)*15.3*MAX(NSEC/(NSEC+2*10^(-7)),PSEC/(PSEC+3.5*10^(-8)))
DIF1 = (3987*(CO2M/75-CO2M2/125)/75)*(TEMP/15.2)^0.5
CO2M2(t) = CO2M2(t - dt) + (DIF1 - DIF2) * dt
INIT CO2M2 = 4110

INFLOWS:
DIF1 = (3987*(CO2M/75-CO2M2/125)/75)*(TEMP/15.2)^0.5
OUTFLOWS:
DIF2 = (3987*(CO2M2-CO2M3)/(125*125))*(TEMP/15.2)^0.5
CO2M3(t) = CO2M3(t - dt) + (DIF2 - DIF3) * dt
INIT CO2M3 = 4110

INFLOWS:
DIF2 = (3987*(CO2M2-CO2M3)/(125*125))*(TEMP/15.2)^0.5
OUTFLOWS:
DIF3 = 3987*((CO2M3-CO2M4)/(125*125))*(TEMP/15.2)^0.5
CO2M4(t) = CO2M4(t - dt) + (DIF3 - DIF4) * dt
INIT CO2M4 = 4110

INFLOWS:
DIF3 = 3987*((CO2M3-CO2M4)/(125*125))*(TEMP/15.2)^0.5
OUTFLOWS:
DIF4 = 3987*((CO2M4-CO2M5)/(125*125))*(TEMP/15.2)^0.5
CO2M5(t) = CO2M5(t - dt) + (DIF4 - DIF5) * dt
INIT CO2M5 = 4110

INFLOWS:
DIF4 = 3987*((CO2M4-CO2M5)/(125*125))*(TEMP/15.2)^0.5

(continued)

TABLE 16.1 (Continued)

The Global Warming Model: Equations in the STELLA Format

OUTFLOWS:
DIF5 = 3987*((CO2M5-CO2M6)/(125*125))*(TEMP/15.2)^0.5
CO2M6(t) = CO2M6(t - dt) + (DIF5 - DIF6) * dt
INIT CO2M6 = 4110

INFLOWS:
DIF5 = 3987*((CO2M5-CO2M6)/(125*125))*(TEMP/15.2)^0.5
OUTFLOWS:
DIF6 = 3987*((CO2M6/125-CO2DS/2800)/(2800))*(TEMP/15.2)^0.5
DETR(t) = DETR(t - dt) + (MOR - DEC) * dt
INIT DETR = 1500

INFLOWS:
MOR = 0.49*ASS
OUTFLOWS:
DEC = DETR*66*1.05^(TEMP-15.2)/1500
DTEM(t) = DTEM(t - dt) + (MOP + MOZ - REM) * dt
INIT DTEM = 3000

INFLOWS:
MOP = PHYT*4.5*1.05^(TEMP-15.2)
MOZ = (4.8*ZOOP+0.2222*ZOOP*PHYT)*1.05^(TEMP-15.2)
OUTFLOWS:
REM = 0.02*DTEM*1.05^(TEMP-15.2)
NITA(t) = NITA(t - dt) + (DET - NIT) * dt
INIT NITA = 3800000

INFLOWS:
DET = 0.000063*DETR
OUTFLOWS:
NIT = PLAL*0.00016
NLIT(t) = NLIT(t - dt) + (PFE + NIT - DET - NRUN) * dt
INIT NLIT = 852

INFLOWS:
PFE = 0.03*PTIME^TIME
NIT = PLAL*0.00016
OUTFLOWS:
DET = 0.000063*DETR
NRUN = NLIT*0.000214
NSEA(t) = NSEA(t - dt) + (NRUN) * dt
INIT NSEA = 1904

INFLOWS:
NRUN = NLIT*0.000214
PHYT(t) = PHYT(t - dt) + (GRO - GRZ - MOP) * dt
INIT PHYT = 5
INFLOWS:
GRO = PHYT*(1-PHYT/20)*1.05^(TEMP-15.2)*15.3*MAX(NSEC/(NSEC+2*10^(-7)),PSEC/
  (PSEC+3.5*10^(-8)))
OUTFLOWS:
GRZ =

(continued)

TABLE 16.1 (Continued)

The Global Warming Model: Equations in the STELLA Format

PHYT*16*TEMP*1.05^(TEMP-15.2)

MOP = PHYT*4.5*1.05^(TEMP-15.2)

PLAL(t) = PLAL(t - dt) + (ASS - MOR - RES) * dt

INIT PLAL = 560

INFLOWS:

ASS = (120*13*CO2C)/(7*(CO2C+300))*(NLIT/865)*(1.05^(TEMP-15.2))*(1-0.00004*(TEMP-15.71))

OUTFLOWS:

MOR = 0.49*ASS

RES = 0.5*ASS

PLIT(t) = PLIT(t - dt) + (PPF - PRUB) * dt

INIT PLIT = 60

INFLOWS:

PPF = PFE*0.1

OUTFLOWS:

PRUB = PLIT*0.00005

PSEA(t) = PSEA(t - dt) + (PRUB) * dt

INIT PSEA = 129.5

INFLOWS:

PRUB = PLIT*0.00005

WATV(t) = WATV(t - dt) + (FFU - WEP) * dt

INIT WATV = 67580

INFLOWS:

FFU = 0.63*FFC+RES

OUTFLOWS:

WEP = 0.0547*(TEMP-15.2)

ZOOP(t) = ZOOP(t - dt) + (GRZ - REZ - MOZ) * dt

INIT ZOOP = 1

INFLOWS:

GRZ =

PHYT*16*TEMP*1.05^(TEMP-15.2)

OUTFLOWS:

REZ = ZOOP*10*1.05^(TEMP-15.2)

MOZ = (4.8*ZOOP+0.2222*ZOOP*PHYT)*1.05^(TEMP-15.2)

ALBE = 0.301+2.1*10^(-6)*(WATV-67580)+1.0*(CLOD-0.5)

BETA = 0.399+118*10^(-6)*(CO2C-350)+0.563*(CLOD-0.5)+2.73*10^(-6)*(WATV-67580)

CLOD = WATV*0.5/(67580)

CO2C = CO2A*29/(12*5.35)

COMC = CO2M*3800/(75*1.36*10^3)

FTIME = 1

Keel = 10+(8.4-pH)*0.7

NSEC = NSEA/1.36*10^9

pH = 8.4 -1.2* COMC/919

PSEC = PSEA/1.36*10^9

PTIME = 1.05

TEMP = (WS*(1-ALBE)/(28840*(1-BETA)))^0.25-273.3

WS = 1.73*10^14

the model developed by Jørgensen and Mejer (1976a, 1976b), and is modified slightly from a similar model presented by Jørgensen and Fath (2011).

The model has the following *state variables*:

Water in the atmosphere (WATV)
Nitrogen in the atmosphere (NITA)
Nitrogen in the lithosphere (NLIT)
Nitrogen in the hydrosphere (NSEA)
Phosphorus in the lithosphere (PLIT)
Phosphorus in the hydrosphere (PSEA)
Carbon as carbon dioxide in the atmosphere (CO2A)
Carbon in plant biomass in the lithosphere (PLAL)
Carbon in detritus in the lithosphere (DETR)
Carbon in the upper layer of the sea (CO2M)
Carbon in phytoplankton in the sea (PHYT)
Carbon in the zooplankton in the sea (ZOOP)
Carbon in detritus in the sea (DTEM)
Carbon in six deeper layers of the sea (CO2M2, CO2M3, . . . , CO2M6, CO2DS)

The *forcing functions* are:

Use of fossil fuel (FFC)
Use of nitrogen fertilizer (PFE)
Use of phosphorus fertilizer (PPF)

Figure 16.6 shows the model results corresponding to a 2% increase in the use of fossil fuels since time = 0, which represents 1990 as the reference year. The starting average global temperature is 15.71°C, and the model simulation shows the temperature increasing to 20.16°C, corresponding to an increase in the global average temperature of about 4.5°C. The temperature in 2010 is 16.32°C or 0.62°C higher than in 1990. This value corresponds very closely to the recorded increase in global average temperature during the past 20 years. The carbon dioxide concentration has increased from about 350 ppm in 1990 to about 390 ppm million in 2010, which also matches the measured carbon dioxide concentration increase in the atmosphere. The model simulation projects that carbon dioxide concentration will be about 720 ppm in 2100. If the temperature increase is to be held at only 2°C during this century, as recommended by the climate panel of the United Nations, then it is necessary to phase out fossil fuel use during the next 30–50 years. A prediction based on a continuous 2% annual increase of fossil fuel use during the next

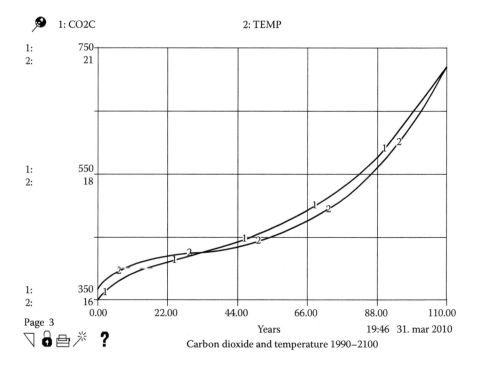

FIGURE 16.6
The model simulation showing atmospheric carbon dioxide concentration and the average global temperature from 1990 to 2100.

10 years, followed by a linear decrease to total cessation of use of fossil fuel by year 2060 shows a temperature increase to 18.1°C. This scenario results in a 1.78°C increase compared with today and 2.4°C higher than that in 1990. It should be realistic and feasible. However, if the gradual phasing out of fossil fuel does not start before year 2030 and it will take the rest of the century to year 2100 before it reaches cessation of use of fossil fuel, the results will be about 3.0°C higher than today later in this century (and before year 2100) and 3.6°C higher than that in 1990—hardly an acceptable scenario. The conclusion is therefore based on the shown model—that from 2020 to 2060, we have 40 years to shift to renewable energy. This leaves us the next 10 years to get prepared for the shift and to start (in the most advanced countries) reducing our use of fossil fuel.

The model has the following features and includes the following processes:

1. The global cycling of N, P, and C is included because nitrogen and phosphorus cycles interact with the carbon cycle due to the need for fertilizer in primary production.

2. Carbon dioxide diffusion in the sea is described by a multilayer model (seven layers).

3. The ability of the oceans to take up carbon dioxide is a function of pH, which is dependent on the carbon dioxide concentration relative to the concentrations of hydrogen carbonate and carbonate. These relationships are considered in the model.

4. The ability of the oceans to take up carbon dioxide is a function of temperature and in accordance with the solubility of carbon dioxide as a function of temperature.

5. Increased photosynthesis by increased carbon dioxide concentration according to a Michaelis–Menten expression; see the equations in Table 16.1

6. Water content in the atmosphere changes when the temperature changes, and atmospheric water is also a greenhouse gas with a known effect.

7. The cloudiness changes when the water content in the atmopshere changes, and this also affects the albedo; see the equations for ALBE and CLOD (Table 16.1).

8. A deforestation, 100,000 km²/year according to FAO, is included.

9. The change in albedo due to decreased ice coverage is estimated and included.

10. The change in primary production in the sea, in wetlands, and in forests due to changed temperature and due to the increased carbon dioxide concentration (see also item (5)) is considered.

11. Permafrost melting for tundra regions is considered. It will decrease the albedo and increase primary production.

12. The primary production is reduced according to the flooding. It is estimated by IPCC that the sea level will rise by about 1 m from 1990 to 2100, which would correspond to an increase of 2.5°C. Geographically, it is possible to calculate that these estimations correspond approximately to an annual decrease in the photosynthesis equal to the observed (modeled) increase of the temperature = (present temperature – 15.71) × 0.00004.

The results of the model should not be considered as accurate prognoses, but should be compared with other model results and used as a first approximate indicator for various political decisions about the energy policy. The temperatures for the three mentioned scenarios are the maximum temperature. The carbon dioxide will decrease as soon as the emission from the use of fossil fuel stops, because the carbon dioxide continues to diffuse to the deep sea. The uncertainty of the prognoses are, however, rooted in the political decisions. They are hardly possible to predict, because they are not necessarily rational.

16.4 Models of Global Warming Consequences for Nature

It is, of course, possible for models of ecosystems to change the climatic forcing functions according to the changes that are determined by use of global warming models presented in this chapter. We know, however, that organisms are adaptable and that the species' composition may change when the conditions (forcing functions) are changed. Examples of structurally dynamic models (SDMs) and how to use them to account for adaptation and shifts in species' composition are presented in Sections 2.7, 12.4, and 14.4. It cannot be ruled out that some of the models mentioned below could yield even more accurate predictions by using the SDM approach.

Zlatev and Moseholm (2008) have used a Eulerian model for Denmark, which they call UNI-DEM, to assess the influence of climatic changes on air pollution. They find that the concentrations of the major pollutants do not depend very much on climatic changes, but that quantities related to high ozone levels may increase significantly. The changes in biogenic emissions are the core issue here. The model shows that the interacting chemical processes may be able to cause changes that are more or less unexpected, and it is therefore recommended to develop climate change models to be able to capture the results of many interacting processes, which is impossible without the use of a model.

Global warming changes the water cycle and the regional precipitation level. Holsten et al. (2009) have looked into how global climate changes affect water flows and soil moisture dynamics. They have modeled the soil moisture in Brandenburg (Berlin region), and they found that *a* significant decrease of the soil moisture can be expected in the region during the coming years, indicating a high-level risk for many wetland areas. The applied model was a spatial model that can indicate the changes in soil moisture spatially. The focal problem—the soil moisture—requires the use of a modeling approach to be able to account for the many interacting processes.

Caetano et al. (2009) have developed a model that focuses on the efficient allocation of resources to reduce the greenhouse effect. The model estimates the investments needed in land reforestation and in adoption of renewable energy sources. This model is therefore different from the other models presented in this chapter, as it looks into the optimum use of economic resources. It is, however, included in this overview, because the modeling approach has a potential to find a much wider application than the period 1996–2014 (for which it was used). It could be used in the general discussion on how much it will cost to shift to a cleaner energy technology. The economic parameters, that is, the cost of carbon dioxide reduction, discussed in the model are probably not realistic, and further discussion is required to assess these parameters. The main ideas behind the model, that is, the conceptualizations of the problem, are, on the other hand, very innovative and could be considered in the development of other global warming—economic models.

Tatarinov and Cienciala (2009) have focused on the effect of climate changes on the growth of forest tree species—probably one of the most significant climate change problems in many parts of the world. The model is straight-forward as it looks into the consequences of the changed carbon dioxide concentration in the atmosphere and the increased nitrogen deposition resulting from climate changes. The model is based on a rich literature about the response of trees and forest growth to climatic changes. Therefore, it probably represents a synthesis of what we know about the problem. The model shows a decrease in wood carbon and buildup of litter carbon. For other models that look into the consequences of changed carbon dioxide concentration on forests, see Ogris and Jurc (2010) and Rammig et al. (2010) (see also Chapter 6).

Another very relevant question about the consequences of climate change is how the crop growth is changed. Lenz-Wiedemann, Klar, and Schneider (2010) have developed a comprehensive crop growth model, similar to the agriculture model presented in Chapter 8. Although it only includes carbon, water, and nitrogen cycles, the growth of all plant parts is covered in the model, which is an object-oriented generic model, named DANUIBA. It is used on the growth of sugar beets, maize, winter wheat, potato, and spring barley. The conclusion of the model application to changed climatic conditions is that it is a suitable tool to assess the consequences of global change on biomass production, water, and nitrogen demand.

References

Ackerman, Frank, Stephen J. DeCanio, Richard B. Howarth, and Kristen Sheeran (2009). Limitations of Integrated Assessment Models of Climate Change. *Climatic Change* 95: 3–4, 297–315.

Caetano, M. A. L. et al. 2009. Reduction of carbon dioxide emission by optimally tracking a pre-defined target. *Ecological Modelling* 220: 2525–2541.

Holsten, A. et al. 2009. Impact of climate change on soil moisture dynamics in Brandenburg with a focus on nature conservations areas. *Ecological Modelling* 220: 2076–2087.

Jørgensen, S. E., and H. Mejer. 1976a. Modelling the global cycle of carbon, nitrogen and phosphorus and their influence on the global heat balance. *Ecological Modelling* 2: 19–31.

Jørgensen, S. E., and H. Mejer. 1976b. Modelling the global heat balance. *Ecological Modelling* 2: 273–277.

Jørgensen, S. E., and B. Fath. 2011. *Fundamentals of Ecological Modelling, 4th ed: Ecological Modelling in Research and Environmental Management.* Amsterdam, The Netherlands: Elsevier.

Lenz-Wiedemann, V. I. S., C. W. Klar, and K. Schneider. 2010. Development and test of a crop growth model for application within a Global Change decision support system. *Ecological Modelling* 221: 314–329.

Manabe, S., and R. T. Wetherald. 1975. The effects of doubling the carbon dioxide concentration on the climate of a General Circulation Model. *Journal of the Atmospheric Sciences* 32: 3–15.

Ogris, N., and M. Jurc. 2010. Sanitary felling of Norway spruce due to spruce beetles in Slovenia: A model and projections for various climate change scenarios. *Ecological Modelling* 221: 290–302.

Ramanathan, V., M. S. Lian, and R. D. Cess. 1979. Increased atmospheric carbon dioxide: Zonal and seasonal estimates of the effect on the radiation energy balance and the surface temperature. *Journal of Geophysical Research* 84: 4949–4958.

Ramming, A. et al. 2010. Impacts of changing forest regimes on Swedish forests: Incorporating cold hardiness in a regional ecosystem model. *Ecological Modelling* 221: 303–313.

Tatarinov, F. A., and E. Cienciala. 2009. Long-term simulation of the effect of climate changes on the growth of main Central-European forest tree species. *Ecological Modelling* 220: 3018–3088.

Zlatev, Z., and L. Moseholm. 2008. Impact of climate change on pollution levels in Denmark. *Ecological Modelling* 217: 305–321.

17

Landscape-Scale Resource Management: Environmental Modeling and Land Use Optimization for Sustaining Ecosystem Services

Ralf Seppelt

CONTENTS

17.1 Introduction 457
17.2 Environmental Modeling, Scenarios, and Management 459
    17.2.1 Introduction: Conceptual, Physical, and Mathematical
        Models 459
    17.2.2 Aspects of Scale in Model Definition 460
    17.2.3 Integrated Models for Ecosystem Service Quantification 461
    17.2.4 Scenario-Based Analysis of Environmental Impacts 461
    17.2.5 Optimization Framework: Analysis of Resource Use
        Options 462
    17.2.6 Tools and Applications 464
17.3 Examples on Model-Based Land Use Management 465
    17.3.1 Catchment Management 465
    17.3.2 Ecosystem Services and Trade-Off Analysis 467
    17.3.3 Optimal Configuration of Species Habitats 469
    17.3.4 Pareto-Frontiers for Optimum Land Use Configurations 471
17.4 Discussion and Recommendations 473
Acknowledgments 475
References 475

17.1 Introduction

Land can provide a multitude of goods and services. It provides space for the production of goods such as food and fiber. Landscapes and their ecosystem act as a filter for water and storage for carbon and habitats that host various species. Altogether, landscapes and the embedded ecosystems support the existence of life on earth. However, land use for one purpose often excludes others: bioenergy production can compete with food production, settlement development with biodiversity conservation, and enlargement of recreation areas with the production of renewable resources. Thus, the term of multifunctional use of landscapes was introduced (Hector and Bagchi 2007). However, a multifunctional approach for land management does not avoid a profound quantitative description of trade-offs. Trade-offs may even be aggravated under conditions of climate of demographic change when demands for goods and services are increased due to growth of population. At the same time, land use management offers a broad range of options for adapting to and mitigating these change processes, both technically and in terms of institutional arrangements. Thus, there is a need to use or manage landscapes in an optimum manner, considering a multitude of different natural goods and services. These goods and services are summarized as *ecosystem services* (see Millennium Ecosystem Assessment (MA), 2003, 2005, for a widely used list of these goods and services). Services provided by abiotic parts of the ecosystems are sometimes distinguished as environmental services—we will use the term ecosystem services here in a more general way, which covers biotic as well as abiotic aspects of services. As these environmental processes get their value via the benefit people obtain from them (Diaz et al. 2007), there is a clear anthropocentric concept: Without a benefit there is no service. This involvement of beneficiaries for human life is the key feature that distinguishes these services from ecosystem functions or processes (Chan et al. 2006).

Since the value of ecosystem services is typically external to the valuation framework of decision-makers, suboptimal decisions and allocations of sparse resources result. Ecosystem service assessments can help to incorporate the value of ecosystem, services into decisions, and thereby help in the formulation of better decisions. Although process studies at the field and landscape levels are important tools, methods for regional ecosystem service assessments are an important complement. Because of the multifunctionality of the landscape, ecosystem services, assessments should aim at studying several ecosystem services in parallel (Seppelt et al. 2011). To be most cost-efficient, regional assessments should build on indicators derived from available data. Indicators based on land use or land cover data seem to be a good fit for this purpose, and have been recently the focus of interest among certain sectors of the research community. Several studies were conducted at the regional scale. Frequently, only a limited number of ecosystem services were analyzed. Individual ecosystem services were modeled via land use data, and results were successfully compared to field observations (Costanza et al. 1997).

Thus there are two strands that guide us to the concept of applying environmental models within optimization frameworks: (1) we are to solve resource management problems, for example, the need to identify optimum pattern of land resource allocations in an optimum fashion. And—in doing so—the aspired solution needs to cope with a multicriteria assessment process and the consideration of externalities.

17.2 Environmental Modeling, Scenarios, and Management

17.2.1 Introduction: Conceptual, Physical, and Mathematical Models

Environmental models are tools that help us to understand how ecological processes function and allow testing hypotheses about ecological processes in a systematic manner. Thus, simulation models are the bases for the analysis of a complex interacting system such as our environment with the ecosystem services provided. However, setting up an environment requires detailed system analysis of the processes of interest. This system analysis, however, encompasses conceptual work, biophysical measurements, and final model coding using math or computer science. All three steps belong to the same process, and none of the parts can be neglected in compiling a well-defined simulation model. Table 17.1 illustrates this condition.

The table contains five sentences in the rows with three different options in the columns. Start reading with the sentence beginning the first column and complete the sentence with the completion in the column (conceptual, physical, or mathematical) of your choice, redrawn from (Seppelt 2003).

Note that compiling a mathematical model necessitates consideration of physical models and conceptual models. Without a clear specification of system boundaries of the parameters of interest, without knowledge of

TABLE 17.1

Conceptual, Physical, and Mathematical Modeling

Models are...	...mental or conceptual	...biophysical	...mathematical
System identification encompasses...	the definition of system boundary, components, interactions		
The model is...	a conceptual verbal description of systems behavior	a scaled reproduction of a real system	built by coupling of functions, rules, equations
Elements of models are...	premises, conclusions, syllogisms	physical objects	mathematical functions and (state) variables
With the plausibility check...	conclusions are tested (for well-known situations)	an experiment in a well-known environment is performed	model behavior is analyzed using different methodologies (stability or sensitivity analysis)
Finally, a simulation is...	a *Gedankenexpriment*	a physical measurement by a given boundary conditions	a (numerical) solution of the mathematical equations of rules under given initial and boundary conditions

important and less important processes, and without well-educated guesses of the values of sensitive parameters, an environmental model cannot be set up. On top of it all sits the purpose of the model since this has an important influence on how system borders are drawn, which processes are considered important, and which spatial and temporal scale seem appropriate.

17.2.2 Aspects of Scale in Model Definition

In performing such a modeling and simulation study, we most frequently focus on a specific focal scale (cf. Figure 17.1), which is relevant for the problem at hand. This can be located within well-known space–time diagrams (Delcourt and Delcourt 1988). In doing so, two aspects are to be noted: First, processes in the sense of dynamics of spatial extent or grain with a finer resolution than appropriate are neglected, *aggregated*, or replaced by *effective parameters*. By this, *nonlinear interdependencies* are incorporated frequently. Second, we deal with *boundary conditions* and *constraints* from the level above, most frequently using a scenario approach for analyzing the impacts and consequences of these scenarios to the system of interest. These constraints, however, may also be interpreted as *control variables*. These variables need to fulfill certain constraints, mainly given by economic constraints of by certain scenarios.

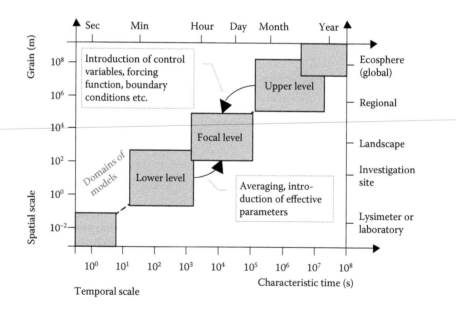

FIGURE 17.1
Space–time illustration of model exercises. (From Seppelt, R. *Computer-based Environmental Modelling*. Wiley-VCH, New York, NY, 2003. With permission.)

17.2.3 Integrated Models for Ecosystem Service Quantification

Arguably, model-based assessments of ecosystem functions and services provide more consistent insights into the ecosystem impacts of human actions than expert knowledge. The measurement, modeling, and monitoring of ecosystem function in a landscape, and thus the potential ecosystem services provided by the landscape, are the foundation for sustainable use of biodiversity, ecosystems, and natural resources in general (Holling 2001; Carpenter et al. 2006, 2009). This integrated concept of ecosystem services requires us to consider a wide range of ecosystem functions, which do not necessarily perform and function on the same temporal and spatial scale. Thus, this integration requires a harmonization or simplification of models needed. Generally speaking, disciplinary simulation models excel at realistic descriptions of ecosystem processes but lack feedback between ecosystem processes, with the reverse being true for interdisciplinary models. A good example of the disciplinary approach to modeling ecosystem services is the ATEAM project (Schröter et al. 2005). This project was aimed at combining complex models of ecosystem services in a fairly complete description of carbon sequestration, food and timber production, hydrology, and biodiversity, but it was lacking at interactions between these individual models. Another extreme example is the development of the GUMBO model, built from the start to embrace the interrelatedness of system processes, but whose description of ecosystem processes is very simple (Boumans et al. 2002).

Two aspects of realism in modeling seem to be crucial: adequate representation of environmental processes, and an integrated system's perspective, where isolated descriptions of system parts are to be avoided. This necessitates a comprehensive understanding of the system under consideration, and hence a representation of the relevant biophysical processes in a realistic way (Clark et al. 2001). Thus, the nonlinearities as mentioned above become crucial for resource management questions: Thresholds, irreversibility, nonlinearity and uncertainty of system configurations are pervasive in modern system description, but typically not in the models used to investigate ecosystem services under climate or land use change. Models are needed that allow for environmental interactions, but such integration comes at the cost of increased uncertainties, since more parameters have to be calibrated. Using an experimental approach, several biophysical monitoring methods can be combined to produce monetary valuations of the ecosystem services provided by different types of arable land (Sandhu et al. 2008). The study offers a sound basis for considering uncertainties in evaluations of multiple ecosystem services.

17.2.4 Scenario-Based Analysis of Environmental Impacts

Probably the most tempting application of environmental modeling is to develop scenarios for future change projections. The basic assumption behind projections is that all functional relationships that depend on time, whether directly or indirectly (e.g., by definition of rates and initial conditions), remain true throughout the considered time of simulation. Furthermore, the intrinsic uncertainty which might be acceptable at present is assumed to be tolerable in the future. One prominent field of dynamics in spatially explicit model projections is the ongoing and anticipated climate change. In the context of ecosystem services, the ATEAM scenarios or the scenarios of the MA (2002, 2003) are important examples of scenario application.

Scenario analysis is a powerful instrument to inform stakeholders about possible futures. However, the development of scenarios necessitates specification and, much more importantly, the agreement on certain parameters of scenarios (e.g., population growth, expected temperature change). This process of consensus building, which could be (and frequently is) policy-driven, is further influenced by the fact that only a limited number of scenarios can be studied within a modeling exercise. As a consequence, one might expect that the space of possible futures or scenarios is uncovered, which might hold excellent solutions, here in terms of land management and sustaining of ecosystem services.

17.2.5 Optimization Framework: Analysis of Resource Use Options

The question as to how much impact nature can bear without harming our environment is quite old. Ideas such as "most sustainable yield" were introduced in the late 1960s. In this context, a model consists of a *simulator* that represents the environmental processes and connects *control variables* (also known as forcing functions or inputs) to output variables. The former, the control parameters, may be used to define management strategies in terms of the model variables. The latter, the output variables, are used to describe the system behavior and to assess the management strategies being studied. One or more of these output variables are used to assess the simulation, the results of the modeling experiment. Because of this, these variables are often called *indicators*.

Figure 17.2 illustrates this concept. In contrast to scenario analysis, the variations of parameters and functions in the model input (control) are performed automatically, as well as the processing of the output. Control parameters are changed in each step and the changes in the performance criterion is recorded. The procedure stops if no further improvement of the performance criterion can be achieved. This optimization procedure connects the scenarios, the simulation process, and the performance criterion. The basic difference between optimization and scenario analysis is the closed feedback loop in the figure. This feedback loop is closed by the condition of the performance criterion and the modification of control parameters.

Thus, this approach is entirely different from a scenario analysis. However, scenario analysis and this optimization concept do relate to each other, and thus can cross-pollinate each other. The formulation of a management

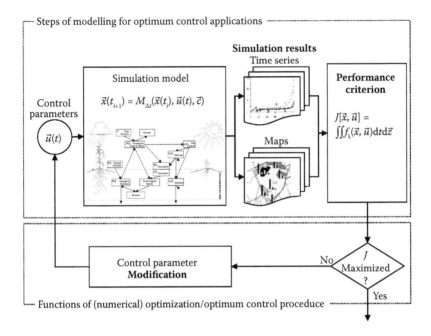

FIGURE 17.2

Concept of applications of environmental models in numerical optimum control procedures. Basic elements of the modeling part are simulation models—displayed by the conceptual model—and the performance criterion. Important part of the optimization procedure is an intelligent modification of the control vector driven by the estimation of the performance criterion. Note that compared to the scenario analysis methodology, which could compare with an open control system, this is a control loop modified from Seppelt and Voinov. (Reprinted from Seppelt, R., and Voinov, A., *Ecological Modelling*, 151, 125–142. Copyright 2002, with permission from Elsevier.)

scenario, assessment, and comparison of results requires considerable effort. Scenario definition requires knowledge of the modeled system, which probably may not be available due to the complexity of the investigated system. Instead of running numerous scenarios through the model and then comparing the results, one may formulate a certain goal that be achievable, and then let the computer sort through the numerous parameter and pattern combinations to reach that goal. Optimization automatically performs a systematic search over the whole control space, which is the combination of all—spatially and temporarily varying—input variables of the model. Thus, setting up optimization tasks requires much less effort in preprocessing, formulating, and consensus finding of the numerous scenario options. The major drawback of this approach is that their computational complexity is high and depends on two factors: the complexity of the process model (number of state variables, degree of nonlinearity, etc.) and the spatial complexity (size of study area, grid cell size, number of spatially interacting processes). The more complex the simulation model and the larger the number of spatial relationships, the lower the chances of success in the optimization. For very complex models, scenario analysis is usually the only feasible option.

17.2.6 Tools and Applications

In the past few decades, the potentials of spatial optimization in the field of conservation management were discovered. Various software packages have been developed that apply spatial optimization for finding optimum allocations for reserve sites [e.g., SITES, SPOT, MARXAN (Ball et al. 2009), C-Plan by Pressey, Ferrier, and Watts (Finkel 1998)]. The results presented here have been derived using the software package: LUPOlib (Land Use Pattern Optimisation library), a generic library for grid-based optimization of spatial landscape configurations with respect to a user-defined optimization goal (Holzkämper and Seppelt 2007b). LUPOlib can be applied to a variety of spatial planning problems (e.g., finding trade-offs between ecological and economic objectives, optimum allocation of management actions, reserve sites or roads). Like most programs dealing with complex spatial optimization problems, LUPOlib utilizes a metaheuristic search algorithm—a genetic algorithm by Wall (1996). This optimization algorithm approaches a global optimum solution in an iterative directed search (Goldberg 1989). LUPOlib utilizes a steady-state genetic algorithm with one-point crossover and flip-mutation as genetic operators. One innovative feature is that changes are performed based on a user-defined patch topology that allows an integration of two different types of land use changes (e.g., areal and linear) (see Figure 17.3). Land use in designated patches (of areas or lines) is changeable, whereas the remaining landscape persists. Within the optimization, the patch topology is represented by two arrays (one for areal patches and one for line patches). For evaluating the goal function, the landscape representation

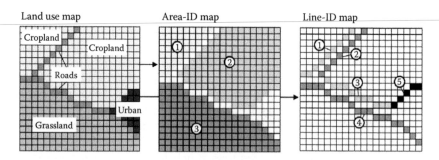

FIGURE 17.3
LUPOlib patch topology: area units and line units are identified in an area- and line-ID map, respectively. Patches of crop and grassland are assigned to area units; urban areas and roads are excluded from land use changes. Edge cells of crop and grassland are assigned to five different line units according to the bordering land-use types. (Reprinted from Holzkämper, A., and Seppelt, R., *Environmental Modelling & Software*, 22, 1801–1804. Copyright 2007b, with permission from Elsevier.)

is transformed to a two-dimensional grid. This allows the user to explicitly consider neighborhood dependencies (e.g., for evaluating habitat suitability, landscape metrics, or lateral flows of water and nutrients). The objective function needs to be specified by the user to solve a designated spatial optimization problem.

17.3 Examples on Model-Based Land Use Management

17.3.1 Catchment Management

Model-based assessment of water and matter dynamics in a landscape as functions of spatial location of habitat structures and matter input necessitates a dynamic model with special interactions. The model needs to consider dynamics, since climate variables change through time and, for instance, crop growth and nutrient uptake strongly depends on these variables. Second, topography and soil parameter determine horizontal fluxes of substances through the landscape. Thus, a simulation model for these processes needs to account for this by considering spatial interactions. Seppelt and Voinov (2002, 2003) studied in a mainly agricultural region the nutrient (N) balance as a function of different land use and land cover schemes. The landscape model uses a grid structure to calculate water and matter dynamics in a spatially explicit way. In other words, flow of water and matter is calculated from cells to neighboring cells for surface, subsurface, and groundwater according to the flow network and conductivities, to soil properties and land use. The task of the case study was to calculate optimum land use maps and fertilizer application maps for the three different nested investigation areas: Patuxent watershed (2365 km^2), Hunting Creek (77.8 km^2), and a subwatershed of Hunting Creek (20.5 km^2) in the Chesapeake Bay region, USA (Seppelt 2003).

For this purpose, optimization tasks were formulated. This required the definition of performance criteria, which compare economic aspects, such as farmer's income from harvest A, costs for fertilization B, with ecologic aspects, such as nutrient loss out of the watershed C. Because A and B can be quantified by monetary units and C is given in nonmonetary terms, for instance, by mass per area, a weight c_N (shadow price) was introduced in the performance criterion J, which is to be maximized:

$$J = A - B - c_N C \tag{17.1}$$

The maxima of the criteria J were calculated based on numerical optimization in spatially explicit dynamic ecosystem simulation models and tests performed via Monte Carlo simulations and gradient-free optimization procedures. Here, we focus on the optimizing nitrogen loss out of the watershed, which are summarized. In other words, we study the trade-off between agricultural production and nutrient retention capability of a landscape including the innerregional dependencies, given by the topography and the spatial configuration of the landscape. The core idea of the investigation is to study optimum land use patterns as a function of the unknown weight c_N in Equation (17.1). Increasing c_N, nutrient loss out of the watershed is "punished" (compared to economic income by agriculture). As a result, the number of agricultural areas decreases by increasing c_N. Important findings, illustrated in Figure 17.4, are as follows:

1. Although Equation (17.1) describes a simple linear performance criterion, the resulting patterns are not linear at all.
2. These nonlinear patterns are similar at different spatial scales.

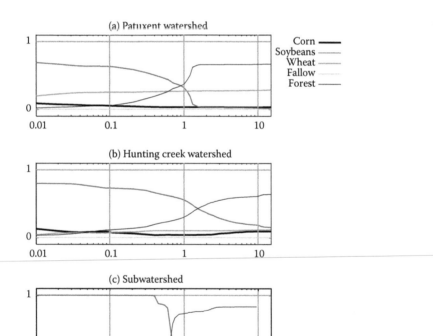

FIGURE 17.4
Comparison of optimum land use distribution patterns optimizing yield against nutrient loss at three different spatial scales: (a) Patuxent study area; Hunting Creek (b) watershed and (c) the subwatershed. (Modified from Seppelt, R., *Computer-based Environmental Modeling*, Wiley-VCH, New York, NY, 2003. With permission.) The x-axis shows the shadow price c_N for which the land use distribution has been optimized.

In addition to the aggregated results, managers are interested in the spatially explicit results. Figure 17.5 displays optimum land use patterns for different c_N values chosen. The striking result is that sensitive regions can be identified based on the presented method. For instance, the maps with increasing c_N from 0 on show that forest habitat near the rivers and creeks supports the retention capability of the ecosystem in terms of nutrients. As a result, important areas with high retention capabilities were identified and a spatially explicit fertilization scheme was computed that depends on soil properties and topological relations to neighboring cells. This shows that optimization methods even in complex simulation models can be a useful tool for a systematic analysis of management strategies of ecosystem use.

17.3.2 Ecosystem Services and Trade-Off Analysis

Although the performance criterion contained only economic yield of crop production and damage costs for nitrogen loss, we are able to include additional ecosystem services in a trade-off analysis:

- Production service: net primary production
- Nitrogen loss: potential loss of retention capability
- Forest area: habitat function
- Forest edge: habitat function for pollination services

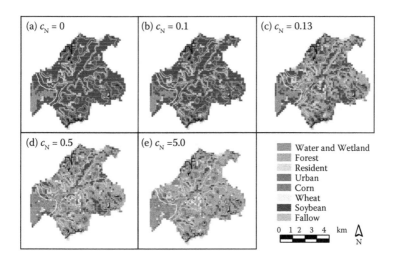

FIGURE 17.5
Optimum land use maps derived for optimization using the weighting c_N. Lower areas near rivers and creeks can be identified by contour lines. (Redrawn from Seppelt, R., *Computer-based Environmental Modelling*, Wiley-VCH, New York, NY, 2003. With permission.)

This can be compared with the monetary value of the crop yield, which related the ecological function to the beneficial services obtained from the region. Using the external cost variable c_N, we calculated a sequence of optimum landscapes configuration and plotted the above-mentioned variables. Thus, in Figure 17.6, each of the dots characterizes one optimized landscape for an assumed value of external costs of nitrogen loss between 0 and US$18/ kg N emitted out of the catchment. Variability has been introduced by using climate data from 1985 to 1995. This variability is most obvious in the plot Forest edge over Nitrogen loss.

For the purpose of simplicity, we assumed a proportional link between land use and the ecosystem services under consideration. The next step would be to include the additional ecosystem services and their related ecosystem functions in the performance criterion. The results of the trade-off analysis, shown in Figure 17.6, show some obvious relationships, such as the strong correlation between net primary production and yield. Here, however, for low yields variation can be identified, which is due to the diversity of crops.

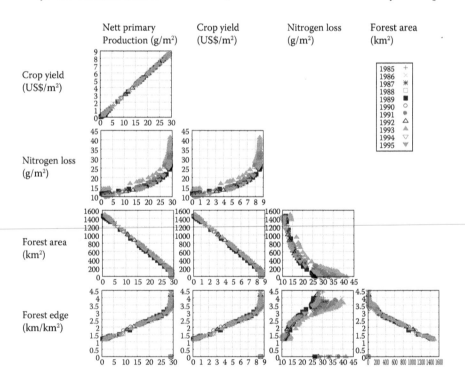

FIGURE 17.6
Trade-off analysis of the ecosystem function retention, production, and habitat based on optimized landscapes. Plotted are final results for the selected biophysical indicators of potential provided ecosystem services for 10 years. Note that each point denotes the results of an optimization derived from different assumptions on shadow prices, assuming that a reconfiguration of the landscape could be achieved. This figure illustrates the overall space of trade-offs in this landscape.

All other plots show strong nonlinearities between the analyzed variables. These are attributed to either the nonlinearities of the processes, the structure of the landscape, and spatial interrelations. Most of these relationships can be clearly explained. For instance, yield and net primary production cannot be increased by further fertilization (if external cost of nitrogen loss is ignored). This just increases nitrogen loss. Because net primary production in forest is not considered as an ecosystem service, an increase in forest area leads to a decrease in yield. Finally, the more structure the landscape has (e.g., a medium amount of forest area), the more habitat for pollinators is available (forest edge). Finally, interannual variability can be derived from the different trajectories, including the sensitivity of trade-off curves on climatic conditions, which is most obvious for nitrogen loss.

17.3.3 Optimal Configuration of Species Habitats

Heterogeneity of agricultural landscapes is supposed to be of significant importance for species diversity in (agro-)ecosystems. Land use pattern changes may lead to an increase in suitable habitat for some species, but could mean habitat deterioration for other species with opposing habitat requirements. To investigate the effects of land use changes on different species' habitat suitability and to allow a trade-off between management objectives, a spatial optimization model is applied. The approach integrates a neighborhood-dependent multispecies evaluation of land use patterns into an optimization framework for generating goal-driven scenarios aims at maximizing habitat suitability of three selected bird species (middle-spotted woodpecker, wood lark, red-backed shrike) by identifying optimum agricultural land use patterns. Thus, for these examples, the performance criterion is to maximize habitat suitability of species i denoted by $HSI_i(x)$ depending on environmental site conditions x, including land use. The evaluation of habitat suitability is based on landscape metrics calculated within the species' home ranges to incorporate the effects of species responses to landscape pattern on a territorial scale. Because different species show different habitat preference, we assume trade-offs between different habitat requirements, such as forest versus open agricultural land or hedgerows for nesting. The performance criterion is as follows

$$J = \sum c_i \, HSI_i(x) \text{ with } \sum c_i = 1 \qquad (17.2)$$

where c_i denotes arbitrary weighting, for example, importance of habitat characteristics of a given species i. Against different combination of c vectors, optimization results can be analyzed.

The main focus of this study was to explore the potential of this approach for conservation management on the basis of a case study. One can investigate where habitat requirements oppose, where they coincide, and how a landscape optimized simultaneously for all target species should be charac-terized. Habitat suitability for all three target species improved during almost all optimization runs, except for the runs optimizing habitat suitability for the wood lark and the middle-spotted woodpecker, where red-backed shrike habitat suitability decreased. We also observed a slight decrease in habitat suitability for wood lark in the runs optimizing habitat suitability for middle-spotted woodpecker and red-backed shrike (see Figure 17.7). The highest mean

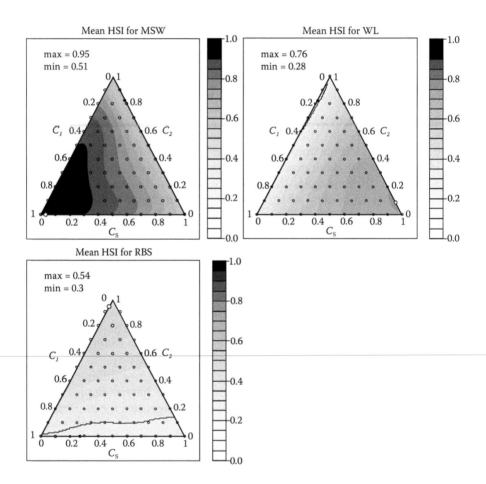

FIGURE 17.7

Mean habitat suitability indices of the focal species (MSW, middle-spotted woodpecker; WL, wood lark; RBS, red-backed shrike) depending on species' weightings in performance criterion, Equation 17.2 (c_1, weight for middle-spotted woodpecker; c_2, weight for red-backed shrike; c_3, weight for wood lark). White dots indicate maximum values; black dots indicate minimum values; lines indicate mean HSI values derived from the initial landscape. For MSW, the initial value is below the values in the plot. (Reprinted from Holzkämper, A., et al. *Ecological Modelling*, 198, 277–292. Copyright 2006, with permission from Elsevier.)

habitat suitability values were recorded for the middle-spotted woodpecker. In almost all optimization runs, the mean HSI for the middle-spotted woodpecker exceeded those of the other species. The optimization was least successful for the red-backed shrike. As the initial mean habitat suitability index was 0.40 for the middle-spotted woodpecker, 0.30 for the wood lark, and 0.32 for the red-backed shrike, the improvement was best for the middle-spotted woodpecker and worst for the red-backed shrike. The minimum values of habitat suitability of the middle-spotted woodpecker and the wood lark are reached when the weight for red-backed shrike is high (0.9).

We find that all species would benefit from an increase in deciduous and coniferous forest, a decrease in cropland and grassland in the study area, and more heterogeneous land use patterns (smaller patches, more diversity of land use types). Habitat requirements of red-backed shrike contrast most to those of the other two species with respect to landscape composition and configuration (Holzkämper et al. 2006).

17.3.4 Pareto-Frontiers for Optimum Land Use Configurations

Land use intensification and spatial homogeneity are major threats to biodiversity in these landscapes. Thus, cost-effective strategies for species conservation in large-scale agricultural landscapes are required. Based on the results discussed in the preceding section, the process for a change in land use can be incorporated to derive cost–benefit function of habitat improvement used in working landscapes. Thus, Equation (17.2) can be modified in different ways:

$$J = \text{HSI}_i(x) - \lambda A \qquad (17.3)$$

considers a single species i, and

$$J = \sum c_i^* \, \text{HSI}_i(x) - \lambda A \text{ with } \sum c_i^* = 1 \qquad (17.4)$$

defines a performance criterion for identifying optimum land use configuration, taking into account habitat demand for all three species (discussed in the previous section). Here, c^* denotes the weighting derived from the results mentioned in the previous section, and denotes the compromises of habitat enhancement for all three species. λA summarizes the total costs for changing a parcel of land, for example, A is the total area of land changed and λ denotes the related costs (see Holzkämper and Seppelt 2007a for details).

Spatial optimization methods are applied to identify the most effective allocation of a given budget for conservation. The optimization of spatial land use patterns in real landscapes on a large spatial scale is often limited by computational power. The solution here makes use of a simplifying methodology for analyzing cost-effectiveness of management actions on a regional scale. A spatially explicit optimization approach is employed to identify optimum agricultural land use patterns with respect to an ecological–economic goal function. Figure 17.8 shows the examples for single species optimization based on Equation (17.3). Depending on the assumed available costs for land use change, landscape structure change according required habitat. Based on the optimization results using Equation (17.4) for small-scale landscape samples, we derive a target- and site-specific cost–benefit function that can be applied to predict ecological improvement as a function of costs and local conditions on a large spatial scale.

Optimization results show that the cost–benefit relationships differed not only between species, but also between study sites. The relations between habitat suitability and assumed available land use change for each of the three

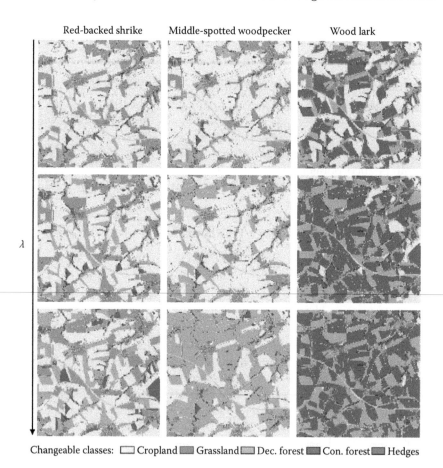

Changeable classes: ☐ Cropland ▨ Grassland ☐ Dec. forest ■ Con. forest ▨ Hedges

FIGURE 17.8
Exemplary set of optimum trade-off solutions for the three species given different values for λ. (Reprinted from Holzkämper, A., and R. Seppelt. *Biological Conservation*, 136, 117–127. Copyright 2007a, with permission from Elsevier.)

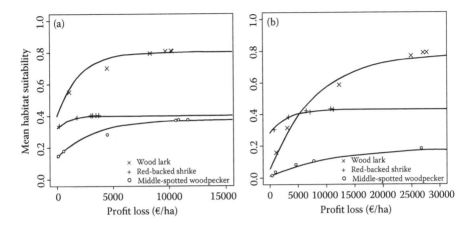

FIGURE 17.9
Empirical relationship of profit–loss by income over 10 years for three species. This empirical function was fitted based on optimization results and does depend only on species habitat requirements, the structure of the landscape. (a) Cost–benefit functions for low fertility score values and graph; (b) comparable high fertility scores. (Reprinted from Holzkämper, A., and Seppelt, R., *Biological Conservation*, 136, 117–127. Copyright 2007a, with permission from Elsevier.)

target species could be described by three exponential saturation functions. For all three species, the optimum habitat suitability as a function of money invested for land use change as an approximation depended on initial mean habitat suitability, maximum possible mean habitat suitability, profit loss P, and mean soil fertility (see Figure 17.9). By doing this, it is possible to identify areas where management actions for ecological improvement are most efficient with respect to a certain conservation goal. The fitted function is validated independently. In a case study, we analyze cost-effectiveness of management actions to enhance habitat suitability for three different target species.

17.4 Discussion and Recommendations

Application of environmental models within frameworks of optimization or optimum control is still a challenging task because it is computationally exhaustive. This goes along the lines of applying environmental modeling in data based on inverse modeling approaches. Not explicitly mentioned assumptions are, for instance, that the parameter space of independent variables can be controlled, either by intelligent search techniques or by sufficient knowledge on the behavior of the system. Second, this requires sufficient robustness of the model, which means stability for these domains in the parameter space, where the optimum solution is assumed.

The advantage of these approaches are manifold: With this systematic search for goal-oriented scenarios, much more can be learned about the system of interest than with a simple comparison of distinct scenarios. Second, the tools developed to tackle these tasks are suitable for management problems on nature protection and give planners insights into how landscapes function. The results indicate several trade-offs between different ecosystem services. These trade-offs are valuable information for decision makers and regional planners, because they reveal otherwise undetected options for land use planning. Without the combined use of simulation models and optimization techniques, this information would be unavailable. Although static GIS-based land use analysis approaches have their merits, this type of analysis would be unable to reveal this type of trade-offs. It is the combination of simulation models with optimization techniques that enables decision makers to take a deeper look into the causalities between land use and ecosystem services.

Note that these results are always based on the assumption that the landscape is used in an optimal manner, assuming external costs arbitrarily. By this method, we can systematically study the interrelationships of these processes and derive a pretty sharp Pareto curve, which gives a hint on the trade-offs of different services. The real use of the landscape usually differs from this optimal land use pattern. Nevertheless, the analysis reveals hidden potentials and offers possible ways of improving the allocation of sparse resources. In this context, it is interesting to note that the real land use pattern in the case study area is fairly similar to some optimized land use patterns.

From a model building viewpoint, we have to recall that the model was built to fulfill a specific purpose. The model for the Patuxent catchment, for instance, was designed to detect different optimal land use patterns and not to analyze, for example, optimal crop rotation systems or optimal crop management practice. A detailed analysis of the pollination service would, for example, enforce the incorporation of crops that benefit from pollination such as almonds, apples, or rapeseed. The model also was designed for the regional scale and therefore does not pay too much attention to details of local scale processes. For the analysis of trade-offs in a multifunctional landscape, optimization is an important tool because it enables the scientist to compare results between well-managed land use systems. In a way, these optimization runs enable us to compare the outcome if the value of ecosystem service would be incorporated in land use decisions. There is still a lack of consistent methods and criteria to help us decide which degree of model aggregation is beneficial without losing too much of the complexity of ecological systems (Jakeman et al. 2006; Koch et al. 2009). Since off-the-shelf solutions to ecosystem service modeling do not exist, careful consideration of local conditions is required. Support research on environmental resource management based on the ecosystem service concept by a concise modeling environment benefits science, society, and decision makers.

Acknowledgments

Special thanks are due to Annelie Holzkämper and Sven Lautenbach for excellent cooperation discussion and support. This work was funded by the Helmholtz Programme "Terrestrial Environmental Research" (Seppelt et al. 2009). The funders had no role in study design, data collection and analysis, decision to publish, or preparation of the manuscript.

References

Ball, I. R., H. P. Possingham, and M. Watts. 2009. Marxan and relatives: Software for spatial conservation prioritisation. In: Spatial conservation prioritisation: Quantitative methods and computational tools, ed. A. Moilanen, K. A. Wilson, and H. P. Possingham, 185–195. Oxford, UK: Oxford University Press.

Boumans, R., R. Costanza, J. Farley, M. A. Wilson, R. Portela, J. Rotmans, F. Villa, and M. Grasso. 2002. Modeling the dynamics of the integrated earth system and the value of global ecosystem services using the GUMBO model. *Ecological Economics* 41: 529–560.

Carpenter, S. R., E. M. Bennett, and G. D. Peterson. 2006. Scenarios for ecosystem services: An overview. *Ecology and Society* 11: http://www.ecologyandsociety.org/vol11/iss11/art29/.

Carpenter, S. R., H. A. Mooney, J. Agard, D. Capistrano, R. S. DeFries, S. Díaz, T. Dietz, A. K. Duraiappah, A. Oteng-Yeboah, H. M. Pereira, C. Perrings, W. V. Reid, J. Sarukhan, R. J. Scholes, and A. Whyte. 2009. Science for managing ecosystem services: Beyond the Millennium Ecosystem Assessment. *Proceedings of the National Academy of Sciences of the United States of America* 106: 1305–1312.

Chan, K. M. A., M. R. Shaw, D. R. Cameron, E. C. Underwood, and G. C. Daily. 2006. Conservation planning for ecosystem services. *PLoS Biology* 4: 2138–2152.

Clark, J. S., S. R. Carpenter, M. Barber, S. Collins, A. Dobson, J. A. Foley, D. M. Lodge et al. 2001. Ecological forecasts: An emerging imperative. *Science* 293: 657–660.

Costanza, R., R. d'Arge, R. de Groot, S. Farber, M. Grasso, B. Hannon, K. Limburg et al. 1997. The value of the world's ecosystem services and natural capital. *Nature* 387: 253–260.

Delcourt, H. R., and P. A. Delcourt. 1988. Quaternary landscape ecology: Relevant scales in space and time. *Landscape Ecology* 2: 23–44.

Diaz, S., S. Lavorel, F. de Bello, F. Quetier, K. Grigulis, and M. Robson. 2007. Incorporating plant functional diversity effects in ecosystem service assessments. *Proceedings of the National Academy of Sciences of the United States of America* 104: 20684–20689.

Finkel, E. 1998. Software helps Australia manage forest debate. *Science* 281: 1789–1791.

Goldberg, D. E. 1989. *Genetic algorithms in search, optimization and machine learning.* Boston, MA: Addison Wesley.

Hector, A., and R. Bagchi. 2007. Biodiversity and ecosystem multifunctionality. *Nature* 448: 188–186.

Holling, C. S. 2001. Understanding the complexity of economic, ecological, and social systems. *Ecosystems* 4: 390–405.

Holzkämper, A., A. Lausch, and R. Seppelt. 2006. Optimizing landscape configuration to enhance habitat suitability for species with contrasting habitat requirements. *Ecological Modelling* 198: 277–292.

Holzkämper, A., and R. Seppelt. 2007a. Evaluating cost-effectiveness of conservation management actions in an agricultural landscape on a regional scale. *Biological Conservation* 136: 117–127.

Holzkämper, A., and R. Seppelt. 2007b. A generic tool for optimising land-use patterns and landscape structures. *Environmental Modelling & Software* 22: 1801–1804.

Jakeman, A. J., R. A. Letcher, and J. P. Norton. 2006. Ten iterative steps in development and evaluation of environmental models. *Environmental Modelling & Software* 21: 602–614.

Koch, E. W., E. B. Barbier, B. R. Silliman, D. J. Reed, G. ME Perillo, S. D. Hacker, E. F. Granek et al. 2009. Non-linearity in ecosystem services: Temporal and spatial variability in coastal protection. *Frontiers in Ecology and the Environment* 7: 29–37.

MA. 2003. *Ecosystems and human well-being: A framework for assessment*. Washington, DC: Island Press.

MA. 2005. Ecosystems and human well-being: Synthesis. Washington, DC: Island Press.

Sandhu, H. S., S. D. Wratten, R. Cullen, and B. Case. 2008. The future of farming: The value of ecosystem services in conventional and organic arable land. An experimental approach. *Ecological Economics* 64: 835–848.

Schröter, D., W. Cramer, R. Leemans, I. C. Prentice, M. B. Araújo, N. W. Arnell, A. Bondeau et al. 2005. Ecosystem service supply and vulnerability to global change in Europe. *Science* 310: 1333–1337.

Seppelt, R. 2003. *Computer-based environmental modelling*. New York, NY: Wiley-VCH.

Seppelt, R., F. V. Eppink, S. Lautenbach, S. Schmidt, and C. F. Dormann. 2011. A quantitative review of ecosystem service studies: Approaches, shortcomings and the road ahead. *Journal for Applied Ecology*, in review.

Seppelt, R., I. Kühn, S. Klotz, K. Frank, M. Schloter, A. Auge, S. Kabisch, C. Görg, and K. Jax. 2009. Land use options—Strategies and adaptation to global change. *GAIA* 18: 77–80.

Seppelt, R., and A. Voinov. 2002. Optimization methodology for land use patterns using spatially explicit landscape models. *Ecological Modelling* 151: 125–142.

Seppelt, R., and A. Voinov. 2003. Optimization methodology for land use patterns—evaluation based on multiscale habitat pattern comparison. *Ecological Modelling* 168: 217–231.

Wall, M. 1996. GALib: A C++ Library of Genetic Algorithm Components. Version 2.4. in. http://lancet.mit.edu/ga/.

18

Simulation of Fire Spread Using Physics-Inspired and Chemistry-Based Mathematical Analogs

Alexandre Muzy, Dominique Cancellieri, and
David R. C. Hill

CONTENTS

18.1 Introduction..477
18.2 Thermal Degradation of Fuels in Chemistry...478
18.3 Simulation Model and Mathematical Analogs of Physics-Based
     Models 481
     18.3.1 Geometry of Fire Spread..482
     18.3.2 Stochastic Wind Effects...483
     18.3.3 Heat Influence: Physics-Inspired and Chemistry-Based
            Mathematical Analogs 485
     18.3.4 Fire Brands..489
18.4 Simulation Results 491
18.5 Discussion and Perspectives 492
References..495

18.1 Introduction

Fire spread research is multiobjective. Foresters are usually interested in obtaining reusable pragmatic information on: fire dynamics, vegetation inflammability, the impacts of forest fires on vegetation, etc. Physicists (Curry and Fons 1938) rigorously study the physics-based mechanisms involved in fire spreading. Using the terms and the models of thermodynamics, their aim is to fully understand *what* is a fire and *how* it spreads. Chemists focus on fuel and gas properties (not on the dynamics of fire spread). Computer scientists usually implement the numerical solutions of mathematical continuous equations (Grishin 1996) and/or manage databases of experiments and simulation results (McRae 1990). They can also build models on their own through mathematical analogs without any physical or chemical foundations [mainly ellipses (Anderson et al. 1982; Glasa and Halada 2008) or cellular automata (Alexandridis et al. 2008)].

Every discipline develops idiosyncratic models, which avoid (or make difficult) interdisciplinary collaborations and lead to a dispersal of research efforts. Our scope here is to discuss the modeling contribution of each discipline involved in this study. The main mechanisms (mass loss in chemistry and heat transfer in physics) are embedded in an original simulation model. Mass loss is studied in detail and an original model is presented. The modeling of heat transfers is achieved through mathematical analogs inspired from physics. In ecological modeling, occurrence risks and impacts of fires on vegetation are usually statistically analyzed (Lasaponara 2010). Here, both structure and behavior of fire spreading are explicitly described in time through a simulation model.

18.2 Thermal Degradation of Fuels in Chemistry

Thermal decomposition kinetics of biomass is an important key in thermochemical conversion processes. Thermal decomposition kinetics of biomass aims at studying the production of energy and chemical products (Soltes 1988; Alves and Figueiredo 1989; Di Blasi, Branca and Galgano 2008). In wildland fire, the rate of mass loss (due to thermal decomposition) determines the available volatile fuel in the flaming zone. To a lesser extent, the mass loss rate also determines the heat release rate (product of combustion heat and mass of fuel burned). Therefore, the analysis of the thermal degradation of lignocellulosic fuels is decisive for wildland fire modeling and fuel hazard studies (Sofer and Zaborsky 1981; Dimitrakopoulos 2001; Alén, Kuoppala, and Oesch 1996; Balbi, Santoni, and Dupuy 1998).

At an experimental level, many studies led to different degradation schemes in inert environment (Grishin 1997), but only a few were monitored in air atmosphere (Statheropoulos et al. 1997). The whole process is complex and concerns solid and gaseous reactions. In solid state, a variation in apparent activation energy may be observed for an elementary reaction. This is due to the heterogeneous nature of the solid or due to a complex reaction mechanism. This variation can be detected by isoconversional or model-free methods. The isoconversional analysis provides a fortunate compromise between the oversimplified but widely used single-step Arrhenius kinetic treatment and the prevalent occurrence of processes whose kinetics are multistep and/or non-Arrhenius (Beall and Eickner 1970). These methods allow estimation of the apparent activation energy, at progressive degrees of conversion, for an independent model. Application of model-free methods was highly recommended in order to obtain a reliable kinetic description of the investigated process. The heat released from combustion causes the ignition of adjacent unburned fuel. Therefore, the analysis of the thermal degradation of lignocellulosic fuels is decisive for wildland fire modeling and fuel hazard studies (Liodakis, Bakirtzis, and Dimitrakopoulos 2002).

Physical fire spread models are based on a detailed description of physical and chemical mechanisms involved in fires. Since the pioneering work of Grishin (1997), these models have been incorporating chemical kinetics for the thermal degradation of fuels. However, kinetic models need to be improved. Knowledge of the kinetic triplet (E_a, K_0, and n) and the kinetic scheme can help in predicting the rate of thermal degradation when the collection of experimental data is impossible in classical thermal analysis (high heating rates encountered in fire conditions).

For a mass loss modeling, based on the Arrhenius's law, we denote here the conversion degree defined as:

$$\alpha = \frac{m_0 - m(t)}{m_0 - m_\infty} \quad \text{or} \quad \alpha = \frac{\Delta H(t)}{\Delta H_\infty} \tag{18.1}$$

where α is the conversion degree, ΔH is the enthalpy of the reaction (kJ/g), and m is the mass loss (%).

The rate of heterogeneous solid-state reactions can be described as:

$$\frac{d\alpha}{dt} = Ae^{-E_a/RT} f(\alpha) \tag{18.2}$$

where E_a is the activation energy (kJ/mol), A is a pre-exponential factor (1/s), t is the time (min), R is the gas constant (8.314 kJ/mol), $f(\alpha)$ is the kinetic model reaction, and T is the temperature (K).

The temperature dependence of the constant rate is described by the Arrhenius equation. Brown et al. (1988) presented a theoretical justification for the application of the Arrhenius equation to the kinetics of solid-state reactions. It is now recognized that this empirical equation represents the experimental rate data as a function of temperature, accurately and in both homogeneous and heterogeneous reactions (Prasad, Kanungo, and Ray 1992). More recently, some reports have demonstrated how a complete isoconversional kinetic analysis can be achieved using the dependence on activation energy in association with thermoanalytical data (Ozawa 2000).

The parameters of the kinetics reaction were determined using the following procedure. Under nonisothermal conditions, in which a sample is heated at a constant rate, the explicit temporal dependence in Equation 18.2 is eliminated through the trivial transformation:

$$\frac{d\alpha}{dT} = \frac{A}{\beta} e^{-E_a/RT} f(\alpha) \tag{18.3}$$

We define $g(\alpha)$ as:

$$g(\alpha) = \int_0^{\alpha} \frac{d\alpha}{f(\alpha)} = \frac{A}{\beta} \int_{T_0}^{T} e^{-E_a/RT} dT = \frac{AE_a}{R\beta} p(x) \qquad (18.4)$$

where $x = \dfrac{E_a}{RT}$ and $p(x) = \displaystyle\int_{x_0}^{\infty} \frac{e^{-x}}{x^2} dx$, β is the heating rate (K/min), and T is the temperature (K).

Kinetics analyses are traditionally expected to produce an adequate kinetic description of the process in terms of the reaction model and the Arrhenius parameters. There are many methods for analyzing solid-state kinetic data. These methods may be classified according to the experimental conditions selected and to the mathematical analysis performed. Experimentally, either isothermal or nonisothermal methods are employed. The earliest kinetics studies were performed under isothermal conditions (Wight 1998). However, a major drawback of the isothermal conditions assumption is that a typical solid-state process has its maximum reaction rate at the beginning of the transformation. Several mathematical methods can be used to calculate the kinetics of solid state reactions: model-fitting and isoconversional (model-free) methods. A model-fitting method involves two fits: the first establishes the model that best fits data, whereas the second determines specific kinetic parameters such as activation energy and pre-exponential factor using Arrhenius equation. The model-fitting approach has the advantage that only one TGA measurement is needed. However, almost any $f(\alpha)$ can satisfactorily fit the data by virtue of the Arrhenius parameters compensation effects and only a single pair of Arrhenius parameters results from the model-fitting method. Consequently, researchers give up this kind of method for the benefit of isoconversional methods, which can compute kinetic parameters without modeling assumptions (Vyazovkin and Wight 1999). The isoconversional method has the ability to reveal the complexity of the process in the form of a functional dependence of the activation energy E_a on the extent of conversion α. The basic assumption of these methods is that the reaction rate for a constant extent of conversion, α, depends only on the temperature (Ozawa, Flynn, Wall, and Quick 1966). To use these methods, a series of experiments has to be performed at different heating rates (Vyazovkin and Wight 1997). The knowledge of E_a vs. α allows detection of multistep processes and prediction of the reaction kinetics over a wide temperature range. For this work, we chose the method of Kissinger–Akahira–Sunose (KAS) (Akahira and Sunose 1971) applied without any assumption concerning the kinetic model. The KAS method uses Doyle's approximation present in Equation 18.5:

$$p(x) \approx \frac{e^{-x}}{x^2} \qquad (18.5)$$

Taking into account the approximation, the logarithm of Equation 18.4 gives:

$$\ln\left(\frac{\beta_j}{T_{jk}^2}\right) = \left[\ln\left(\frac{A_\alpha R}{E_a(\alpha_k)}\right) - \ln g(\alpha_k)\right] - \frac{E_a(\alpha_k)}{RT_{jk}} \tag{18.6}$$

where E_a and A_α are respectively the apparent activation energy and the pre-exponential factor at a given conversion degree α_k, and the temperatures T_{jk} are those which the conversion α_k is reached at a heating rate β_j. During a series of measurements the heating rates are $\beta = \beta_1 \ldots \beta_j \ldots$.

The apparent activation energy was obtained from the slope of the linear plot of $\ln\left(\frac{\beta_j}{T_{jk}^2}\right)$ vs. $\frac{1}{T_{jk}}$ performed thanks to a Microsoft® Excel® spreadsheet developed for this purpose.

When the values of the apparent activation energy and the pre-exponential factor were determined, it was possible to reconstruct the reaction model (Kissinger 1957):

$$f(\alpha) = \beta\left(\frac{d\alpha}{dT}\right)_\alpha\left[A(\alpha)e^{\left(-E_a(\alpha)\big/RT(\alpha)\right)}\right]^{-1} \tag{18.7}$$

18.3 Simulation Model and Mathematical Analogs of Physics-Based Models

"Simulation models (. . .) implement (. . .) models in a simulation rather than modelling context. Mathematical analog models are those that utilise a mathematical precept rather than a physical one for the modelling of the spread of wildland fire" (Sullivan 2009). Mathematical analog models (Weber 1991) first introduced the use of *Huygens' wavelet* principle in fire spread modeling. This kind of modeling considers ignition propagations as ellipses. *Cellular automata* approaches (Alexandridis et al. 2008; Karafyllidis and Thanailakis 1997; Berjak and Hearne 2002; Encinas, et al. 2007a, 2007b) compute simple mathematical rules to reproduce fire spreading. *Percolation*, in mathematics, considers transports through randomly distributed media. It is the "simplest example of fractal growth model" (Caldarelli et al. 2001). The latter study considers satellite images of wildland fires to model statistical properties of fire spreading. In artificial neural networks (McCormick, Brander, and Allen

2000), weightings for each connection between the nodes is learned through usual neural network learning rules. Physics focuses on the propagation of fire through heat transfers (Sullivan 2007) (diffusion, convection, and radiation), possibly integrating simplified chemical submodels. However, no single model faithfully reproduces all heat transfers. Another perspective (the one chosen here) is to consider heat transfers as related to the mass loss of ignited fuels. The total heat released by ignited fuels depends on their burning mass (and loss rate). Then, a part of the heat impinges on other fuels according to their distance, their moisture, the wind conditions, and topography (the other part of heat is transferred to the air). This approach allows simplifying the modeling and simulation of heat transfers.

We propose here a mathematical model integrating all these parameter interactions. This model is based on the chemical degradation of fuel and inspired from physics.

18.3.1 Geometry of Fire Spread

As presented in Section 18.2, characteristics (energy released and received, mass loss) in a combustible sample can be studied through chemistry. The experimental protocol consists in submitting the combustible sample to a heating source, until ignition. A weighting apparatus is then used to measure the mass loss. A temperature sensor is used to measure the temperature in the fuel. The experiment (and the corresponding mathematics-based knowledge presented in Section 18.2) is repeated for many species of vegetation. The goal is to calculate the main mass loss characteristics of every species, according to the heating rate they are submitted to.

A difficulty arises for measuring heat transfers in the fuel and in the air. First, because it is difficult to determine the effective boundaries of "a flame." Second, because it is impossible to measure the temperature at every location within and around the fuel. Third, because both slope and wind consequences on fire spread are difficult to model. Physics attempts to achieve this goal (Alves and Figueiredo 1989). However, there is no physical model able to reproduce precisely and faithfully all the mechanisms of heat transfers.

Therefore, it is usually convenient to achieve mathematical and computational analogs (without any physical assumptions) to build experiments of fire spread using discrete and continuous functions. The parameters of these functions do not have any physical foundations. In physics, although the modeling parameters benefit from a long history of mathematical modeling (validated by experiments), it should be noticed that their numerical value was determined under precise experimental conditions. These conditions are difficult (impossible?) to reproduce. They are always slightly different, or even incompatible with actual conditions of fire spreading.

Simulation and mathematics can be used to model the propagation of energy/heat in space. Virtual experiments are implemented. Parameter combinations are successively tested to fit (quantitatively and/or qualitatively) many fire propagations. These mathematical analogs can finally be used to feed physical and/or chemical models.

Another solution (the one investigated here) is to use chemical modeling to describe precisely the mass loss process according to the heat received by fuels. The geometry and heat influence of the fire front depending on many unknown correlated parameters, probabilities are used to provide more flexibility and robustness to fire spread modeling. Probabilities enhance flexibility because a range of many parameters will produce a mean area of propagations. A computational/mathematical model is built here using analogs of physical modeling of heat transfers.

We present here a wind-based ellipse model for the determination of the heat influence zone of burning sites. Inside this influence zone, the fuel mass degradation is considered through radiative influences. Finally, fire brands, stochastically generated, are taken into account.

18.3.2 Stochastic Wind Effects

Wind (and slope) tilts the flames of a fire front thus increasing the heat influence area of the fire front, in the wind direction. The difficulty in modeling wind intensity and directions is attributed to the fact that wind blows in a squally fashion. Wind distribution, speed, and directions are highly erratic. A wind blowing at a speed of 30 km/h is an average of squalls below and above 30 km/h. Moreover, a northeast wind is not, dynamically, blowing northeast all the time. Finally, because of vegetation and topographic shapes, and vegetation porosity, local and global winds have very different directions and speed. More than humidity and slope characteristics, winds need to be modeled stochastically.

The heat influence area of ignited cells can be determined mathematically through ellipses (Glasa and Halada 2008) (Figure 18.1). A physics-based and computational-based model has been presented by Porterie et al. (2007). The notion of impact parameter and ellipses is used through small world

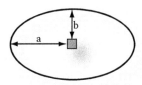

FIGURE 18.1
Elliptical heat influence area.

networks. "The impact parameters can be either identical to the characteristic lengths l_x and l_y (axis of the ellipse) directly related to the radiative/convective impact of the burning site (deterministic case) or generated following a Poisson-like distribution based on these lengths (random case) (Porterie et al. 2005). They are dependent on fire and fuel conditions and expressed in an arbitrary length unit (δl) corresponding to the lattice parameter. A high value of the impact parameter ratio l_y/l_x corresponds to a strong anisotropy of the front shape induced by the terrain slope and/or wind effects in the y-direction.

In our model, the wind lengthens the shape of the ellipse in the wind direction. Width b and length a of the ellipse are considered as follows.

The shape deformation of the ellipse by wind can be modeled through the calculation of parameters a and b:

b is calculated through an exponential probability distribution function $b \sim \exp(\lambda_b)$, for $\lambda_b - 1$ (cf. Figure 18.2).

a is initialized to b value plus a negative exponential law of parameter

$b + \exp(\lambda_a)$, with $\lambda_a = \dfrac{3}{2} - 0.9\dfrac{v}{v_{max}}$, where v_{max} is the maximum wind speed and v is the wind speed, with $\lambda_a \in [0.6;1.5]$ (cf. Figure 18.2). Therefore, the ellipse lengthens with wind strength. Under no wind conditions, it is highly probable that the ellipse will be a circle.

Then, once the ellipse shape is calculated, the ellipse is rotated according to wind direction, and translated according to wind strength (Figure 18.3).

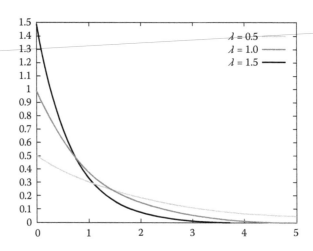

FIGURE 18.2
Density of probabilities of an exponential distribution.

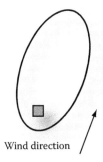

FIGURE 18.3
Ellipse rotated and translated.

18.3.3 Heat Influence: Physics-Inspired and Chemistry-Based Mathematical Analogs

The conversion degree α represents the advancement of the reaction according to the temperature in the fuel. When $\alpha = 100\%$, the reaction is over (all the combustible is burned). When $\alpha > 0\%$, the reaction starts. As experimentally proved by Leroy et al. (2010), $\alpha(T)$ is equivalent to a sigmoid function, depending on the gradient temperature β (cf. Figure 18.4).

If we refer to Equation 18.1, knowing the conversion degree α and the initial mass m_0, the current mass $m(t)$ can be calculated. Therefore, the problem now is to determine the temperature propagation in space.

In the physics-based *point source model* (McGrattan, Baum, and Hamins 2000) (cf. Figure 18.5), flames are modeled as point sources situated in the center of the flame. It is assumed that a fraction χ_r of the *total heat released rate* $\dot{Q}_i$, at position i, is received as a *thermal radiation flux* $\dot{\phi}_j$ by a target fuel (T), of dimensions A_j at position j. The radiated energy is calculated by dividing

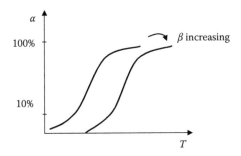

FIGURE 18.4
Simplified curve of conversion degree.

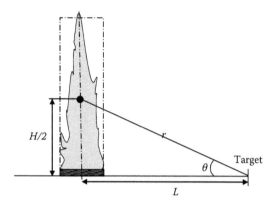

FIGURE 18.5
Geometry of radiation in the point source model.

this fraction by the surface area of a sphere whose radius r is the distance from the middle of flame height to the influenced fuel.

The total heat released rate $\dot{Q}_i$ is a product of the mass derivative in i, $\dfrac{dm_i}{dt}$ (kg/s), by the heat of combustion ΔH_i (kJ/kg), which is constant:

$$\dot{Q}_i = \frac{dm_i}{dt}\Delta H_i \tag{18.8}$$

Thus, we obtain:

$$\dot{\phi}_j = \chi_r \frac{dm_i}{dt}\frac{\Delta H_i}{4\pi r^2}\cos(\theta)A_j \tag{18.9}$$

Equation 18.8 remains valid for vertical flames, that is, without wind and/ or slope. When the wind blows, the flame is inclined. Therefore, a relation needs to be found [cf. Figure 18.6 and reference (Drysdale 1999) for more information], between: the angle formed by normal to the flame and the radiative ray directed to the target (φ_1); the angle formed by the normal to the fuel and the radiative ray directed to the target (φ_2); the distance between the emitter and the receiver; and the surface of emission and reception.

Using such a mathematical description of radiation geometry necessitates describing precisely the flame shape in three dimensions. Empirical formula can be used to determine the flame height (considered as constant) (McCaffrey 1979), such as:

$$H_i = 0.08\dot{Q}_i^{2/5} \tag{18.10}$$

where $\dot{Q}_i$ is the rate of heat released at position i.

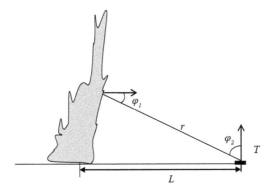

FIGURE 18.6
Geometry of radiation under wind.

Then, wind conditions, using a precise modeling of gas motion, are integrated to determine flame tilt angle in the equation (Morandini et al. 2002). However, this *level of precision* is difficult (computationally possible?) to fit with actual gas and flame motion during a fire spread. Moreover, this requires manipulating physics-specific notions, which avoid proposing a simple propagation simulation model of fire spread. One solution is to use automata to build computational models of fire spread using mathematical analogs (Dunn and Milne 2004). We attempt to go a step further here, building mathematical/computational analogs of physics-inspired processes of fire spread, integrating chemical processes.

We will take advantage of the *point source model* assumption, which considers that a portion of the total heat rate released by an ignited combustible is transmitted to the neighboring combustibles (without considering heat exchanges with air): $\chi_r \dot{Q}_i$. We assume now that a fraction χ_{h_i} of the total heat released at position i is transmitted (by radiation and diffusion) to the neighboring sites j, and that this heat influence decreases with the square distance between the sites according to an exponential law (cf. Figure 18.7).

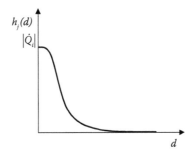

FIGURE 18.7
Heat influence as a function of distance.

The precise geometry of the flame (and flame tilt) is not taken into account. χ_h is assumed to be proportional to the flame height: $\chi_{h_i} \propto H_i$ (cf. Equation 18.10). Areas are considered to be equal at emission and reception (using a propagation cellular domain). We obtain the *heat influence rate* h_j:

$$\frac{dh_j}{dt} = \chi_{h_i} \dot{Q}_i e^{-qd^2} \tag{18.11}$$

where q (m^{-2}) is a constant depending on fuel characteristics.

Considering a distance reduction factor $\delta_{ij} = e^{-qd^2}$ and combining Equation 18.8 with Equation 18.11, we obtain:

$$\frac{dh_j}{dt} = \chi_{h_i} \delta_{ij} \frac{dm_i}{dt} \Delta H_i \tag{18.12}$$

Now, another factor needs to be integrated for modeling heat transfers: slope. We consider a slope reduction factor $\lambda_{ij} = z_j - z_i$, with z_i, the altitude at position i, and z_j, the altitude at position j. We obtain:

$$\frac{dh_j}{dt} = \chi_{h_i} \delta_{ij} \frac{dm_i}{dt} \Delta H_i (1 + \lambda_{ij}) \tag{18.13}$$

Another major parameter influencing a fire spreading is the moisture reduction of heat influence γ_j in the fuel. Moisture decreases the impact of heat on fuels. We obtain:

$$\frac{dh_j}{dt} = \chi_{h_i} \delta_{ij} \frac{dm_i}{dt} \Delta H_i (1 - \gamma_j)(1 + \lambda_{ij}) \tag{18.14}$$

The mass decrease in an ignited fuel (cf. Figure 18.4, Equation 18.3) depends on the temperature in the fuel. Variations of the temperature in the fuel depend mainly on three causes: (1) the heat transfer received, (2) the heat released, and (3) the heat exchanged with the air. Causes (1) and (2) increase the temperature in the ignited fuel. Cause (3) decreases the temperature. Many equations can be written to model this temperature dynamics (Drysdale 1999; Morandini et al. 2002).

Previous equations have been developed for a one-dimension spreading considering two different sites in the fuel (one site ignited and influencing, the other one influenced). In two dimensions, for many sites {i} influencing one site j, we obtain:

$$\frac{dh_j}{dt} = (1 - \gamma_j) \sum_{\{i\}} \chi_{h_i} \delta_{ij} \frac{dm_i}{dt} \Delta H_i (1 + \lambda_{ij}) \tag{18.15}$$

Our aim here is not to reuse these equations. It is more to focus on the transfer and release of heat in fire spread and to combine both with the mass loss rate in combustible. Figure 18.8 sketches the temperature curve in fuels.

Therefore, we do not account explicitly for temperature cooling by the air and consider merely that the temperature in cell j is a proportion τ of the heat received in this fuel:

$$T_j \propto \tau h_j \tag{18.16}$$

Finally, we describe the simulation sequence by the following algorithm:

ALGORITHM 1. MODEL COMPUTATION SEQUENCE

1. For all sites j influenced by ignited sites (and including them):
2. Compute temperature: $T_j = \tau h_j$
3. For all ignited sites i:
4. Compute mass loss: $\dfrac{d\alpha_i}{dT_i} = \dfrac{A}{\beta_i} e^{-E_a/RT_i} f(\alpha_i)$
5. Compute total heat released rate at position i: $\dot{Q}_i = \dfrac{dm_i}{dt} \Delta H_i$
6. Compute axis lengths of wind-based ellipse (area of heat influence of the ignited sites i):
$$\begin{cases} b \sim \exp(\lambda_b = 1) \\ a \sim b + \exp(\lambda_a) \end{cases}, \text{ with } \lambda_a = \frac{3}{2} - 0.9\frac{v}{v_{max}}, \text{ where } v_{max} \text{ is the}$$
maximum wind speed and v the wind speed.
7. For all cells j influenced by ignited sites:
8. Compute heat reception at position j: $\dfrac{dh_j}{dt} = (1-\gamma_j) \sum_{\{i\}} \chi_{h_i} \delta_{ij} \dfrac{dm_i}{dt}$ $\times \Delta H_i (1+\lambda_{ij})$

18.3.4 Fire Brands

Fire brands (when occurring) have a major effect on a fire spread path. This phenomenon receives recent attention by physicists and computer scientists (Porterie 2007). In the work of Porterie et al. (2007), small world networks are used (Watts and Strogatz 1998). The travel distance of firebrands is calculated according to an exponentially decaying probability distribution depending on fuel type, moisture, and wind.

Moisture influence has been previously integrated in the preceding section. We consider here the wind influence on fire brands (through direction and travel distance). According to the fuel type (e.g., merely if it is a pine tree), a simple probability p_{fb} is used for fire brand occurrence (cf. Figure 18.9).

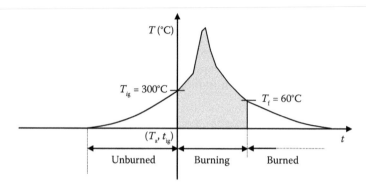

FIGURE 18.8
Simplified temperature curve of fuels.

If a fire brand occurs, the travel distance d_{fb} is determined according to an exponential law, based on wind speed v:

$$d_{fb} \sim \exp\left(\lambda_{d_{fb}} = \frac{1}{v} \right) \qquad (18.17)$$

The emission angle ω is calculated according to a normal law:

$$\omega \sim N(\mu, \sigma^2) \qquad (18.18)$$

With $\mu = v$ and $\sigma^2 = \dfrac{1}{d_{fb}}$.

Hence, the stronger the wind, the more chance a fire brand has to travel a long distance in the wind direction. Conversely, for low speed winds, fire brand distance is smaller and directions more stochastic.

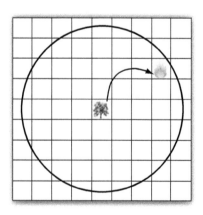

FIGURE 18.9
Fire brand ignition.

FIGURE 18.10
Temperature changes in the propagation domain.

18.4 Simulation Results

We present here qualitative simulation results validating our modeling approach. First, we check the temperature increase and decrease behavior presented in Figure 18.8. Colors are affected to temperatures. Figure 18.10 presents a fire spreading in a homogeneous fuel (dark gray). We can see that colors corresponding to temperatures change gradually from *gray* to *white and* then back (during temperature decrease) to *gray and* finally *black*.

We then defined a simple image analysis filter for detecting vegetation species. Figure 18.11 depicts the original picture.

Figure 18.12 describes a first image analysis of all the vegetation in the propagation domain.

Figure 18.13 depicts the detection of the shrub species (in gray).

Figure 18.14 validates the fire spread model under no wind, no moisture, and no wind conditions.

FIGURE 18.11
Original aerial picture of a possible propagation domain.

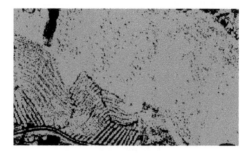

FIGURE 18.12
Detection of all the vegetation.

Figure 18.15 validates the fire spread model under no slope, no moisture, and wind conditions.

Figure 18.16 validates the fire spread model under no wind, no moisture, and slope conditions.

Figure 18.17 validates the fire spread model under no wind, no slope, and moisture conditions.

18.5 Discussion and Perspectives

We have presented a chemistry-based, physics-inspired, computational model using mathematical analogs. Parameters of the model have been determined through virtual experiments, to achieve fire spread qualitative

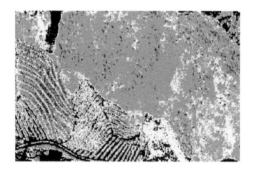

FIGURE 18.13
Detection of shrubs.

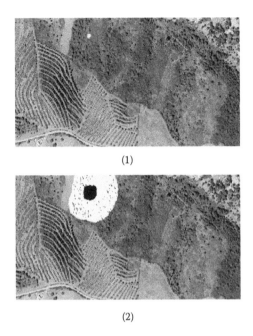

(1)

(2)

FIGURE 18.14
Fire spread simulation under no wind, no moisture, and no slope. Fire spread is punctual and circular, except in areas without vegetation detected (cf. Figure 13).

behaviors. As many fire spread phenomena have not been quantitatively described in physics, this global model tries to fill these modeling gaps. Mass loss of ignited fuel has been chosen as the major driving parameter of fire spread (through its impact on heat influence).

Currently, only two simulation models incorporate physical rules and bidimensional propagation algorithms: those developed by Porterie et al. (2007) and Rothermel (1972). Mathematical analogs and computational tools are fundamental to model and simulate fire spread. A lot of effort is necessary

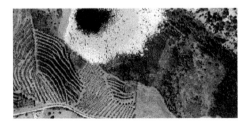

FIGURE 18.15
Fire spread simulation under a wind of 50 km/h blowing west, no moisture, and no slope. Fire spreads elliptically, except in areas without vegetation. Fire spread exhibits ignitions from fire brands.

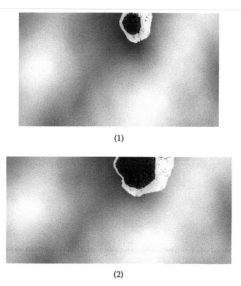

(1)

(2)

FIGURE 18.16
Fire spread simulation under no wind, no moisture, and slope. Increasing slope, in the west direction, is indicated in black. Decreasing slope in the west direction is indicated in white. In (1), fire spreads first elliptically in the black area, regressing in the white area. In (2), fire spreads more in the black area and slows down in the white area.

to develop mathematical and computational structures catching the main physical and chemical mechanisms of fire spread. The difficulty to achieve this goal resides in knowledge (and related vocabulary and models) which is specific to both chemistry and physics disciplines. However, fire spread can still be described at a first simple, rough, modeling level, using a *natural language vocabulary* for entities (wind, mass, heat released, heat influence, etc.) and behaviors (spreading, brand sending, etc.). This kind of canonical model (and related vocabulary) can then be specified to particular domain-specific approaches (e.g., "heat" will be called radiation, convection, or diffusion in physics).

FIGURE 18.17
Fire spread simulation under no wind, no slope, and moisture conditions. No moisture is indicated in black. Fire spreads more in the black area and slows down in the white area.

We believe that a mathematical and computational framework can be developed through a nonlinear fire spread model integrating all interactions (heat with fuel characteristics, topography, climate, etc.), such as the one presented in Equation (18.15). A long distance now remains to be covered for the experimental validation of all the reduction/interaction parameters of such an equation.

References

Akahira, T., and T. Sunose. 1971. Joint convention of four electrical institutes. *Research Report Chiba Institute Technology* 246.

Alexandridis, A., D. Vakalis, C. Siettos, and G. Bafas. 2008. A cellular automata model for forest fire spread prediction: the case of the wildfire that swept through Spetses Island in 1990. *Applied Mathematics and Computation* 204: 191–201.

Alén, R., E. Kuoppala, and P. Oesch. 1996. Formation of the main degradation compound groups from wood and its components during pyrolysis. *Journal of Analytical and Applied Pyrolysis* 36: 137–148.

Alves, S. S., and J. L. Figueiredo. 1989. Kinetics of cellulose pyrolysis modelled by three consecutive first-order reactions. *Journal of Analytical and Applied Pyrolysis* 17: 37–46.

Anderson, D. H., E. A. Catchpole, N. J. De Mestre, and T. Parkes. 1982. Modelling the spread of grass fires. *The ANZIAM Journal* 23: 451–466.

Balbi, J. H., P. A. Santoni, and J. L. Dupuy. 1998. Dynamic modelling of fire spread across a fuel bed. *Internation Journal of Wildland Fire* 9: 275–284.

Beall, F. C., and H. W. Eickner. 1970. WIS. *Thermal Degradation of Wood Components: A Review of the Literature*, Madison, WI: Forest Products Lab.

Berjak, S. G., and J. W. Hearne. 2002. An improved cellular automaton model for simulating fire in a spatially heterogeneous Savanna system. *Ecological Modelling* 148: 133–151.

Brown, M. E., A. K. Kelly, et al. 1988. A thermoanalytical study of the thermal decomposition of silver squarate. *Thermochimica Acta* 127: 139–158.

Caldarelli, G., R. Frondoni, A. Gabrielli, M. Montuori, R. Retzlaff, and C. Ricotta. 2001. Percolation in real Wildfires. *cond-mat/0108011*, Aoû.

Curry, J. R., and W. L. Fons. 1938. Rate of spread of surface fires in the ponderosa pine type of California. *Journal of Agricultural Research* 57: 239–267.

Di Blasi, C., C. Branca, and A. Galgano. 2008. Thermal and catalytic decomposition of wood impregnated with sulfur and phosphorus-containing ammonium salts. *Polymer Degradation and Stability* 93: 335–346.

Dimitrakopoulos, A. P. 2001. Thermogravimetric analysis of Mediterranean plant species. *Journal of Analytical and Applied Pyrolysis* 60: 123–130.

Drysdale, D. 1999. *An Introduction to Fire Dynamics*. 2nd ed. West Sussex, UK: Wiley.

Dunn, A., and G. Milne. 2004. Modelling wildfire dynamics via interacting automata. *Cellular Automata* 3305: 395–404.

Encinas, A. H., L. H. Encinas, S. H. White, A. M. D. Rey, and G. R. Sánchez. 2007a. Simulation of forest fire fronts using cellular automata. *Advances in Engineering Software* 38: 372–378.

Encinas, L. H., S. H. White, A. M. D. Rey, and G. R. Sánchez. 2007b. Modelling forest fire spread using hexagonal cellular automata. *Applied Mathematical Modelling* 31: 1213–1227.

Flynn, J. H., L. A. Wall, and A. Quick. 1966. Direct method for the determination of activation energy from thermogravimetric data. *Journal of Polymer Sciences* 4: 323–328.

Glasa, J., and L. Halada. 2008. On elliptical model for forest fire spread modeling and simulation. *Mathematics and Computers in Simulation* 78: 76–88.

Grishin, A. M. 1996. General mathematical model for forest fires and its applications. *Combustion, Explosion, and Shock Waves* 32: 503–519.

Grishin, A. M. 1997. *Mathematical modeling of forest fires and new methods of fighting them.* trans. M. Czuma. Tomsk: Publishing House of the Tomsk State University.

Karafyllidis, I., and A. Thanailakis. 1997. A model for predicting forest fire spreading using cellular automata. *Ecological Modelling* 99: 87–97.

Khawam, A., and D. R. Flanagan. 2005. Complementary use of model-free and modelistic methods in the analysis of solid-state kinetics. *Journal of Physical Chemistry B* 109: 10073–10080.

Kissinger, H. E. 1957. Reaction kinetics in differential thermal analysis. *Analytical Chemistry* 29: 1702–1706.

Lasaponara, R. 2010. Editorial. *Ecological Modelling* 221: 1.

Leroy, V., D. Cancellieri, E. Leoni, and J. L. Rossi. 2010. Kinetic study of forest fuels by TGA: Model-free kinetic approach for the prediction of phenomena. *Thermochimica Acta* 497 (1–2): 1–6.

Liodakis, S., D. Bakirtzis, and A. Dimitrakopoulos. 2002. Ignition characteristics of forest species in relation to thermal analysis data. *Thermochimica Acta* 390: 83–91.

McCaffrey, B. J. 1979. Purely buoyant diffusion flames: Some experimental results. *National Bureau of Standards, NBSIR* 79–1910.

McCormick, R. J., T. A. Brander, and T. F. H. Allen. 2000. Towards a theory of mesoscale wildfire modeling—a complex systems approach using artificial neural networks. In *Proceedings from the Joint Fire Science Conference and Workshop.* Crossing the Millennium: Integrating Spatial Technologies and Ecological Principles for a New Age in Fire Management, Boise, ID. June 15–1, 3–15.

McGrattan, K. B., H. R. Baum, and A. Hamins. 2000. Thermal radiation from large pool fires. NISTIR-6546, NIST, Gaithersburg, MD, USA.

McRae, R. H. 1990. Use of digital terrain data for calculating fire rates of spread with the PREPLAN computer system. *Mathematical and Computer Modelling* 13: 37–48.

Morandini, F., P. A. Santoni, J. H. Balbi, J. M. Ventura, and J. M. Mendes-Lopes. 2002. A two-dimensional model of fire spread across a fuel bed including wind combined with slope conditions. *International Journal of Wildland Fire* 11: 53–63.

Ozawa, T. 2000. Thermal analysis—review and prospect. *Thermochimica Acta* 355: 35–42.

Soltes, E. J. 1988. Of biomass, pyrolysis and liquids therefrom. In *Pyrolysis Oils from Biomass: Producing, Analyzing, and Upgrading,* 1. ACS Symposium.

Porterie, B., N. Zekri, J. Clerc, and J. Loraud. 2007. Modeling forest fire spread and spotting process with small world networks. *Combustion and Flame* 149: 63–78.

Porterie, B., S. Nicolas, J. L. Consalvi, J. C. Loraud, F. Giroud, and C. Picard. 2005. Modeling thermal impact of wildland fires on structures in the urban interface: Part 1. Radiative and convective components of flames representative of vegetation fires. *Numerical Heat Transfer, Part A: Applications* 47: 471–489.

Porterie, B., N. Zekri, J. Clerc, and J. Loraud. 2007. Modeling forest fire spread and spotting process with small world networks. *Combustion and Flame* 149: 63–78.

Prasad, T. P., S. B. Kanungo, and H. S. Ray. 1992. Non-isothermal kinetics: some merits and limitations. *Thermochimica Acta* 203: 503–514.

Rothermel, R. C. 1972. A mathematical model for predicting fire spread in wildland fuels, USDA Forest Service Research Paper INT (USA).

Sofer, S. S., and O. R. Zaborsky. 1981. *Biomass Conversion Processes for Energy and Fuels*. New York: Plenum Press.

Statheropoulos, M., S. Liodakis, N. Tzamtzis, A. Pappa, and S. Kyriakou. 1997. Thermal degradation of *Pinus halepensis* pine-needles using various analytical methods. *Journal of Analytical and Applied Pyrolysis* 43: 115–123.

Sullivan, A. L. 2009. Wildland surface fire spread modelling, 1990–2007: 3. Simulation and mathematical analogue models. *International Journal of Wildland Fire* 18: 387–403.

Sullivan, A. L. Jun. 2007. A review of wildland fire spread modelling, 1990–present: 1. Physical and quasi-physical models. *0706.3074*.

Vyazovkin, S., and C. A. Wight. 1997. Kinetics in solids. *Annual Review of Physical Chemistry* 48: 125–149.

Vyazovkin, S., and C. A. Wight. 1999. Model-free and model-fitting approaches to kinetic analysis of isothermal and nonisothermal data. *Thermochimica Acta* 340: 53–68.

Watts, D. J., and S. H. Strogatz. 1998. Collective dynamics of 'small-world' networks. *Nature* 393: 440–442.

Weber, R. 1991. Modelling fire spread through fuel beds. *Progress in Energy and Combustion Science* 17: 67–82.

Wight, S. V. 1998. Isothermal and non-isothermal kinetics of thermally stimulated reactions of solids. *International Reviews in Physical Chemistry* 17: 407–433.

19

Air Pollution

Kostadin Ganev

CONTENTS

19.1 Introduction...500
19.2 Basic Approaches to Air Pollution Modeling 502
    19.2.1 Mass Conservation Relation...503
    19.2.2 Eulerian Approach to Air Pollution Modeling 503
    19.2.3 Lagrangian Approach to Air Pollution Modeling 505
    19.2.4 Comparison of Eulerian and Lagrangian Approaches............507
    19.2.5 Boundary Conditions 508
    19.2.6 Trajectory "Puff" and "Plume" Models..509
        19.2.6.1 Puff Models..510
        19.2.6.2 Plume Models 512
19.3 Main Processes that Determine Air Pollution.......................................513
    19.3.1 Gas-Phase Chemistry...514
        19.3.1.1 CB4 Mechanism 515
        19.3.1.2 RADM2 Mechanism 516
        19.3.1.3 SAPRC-97 Mechanism...516
    19.3.2 Aerosol Dynamics 517
    19.3.3 Cloud Dynamics and Chemistry...519
    19.3.4 Atmospheric Processes 519
    19.3.5 Emissions 520
19.4 Evaluation of the Simulation Results 523
19.5 Some Examples of Up-to-Date Air Pollution Models 523
    19.5.1 Community Multiscale Air Quality System...............................528
        19.5.1.1 The CMAQ Structure 528
        19.5.1.2 The CMAQ Chemical Transport Model 529
        19.5.1.3 CMAQ Output..531
    19.5.2 The Fifth-Generation PSU/NCAR MM5 532
        19.5.2.1 The Meteorology Model..532
        19.5.2.2 Meteorology Model Preprocessing................................537
        19.5.2.3 Nesting 539
        19.5.2.4 Four-Dimensional Data Assimilation 539
        19.5.2.5 Data Required to Run the MM5 Modeling System....540

19.5.3 SMOKE Modeling System 540
    19.5.3.1 Summary of the Major Features of SMOKE
          Version 2.0 541
19.5.4 ADMS 4 Air Pollution Model 543
    19.5.4.1 Model Output 545
    19.5.4.2 ADMS 4 Industrial 545
    19.5.4.3 ADMS-Urban 546
    19.5.4.4 ADMS-Roads 547
    19.5.4.5 ADMS-Airport 547
19.5.5 SILAM Air Pollution Model 548
    19.5.5.1 Lagrangian Advection Algorithm with Monte
          Carlo Random Walk 549
19.6 Conclusion 550
References 551

19.1 Introduction

The composition of the atmosphere is rather complex—a large number of gases and atmospheric aerosol (small particles of solid or liquid with various chemical composition, suspended in the atmosphere)—and it has always been that way, even before mankind appeared on Earth. This is why the question "what actually should be considered as *air pollution*" is not pointless. Intuitively, it is clear that compounds with a concentration significantly larger than the that in "clean air" should be treated as pollutants.

From a practical viewpoint, it is more important to consider the effect that different compounds have on human health, ecosystems, or climate. A brief overview of perhaps the most prominent "bad" effects of different atmospheric compounds is given in Table 19.1. As shown in the table, the different types of pollution causing different problems have also different temporal and spatial scales, which are determined by the processes that characterize the different types of pollution.

Air quality (AQ) is a key element for human well-being and quality of life. According to the World Health Organization (WHO), air pollution severely affects the health of European citizens (WHO 2004)—between 2.5% and 11% of the total number of annual deaths are due to air pollution (WHO 2000). There is considerable concern about impaired and detrimental AQ conditions over many areas in Europe, especially in urbanized areas, in spite of about 30 years of legislation and emission reduction. Current legislation (e.g., ozone daughter directive 2002/3/EC) requires informing the public on AQ, assessing air pollutant concentrations throughout the whole territory of member states and indicating exceedances of limit and target values, forecasting potential exceedances, and assessing possible emergency measures to abate exceedances using modeling tools.

TABLE 19.1

Today's Problems of Atmospheric Chemistry, Aerosols, and Clouds

Problems	Compounds	Effects	Scale (Temporal/Spatial)
Winter smog	benzene (carcinogenic), SO_2, aerosol formation	toxic for humans	days/local (inversions)
Summer smog	formation of photo-oxidants from VOCs, CO, and NO_x, and of aerosol particles	harvest reductions, human health	several days/ regional (continental)
Polluted precipitation, acid rain	H_2SO_4 and HNO_3 (from SO_2 and NO_x)	acidification of soils, N-fertilization	days to weeks/ continental
Stratospheric ozone destruction	chlorofluorocarbons (CFCs), halons, H-CFCs (in interaction with CH_4, N_2O, cloud particles)	UV-B increase at ground level	decades/global
Greenhouse effect I	well-mixed greenhouse gasses: CO_2, CH_4, N_2O, CFCs; also H_2O, O_3, etc.	climate change: warming	decades to centuries/global
Greenhouse effect II	aerosols and clouds, direct and indirect effects	partly compensating climate effect: cooling	days to weeks

It should be noted that say 30 years ago the main concern was about the classic acidifying pollutants (SO_2, NO_x), the heavy metals (Hg, Cd, Pb), and the persistent organic pollutants. Gradually, partly because some of the problems with the above "classic" pollutants are to some extent solved, but also because of the new knowledge gained about the biological impact of different admixtures, the photo-oxidant (e.g., ozone), and particulate matter (PM), air pollution has become a major environmental problem. These pollutants are, to a large extent, secondary—products of chemical transformations and aerosol processes of other atmospheric constituents.

In the past, AQ model paradigms typically addressed individual pollutant issues separately. However, it is becoming increasingly evident that when pollutant issues are treated in isolation, the resulting control strategies may solve one set of problems but may lead to unexpected aggravation of other related pollutant issues. Pollutants in the atmosphere are subject to myriad transport processes and transformation pathways that control their composition and levels. Also, pollutant concentration fields are sensitive to the type and history of the atmospheric mixtures of different chemical compounds. Thus, modeled abatement strategies of pollutant precursors, such as volatile organic compounds (VOC) and NO_x, to reduce ozone levels may, under a variety of conditions, cause an exacerbation of other air pollutants such as PM or issues of acidic deposition. A wide range of AQ issues arise all the time including visibility, fine and coarse particles, indirect exposure to toxic pollutants such as heavy metals, semivolatile organic species, nutrient deposition to water bodies, impacts of atmospheric composition on the optic properties of the atmosphere and thus to Earth's health balance, hence to the global climate.

Because of the increased modeling requirements, more comprehensive modeling approaches appear to be needed. With projections for the increasing rapid pace of the development of computational capabilities at the start of the 1990s, the opportunity arose for a strategic review of modeling approaches leading to design of a system that would both meet and keep pace with the increasing requirements of AQ modeling, of incorporating advances in state-of-science descriptions of atmospheric processes. The scope of such a system must be able to process great and diverse information from complicated emissions mixtures and complex distributions of sources, to modeling the complexities of atmospheric processes that transport and transform these mixtures in a dynamic environment that operates on a large range of time scales covering minutes to days and weeks. The corresponding spatial scales are commensurately large, ranging from local to continental scales. On these temporal and spatial scales, emissions from chemical manufacturing and other industrial activities, power generation, transportation, and waste treatment activities contribute to a variety of air pollution issues including visibility, ozone, PM, and acid, nutrient, and toxic deposition. The residence times of pollutants in the atmosphere can extend to multiple days; therefore, transport must be considered on at least a regional scale. The AQ regulation requirements and other goals for a cleaner environment vary over a range of time scales, from peak hourly to annual averages. These challenges suggest that more comprehensive approaches to AQ modeling are needed, and that assessments and pollution mitigation are achieved more successfully when the problems are viewed in a "one atmosphere" context that considers multiple pollutant issues.

Air pollution is most often a result of human activities. There are some natural events, however, such as volcanoes and forest fires, which are substantial sources of air pollution as well. The biogenic VOC are one of the most important ozone precursors.

19.2 Basic Approaches to Air Pollution Modeling

Atmospheric models are, broadly speaking, any mathematical procedure that results in an estimation of ambient AQ entities (i.e., concentrations, deposition, exceedances). In general terms, a distinction between process-oriented models and statistical models can be made. Process-oriented models are based on the description of physical/chemical processes: starting with emissions, atmospheric advection, and dispersion, chemical transformation and deposition is calculated. This type of model is able to give a description of cause–effect relations.

Statistical models are valuable tools in the diagnosis of present AQ via interpolation and extrapolation of measuring data, but they will not be discussed in the present chapter.

19.2.1 Mass Conservation Relation

Each and any process-oriented air pollution model is based on the mass conservation relation. Let us assume that domain D with a lateral boundary S is an arbitrary three-dimensional (3-D) domain and $c(x, y, z, t)$ is the concentration (mass/volume) of a given atmospheric constituent (admixture)—gas or aerosol. Then, the total admixture quantity in D—$C_D(t) = \iiint_D c(x,y,z,t)\,dD$ should fulfill the mass conservation relation:

$$\frac{dC_D}{dt} = Q_D + F_{chem} + F_{aero} - F_{wash} - F_S \tag{19.1}$$

The right-hand terms of 19.1 are as follows:

QD = emissions of the chosen admixture within the chosen domain

F_{chem} = changes in the chosen admixture quantity C_D due to chemical transformations (interaction with other species)

F_{aero} = are the changes in the chosen admixture quantity C_D due to interactions with or within the atmospheric aerosol

F_{wash} = accounts for the possible washout of the chosen admixture by precipitation—absorption by rain, fog, or snow particles

F_S = total quantity of the given admixture that passed (inflow or outflow) through the lateral boundary S due to the arranged motions (wind and gravity deposition) and turbulence in the atmosphere

It is clear that, depending on the chosen constituent (admixture), as well as on the other constituents in the domain D, the terms F_{chem} and F_{aero} have their signs—they may lead to increase or decrease the total quantity C_D. The washout obviously leads to decrease of C_D.

The sign of F_s can also be positive (net outflow) or negative (net inflow).

19.2.2 Eulerian Approach to Air Pollution Modeling

Like the total admixture quantity C_D can be expressed as an integral of the concentration $c(x, y, z, t)$ over D, the terms Q_D, F_{chem}, F_{aero}, and F_{wash} can also be presented as integrals of the respective functions $q, f_{chem}, f_{aero}, f_{wash}$, which account for the emission, chemical transformation, aerosol processes, and washout in each point with coordinate $\mathbf{x} = (x, y, z)$ in a moment t.

If $\mathbf{u}(x, y, x, t)$ is the vector of the arranged motion velocity and $\tau(x, y, z, t)$ is the vector of the turbulent diffusion flux, the total admixture flux in a given point can be defined as

$$\mathbf{f}(x, y, x, t) = \mathbf{u}(x, y, x, t)c(x, y, x, t) + \tau(x, y, z, t). \tag{19.2}$$

Then, F_s is calculated according to the formula:

$$F_S = \oiint\limits_S \mathbf{f}\cdot\mathbf{n}\,dS, \tag{19.3}$$

where $\mathbf{n}$ is the unity vector normal to the surface S. $\mathbf{n}$ is considered positive when directed out of D, so the scalar product $\mathbf{f} \cdot \mathbf{n}$ will be positive when the net admixture flux $\mathbf{f}$ is also directed outward. This explains the sign of F_s in 19.1.

If the Gauss theorem is applied:

$$F_S = \oiint\limits_S \mathbf{f}\cdot\mathbf{n}\,dS = \iiint\limits_D \nabla\cdot\mathbf{f}\,dD, \tag{19.4}$$

where $\nabla = \left(\dfrac{\partial}{\partial x}, \dfrac{\partial}{\partial y}, \dfrac{\partial}{\partial z}\right)$, the mass conservation relation can be also written in the form:

$$\iiint\limits_D \left(\frac{\partial c}{\partial t} = q + f_{chem} + f_{aero} - f_{wash} - \nabla\cdot\mathbf{f}\right) dD. \tag{19.5}$$

The expression 19.5 is valid for an arbitrary domain D. This is possible only if the relation within the brackets is valid for each point in the atmosphere, and that is actually the Eulerian approach for description of the transport and transformation processes in the atmosphere.

Nothing is said yet about the turbulent diffusion in the atmosphere. The atmosphere is always turbulent—due to instabilities in the arranged flows a cascade of eddies of different size is formed. They move and interact with one another and within one another in a random fashion on the background of the basic (arranged) motions, thus causing a stochastic behavior of the atmospheric characteristics, including velocity and composition.

The statistical (over the ensemble of turbulent fluctuations) averaging produces the mean atmospheric characteristics as well as some covariant terms, which characterize the mean turbulent mixing.

The theory of atmospheric turbulence is too voluminous and complex, and so it cannot be even sketchily presented here. There are two things about the turbulent diffusion in the atmosphere that are very important for air pollution modeling and should be noted:

- The turbulent diffusion causes much more intensive mixture than the molecular one, so the molecular diffusion is not accounted for in the air pollution models.
- Because of the stochastic movement of the turbulent eddies, some analogy with the molecular diffusion can be made, and so the turbulent admixture flux in the air pollution models is usually presented in the form:

$$\tau = -\mathbf{K} \times \nabla c, \tag{19.6}$$

where $\mathbf{K}$ is the tensor of turbulent diffusion coefficients, which depend strongly on the meteorological conditions. Pretty often the nondiagonal terms of $\mathbf{K}$ are neglected in the air pollution models so the turbulent flux is presented by the expression:

$$\tau = -\left(K_{xx} \frac{\partial c}{\partial x}, \; K_{yy} \frac{\partial c}{\partial y}, \; K_{zz} \frac{\partial c}{\partial z} \right). \tag{19.7}$$

If τ from 19.6 is inserted in the expression 19.2 for net admixture flux $\mathbf{f}$, and after that $\mathbf{f}$ in 19.5 is substituted by the expression 19.2, having in mind that the relation within the brackets is valid for each point in the atmosphere and obviously for each admixture, the most popular form of the system of equation for transport and transformation of a set of N admixtures in the atmosphere can be obtained:

$$\frac{\partial c_i}{\partial t} + \nabla \cdot (\mathbf{u} c_i) = q_i + f_{i\,chem} + f_{i\,aero} - f_{i\,wash} + \nabla \cdot (\mathbf{K} . \nabla c_i), \quad i = 1, 2, ..., N. \tag{19.8}$$

19.2.3 Lagrangian Approach to Air Pollution Modeling

If the Eulerian approach is based on keeping an eye on the mean admixture concentration changes in each point, the Lagrangian approach is based on tracking the motion of a single fluid particle. Single fluid particle means a volume of fluid large compared to molecular dimensions, but small enough to act as a particle that exactly follows the fluid. The "particle" may contain fluid (air) with a different composition than the carrier fluid.

Let a particle be in a location $x' = (x', y', z')$ at a moment t' and let $X(x', t'; t)$ be the particle trajectory (the change of the particle location with time t) in the turbulent air and $\Psi(x, t)$ the probability density function ($\Psi(x, t)$ dD defines the probability that the particle is within the elementary volume $x + dx$ at moment t). As the particle should be somewhere, it is obvious that

$$\int_{-\infty}^{\infty}\int_{-\infty}^{\infty}\int_{-\infty}^{\infty} \Psi(x,t)\,dD = 1. \tag{19.9}$$

The probability density $\Psi(x, t)$ of finding the particle in point x at moment t can be expressed as a product of two other probability densities:

- The probability density that a particle, which at moment t' has been in point x', will move to point x at moment t—the transition probability density $T(x, t|x', t')$;
- The probability density $\Psi(x', t')$ that the particle is in point x' at moment t', integrated over all the possible starting points x'. Thus:

$$\Psi(x,t) = \int_{-\infty}^{\infty}\int_{-\infty}^{\infty}\int_{-\infty}^{\infty} T(x,t|x',t')\Psi(x',t')\,dD'. \tag{19.10}$$

The probability density function $\Psi(x, t)$ has been defined with respect to a single particle. If an arbitrary number m of particles are tracked and the position of each is given by the density function $\Psi_i(x, t)$, it can be shown that the ensemble mean concentration (in the terms of number of particles) is given by:

$$c(x,t) = \sum_{i=1}^{m} \Psi_i(x,t)'. \tag{19.11}$$

As demonstrated by Sienfeld and Pandis (1998), by expressing the probability density function $\Psi_i(x, t)$ in terms of the initial particle distribution and the spatiotemporal distribution of particle sources $S(x, t)$ (in unit particles per volume per time), and substituting the resulting expressions in 19.10, the following general formula for the mean concentration can be obtained:

$$c(\mathbf{x},t) = \int\limits_{-\infty}^{\infty}\int\limits_{-\infty}^{\infty}\int\limits_{-\infty}^{\infty} T(\mathbf{x},t|\mathbf{x}_0,t_0)c(\mathbf{x}_0,t_0)\,\mathrm{d}D_0 + \int\limits_{-\infty}^{\infty}\int\limits_{-\infty}^{\infty}\int\limits_{-\infty}^{\infty}\int\limits_{t_0}^{t} T(\mathbf{x},t|\mathbf{x}',t')S(\mathbf{x}',t')\,\mathrm{d}t'\,\mathrm{d}D'.$$

(19.12)

The first term on the right-hand side of 19.12 presents the contribution of the particles present at the initial moment t_0, and the second term accounts for the particles emitted by particle sources during the time interval (t_0, t).

Equation 19.12 is the fundamental Lagrangian relation for mean concentrations. It is clear, however, that although the initial concentration and particle sources may be known, the transition probability density could be known only in case of complete knowledge of the turbulence properties, hence the transition probability density can be evaluated only in the simplest circumstances (which is actually valid for the solution of Equations 19.8 as well).

One cannot help but notice also that all the considerations in the Lagrangian approach do not explicitly account for chemical reactions.

19.2.4 Comparison of Eulerian and Lagrangian Approaches

As already stated, the techniques for describing the statistical properties of the concentrations in a turbulent fluid can be divided into two categories: Eulerian and Lagrangian. The Eulerian methods attempt to formulate the concentration statistics in terms of the statistical properties of the Eulerian fluid velocities, that is, the velocities measured at fixed points in the fluid. A formulation of this type is very useful not only because the Eulerian statistics are readily measurable (as determined from continuous-time recordings of the wind velocities by a fixed network of instruments) but also because the mathematical expressions are directly applicable to situations in which chemical reactions are taking place. Unfortunately, the Eulerian approaches lead to a serious mathematical obstacles known as the "closure problem" (explicit formulation of the turbulent diffusion coefficients), for which no generally valid solution has yet been found.

In contrast, the Lagrangian techniques attempt to describe the concentration statistics in terms of the statistical properties of the displacements of groups of particles released in the fluid. The mathematics of this approach is more tractable than that of the Eulerian methods, in that no closure problem is encountered, but the applicability of the resulting equations is limited because of the difficulty of accurately determining the required particle statistics. Moreover, the equations are not directly applicable to problems involving nonlinear chemical reactions.

It should be once again noted that exact solution for the mean concentrations $c(x, y, z, t)$ even of inert species in a turbulent fluid is not possible in general by either the Eulerian or Lagrangian approaches.

All the existing air pollution models are either numerical or/and based on simplifying assumptions and approximations. It seems that most of the current air pollution models are based on the Eulerian approach. There are, however, some so-called "particle" models, which are essentially Lagrangian—they are based on following the movement of a large number of particles (released by the sources or entering through the model domain boundaries), then at some steps concentration fields for the different admixtures are constructed by the particle ensemble, the transformation processes are accounted for (change of the composition of the particles), and then again the particle ensemble is treated in Lagrangian manner.

The "puff" and "plume" models, which will be considered below, are sometimes also called Lagrangian, but this is not quite precise. They should be better called "trajectory" models.

19.2.5 Boundary Conditions

The admixture concentrations in a given domain depend not only on the admixture sources, the transport and transformation within the domain, but also on the admixture fluxes through the domain boundaries. Even the global models are formulated in a semispace, so they need a boundary condition on Earth's surface. The models formulated for limited domains should also have boundary conditions for the lateral and upper boundary.

The boundary conditions are usually formulated as a relation between normal to the boundary admixture fluxes and admixture concentrations at the boundaries:

$$\alpha \mathbf{f} \cdot \mathbf{n} = \beta c + m. \tag{19.13}$$

Formula 19.13 is general enough to describe a large variety of regimes at the boundaries. Depending on the coefficients α, β, and m, it could obviously mean prescribed concentrations or fluxes (in particular, zero concentrations or fluxes), admixture sources, or sinks at the boundary, etc.

The pollution concentrations are hardly ever measured with spatial and temporal resolution good enough, so that this data can be used as boundary conditions. This is why the up-to-date numerical models apply "nesting" procedures for obtaining the boundary concentrations for the lateral boundaries. The nesting means that the air pollution problem is solved for a chain of successive domains each smaller domain "nested" and taking its lateral boundary conditions from the simulations within the preceding one, thus downscaling the problem to a desired resolution. The lateral boundary concentrations for the outermost domain are usually taken from some global models or from some typical climatic concentration profiles.

The boundary condition at Earth's surface also has the form 19.13. In this case, it reflects the interaction of the admixture with the underlying surface—mechanical and/or chemical absorption by soil, water surfaces, and vegetation, possible admixture reemissions, as well as surface pollution sources at the boundary. The coefficients α, β, and m in this case depend on the nature of the admixture and the underlying surface and on the near surface dynamic (turbulent) properties of the atmosphere. The explicit form of the coefficients could be rather complex as can be seen, for example, in the work of Baldocchi et al. (1987), Erisman and Wyers (1994), Ganev and Yordanov (2005), Hicks et al. (1987), Joffre (1988), Sienfeld and Pandis (1998), Seland et al. (1995), Venckatram and Pleim (1999), and Wesley (1989).

In the Lagrangian models, the interaction (absorption, reflection) of fluid particles with the underlying surface should be accounted for by the transition probability density.

19.2.6 Trajectory "Puff" and "Plume" Models

As shown in the report of Sienfeld and Pandis (1998), the concentration of an instantaneous point source (let in 19.12, $S(\mathbf{x}, t) = 0$, $c(\mathbf{x}_0, t_0) = \delta(\mathbf{x}_0)$, $\mathbf{x}_0 = (0, 0, 0)$, $t_0 = 0$, δ—the Dirac function) in a flow, where the wind is a stationary Gaussian random process (say with mean wind velocity $\mathbf{u} = (u, 0, 0)$), has a Gaussian distribution itself:

$$c(\mathbf{x},t) = \frac{1}{(2\pi)^{3/2}\sigma_x\sigma_y\sigma_z}\exp\left[-\frac{(x-ut)^2}{2\sigma_x^2(t)} - \frac{y^2}{2\sigma_y^2(t)} - \frac{z^2}{2\sigma_z^2(t)}\right]. \tag{19.14}$$

Formula 19.14 actually describes the movement along the mean wind (($x - ut, y, z$) describes the trajectory of the puff center) and the spread in all three directions of an admixture cloud ("puff").

It can be shown that the dispersions in the three directions σ_x, σ_y, σ_z, which are due to the atmospheric turbulence (wind velocity fluctuations), vary with time in the following way:

$$\sigma^2 \sim t^2 \text{ for small } t \text{ and } \sigma^2 \sim t \text{ for large } t. \tag{19.15}$$

Again, in the work of Sienfeld and Pandis (1998), the solution for the case of continuous point source ($S(\mathbf{x}, t) = q\delta(\mathbf{x}_0)$) under the same dynamic conditions is shown:

$$c(\mathbf{x}) = \frac{q}{2\pi u\,\sigma_y\sigma_z}\exp\left[-\frac{y^2}{2\sigma_y^2} - \frac{z^2}{2\sigma_z^2}\right], \tag{19.16}$$

where the dispersions σ_y and σ_z change with x as in 19.14, only instead on t, they depend on $\frac{x}{u}$ (a ratio, which also has the dimension of time).

It should be noted that in obtaining 19.16, the turbulent diffusion in wind direction is neglected—considered small compared to the advection (transport by the mean wind).

Formula 19.16 describes a stationary "plume" with an axis oriented along the wind, which spreads from the stationary source in vertical and crosswind direction.

The same puff and plume formula can be obtained from 19.8 if the Eulerian approach is applied. The restricting conditions are of stationary horizontally homogeneous wind and turbulence, diagonal tensor of turbulent diffusion coefficients, and no transformation of the admixture. The respective formulae for puff and plume have the same form as 19.14 and 19.16. The dispersions in all directions depend on the respective diagonal elements in the following way:

$$\sigma = \sqrt{2Kt} \text{ for puff or } \sigma = \sqrt{2K\frac{x}{u}} \text{ for plume.} \tag{19.17}$$

The "puff" (19.14) and "plume" (19.16) solutions are obtained under very restrictive conditions—homogeneous and stationary meteorological fields, which was already mentioned above, and what have not been mentioned yet—both solutions do not account for the interaction of the admixtures with underlying surface and for the admixture transformations. The "puff" and "plume" constructions, however, are widely used as approximate solutions, which are in the basis of a large class of trajectory models, called "puff" or "plume" models.

19.2.6.1 Puff Models

The puff models again consider the concentration from an instantaneous point source—a release of admixture quantities C_{0i} in point $x_0 = (x_0, y_0, z_0)$ in moment $t = t_0$ ($c_i(x, t_0) = C_{0i}\delta(x - x_0)$). It is assumed that the center of masses $X = (X, Y, Z)$ of this puff moves with the air flow according to the law:

$$\frac{dX}{dt} = U, \ X(t = t_0) = x_0, \ U(t) = u(X(t)) \tag{19.18}$$

It is also assumed that the whole puff moves with the center of masses without deformations, with Gaussian distribution of concentrations in horizontal direction, although the wind velocity could not be horizontally homogeneous. Under such assumptions, the concentration fields for an ensemble of pollutants from instantaneous point source could be approximately described in the form:

$$c_i(x, y, z, t) = c_{i00}(z, t)c_{hor}(x, y, t),$$

$$c_{hor}(x,y,t) = \frac{1}{2\pi\sigma_x\sigma_y}\exp\left[-\frac{(x-X(t))^2}{2\sigma_x^2(t)} - \frac{(y-Y(t))^2}{2\sigma_y^2(t)}\right], \qquad (19.19)$$

where $c_{hor}(x, y, t)$ accounts for the pollutants horizontal transport and spread and $c_{i00}(z, t)$ accounts for the vertical transport, admixture transformations of any kind, and for the interaction of the underlying surface:

$$\frac{\partial c_{i00}}{\partial t} + \frac{\partial}{\partial z}\left(wc_{i00} - K_{zz}\frac{\partial c_{i00}}{\partial z}\right) = f_{i\,chem} + f_{i\,aero} - f_{i\,wash},$$

$$c_{i00}(z,t_0) = C_{0i}\delta(z - z_0), \quad i = 1,2,\ldots,N. \qquad (19.20)$$

Thus, by dividing the processes into "horizontal transport and diffusion" and "all the others," a much simpler approximate solution of 19.8 can be obtained. In some of the existing models, things are further simplified if an analytic expression for the admixtures vertical profile $c_{ivert}(z, t)$ is introduced. To be consistent with the problem, the function $c_{ivert}(z, t)$ should obviously fulfill the equation:

$$\frac{\partial c_{ivert}}{\partial t} + \frac{\partial}{\partial z}\left(wc_{ivert} - K_{zz}\frac{\partial c_{ivert}}{\partial z}\right) 0. \qquad (19.21)$$

Depending on the boundary condition at the lower boundary at which 19.21 is solved, the vertical profile can partially account for the absorption by underlying surface, so it could be admixture specific. Then the component $c_{i00}(z, t)$ could be presented in the form:

$$c_{i00}(z, t) = c_{ivert}(z, t)C_i(t), \quad i = 1, 2, \ldots, N, \qquad (19.22)$$

where $C_i(t)$ accounts for all the admixture transformations. They give, actually, the total admixture contained in the puff, and so should be a subject of relations similar to 19.1, where F_S accounts for the absorption by the underlying surface and eventually for the flux trough the upper boundary.

Equations 19.20 and 19.21 define an approximate solution for an instantaneous point source. However, the "puff" approximation could also be applied for describing the pollution fields from arbitrary sources. It is clear that the emission terms $q_i(x, y, z, t)$ in 19.8 can be presented as a superposition of elementary instantaneous point sources:

$$q_i(x,y,z,t)= \int_0^t dt_0 \delta(t-t_0) \iiint_D C_{0i}(x_0,y_0,z_0,t_0) \delta(x-x_0)\delta(y-y_0)\delta(z-z_0)dx_0 dy_0 dz_0,$$

(19.23)

Thus, according to the superposition principle (which is strictly valid only in case of linear admixture transformations), the admixture fields could be present as a superposition of the "puffs" $c_i(x, y, z, t|x_0, y_0, z_0, t_0)$ from these elementary point sources:

$$c_i(x,y,z,t)= \int_0^t dt_0 \iiint_D c_i(x,y,z,t|x_0,y_0,z_0,t_0)dx_0 dy_0 dz_0, \qquad (19.24)$$

19.2.6.2 Plume Models

Following the same considerations and with the same rate of inaccuracy the exact "plume" solution, 19.16 can be used to construct approximate solution for the admixture fields from 19.16. It is assumed that even in heterogeneous wind field the admixture concentrations retain the shape of a plume (Figure 19.1) and again the processes could be split in transport and transformation along the plumes axis and Gaussian turbulent spread across the axis.

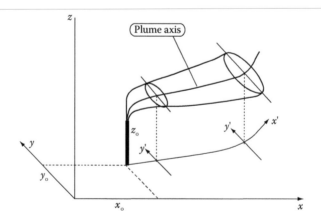

FIGURE 19.1
A scheme of a "plume."

Thus, in the coordinate system (x', y', z) (x' directed along the plume axis, and y' across x'), the concentration field has the form (Gifford 1959):

$$c_i(x', y', z) = c_{i0}(x', z)c_{hor}(y'), \quad c_{hor}(y') = \frac{1}{\sqrt{2\pi}\sigma_y(x')} \exp\left[-\frac{y'^2}{2\sigma_y^2}\right], \quad (19.25)$$

where $c_{hor}(y')$ accounts for the Gaussian turbulent spread across the plume axis and $c_{i0}(x', z)$ accounts for admixture advection along the axis and all the other processes (vertical spread, transformations, washout, absorption by underlying surface):

$$U\frac{\partial c_{i0}}{\partial x'} + \frac{\partial}{\partial z}\left(wc_{i00} - K_{zz}\frac{\partial c_{i00}}{\partial z}\right) = f_{i\text{chem}} + f_{i\text{aero}} - f_{i\text{wash}}, \quad c_{i0}(0, z) = \frac{Q_i \delta(z - z_0)}{U},$$

(19.26)

where z_0 is the source height, $i = 1, 2, \ldots, N$, and U is the wind velocity along the plume axis.

Further simplifications can be made in the same style as in the puff models. The superposition principle can be applied as well, and the admixture fields can be presented as superposition of plumes.

The puff and plume trajectory models are widely used, because they have some important advantages:

- These models can use large, explicit mechanisms, and the mechanism is easily modified.
- They are easy to code, run, and analyze.
- They are well suited for source apportionment because puffs can be tracked from source area to receptor site.
- They can use empirical expressions for the dispersions, which can improve the results in local scales (near the pollution source).

They also have some disadvantages though, the most serious of which is that they are based on the principle of superposition and so do not account for the interaction between puffs/plumes, which could be a serious defect in the case of nonlinear chemistry.

19.3 Main Processes That Determine Air Pollution

As can be best seen in 19.8, the main processes that determine the evolution of the concentration fields are the transport (advection by the arranged air motions and diffusion by the atmospheric turbulence), the chemical transformations, the aerosol processes, and the pollution washout by precipitation. A very important factor of air pollution is, of course, the emissions. The boundary conditions, already briefly discussed above, also can have substantial influence on the concentration patterns.

19.3.1 Gas-Phase Chemistry

Since atmospheric chemistry plays a major role in many air pollution problems, the representation of chemical interactions among atmospheric constituents is often an essential element of an AQ model. All important chemical transformations relevant to the problem being studied must be included to make accurate predictions of ambient pollutant concentrations.

Many atmospheric pollutants or their precursors are emitted as gases and interact primarily in the gaseous phase. However, some important atmospheric processes such as acid deposition and the formation of aerosols involve the interaction of constituents in the gas, liquid, and solid phases, so transformations taking place in all three phases often need to be represented. For computational efficiency, these processes are usually modeled separately.

Interactions in the gas phase are represented in AQ models via chemical mechanisms. A chemical mechanism is a collection of reactions that transforms reactants into products, including all important intermediates. Chemical mechanisms developed for AQ modeling are highly condensed, parameterized representations of a true chemical mechanism. They include artificial species and operators, and many of the mechanism reactions are parameterizations of a large set of true atmospheric reactions.

Mechanism species can be divided into two categories: inorganic and organic. The number of important inorganic species is relatively small, and they are almost always represented explicitly in chemical mechanisms. The important inorganic species included in these mechanisms are ozone, nitric oxide, nitrogen dioxide, nitric acid, nitrous acid, hydrogen peroxide, sulfur dioxide, and several radicals formed through their interactions with other species. Although most of the chemical reactions involving these species are common to all mechanisms, some differences do exist. For example, some of the mechanisms omit a few reactions because they are normally minor pathways and thus do not affect modeling results significantly. Also, different rate constants may be used for some reactions, especially those that are photolytic.

The representation of organic species usually differs more substantially, however. Some species in the mechanism represent real organic compounds, but others represent a mixture of several different compounds. The manner in which the grouping of organic compounds is carried out typically distinguishes one mechanism from another.

Although explicit mechanisms have been developed for many organic compounds, the resultant number of reactions and species needed to represent their atmospheric chemistry is too large to model efficiently in photochemical grid models. In addition, explicit mechanisms have not yet been developed for most organic compounds, thereby requiring that some reaction pathways be postulated. Thus, both compression and generalization are necessary when depicting organic reactions. Although chemical mechanisms differ in the manner in which organic species are represented, the mechanism developer usually chooses some distinguishable organic property to group similar organics into classes that reduce both the number of mechanism species and reactions. The three most common representations include the lumped structure technique, the surrogate species approach, and the lumped species method.

In the lumped structure approach, organic compounds are apportioned to one or more mechanism species on the basis of chemical bond type associated with individual carbon atoms (Whitten et al. 1980). In the surrogate species method, the chemistry of a single species is used to represent compounds of the same class (e.g., Lurmann et al. 1987). Generalized reactions are then written based on the hypothetical model species. The lumped species method is very similar to the surrogate species approach, but various mechanism parameters associated with a particular surrogate are adjusted to account for variations in the composition of the compounds being represented by the surrogate species (e.g., Carter 1990; Stockwell et al. 1990).

It seems that the most popular chemical mechanisms currently used in AQ models are the CB4 (Gery et al. 1989), the RADM2 (Stockwell et al. 1990), and the SAPRC-97 mechanism (Carter et al. 1997).

19.3.1.1 CB4 Mechanism

The CB4 (Gery et al. 1989) mechanism is a lumped structure type that is the fourth in a series of carbon-bond mechanisms, and differs from its predecessors notably in the detail of the organic compound representation.

The CB4 mechanism uses nine primary organic species (i.e., species emitted directly to the atmosphere as opposed to secondary organic species formed by chemical reaction in the atmosphere). Most of the organic species in the mechanism represent carbon–carbon bond types, but ethene (ETH), isoprene (ISOP), and formaldehyde (FORM) are represented explicitly. The carbon-bond types include carbon atoms that contain only single bonds (PAR), double-bonded carbon atoms (OLE), 7-carbon ring structures represented by toluene (TOL), 8-carbon ring structures represented by xylene (XYL), the carbonyl group and adjacent carbon atom in acetaldehyde and higher molecular weight aldehydes represented by acetaldehyde (ALD2), and nonreactive carbon atoms (NR). Many organic compounds are apportioned to the carbon-bond species simply on the basis of molecular structure. For example, propane is represented by three PARs since all three carbon atoms have only single bonds, and propene is represented as one OLE (for the one carbon–carbon double bond) and one PAR (for the carbon atom with all single bonds). Some apportionments, however, are based on reactivity considerations. For example, olefins with internal double bonds are represented as ALD2s and PARs rather than OLEs and PARs. Furthermore, the reactivity of some compounds may be lowered by apportioning some of the carbon atoms to the nonreactive class NR (e.g., ethane is represented as 0.4 PAR and 1.6 NR).

19.3.1.2 RADM2 Mechanism

The RADM2 mechanism is a lumped species type that uses a reactivity based weighting scheme to adjust for lumping (Stockwell et al. 1990). It has evolved from the original RADM1 mechanism (Stockwell 1986). The base mechanism as implemented in the CMAQ system contains 57 model species and 158 reactions, of which 21 are photolytic. In RADM2, the primary organics are represented by 15 mechanism species, five of which are explicit because of their high emission rates or because of special reactivity considerations (methane, ethane, ethene, isoprene, and formaldehyde). The other 10 represent groups of organic compounds aggregated on the basis of their reactivity with the hydroxyl radical (HO) and/or their molecular weights.

19.3.1.3 SAPRC-97 Mechanism

The SAPRC-97 mechanism (Carter 1997) employs the lumped surrogate species approach, but offers the capability to incorporate semiexplicit chemistry of selected organics. The SAPRC series of mechanisms evolved from the "ALW" mechanism of Atkinson et al. (1982), but incorporates improvements to aromatic chemistry and updates to reactions of many individual organic compounds. Although many of the reactions for organic compounds are generalized and incorporate nonexplicit species, product yield coefficients and rate constants are tabulated for more than 100 individual organic compounds. Thus, each of these organics can be modeled individually by including their semiexplicit chemistry in the mechanism. Due to computational constraints, however, the full set of organic compounds cannot be incorporated in an Eulerian model. For this situation, the mechanism is condensed by lumping individual organic compounds into groups with corresponding rate constants and product yield coefficients that have been weighted by mole fractions of the individual organics. The mole fractions are typically derived from emission inventory data used in the model simulation. Thus, unlike the previous mechanisms, the SAPRC-97 mechanism can potentially change with each application since new rate constants and product yield coefficients can be computed for each application.

19.3.2 Aerosol Dynamics

Inclusion of aerosol particles in an AQ model presents several challenges. Among these are the differences between the physical characteristics of gases and particles. In treating gases in an AQ model, the size of the gas molecules is not usually of primary importance. In contrast, particle size is of primary importance. The interaction between condensing vapors and the target particle depends in an important way on the particle size in relation to the mean free path in the atmosphere. For gases, once the concentration is known, the corresponding number of molecules is known. This is not the case for particles. Thus, including aerosol particles in an AQ model means choosing how the total number, total mass, and size distribution of the particles is represented. Once this choice is made, then important physical and chemical processes involving particles must be represented.

Particles may be emitted into the air by natural processes such as wind blowing dust from a desert. Human activities may disturb the soil to allow wind to blow soil particles off the ground. Sea salt particles come into the atmosphere by wind-driven waves on the sea surface. Volcanic activity is another source of particles for both the troposphere and the stratosphere. Particles can be made in the atmosphere directly from chemical reaction. The most important example of this is the transformation of sulfur dioxide, a by-product of fossil fuel combustion, into sulfate particles. Hydroxyl radicals attack the sulfur dioxide, and make sulfuric acid that then may nucleate in the presence of water vapor and ammonia to produce new particles. If there are particles already present in the atmosphere, the new sulfate may condense on the existing particles or nucleate to form new particles depending on conditions that are only recently beginning to be understood. Reactions of organic precursors such as natural monoterpenes and anthropogenic organic species with ozone and other oxidants or radicals make new species that condense on existing particles or make new particles depending on conditions.

Combustion sources emit particles composed of mixtures of organic carbon and elemental carbon. The exact mixture of organic and elemental carbon is a strong function of the conditions of combustion. Once these particles are in the air, they may grow by condensing of species upon them as has already been mentioned. For a large group of particles made in the air, that is, secondary particles, growth may be related to relative humidity because of water condensing on the particles.

Another gas–particle interaction is the chemical equilibration of species within or on the surface of a particle with gases and vapors within the air. Unlike gases, particles coagulate, for example, collide and form a particle whose mass and volume are the sums of the masses and volumes of the colliding particles. Thus, adding particles to an AQ model means adding a new set of physical processes.

There are perhaps two methods recently used for aerosol description in the contemporary models. The first method is to model particle behavior in sets of bins of increasing size. This approach is quite popular and is described originally by Gelbard et al. (1980) and more recently by Jacobson (1997). The second approach is the one suggested by Whitby (1978), and represent the particles as a superposition of lognormal subdistributions called modes.

The sectional method using the discrete size bins requires a large number of bins to capture the size distribution. If one wishes to model several chemical components, then the number of components is multiplied by the number of bins. This leads to a very large number of variables that must be added to an AQ model to capture particle behavior.

In the modal approach, using the three modes suggested by Whitby (1978), only three integral properties of the distribution—the total particle number concentration, the total surface area concentration, and the total mass concentration of the individual chemical components in each of the modes—have to be followed.

In the most popular currently used models, the particles are divided into two groups: fine particles and coarse particles (modal approach). These groups generally have separate source mechanisms and chemical characteristics. The fine particles result from combustion processes and chemical production of material that then condenses upon existing particles or forms new particles by nucleation. The coarse group is composed of material such as wind-blown dust and marine particles (sea salt). The anthropogenic component of the coarse particles is most often identified with industrial processes.

The regulation standards in AQ refer to PM2.5 (particles with diameters less than 2.5 µm) and PM10 (particles with diameters less than 10 µm). Note that PM10 includes PM2.5. Thus, in the present context, coarse particles are those with diameters between 2.5 and 10 µm. Then, the mass of the coarse particles is the difference between the masses in PM10 and PM2.5.

Usually in the advanced models, applying the modal approach PM2.5 is treated by two interacting subdistributions or modes. The coarse particles form a third mode. Conceptually within the fine group, the smaller (nuclei or Aitken) mode represents fresh particles either from nucleation or from direct emission, whereas the larger (accumulation) mode represents aged particles. Primary emissions may also be distributed between these two modes. The two modes interact with each other through coagulation. Each mode may grow through condensation of gaseous precursors; each mode is subject to wet and dry deposition. Finally, the smaller mode may grow into the larger mode and partially merge with it. The chemical species treated in the aerosol component are fine species sulfates, nitrates, ammonium, water, anthropogenic and biogenic organic carbon, elemental carbon, and other unspecified material of anthropogenic origin. The coarse-mode species include sea salt, wind-blown dust, and other unspecified material of anthropogenic origin.

The dynamics (interaction) of the different modes and the aerosol chemistry are too voluminous material to be discussed here in detail.

19.3.3 Cloud Dynamics and Chemistry

Clouds play an important role in boundary layer meteorology and AQ. Convective clouds transport pollutants vertically, allowing an exchange air between the boundary layer and the free troposphere. Cloud droplets formed by heterogeneous nucleation on aerosols grow into rain droplets through condensation, collision, and coalescence. Clouds and precipitation scavenge pollutants from the air. Once inside the cloud or rain water, some compounds dissociate into ions and/or react with one another through aqueous chemistry (i.e., cloud chemistry is an important process in the oxidation of sulfur dioxide to sulfate). Another important role for clouds is the removal of pollutants trapped in rain water and its deposition onto the ground. Clouds can also affect gas-phase chemistry by attenuating solar radiation below the cloud base, which has a significant impact on the photolysis reactions.

The cloud models within the up-to-date air pollution models incorporate many of these cloud processes. The models includes parameterizations for several types of clouds, including subgrid convective clouds (precipitating and nonprecipitating) and grid-scale resolved clouds. They also include an aqueous chemistry model for sulfur and mechanisms for scavenging.

19.3.4 Atmospheric Processes

Air pollution dispersion phenomena are decisively influenced by atmospheric processes—wind and atmospheric turbulence determine admixture transport; temperature, humidity, and solar radiation could have influence on chemical reactions rate and aerosol dynamics and chemistry; precipitation intensity influences the pollution washout.

Atmospheric processes are commonly classified with regard to their spatial scale. The latter is, in turn, related to the characteristic time of the individual process. Orlanski (1975) recommends distinguishing the following scales.

- **Macroscale**

  At this scale (characteristic lengths exceeding 1000 km), the atmospheric flow is mainly associated with synoptic phenomena, that is, the geographical distribution of pressure systems. Such phenomena are mainly due to large-scale inhomogeneities of the surface energy balance. Global and the majority of regional-to-continental scale dispersion phenomena are related to macroscale atmospheric processes, for which the hydrostatic approximation can be considered valid.

- **Microscale**

  In general, air flow is very complex at this scale (characteristic lengths below 1 km), as it depends strongly on the detailed surface characteristics (i.e., form of buildings, their orientation with regard to the wind direction). Although thermal effects may contribute to the generation of these flows, they are mainly determined by hydrodynamic effects (e.g., flow channeling, roughness effects), which have to be described well in an appropriate simulation model. In view of the complex nature of such effects, local scale dispersion phenomena (which are to a large extent associated with microscale atmospheric processes) are mainly described with robust "simple" models in the case of practical applications, such as street canyon models.

- **Mesoscale**

  The flow configuration in the mesoscale (characteristic lengths between 1 and 1000 km) is dependent both on hydrodynamic effects (e.g., flow channeling, roughness effects) and inhomogeneities of the energy balance (mainly due to the spatial variation of area characteristics (e.g., land use, vegetation, water), but also as a consequence of terrain orientation and slope). From the air pollution viewpoint, thermal effects are the most interesting, as they are of particular importance at times of a weak synoptic forcing, that is, bad ventilation conditions. As a minimum requirement, mesoscale meteorological models should be capable of simulating local circulation systems, as for instance sea and land breezes. Mesoscale atmospheric processes affect primarily local-to-regional scale dispersion phenomena, for which urban studies are the most important examples. The description of such phenomena requires, even for practical applications, the utilization of fairly complex modeling tools.

The proper atmospheric processes treatment is one of the main challenges of AQ modeling. These are linked to the very high variability of the concentration fields, combined with the sufficiently long lifetime of some species in the atmosphere that require multiscale simulations in order to take into account the inputs from both remote and local sources. There are also significant differences in the relevant most important pollutants from region to region (e.g., near-surface ozone problem in central and southern Europe vs. aerosol pollution in the Nordic countries). Another problem is the strong dependence of concentrations on fluctuations of local and regional meteorological conditions.

19.3.5 Emissions

Emission data are of crucial importance for AQ assessment/modeling. All the other activities that are carried out for this purpose are actually nothing more than emission data processing.

There are many types of sources of atmospheric emissions and many examples (often millions) of each type, for example: power plants, refineries, incinerators, factories, domestic households, offices and public buildings, cars and other vehicles, fossil fuel extraction and production sites, fossil fuel distribution pipelines, animals and humans, fertilized land, trees and other vegetation, and vegetation fires.

It is not possible to measure emissions from all of the individual examples of these sources or, in the short term, from all the different source types. In practice, atmospheric emissions are estimated on the basis of measurements made at selected or representative samples of the (main) sources and source types.

The basic model for an emission estimate is the product of (at least) two variables, for example:

- An activity statistic (fuel consumption for a given period) and a typical average emission factor for the activity
- An emission measurement over a period of time and the number of such periods emissions occurred in the required estimation period

For example, to estimate annual emissions of sulfur dioxide in grams per year from an oil-fired power plant you might use, either:

- Annual fuel consumption (in tons fuel/year) and an emission factor (in grams SO_2 emitted/tons fuel consumed) or
- Measured SO_2 emissions (in grams per hour) and number of operating hours per year

In practice, the calculations tend to be more complicated, but the principles remain the same.

Emission estimates are collected together into inventories or databases, which usually also contain supporting data on, for example: the locations of the sources of emissions; emission measurements where available; emission factors; capacity, production, or activity rates in the various source sectors; operating conditions; methods of measurement or estimation, etc.

Emission inventories may contain data on three types of source: point, area, and line. However, in some inventories, all of the data may be on area basis—region, country, subregion, etc.

> *Point sources*—Emission estimates are provided on an individual plant or emission outlet (usually large) usually in conjunction with data on location, capacity or throughput, operating conditions, etc. The tendency is for increased amounts of information on point source

emissions to become available as legislative requirements extend to more source types and pollutants, and "access to information" initiatives increase the availability of such data.

Area sources—Smaller or more diffuse sources of pollution are provided on an area basis either for administrative areas, such as counties, regions, etc., or for regular grids.

Line sources—In some inventories, vehicle emissions from road transport, railways, inland navigation, shipping, or aviation, etc. are provided for sections along the line of the road, railway track, sea lane, etc.

At present, the methodology generally accepted for emission evaluation in Europe is the Core Inventory of Air Emissions in Europe (CORINAIR; see http://www.eea.europa.eu/publications/EMEPCORINAIR5). Inventory recognizes 11 main source sectors (Selected Nomenclature for Air Pollution (SNAP)):

- Public power, cogeneration, and district heating plants (energy sector, utilities, refineries)
- Commercial, institutional, and residential combustion plants
- Industrial combustion
- Production processes (mining)
- Extraction and distribution of fossil fuels
- Solvent use
- Road transport
- Other mobile sources and machinery
- Waste treatment and disposal
- Agriculture
- Nature

Data were provided on large point sources on an individual basis and on other smaller or more diffuse sources on an area basis, usually by administrative boundary at the county, department, or municipality level.

As for some of the SNAP categories (e.g., road transport), the emission inventory is sometimes made for the whole country on the basis of total fuel consumption. In order to prepare inventory with good enough spatial resolution (in a desired computational grid), it is necessary for the emissions to be disaggregated—some surrogates (population density, road density in a given grid cell, etc.) are used and the emissions are distributed in the grid proportionally to the surrogate values in a grid cell.

As the emission inventories are usually made on an annual basis, some temporal profiles (diurnal, weekly, and annual) have to be allocated to the different SNAPs in order to obtain sufficient temporal details.

It seems that currently the most detailed generally available emission inventory for Europe is the one presented by Visschedijk et al. (2007). The inventory spatial resolution is $0.25° \times 0.125°$ longitude–latitude, that is on average 15×15 km for midlatitudes. The inventory is elaborated for area and large point sources separately, distributed over the first 10 SNAPs (without the natural emissions). The inventory contains eight pollutants: CH_4, CO, NH_3, NMVOC (VOC), NO_x, SO_x, PM10, and PM2.5. The temporal allocation is made on the basis of daily, weekly, and monthly profiles, provided by Builtjes et al. (2003). The temporal profiles are country-, pollutant-, and SNAP-specific.

19.4 Evaluation of the Simulation Results

If data from air pollution measurements are used, it is highly recommended for model simulations to be compared with the data. By such verification, the model's performance can be evaluated and its results validated.

If a set of observations O_i is available (possibly at different times and locations $i, i - 1, 2, \ldots, N$) and P_i are the corresponding model results for the respective times/location, several criteria have been elaborated for evaluation of the models by comparison with measurements. They are given in Table 19.2.

The model performance in each specific case depends not only on the model quality, but also on the quality of input data (meteorology, emissions). This is why it is good for simulation evaluations to be performed even in models that have been generally proven to have good simulation abilities.

19.5 Some Examples of Up-to-Date Air Pollution Models

At present, there is a huge number of air pollution models of various types and classes. Combined with various meteorological and emission preprocessors, this gives a set of numerous model applications.

A meta database of up-to-date and widely used meteorological and AQ models was created by COST Action 728 (http://www.cost728wg1.org.uk/). The Action addresses key issues concerning the development of mesoscale modeling capability for air pollution and dispersion applications. This project, which coordinates the efforts and gathers the expertise of more than 50 research institutions in Europe, is one of the most comprehensive databases of its kind.

TABLE 19.2

Statistical Measures Used in Model Performance Evaluation (P_i and $\bar{P} = \sum_{1=1}^{N} P_i$ Are the Arithmetic Averages of O_i and P_i, Respectively)

Measure	Mathematical Expression	Meaning
Coefficient of determination	$$\frac{\left[\sum_{i=1}^{N}(P_i - \bar{P})(O_i - \bar{O})\right]^2}{\sum_{i=1}^{N}(P_i - \bar{P})^2 \sum_{i=1}^{N}(O_i - \bar{O})^2}$$	Measures association between model prediction and observations.
Normalized Mean Error (NME) (Reported as %)	$$\frac{\sum_{i=1}^{N}|P_i - O_i|}{\sum_{i=1}^{N}O_i}$$	Relative differences between the model and observations.
Root Mean Square Error (RMSE)	$$\left[\frac{1}{N}\sum_{i=1}^{N}(P_i - O_i)^2\right]^{1/2}$$	Average magnitude of the forecast errors. RMSE puts greater influence on large errors than small errors.
Mean Absolute Gross Error (MAGE)	$$\frac{1}{N}\sum_{i=1}^{N}|P_i - O_i|$$	Average magnitude of the forecast errors.
Fractional Gross Error (FE) (Reported as %)	$$\frac{2}{N}\sum_{i=1}^{N}\left|\frac{P_i - O_i}{P_i + O_i}\right|$$	
Mean Normalized Gross Error (MNGE) (Reported as %)	$$\frac{1}{N}\sum_{i=1}^{N}\frac{|P_i - O_i|}{O_i}$$	It quantifies the mean absolute deviation of the residuals. It indicates the average unsigned discrepancy between estimates and observations and is calculated for all pairs.
Mean Bias (MB)	$$\frac{1}{N}\sum_{i=1}^{N}(P_i - O_i)$$	Average forecast error.
Mean Normalized Bias (MNB) (Reported as %)	$$\frac{1}{N}\sum_{i=1}^{N}\frac{P_i - O_i}{O_i}$$	Relative differences between model prediction and observations.
Mean Fractionalized Bias (Fractional Bias, MFB) (Reported as %)	$$\frac{2}{N}\sum_{i=1}^{N}\frac{P_i - O_i}{P_i + O_i}$$	It bounds maximum bias and gives equal weight to underestimations and overestimations.

(continued)

TABLE 19.2 (Continued)

Statistical Measures Used in Model Performance Evaluation (P_i and $\bar{P} = \sum\limits_{1=1}^{N} P_i$ Are the Arithmetic Averages of O_i and P_i, Respectively)

Measure	Mathematical Expression	Meaning
Normalized Mean Bias (NMB) (Reported as %)	$$\dfrac{\sum\limits_{i=1}^{N} P_i - O_i}{\sum\limits_{i=1}^{N} O_i}$$	Relative differences between the model and observations.
Bias Factor (BF) Reported as ratio notation (prediction/observation)	$$\dfrac{1}{N} \sum\limits_{i=1}^{N} \dfrac{P_i}{O_i}$$	BF = 1 + MNB

As shown in Table 19.3, the models can be categorized by the spatial/temporal scales they are relevant for (microscale, mesoscale, macroscale) as well as by their functions and formulation (meteorological, AQ, and integrated meteorological and AQ models).

The site provides extensive information about the models as well as links to model sites and model-using groups.

Other types of models will be briefly described in the following subsections. Although these sketchy descriptions do not provide a deep and detailed information about these models, our aim is to demonstrate the wide range of processes that these models are able to account for and the richness of the AQ output they produce.

The first modeling system to be mentioned is the US EPA Models 3 system, an advanced AQ modeling system that addresses AQ from the "one atmosphere" multipollutant perspective integrating meteorological, air pollution, and emission models. The system is fully based on the Eulerian approach and is widely used all over the world. It consists of three models:

- **CMAQ** (Community Multi-scale Air Quality model; http://www.cmaq-model.org/, http://www.cmascenter.org/index.cfm?temp_id=99999&model=cmaq)
- **MM5** (the Fifth-Generation Pennsylvania State University (PSU)/National Center for Atmospheric Research (NCAR) Meso-Meteorological Model; http://www.mmm.ucar.edu/mm5/);
- **SMOKE** (Sparse Matrix Operator Kernel Emissions Modelling System; http://www.smoke-model.org/, http://www.cmascenter.org/index.cfm?temp_id=99999&model=smoke)

TABLE 19.3

Classification of Models

	Meteorology	AQ	Integrated AQ and Meteorology
Microscale	ADREA	ADREA	GEM-AQ
	Chensi	AERMOD	M-SYS
	GEM-AQ	Chensi	Meso-NH
	LESNIC	FINFLO	MIMO
	M-SYS	GEM-AQ	RCG
	M2UE	M-SYS	
	MERCURE	M2UE	
	Meso-NH	MERCURE	
	MIMO	Meso-NH	
	MITRAS	MICTM	
	RCG	MIMO	
	STAR-CD	MITRAS	
	VADIS	NAME	
		RCG	
		STAR-CD	
		VADIS	
Mesoscale	ADREA	ADREA	BOLCHEM
	ALADIN/A	AERMOD	CALMET/CALPUFF
	ALADIN/PL	ALADIN-CAMx	CALMET/CAMx
	ARPS	AURORA	COSMO-MUSCAT
	BOLCHEM	BOLCHEM	GEM-AQ
	CALMET/CALPUFF	CAC	M-SYS
	CALMET/CAMx	CALGRID	MC2-AQ
	CLM	CALMET/CALPUFF	MCCM
	COSMO-2	CALMET/CAMx	Meso-NH
	COSMO-7	CAMx	RCG
	COSMO-MUSCAT	CHIMERE	TAPM
	ENVIRO-HIRLAM	CHIMERE (ARPA-IT)	WRF/Chem
	GEM-AQ	CMAQ	
	GESIMA	CMAQ(GKSS)	
	GME	COSMO-MUSCAT	
	Hirlam	EMEP	
	LAMI	ENVIRO-HIRLAM	
	LME	EPISODE	
	LME_MH	EURAD-IM	
	M-SYS	FARM	
	MC2-AQ	FLEXPART	
	MCCM	FLEXPART V6.4	
	MEMO (UoT-GR)	FLEXPART/A	
	MEMO (UoA-PT)	GEM-AQ	
	MERCURE	LOTOS-EUROS	
	Meso-NH	LPDM	
	METRAS	M-SYS	
	MM5 (met.no)	MARS (UoT-GR)	
	MM5 (UoA-GR)	MARS (UoA-PT)	
	MM5 (UoA-PT)	MATCH	
	MM5 (UoH-UK)	MC2-AQ	

(continued)

TABLE 19.3 (Continued)

Classification of Models

	Meteorology	AQ	Integrated AQ and Meteorology
	MM5(GKSS-D)	MCCM	
	NHHIRLAM	MECTM	
	RAMS	MEMO (UoT-GR)	
	RCG	MERCURE	
	SAIMM	Meso-NH	
	TAPM	MOCAGE	
	UM	MUSE	
	WRF-ARW	NAME	
	WRF/Chem	OFIS	
		Polyphemus	
		RCG	
		SILAM	
		TAPM	
		TCAM	
		TREX	
		WRF/Chem	
Macroscale	GME	CAM-CHEM	ECHMERIT
	Hirlam	CHIMERE	EMAC
	UM	CHIMERE (ARPA-IT)	GEM-AQ
		ECHAM5-MOZ	STOC_HadAM3
		EMEP	
		EURAD-IM	
		FLEXPART	
		FLEXPART V6.4	
		FLEXPART/A	
		FRSGC/UCI	
		GEOS-Chem	
		GISS-PUCCINI	
		GLEMOS	
		GOCART	
		IMPACT	
		LPDM	
		MATCH	
		MOCAGE	
		MOZART-2	
		MSCE-HM-Hem	
		MSCE-POP	
		NAME	
		OsloCTM2	
		SILAM	
		TM5	
		TOMCAT	
		UKCA	

19.5.1 Community Multiscale Air Quality System

CMAQ is a multipollutant, multiscale AQ model that contains state-of-science techniques for simulating atmospheric and land processes that affect the transport, transformation, and deposition of atmospheric pollutants and/or their precursors on both regional and urban scales. It is designed as a science-based modeling tool for handling all the major pollutant issues (including photochemical oxidants, PM, acidic, and nutrient deposition) holistically.

Models-3 CMAQ was released to the public in June 1998. The science and model engineering concepts and progress of the project have been described in several studies (Byun et al. 1995a, 1995b, 1996, 1998; Coats et al. 1995; Ching et al. 1995). The model is in constant development and had several newer versions released since.

19.5.1.1 The CMAQ Structure

The CMAQ core is the Chemical Transport Model (CCTM). The CMAQ modeling system also includes interface processors to incorporate the outputs of the meteorology and emissions processors and to prepare the requisite input information for initial and boundary conditions and photolysis rates to the CCTM. Figure 19.2 illustrates the relationship and purpose of each of the CMAQ processors (and requisite interfaces) and their relation to the chemical

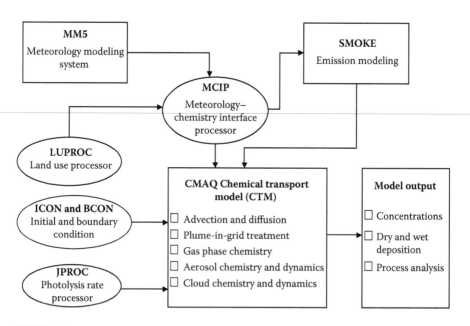

FIGURE 19.2
SMAQ organization.

transport modeling system. The arrows show the flow of data through the modeling system. Two additional functional features of the CMAQ system are included, one for process analysis, which is primarily for model diagnostic analyses, and a second one that is an aggregation methodology for estimating longer-term averaged fields. Each of these processors is described briefly below.

The Meteorology–Chemistry Interface Processor (MCIP) translates and processes model outputs from the meteorology model for the CCTM. MCIP interpolates the meteorological data if needed, converts between coordinate systems, computes cloud parameters, and computes surface and planetary boundary layer (PBL) parameters for the CCTM. MCIP uses land use information from the land use processor (LUPROC) to calculate the PBL and surface parameters.

Initial Conditions and Boundary Conditions (ICON and BCON) provide concentration fields for individual chemical species for the beginning of a simulation and for the grids surrounding the modeling domain, respectively. The ICON and BCON processors use data provided from previous 3-D model simulations or from clean-troposphere vertical profiles. Both the vertical profiles and modeled concentration fields have a specific chemical mechanism associated with them, which are a function of how these files were originally generated.

The photolysis processor (JPROC) calculates temporally varying photolysis rates. JPROC requires vertical ozone profiles, temperature profiles, a profile of the aerosol number density, and Earth's surface albedo to produce the photolysis rates for the CCTM. JPROC uses this information in radiative transfer models to calculate the actinic flux needed for calculating photolysis rates. JPROC generates a lookup table of photodissociation reaction rates.

19.5.1.2 The CMAQ Chemical Transport Model

The CMAQ CCTM simulates the relevant and major atmospheric chemistry, transport, and deposition processes involved throughout the modeling domains. A general overview of the CMAQ abilities and functions is provided below.

19.5.1.2.1 Advection and Diffusion

Several advection methods are implemented in the CMAQ; these include a scheme by Bott (1989), a piecewise parabolic method (PPM) (Colella and Woodward 1984), and the Yamartino–Blackman cubic algorithm. Options for computing subgrid vertical transport include eddy diffusion, and the Asymmetric Convective Model (ACM) (Pleim and Chang 1992) applicable to convective conditions.

Horizontal diffusion is modeled using a constant eddy diffusion coefficient. Numerical methods differ in the handling of advection of concentration fields.

19.5.1.2.2 Gas-Phase Chemistry

CMAQ includes the RADM2, CB4, and SAPRC-97 gas-phase chemical mechanisms. The CMAQ version of CB4 includes the most recent representation of isoprene chemistry, and two additional variants of the RADM2 mechanism also contain the newer isoprene chemistry at two levels of detail. In addition, CMAQ provides the capability to edit these mechanisms or to import a completely new mechanism by means of a generalized chemical mechanism processor. CMAQ also accounts for the formation of secondary aerosols and the reactions of pollutants in the aqueous phase, and aqueous reactions are simulated by means of the aqueous chemical mechanism incorporated in RADM. All CMAQ gas-phase mechanisms are linked to these processes to provide the capability to simulate multiphase interactions.

Two chemistry solvers are available—the Sparse Matrix Vectorized Gear (SMVGEAR) algorithm developed by Jacobson and Turco (1994) and the Quasi-Steady State Approximation (QSSA) method used in the Regional Oxidant Model. SMVGEAR is generally recognized as the more accurate of the two, but it is much slower than QSSA on nonvector computers.

A chlorine chemical mechanism has been added to CMAQ based on Tanaka et al. (2003). Twenty gas-phase reactions were combined with the CB05V mechanism and incorporated into the model CB05CLTX mechanism. Six species track emissions from anthropogenic and biogenic sources. Other species allow simulating the fate and transport of molecular chlorine and hydrogen chloride emissions. The additional reactions not only simulate the photochemical destruction of these eight compounds but also simulate how the chlorine compounds affect ozone photochemistry. To simulate the effect, CB05CLTX includes species representing the daughter products of molecular chlorine and hydrogen chloride.

19.5.1.2.3 Plume-in-Grid Modeling

CMAQ includes algorithms to treat subgrid-scale physical and chemical processes impacting pollutant species in plumes released from selected Major Elevated Point Source Emitters (MEPSEs). The Plume-in-Grid (PinG) modules simulate plume rise and growth, and the relevant dynamic and chemical reaction processes of subgrid plumes. PinG can be used for the simulations at 36- and 12-km resolutions. PinG is not invoked at 4 km resolutions and the MEPSE emissions are directly released into the CMAQ 3-D grid cells.

19.5.1.2.4 Particle Modeling and Visibility

One of the major advancements in CMAQ is the modeling of fine and coarse mode particles, with the use of the fine particle model described by Binkowski and Shankar (1995). CMAQ predicts hourly gridded concentrations of fine particle mass whose size is equal to or less than 2.5 µm in diameter (PM 2.5), speciated to sulfate, nitrate, ammonium, organics, and aerosol water.

Secondary sulfate is produced when hydroxyl radicals react with sulfur dioxide to produce sulfuric acid that either condenses to existing particles or nucleates to form new particles. CMAQ model output includes number densities for both fine and coarse modes. The modeling of aerosols in CMAQ also provides the capability to handle visibility, which is another CMAQ output. In another potential application, CMAQ can provide the basis for modeling the atmospheric transport and deposition of semivolatile organic compounds (SVOC) with parameterizations for their rates of condensation to and/or volatilization from the modeled particles.

19.5.1.2.5 Cloud Processes
Proper descriptions of clouds are essential in AQ modeling due to their critical role in atmospheric pollutant transport and chemistry processes. Clouds have both direct and indirect effects on the pollutant concentrations: they directly modify concentrations via aqueous chemical reactions, vertical mixing, and wet deposition removal processes, and they indirectly affect concentrations by altering radiative transmittances, which affect photolysis rates and biogenic fluxes. CMAQ models deep convective clouds (Walcek and Taylor 1986) and shallow clouds using the algorithms as implemented in RADM (Dennis et al. 1993) for 36 and 12 km resolutions. At 4 km resolution, the clouds are generally resolved, and explicit type cloud dominates.

19.5.1.2.6 Photolysis Rates
The photochemistry of air pollutants is initiated by photodissociation of smog precursors, which are driven by solar radiation. The amount of solar radiation is dependent on sun angle (time of day), season, latitude, and land surface characteristics, and is greatly affected by atmospheric scatterers and absorbers. Photolytic rates are also wavelength- and temperature-dependent. Within CMAQ, temporally resolved 3-D gridded photolysis rates are interpolated from a lookup table generated by JPROC processor and corrected for cloud coverage.

19.5.1.3 CMAQ Output

The CMAQ output consists of 3-D concentration fields of all the admixtures the model accounts for, delivered at chosen equidistant time periods. In addition, the 2-D fields of dry and wet deposition fields for chosen time periods. This is a kind of output that all the models produce. The CMAQ output also optionally contains other fields—those of the contribution of different processes to the concentration evolution.

AQ modeling simulations arise from modeling of complex atmospheric processes. It is important to assure and to understand the model results. Sensitivity tests are needed to detect problems in model formulations and to determine if the model is credible for assessing emission control strategies.

A very powerful sensitivity analysis tool called process analysis is provided with CMAQ.

19.5.1.3.1 Process Analysis

Sensitivity analyses are needed to detect errors and uncertainties introduced into a model by the parameterization schemes and the input data. Results must also be analyzed to ensure that realistic values are obtained for the right reasons rather than through compensating errors among the science processes. Process analysis techniques quantify the contributions of individual physical and chemical atmospheric processes to the overall change in a pollutant's concentration, revealing the relative importance of each process. Process analysis is particularly useful for understanding the effects from model or input changes. CMAQ provides the capability to perform process analyses using two different pieces of information: Integrated Process Rates (IPRs) and Integrated Reaction Rates (IRRs).

- The IPRs are obtained during a model simulation by computing the change in concentration of each species caused by physical processes (e.g., advection, diffusion, emissions), chemical reaction, aerosol production, and aqueous chemistry. Values provide only the net effect of each process. IPRs are particularly useful for identifying unexpectedly low or high process contributions, which could be indicative of model errors.

- The IRR analysis involves the details of the chemical transformations. For gas-phase chemistry, the CCTM has been designed to compute not only the concentration of each species, but also the integral of the individual chemical reaction rates. IRR analyses have typically been used to understand the reasons for differences in model predictions obtained with different chemical mechanisms.

19.5.2 The Fifth-Generation PSU/NCAR MM5

The Fifth-Generation PSU/NCAR Mesoscale Model (MM5) was developed in cooperation with Penn State and the University Corporation for Atmospheric Research (UCAR). It is the latest in a series that developed from a mesoscale model used by Anthes at Penn State in the early 1970s that was later documented by Anthes and Warner (1978).

19.5.2.1 The Meteorology Model

Thorough descriptions of the standard MM5 options are given by Grell et al. (1994) and Dudhia et al. (1998). The source code for an early release of MM5 is documented by Haagenson et al. (1994).

The coordinate system for MM5 is $(x, y, \sigma - p)$. The x and y dimensions are a regular lattice of equally spaced points ($\Delta x = \Delta y$ = horizontal grid spacing, in kilometers) forming rows and columns. Sigma is a terrain-following vertical coordinate that is a function of the pressure at the point on the grid (in hydrostatic runs) or the reference state pressure (in nonhydrostatic runs), the surface pressure at the grid point, and the pressure at the top of the model. Sigma varies from 1 at the surface to 0 at the top of the model. The influence of the terrain on the sigma structure diminishes with height, such that the sigma surfaces near the top of the model are nearly parallel. The horizontal grid in MM5 has an Arakawa-B staggering of the velocity vectors with respect to the scalars (Arakawa and Lamb 1977). The momentum variables (u and v components of wind and the Coriolis force) are on "dot" points, whereas all other variables (e.g., mass and moisture variables) are on "cross" points. The dot points form the regular lattice for the simulation domain, whereas the cross points are offset by 0.5 grid point in both the x- and y-directions. The interpolation of the variables to the staggered grid is done automatically within the INTERP program.

MM5 is based on primitive physical equations of momentum, thermodynamics, and moisture. The state variables are temperature, specific humidity, grid-relative wind components, and pressure.

19.5.2.1.1 Prognostic Equations

In the prognostic equations, the state variables are mass-weighted with a modified surface pressure. MM5 can be run as either a hydrostatic or nonhydrostatic model. In the hydrostatic model, the state variables are explicitly forecast. In the nonhydrostatic model (Dudhia 1993), pressure, temperature, and density are defined in terms of a reference state and perturbations from the reference state. MM5 is not mass-conserving in the nonhydrostatic mode. The vertical (sigma) coordinate is defined as a function of pressure. The model's prognostic equations are thoroughly discussed by Grell et al. (1994) and Dudhia et al. (1998).

19.5.2.1.2 Lateral Boundary Conditions

There are five options for lateral boundary conditions in MM5: fixed, relaxation, time dependent, time and inflow/outflow dependent, and sponge. The lateral boundaries in MM5 consist of either the outer five grid points (relaxation and sponge options) or the outer grid point (all other options) on the horizontal perimeter of the simulation domain. (The outer four grid points are used for boundary conditions for "cross" point variables for the relaxation and sponge options.) The lateral boundary conditions for the coarse domain are derived from the background fields processed in DATAGRID and INTERP. When the one-way nest option is selected, the lateral boundary conditions for nested domains are interpolated from the simulation on the parent domain.

19.5.2.1.3 Model Physics

Several model physics options in MM5 are briefly noted below. The model physics options are further discussed and compared by Dudhia et al. (1998).

19.5.2.1.4 Radiation

There are five atmospheric radiation cooling schemes available in MM5.

- **The "None" option** applies no mean radiative tendency to the atmospheric temperature. This scheme is unrealistic for long-term simulations.
- **The Simple Cooling scheme** sets the atmospheric cooling rate strictly as a function of temperature. There is no cloud interaction or diurnal cycle.
- **The Surface Radiation scheme** is used with the "none" and "simple cooling" schemes. This scheme includes a diurnally varying short-wave and longwave flux at the surface for use in the ground energy budget. These fluxes are calculated based on atmospheric column-integrated water vapor and low/middle/high cloud fraction esti-mated from relative humidity.
- **The Dudhia Longwave and Shortwave Radiation scheme** is sophis-ticated enough to account for longwave and shortwave interactions with explicit cloud and clean air. This scheme includes surface radia-tion fluxes and atmospheric temperature tendencies. This scheme requires longer CPU time, but not much memory. (See Dudhia 1989.)
- **The CCM2 Radiation scheme** includes multiple spectral bands in shortwave and longwave, but the clouds are treated simply as func-tions of relative humidity. This scheme is suitable for larger grid scales and probably more accurate for long time integration (e.g., cli-mate modeling). It also provides radiative fluxes at the surface. (See Hack et al. 1993.)

19.5.2.1.5 Convective Parameterization

There are currently six convective parameterization schemes in MM5. There is also the option for no convective parameterization and an independent option for shallow convection. The convective parameterization schemes have been designed for use at various simulation scales, and they are not entirely interchangeable. For example, each scheme uses different assump-tions for convective coverage on the subgrid-scale and for the convective trig-ger function. The convective parameterization schemes also differ greatly in CPU usage and memory requirements.

The Anthes–Kuo scheme is based on moisture convergence and is mostly applicable to larger grid scales (i.e., greater than 30 km). This scheme tends to produce more convective rainfall and less resolved-scale precipitation. This scheme uses a specified heating profile where moistening is dependent on relative humidity. (See Anthes 1977.)

- **The Fritsch–Chappel scheme** is based on relaxation to a profile due to updraft, downdraft, and subsidence region properties. The convective mass flux removes 50% of the available buoyant energy in the relaxation time. There is a fixed entrainment rate. This scheme is suitable for 20–30 km scales due to the single cloud assumption and local subsidence. (See Fritsch and Chappel 1980.)

- **The Arakawa–Schubert scheme** is a multicloud scheme that is otherwise similar to the Grell scheme (described below). This scheme is based on a cloud population, and it allows for entrainment into updrafts and the existence of downdrafts. This scheme is suitable for larger grid scales (i.e., greater than 30 km). This scheme can be computationally expensive compared to the other available schemes. (See Arakawa and Schubert 1974.)

- **The Kain–Fritsch scheme** is similar to the Fritsch–Chappel scheme, but it uses a sophisticated cloud-mixing scheme to determine entrainment and detrainment. This scheme also removes all available buoyant energy in the relaxation time. (See Kain and Fritsch 1990, 1993.)

- **The Betts–Miller scheme** is based on relaxation adjustment to a reference postconvective thermodynamic profile over a given period. This scheme is suitable for scales larger than 30 km. However, there is no explicit downdraft, so this scheme may not be suitable for severe convection. (See Betts 1986; Betts and Miller 1986, 1993; Janjic 1994.)

- **The Grell scheme** is based on the rate of destabilization or quasi-equilibrium. This is a single-cloud scheme with updraft and downdraft fluxes and compensating motion that determines the heating and moistening profiles. This scheme is useful for smaller grid scales (e.g., 10–30 km), and it tends to allow a balance between the resolved scale rainfall and the convective rainfall. (See Grell et al. 1991; Grell 1993.)

- **The "no convective parameterization" option** (e.g., explicitly resolved convection on the grid scale) is also available. This option is generally used for simulations on domains with horizontal grid spacing smaller than 10 km.

- **The Shallow Convection scheme** is an independent option that handles nonprecipitating clouds that are assumed to be uniform and to have strong entrainment, a small radius, and no downdrafts. This scheme is based on the Grell and Arakawa–Schubert schemes. There is also an equilibrium assumption between cloud strength and subgrid boundary layer forcing.

19.5.2.1.6 PBL Processes

Four PBL parameterization schemes are available in MM5. These parameterizations are most different in the turbulent closure assumptions that are used. The PBL parameterization schemes also differ greatly in CPU usage.

- **The Bulk Formula scheme** is suitable for coarse vertical resolution in the PBL (i.e., greater than 250 m vertical grid sizes). This scheme includes two stability regimes.

- **The Blackadar scheme** is suitable for "high-resolution" PBL (e.g., five layers in the lowest kilometer and a surface layer less than 100 m thick). This scheme has four stability regimes; three stable and neutral regimes are handled with a first-order closure, whereas a free convective layer is treated with a nonlocal closure. (See Blackadar 1979.)

- **The Burk–Thompson scheme** is suitable for coarse and high-resolution PBL. This scheme explicitly predicts turbulent kinetic energy for use in vertical mixing, based on a 1.5-order closure derived from the Mellor–Yamada formulas. (See Burk and Thompson 1989.)

- **The Medium Range Forecast (MRF) model scheme** is suitable for high-resolution PBL. This scheme is computationally efficient. It is based on a Troen–Mahrt representation of the countergradient term and a first-order eddy diffusivity (K) profile in the well-mixed PBL. This scheme has been taken from the National Centers for Environmental Prediction's (NCEP) MRF model. (See Hong and Pan 1996.)

19.5.2.1.7 Surface Layer Processes

The surface layer processes with the Blackadar and MRF PBL schemes have been parameterized with fluxes of momentum, sensible heat, and latent heat, following Zhang and Anthes (1982). The energy balance equation is used to predict the changes in ground temperature using a single slab and a fixed-temperature substrate. The slab temperature is based on an energy budget, and the depth is assumed to represent the depth of the diurnal temperature variation (~10–20 cm). The 13 land use categories are used to seasonally define the physical properties at each grid point (e.g., albedo, available moisture, emissivity, roughness length, and thermal inertia).

A five-layer soil temperature model (Dudhia 1996) is also available as an option in MM5. In this model, the soil temperature is predicted at layers of approximate depths of 1, 2, 4, 8, and 16 cm, with a fixed substrate below using a vertical diffusion equation. This scheme vertically resolves the diurnal temperature variation, allowing for more rapid response of surface temperature. This model can only be used in conjunction with the Blackadar and MRF PBL schemes.

In a subsequent release, the Pleim–Xiu land–surface scheme (Pleim and Xiu 1995) was included in MM5. The Pleim–Xiu scheme has been developed to address surface processes that can significantly impact AQ modeling, including evapotranspiration and soil moisture. Notable in the Pleim–Xiu scheme is the more careful treatment of the surface characteristics (particularly vegetation parameters) that are currently assigned to grid points based on land use specification, as well as a more detailed land use and soil type classification database.

19.5.2.1.8 Resolvable-Scale Microphysics Schemes

There are six resolvable-scale (explicit grid-scale) microphysics schemes in MM5. There is also an option for a "dry" model run. The microphysics schemes have been designed with varying degrees of complexity for different applications of the model. In addition, there are new prognostic output variables that are generated by the more sophisticated schemes. These microphysics schemes also differ greatly in CPU usage and memory requirements.

- **The Dry scheme** has no moisture prediction. Water vapor is set to zero. This scheme is generally used for sensitivity studies.

- **The Stable Precipitation scheme** generates nonconvective precipitation. Large-scale saturation is removed and rained out immediately. There is no evaporation of rain or explicit cloud prediction.

- **The Warm Rain scheme** uses microphysical processes for explicit predictions of cloud and rainwater fields. This scheme does not consider ice phase processes. (See Hsie and Anthes 1984.)

- **The Simple Ice scheme** adds ice phase processes to the warm rain scheme without adding memory. This scheme does not have supercooled water, and snow is immediately melted below the freezing level. (See Dudhia 1989.)

- **The Mixed-Phase scheme** adds supercooled water to the simple ice scheme, and slow melting of snow is allowed. Additional memory was added to accommodate the ice and snow. This scheme does not include graupel or riming processes. (See Reisner et al. 1993, 1998.)

- **The Mixed-Phase with Graupel scheme** adds graupel and ice number concentration prediction equations to the mixed-phase scheme. This scheme is suitable for cloud-resolving scales. (See Reisner et al. 1998.)

- **The NASA/Goddard Microphysics scheme** explicitly predicts ice, snow, graupel, and hail. This scheme is suitable for cloud-resolving scales. (See Tao and Simpson 1993.)

19.5.2.2 Meteorology Model Preprocessing

Before the meteorology model can be run, several smaller ("preprocessing") programs must be run to set up the domain for the simulation and to generate a set of initial and boundary conditions for the meteorology model. The preprocessing programs are briefly described in this section.

19.5.2.2.1 *Defining the Simulation Domain (TERRAIN)*

Domains for the meteorology simulations are defined by several primary parameters: number of grid points in each horizontal dimension, grid spacing, center latitude, center longitude, map projection (Mercator, Lambert conformal, or polar stereographic), and number of "nested" domains and their horizontal dimensions. These parameters are processed by a program that is executed only when a new domain location is required. This program, TERRAIN, makes use of high-resolution global terrain and land use data sets to create "static files" for the domain. The static files currently include values for each grid point for terrain height and land use specification (e.g., deciduous forest, desert, water).

19.5.2.2.2 *Processing the Meteorological Background Fields (DATAGRID)*

After the simulation domain has been established, the program DATAGRID is run to process the meteorological background fields. DATAGRID generates first-guess fields for the model simulation by horizontally interpolating a larger-scale data set (global or regional coverage) to the simulation domain. DATAGRID interpolates the background fields to the simulation domain for times throughout the simulation period; these files are used ultimately to generate lateral boundary conditions for the coarse-domain simulations. (Nested domains obtain lateral boundary conditions from the coarse domain.) DATAGRID also processes the sea-surface temperature and snow files by interpolating the analyses to the simulation domain. Lastly, DATAGRID calculates map-scale factors and Coriolis parameters at each grid point to be used by MM5.

19.5.2.2.3 *Objective Analysis (RAWINS)*

The program RAWINS performs an objective analysis by blending the first-guess fields generated by DATAGRID with upper-air and surface observations. There are four objective analysis techniques available in RAWINS. (See Dudhia et al. 1998.)

19.5.2.2.4 *Setting the Initial and Boundary Conditions (INTERP)*

The "standard" INTERP program sets the initial and boundary conditions for the meteorology simulation. The analyses from RAWINS are interpolated to MM5's staggered grid configuration and from their native vertical coordinate (pressure) to MM5's vertical coordinate (a terrain-following, pressure-based "sigma" coordinate). In addition, the state variables are converted as necessary, for example, relative humidity to specific humidity. The analyses from one time (generally the first time analyzed by RAWINS) are interpolated by INTERP to provide MM5's initial conditions, whereas analyses from all times are interpolated to generate MM5's lateral boundary conditions.

INTERP can also be used for "one-way nesting," where the MM5 output from one model run is interpolated to provide initial and boundary conditions for the nested run. This is a "nonstandard" use of the INTERP program, but it is commonly used to support AQ modeling.

19.5.2.3 Nesting

MM5 can simulate nested domains of finer resolutions within the primary simulation domain. In MM5, the software is configured to enable up to nine nests (10 domains) within a particular run. However, due to current hardware resources, the state of the science, numerical stability, and practicality, the number of domains in a simulation is generally limited to four or fewer.

Nesting can be accomplished by either a "one-way" or a "two-way" method. In one-way nesting, the coarse-resolution domain simulation is run independently of the nest. The coarse domain can then provide the initial and boundary conditions for its nest. In one-way nesting, each domain can be defined with independent terrain fields, and there are no feedbacks to the coarse domain from its nest. Note that the simulated meteorology at the same grid point in the coarse-resolution domain is likely to be different (if only slightly) from the nest in a one-way nest simulation.

Two-way interactive nesting (Zhang et al. 1986; Smolarkiewicz and Grell 1992) allows for feedback to occur between the coarse-resolution domain and the nest throughout the simulation. The two domains are run simultaneously to enable this feedback, and terrain in the overlapping regions must be compatible to avoid mass inconsistencies and generation of numerical noise. The TERRAIN program automatically defines the terrain compatibility when the user specifies the two-way nesting interaction. When two-way nesting is used, the portion of the coarse-resolution domain that is simulated in the nest may reflect too much smaller-scale detail from the nest to be useful for the CMAQ simulations.

19.5.2.4 Four-Dimensional Data Assimilation

The four-dimensional data assimilation (FDDA) scheme included in MM5 is based on Newtonian relaxation or "nudging." Nudging is a continuous form of FDDA where artificial (nonphysical) forcing functions are added to the model's prognostic equations to nudge the solutions toward either a verifying analysis or toward observations. The artificial forcing terms are scaled by a nudging coefficient that is selected so that the nudging term will not dominate the prognostic equations. The nudging terms tend to be one order of magnitude smaller than the dominant terms in the prognostic equations and represent the inverse of the e-folding time of the phenomena captured by the observations.

There are two types of nudging in MM5: analysis nudging and observation nudging. Analysis nudging gently forces the model solution toward gridded fields. Analysis nudging can make use of 3-D analyses and some surface analyses. Analysis nudging is generally used for scales where synoptic and mesoalpha forcing are dominant. Observation nudging gently forces the model solution toward individual observations, with the influence of the observations spread in space and time. It is better suited for assimilating high frequency, asynoptic data that may not otherwise be included in an analysis.

Nudging in MM5 is extensively discussed by Stauffer and Seaman (1990, 1994) and Stauffer et al. (1991).

19.5.2.5 Data Required to Run the MM5 Modeling System

Since the MM5 modeling system is primarily designed for real-data studies/simulations, it requires the following data sets to run:

- Topography and land use (in categories)—a TERRAIN own database is available for the MM5 users.
- Gridded atmospheric data that have at least these variables: sea-level pressure, wind, temperature, relative humidity, and geopotential height; and at these pressure levels: surface, 1000, 850, 700, 500, 400, 300, 250, 200, 150, and 100 hPa. The NCEP Final Analysis (FNL) data with $1° \times 1°$ resolution has been available since July 1999. It can be freely downloaded from http://dss.ucar.edu/datasets/ds083.2/data/.
- Observation data that contain soundings and surface reports (in case of observation nudging).

19.5.3 SMOKE Modeling System

The SMOKE Modeling System was originally developed to integrate emissions data processing with high-performance computing (HPC) sparse-matrix algorithms. SMOKE is now under active development at Carolina Environmental Program (CEP), and is partially supported by the Community Modeling and Analysis System (CMAS).

The primary purposes of the first SMOKE redesign were support of emissions processing with user-selected chemical mechanisms and emissions processing for reactivity assessments. In 2002, SMOKE was enhanced to support driving the MOBILE6 model used to create on-road mobile emission factors and to support on-road and nonroad mobile toxics inventories. In 2003, SMOKE version 2.0 was created to include all toxic inventories, including point and nonpoint (stationary sources reported at the county level) sources. It can process criteria gaseous pollutants such as carbon monoxide (CO), nitrogen oxides (NO_x), VOC, ammonia (NH_3), sulfur dioxide (SO_2); PM pollutants such as PM2.5 and PM10; as well as a large array of toxic pollutants, such as mercury, cadmium, benzene, and formaldehyde. In fact, SMOKE has no limitation regarding the number or types of pollutants it can process.

The purpose of SMOKE is to convert the resolution of the emission inventory data to the resolution needed by an AQ model. Emission inventories are typically available with an annual total emissions value for each emissions source, or perhaps with an average-day emissions value. The air pollution models, however, typically require emissions data on an hourly basis, for each model grid cell (and perhaps model layer), and for each model species. Consequently, emissions processing involves transforming an emission inventory through temporal allocation, chemical speciation, and spatial allocation, to achieve the input requirements of the air pollution models.

SMOKE is primarily an emissions processing system and not an emission inventory preparation system. This means that its main purpose is to provide an efficient tool for converting emission inventory data into the formatted emission input files required by an AQM. However, for mobile and biogenic sources, SMOKE does offer emission inventory preparation functions. For mobile sources, SMOKE computes an emission inventory from mobile-source activity data, using emission factors from the MOBILE6 model. Previous versions of SMOKE (version 1.4 and below) have also supported the MOBILE5b model. For biogenic sources, SMOKE includes both the BEIS2 and BEIS3 models for computation of hour-specific, meteorology-based biogenic emissions from vegetation and soils.

19.5.3.1 Summary of the Major Features of SMOKE Version 2.0

- It supports both gridded and county-total land use for biogenic emissions modeling.
- It includes the BEIS2 or BEIS3 system (and the user can choose which he or she wants).
- It includes a driver for MOBILE6 runs and features to improve run time when using MOBILE6 for large domains and long periods (can support annual, national runs using MOBILE6).
- It has a multicountry capability (up to 10 countries).
- It has lower disk space requirements over other emissions processing software for the same level of detail in the emissions.
- Any pollutant can be processed by the system.
- Any chemical mechanism can be used to partition pollutants to model species, as long as the appropriate input data are supplied.
- It has improved control strategy input formats and design over previous SMOKE versions.

- Control strategies can include changes in the reactivity of emitted pollutants. This is useful, for example, when a solvent is changed in an industrial process.
- Run-time memory allocation, eliminating any need to recompile the source code for different inventories, grids, or chemical mechanisms.
- No third-party software is required to run SMOKE, although some input file preparation, such as gridded land use or spatial allocation surrogates, does require other software.
- Supports plume in grid (PinG) processing.
- Improved data file formats from previous SMOKE version.
- It features integrated use of Continuous Emissions Monitoring (CEM) data, an hour-specific data set of emissions from electric generating facilities. It also supports other day-specific and hour-specific point-source data.
- It supports hour-specific point-source stack parameters (exit temperature, exit velocity, and exit flow rate).
- It supports externally computed hourly plume rise (used for processing wildfires as point sources with a different plume rise algorithm than is available in SMOKE).

The sparse matrix approach used throughout SMOKE permits rapid and flexible processing of emissions data. Rapid processing is possible because SMOKE uses a series of matrix calculations rather than a less-efficient sequential approach used by previous systems. Flexible processing comes from splitting the processing steps of inventory growth, controls, chemical speciation, temporal allocation, and spatial allocation into independent steps whenever possible. The results from these steps are merged together in the final stage of processing using vector–matrix multiplication. This means that individual steps (such as adding a new control strategy, or processing for a different grid) can be performed and merged without having to redo all of the other processing steps.

SMOKE is written in Fortran 90 and is designed to run on a variety of UNIX platforms. Executables for Linux and IRIX are provided and the source code is available for download and can easily be compiled for a particular system. Running SMOKE on Windows is not supported, due to the inherent limitations of that system. The current version of SMOKE is version 2.1, although previous versions are still available for download.

It should be noted that SMOKE refers to a very extensive database of emissions, temporal and speciation profiles, which is available only for North America. This is why, for applications in the other parts of the world, only a part of SMAQ functions can be used—calculation of biogenic emissions, plume rise calculations, etc.

19.5.4 ADMS 4 Air Pollution Model

ADMS is an example of advanced trajectory model. It is an air pollution modeling software for industrial and urban AQ management. The model was developed by Cambridge Environmental Research Consultants (CERC; http://www.cerc.co.uk/environmental-software.html), with some of the model options written in collaboration with other research organizations, such as the UK Meteorological Office, University of Surrey, and National Power (now Innogy). It is a new-generation Atmospheric Dispersion model, which means that:

- The atmospheric boundary layer properties are described by two parameters—the boundary layer depth and the Monin–Obukhov length—rather than in terms of a single parameter (Pasquill Class).
- Dispersion under convective meteorological conditions uses a skewed Gaussian concentration distribution (shown by validation studies to be a better representation than a symmetric Gaussian expression).

ADMS 4 has a built-in meteorological preprocessor that allows a variety of input meteorological data. Both hourly sequential and statistical data can be processed, and all input and output meteorological variables are written to a file after processing. The model has links to the Surfer contour-plotting package, in addition to ArcView and MapInfo Geographical Information System (GIS) software. The GIS links can be used to enter and display input data, and display output, usually as color contour plots.

The ADMS 4 model includes dry and wet deposition, NO_x chemistry, hills, buildings, puffs, fluctuations, odors, radioactivity (and gamma dose), plume visibility, coastline, and time varying sources.

ADMS 4 can model both continuous releases, that is, "plumes," in addition to instantaneous and time-dependent releases, that is, "puffs." It uses a Runge–Kutta method to solve the conservation equations to estimate plume rise, rather than using empirical expressions (as used by a number of other models). The ADMS 4 method takes into account:

- The effect of plume buoyancy and momentum.
- It also includes the penetration of boundary layer inversions.

Dry deposition is assumed to be proportional to the near-surface concentration, and deposition velocities can either be entered by the user, or estimated by the model.

Wet deposition is modeled through a washout coefficient; irreversible uptake is assumed, and plume strength following wet deposition decreases with downwind distance.

A simple NO_x chemistry scheme is included, involving the conversion of nitrogen dioxide (NO_2) to nitrous oxide (NO) and ozone (O_3) in daylight, and a reverse reaction that occurs both day and night.

There is a *complex terrain option* that can be used in regions where the gradient exceeds 1:10, but is less than 1:3.

The dispersion of air pollution around buildings is complicated to model. The building effects module in ADMS 4 includes the following features (see Figure 19.3):

- Up to 25 buildings can be included in each model run with a "Main Building" being defined for each source.
- For each wind direction, a single effective wind-aligned building is defined, around which the flow is modeled.
- The flow field consists of a recirculating region (or cavity), with a diminishing turbulent wake downstream.
- Concentrations within the cavity, CR, are uniform, and based on the fraction of the release that is entrained.
- The concentration at a point further downwind is the sum of contributions from two plumes: a ground-based plume from the recirculating flow region and an elevated plume from the nonentrained remainder.
- The concentration and deposition are set to zero within the user-defined buildings.

ADMS 4 is the only model of its kind to model short time scale *fluctuations*. These are particularly important for the modeling of odors, and for calculation of a 15-min average for comparison with the UK National Air Quality

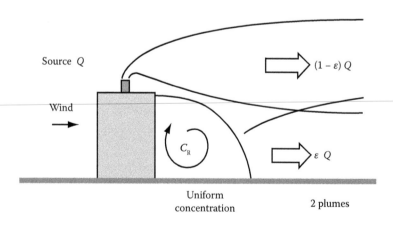

FIGURE 19.3
Plume–building interactions in ADMS 4.

Standard objective for SO_2. This module takes into account variations due both to turbulence and changes in meteorology.

ADMS 4 includes a radioactivity module that predicts the decay of radio-active species released from a source. Users may enter up to 10 "parent" iso-topes in any model run, and up to 50 isotopes ("parents" and "daughters") will be output. Half-lives of over 800 isotopes are included in the model and ADMS 4 can also calculate the associated levels of *gamma ray dose*.

For air dispersion modeling in coastal areas, ADMS 4 includes a coastline module (see Figure 19.4) that may be invoked when the following conditions are satisfied:

- The sea is colder than the land.
- There are convective meteorological conditions on land.
- There is an onshore wind.

19.5.4.1 Model Output

The user chooses the pollutant, the averaging time (which may be an annual average or a shorter period), any percentiles and exceedence values that are of interest, and whether a rolling average is required. The output options are designed to be flexible because AQ limits can vary from country to country and over time.

ADMS 4 has several versions designed for studying several specific air pollution issues.

19.5.4.2 ADMS 4 Industrial

This version is used to model the impact of existing and proposed industrial installations. Current and future AQ can be assessed with respect to the AQ standards. Typical applications include:

- IPPC authorizations
- Stack height determination
- Odor modeling

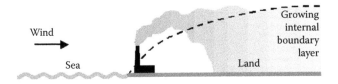

FIGURE 19.4
Treatment of sea–land internal boundary layer in ADMS 4.

- Environmental impact assessments
- Safety and emergency planning

The ADMS 4 Industrial accounts for several types of sources, as shown in Table 19.4.

The maximum number of sources that can be modeled in ADMS 4 is 300. Of these:

- Up to 300 may be point or jet sources.
- Within the limit of 300, up to 30 line sources, 30 area sources, and 30 volume sources may be modeled simultaneously.

19.5.4.3 ADMS-Urban

This version is used in most of the urban areas in the UK, because of its distinctive ability to describe in detail what happens on a range of scales, from the street scale to the citywide scale, taking into account a whole range of relevant emission sources: traffic, industrial, commercial, domestic, and other less well-defined sources.

It can be used to examine emissions from 6000 sources simultaneously, including 1500 road sources, up to 1500 point, line, area, or volume sources, and up to 3000 aggregated sources.

Modeling roads in urban areas is more complex than just modeling the emissions from traffic as a line source. Both the effect of *street canyons* and *traffic-induced turbulence* are included when roads are modeled in ADMS-Urban. In urban areas, it is also important to include the aggregated emissions from sources that may be too small to define explicitly, but whose aggregate emissions contribute to overall pollution levels. For example, domestic emissions of NO_x from an individual household may not be known, but the aggregated emissions could be calculated using areawide figures for fuel consumption.

Source and emissions data can be imported from a Microsoft Access database created by the user or exported from CERC's Emissions Inventory, which

TABLE 19.4

Industrial Source Types in ADMS 4 Industrial

Source Type	Example Sources
Point	Emissions from a stack or vent
Area	Evaporative emissions from a tank
Line	Emissions from a conveyor belt at a quarry
Volume	Fugitive emissions
Jet (directional releases)	Emissions from a ruptured pipe

contains current and future emission factors including those for vehicles, industrial processes, and fuel consumption.

Source parameters include:

- Source location data
- Road widths and canyon heights for road sources
- Stack heights, diameters, exit velocities, etc., for industrial sources
- Grid dimensions for aggregated emissions data

19.5.4.3.1 Modules in ADMS Urban

- NO_x–NO_2 chemistry.
- Sulfate chemistry—The reactions between SO_2 and other compounds in the air to produce particulates are based on those used in the EMEP model.
- Street canyons—This module is based on the Danish Operational Street Pollution Model (OSPM; Hertel and Berkowicz 1990).
- Complex terrain—This module is based on FLOWSTAR advanced airflow model, which calculates the change in mean flow and turbulence due to terrain and changes in surface roughness (land use).
- Buildings—Users can include up to 10 dominant buildings. ADMS-Urban creates an effective building for each point source from the user-defined buildings and models the recirculating flow in the lee of the building, the cavity region, as well as the building main wake.

19.5.4.4 ADMS-Roads

The ADMS-Roads pollution model is used for investigating air pollution problems due to small networks of roads that may be in combination with industrial sites, for instance, small towns or rural road networks. ADMS-Roads combines many of the scientific capabilities of ADMS-Urban with an easy-to-use interface. For large towns and cities, ADMS-Urban is the preferred tool.

ADMS-Roads can be used to examine emissions from many sources simultaneously, including 150 road sources, and up to 3 point, 3 line, 4 area, and 25 volume industrial sources.

19.5.4.5 ADMS-Airport

ADMS-Airport is a tool for managing AQ at airports. It is an extension of the ADMS-Urban model, designed to model the concentration of pollutants at airports in rural or complex urban environments. It is able to take into account the whole range of relevant emission sources: aircraft traffic, auxiliary power units, ground support equipment, road traffic, industrial, commercial, domestic, and other less well-defined sources. ADMS-Airport has been used to model AQ at London's Heathrow airport for the 2002 base case and future year scenarios. It can be used to examine emissions from 6500 sources simultaneously, including up to 500 aircraft jet sources, more than 1500 road sources, up to 1500 point, line, area, or volume industrial sources, and up to 3000 aggregated sources.

ADMS-Airport makes use of the ADMS jet model to calculate the impact of aircraft exhausts. The jet model calculates an integral solution to the equations of conservation of mass, momentum, heat, and species, capturing the effect of the movement of the jet engine source in reducing the effective buoyancy of the exhaust. This is particularly important in capturing the near-field dispersion from the high-momentum, buoyant takeoff ground roll sources.

To model the airport's flight schedule in detail, users can construct up to 500 annual hourly profiles. These detailed schedules can also be used for detailed modeling of non-airport sources, such as the effect of school terms and public holidays on road traffic. For less-detailed, time-dependent modeling, ADMS-Airport allows up to 50 diurnal and 50 monthly profiles plus wind direction dependence for any source.

19.5.5 SILAM Air Pollution Model

SILAM is a model with well-proven simulation abilities. It has been used in international emergency exercises and EU projects for model intercomparisons. It treats the chemical, aerosol, and cloud processes in a quite advanced manner. The model also produces a very extensive output. Depending on the user request, it can include 4-D concentration of the pollutants, 3-D dry and wet deposition, 4-D particle counts, 3-D vertically integrated particle counts, and any meteorological variable available from the meteorological preprocessor, whether or not it is used for the simulations. The last feature enables one to use the SILAM framework for meteorological analysis or data preparation without running any dispersion. The two nonstandard output variables—the particle counts (number of particles per cubic meter) and the vertically integrated value of this (number of particles per square meter) are convenient for the probability analysis of territories affected by the release.

The model also accounts for radioactive pollution including relatively detailed radiology—496 nuclides, 80 dose pathways, 23 human target organs—and allows computation of the environmental removal (migration) of nuclides after deposition, and of external and internal exposure doses. The output variables include air concentrations (Bq/m^3), deposited amounts of radioactive nuclides (Bq/m^2), external dose rates (Sv/s), and external and internal radiation doses (Sv) (the last two parameters being available in the separate mode only).

However, the main reason why this model is particularly mentioned in this very brief survey among numerous other models, is that SILAM implements the Lagrangian approach to describing advection and turbulent diffusion.

19.5.5.1 Lagrangian Advection Algorithm with Monte Carlo Random Walk

The advection–diffusion scheme of SILAM considers the pollution plume as a set of many Lagrangian particles, each carrying part of the total mass. These particles are advected by a 3-D synoptic-scale wind field and mixed by small-scale turbulent eddies. Thus, the admixture advection and diffusion is described by the motion of individual particles. Each particle moves according to the law:

$$\frac{d\mathbf{X}}{dt} = \mathbf{U}, \ \mathbf{U}(t) = \mathbf{u}(\mathbf{X}(t)), \tag{19.27}$$

where $\mathbf{X}$ is the radius vector (the coordinates) of the particle and $\mathbf{u}$ is the wind, which is composed of the grid-scale wind $\bar{\mathbf{u}}$, the turbulent wind fluctuations $\mathbf{u}_t$, and the mesoscale wind fluctuations $\mathbf{u}_m$ (Zannetti 1992), that is, $\mathbf{u} = \bar{\mathbf{u}} + \mathbf{u}_t + \mathbf{u}_m$.

The turbulent components of the wind velocity are parameterized assuming a Markov process based on the Langevin equation (Thomson 1987) for the individual wind components. Assuming Gaussian diffusion and neglecting cross-wind correlations, the approximation of three independent turbulent diffusion processes (Stohl and Thomson 1999) can be reached:

$$\frac{du_i}{dt} = -u_i \frac{dt}{d\tau_i} + \sqrt{\frac{2}{\tau_i}} \sigma_i dW_i, \ i = 1, 2,$$

$$\frac{du_3}{dt} = -u_3 \frac{dt}{d\tau_3} + \frac{\partial \sigma_3^2}{\partial x_3} dt + \frac{\sigma_3^2}{\rho} \frac{\partial \rho}{\partial x_3} dt + \sqrt{\frac{2}{\tau_3}} \sigma_3 dW_3 \tag{19.28}$$

where τ_i are Lagrangian time scales and σ_i are standard deviations of the turbulent wind components (parameterized from meteorological and similarity theory parameters following Nieuwstadt and Van Dop 1982, and Ryall and Maryon 1997). W_i are incremental components of a zero-mean Wiener process with a variance dt.

Equation 19.28 takes into account the variation of air density with height ($i = 3$ denotes the vertical direction), but neglects possibly skewed turbulence in the case of strong convection. The last effect, however, is believed to be not too large in the general case.

Solving Equation 19.28 is too time-consuming and is not fully tested yet, so the operational SILAM algorithm for computation of the random relocation of Lagrangian particles is based on the so-called fully mixed boundary layer assumption (Thomson 1987; Rodean 1996; Saltbones et al. 1996). According to this assumption, a stochastic displacement of the particle is given by:

$$x_i^{new} = x_i + r \cdot l_i,$$ (19.29)

where l_i denotes characteristic turbulent length scales and the random numbers r have a Gaussian (or uniform) distribution with zero mean and unit standard deviation. The length scales for horizontal diffusion are computed following Saltbones et al. (1996) and Sutton (1953):

$$l_i = a \Delta x_i^b,$$ (19.30)

where Δx is the distance traveled by the particle during the time step, $a = 0.5$ and 0.25 inside the boundary layer and in the free troposphere, respectively, and $b = 0.875$.

A common area of compromise for Lagrangian models is the number of Lagrangian particles representing the pollution cloud. In SILAM, this parameter is limited only by computer resources and time requirements, and should be set by the user. Typically, it ranges from 10^4 to 10^5 t for operational runs with a short period covered, and reaches 10^6 t for more detailed investigations. In some cases, where only a crude evaluation is required, the model can be run with a few thousand or even some hundreds of particles, storing the results in ASCII files in the form of random-walk trajectories. Such an output is widely used for various scenario assessments, in which only the composition of the release varies while the dispersion pattern is assumed to be fixed.

19.6 Conclusion

Models describing the dispersion and transport of air pollutants in the atmosphere can be distinguished on many grounds, for example:

- On the spatial scale (global, regional-to-continental, local-to-regional, local)

- On the temporal scale [episodic models, (statistical) long-term models]

- On the treatment of the transport equations (Eulerian, Lagrangian models)
- On the treatment of various processes (chemistry, aerosol, wet, and dry deposition)
- On the complexity of the approach

The modeling objectives can also be different: regulatory purposes, policy support, public information, scientific research.

The model choice for each particular case should depend both on the goals and the typical spatial/temporal scales of the studied air pollution phenomenon. The limitations due to data gaps and uncertainties, model restrictions, and imperfections should also be evaluated and kept in mind.

An accurate, comprehensive, reliable, and efficient air pollution modeling requires various types of data: meteorological and chemical fields and measurements (AQ data), emission inventories, physiographic, and even demographic data. The data are necessary as an input to the model, as well as for model evaluation and validation. The complete coherence of models and data is of crucial importance for successful modeling (i.e., sophisticated models will be useless without adequate input data quality and quantity).

References

Anthes, R. A., and T. T. Warner. 1978. Development of hydrodynamic models suitable for air pollution and other mesometeorological studies. *Monthly Weather Review* 106: 1045–1078.

Arakawa, A., and V. R. Lamb. 1977. Computational design of the basic dynamical process of the UCLA general circulation model. *Methods in Computational Physics* 17: 173–265.

Arakawa, A., and W. H. Schubert. 1974. Interaction of a cumulus cloud ensemble with the large scale environment. Part I. *Journal of the Atmospheric Sciences* 31: 674–701.

Atkinson R., A. C. Lloyd, and L. Winges. 1982. An updated chemical mechanism for hydrocarbon/NO_x/SO_2 photooxidations suitable for inclusion in atmospheric simulation models. *Atmospheric Environment* 16: 1,341–1,355.

Baldocchi, D. D., B. B. Hicks, and P. Camara. 1987. A canopy stomatal resistance model for gaseous deposition to vegetation surfaces. *Atmospheric Environment* 21 (1): 91–101.

Betts, A. K. 1986. A new convective adjustment scheme: Part I. Observational and theoretical basis. *Quarterly Journal of the Royal Meteorological Society* 112: 677–692.

Betts, A. K., and M. J. Miller. 1986. A new convective adjustment scheme: Part II. Single column tests using GATE wave, BOMEX, ATEX and Arctic air-mass data sets. *Quarterly Journal of the Royal Meteorological Society* 112: 693–709.

Betts, A. K., and M. J. Miller. 1993. The Betts–Miller scheme. In *The Representation of Cumulus Convection in Numerical Models*, 246 pp. eds. K. A. Emanuel and D. J. Raymond. American Meteorological Society.

Binkowski, F. S., and U. Shankar. 1995. The Regional Particulate Model: Part I. Model description and preliminary results. *Journal of Geophysical Research* 100 (D12): 26,191–26,209.

Blackadar, A. K. 1979. High resolution models of the planetary boundary layer. In *Advances in Environmental Science and Engineering*, 1 (1), eds. J. Pfafflin and E. Ziegler, 50–85. New York, NY: Gordon and Briech Sci.

Bott, A. 1989. A positive definite advection scheme obtained by nonlinear renormalization of the advective fluxes. *Monthly Weather Review* 117: 1006–1015.

Builtjes, P. J. H., M. van Loon, M. Schaap, S. Teeuwisse, A. J. H. Visschedijk, and J. P. Bloos. 2003. Project on the modelling and verification of ozone reduction strategies: contribution of TNO-MEP, TNO-report, MEP-R2003/166, Apeldoorn, the Netherlands.

Byun D. W., A. Hanna, C. J. Coats, and D. Hwang. 1995a. Models-3 air quality model prototype science and computational concept development. *Transactions of Air & Waste Management Association Specialty Conference on Regional Photochemical Measurement and Modeling Studies*, Nov. 8–12, 1993, 197–212. San Diego, CA.

Byun, D. W., C. J. Coats, D. Hwang, S. Fine, T. Odman, A. Hanna, and K. J. Galluppi. 1995b. Prototyping and implementation of multiscale air quality models for high performance computing. Mission Earth Symposium, Phoenix, AZ, April 9–13, 1993, 527–532.

Byun, D. W., D. Dabdub, S. Fine, A. F. Hanna, R. Mathur, M. T. Odman, A. Russell et al. 1996. Emerging air quality modeling technologies for high performance computing and communication environments. In *Air Pollution Modeling and Its Application XI*, eds. S. E. Gryning and F. Schiermeier, 491–502.

Byun, D. W., J. K. S. Ching, J. Novak, and J. Young. 1997. Development and implementation of the EPA's Models-3 initial operating version: Community Multi-scale Air Quality (CMAQ) Model, 1998: Twenty-Second NATO/CCMS International Technical Meeting on Air Pollution Modelling and Its Application, 2–6 June, 1997. In *Air Pollution Modeling and Its Application XII*, ed. S. E. Gryning and N. Chaumerliac, 357–368. Plenum Publishing Corp.

Byun, D. W., J. Young., G. Gipson., J. Godowitch., F. Binkowsk, S. Roselle, B. Benjey, et al. 1998. Description of the Models-3 Community Multiscale Air Quality (CMAQ) model. *Proceedings of the American Meteorological Society 78th Annual Meeting*, Jan. 11–16, 1998, 264–268. Phoenix, AZ.

Carter, W. P. L. 1990. A detailed mechanism for the gas-phase atmospheric reactions of organic compounds. *Atmospheric Environment* 24A: 481–515.

Carter W. P. L., D. Luo, and I. L. Malkina. 1997. Environmental Chamber Studies for Development of an Updated Photochemical Mechanism for Reactivity Assessment. Final Report for California Air Resources Board Contract No. 92–345, Coordinating Research Council, Inc., Project M-9 and National Renewable Energy Laboratory, Contract ZF-2-12252-07.

Ching, J. K. S., D. W. Byun, A. Hanna, T. Odman, R. Mathur, C. Jang, J. McHenry, and K. Galluppi. 1995. Design requirements for multiscale air quality models. *Mission Earth Symposium*, April 9–13, 532–538. Phoenix, AZ.

Coats, C. J., A. H. Hanna, D. Hwang, and D. W. Byun. 1995. Model engineering concepts for air quality models in an integrated environmental modeling system.

Transactions of Air & Waste Management Association Specialty Conference on Regional Photochemical Measurement and Modeling Studies, Nov. 8–12, 1993, 213–223. San Deigo, CA.

Colella, P., and P. L. Woodward. 1984. The Piecewise Parabolic Method (PPM) for gas-dynamical simulations. *Journal of Computational Physics* 54: 174–201.

Dudhia, J. 1989. Numerical study of convection observed during the winter monsoon experiment using a mesoscale two-dimensional model. *Journal of Atmospheric Science* 46: 3077–3107.

Dudhia, J. 1996. A multi-layer soil temperature model for MM5. Preprints. *The Sixth PSU/NCAR Mesoscale Model Users' Workshop.* Boulder, CO: National Center for Atmospheric Research.

Dudhia, J., D. Gill, Y.-R. Guo, D. Hansen, K. Manning, and W. Wang. 1998. PSU/NCAR mesoscale modeling system tutorial class notes (MM5 modeling system version 2). [Available from the National Center for Atmospheric Research, P. O. Box 3000, Boulder, CO 80307.]

Erisman, J. W., A. J. van Pul, and G. P. Wyers. 1994. Parameterization of surface resistance for the quantification of the atmospheric deposition of acidifying pollutants and ozone. *Atmospheric Environment* 28 (16): 2595–2607.

Ganev, K., and D. Yordanov. 2005. Parameterization of dry deposition processes in the surface layer for admixtures with gravity deposition. *International Journal of Environment and Pollution* 25: 1–4, 60–70.

Gelbard, F., Y. Tambour, and J. H. Seinfeld. 1980. Sectional representations for simulating aerosol dynamics. *Journal of Colloid and Interface Science* 76: 541–556.

Gery, M. W., G. Z. Whitten, J. P. Killus, and M. C. Dodge. 1989. A photochemical kinetics mechanism for urban and regional scale computer modeling. *Journal of Geophysical Research* 94 (12): 925–12,956.

Gifford, F. 1959. Statistical properties of a fluctuating plume dispersal model. *Advances in Geophysics* 6: 117–137.

Grell, G. A., Y.-H. Kuo, and R. Pasch. 1991. Semi-prognostic tests of cumulus parameterization schemes in the middle latitudes. *Monthly Weather Review* 119: 5–31.

Grell, G. A. 1993. Prognostic evaluation of assumptions used by cumulus parameterizations. *Monthly Weather Review* 121: 764–787.

Grell, G. A., J. Dudhia, and D. R. Stauffer. 1994. A description of the fifth-generation Penn State/NCAR mesoscale model (MM5). NCAR Tech. Note NCAR/TN-398+STR, 117 pp. [Available from the National Center for Atmospheric Research, P. O. Box 3000, Boulder, CO 80307.]

Hack, J. J., B. A. Boville, B. P. Briegleb, J. T. Kiehl, P. J. Rasch, and D. L. Williamson. 1993. Description of the NCAR community climate model (CCM2). NCAR Tech. Note, NCAR/TN-382+STR, 120 pp. [Available from the National Center for Atmospheric Research, P.O. Box 3000, Boulder, CO 80307.]

Hertel, O., and R. Berkowicz. 1990. Vurdering af spredningsmodellen i den Nordiske Beregningsmetode for Bilavgasser. Sammenfattende rapport, DMU Luft A-136, 27 pp. (in Danish).

Hicks, B. B., D. D. Baldocchi, T. P. Mayers, R. P. Hosker Jr., and D. R. Matt. 1987. A preliminary multiple resistance routine for deriving dry deposition velocities from measured quantities. *Water Air Soil Pollution* 36: 311–330.

Hsie, E.-Y., and R. A. Anthes. 1984. Simulations of frontogenesis in a moist atmosphere using alternative parameterizations of condensation and precipitation. *Journal of Atmospheric Sciences* 41: 2701–2716.

Hong, S.-Y., and H.-L. Pan. 1996. Nonlocal boundary layer vertical diffusion in a medium-range forecast model. *Monthly Weather Review* 124: 2322–2339.

Jacobson, M., and R. P. Turco. 1994. SMVGEAR: A sparse-matrix, vectorized gear code for atmospheric models. *Atmospheric Environment* 28: 273–284.

Jacobson, M. Z. 1997. Development and application of a new air pollution modeling system: II. Aerosol module structure and design. *Atmospheric Environment* 31: 131–144.

Janjic, Z. I. 1994. The step-mountain eta coordinate model: Further development of the convection, viscous sublayer, and turbulent closure schemes. *Monthly Weather Review* 122: 927–945.

Joffre, S. M. 1988. Modelling the dry deposition velocity of high soluble gases to the sea surface. *Atmospheric Environment* 22 (6): 1137–1146.

Kain, J. S., and J. M. Fritsch. 1990. A one-dimensional entraining/detraining plume model. *Journal of Atmospheric Sciences* 47: 2784–2802.

Kain, J. S., and J. M. Fritsch. 1993. Convective parameterization for mesoscale models: The Kain–Fritsch scheme. In *The Representation of Cumulus in Mesoscale Models*, eds. K. A. Emanuel and D. J. Raymond, 246 pp. American Meteorological Society.

Lurmann F. W., W. P. L. Carter, and L. A. Coyner. 1987. A Surrogate Species Chemical Reaction Mechanism for Urban Scale Air Quality Simulation Models Volume 1. EPA-600/3-87-014a.

Nieuwstadt, F. T. M., and H. Van Dop, eds. 1982. *Atmospheric Turbulence and Air Pollution Modelling*, 360 pp. Dordrecht, the Netherlands: D. Reidel Publishing Company.

Orlanski, J. 1975. A rational subdivision of scales for atmospheric processes. *Bulletin of the American Meteorological Society* 56: 527–530.

Pleim, J. E., and J. S. Chang. 1992. A non-local closure model in the convective boundary layer. *Atmospheric Environment. Part A. General Topics* 26: 965–981.

Pleim, J. E., and A. Xiu. 1995. Development and testing of a surface flux and planetary boundary layer model for application in mesoscale models. *Journal of Applied Meteorology* 34: 16–32.

Reisner, J., R. T. Bruintjes, and R. J. Rasmussen. 1993. Preliminary comparisons between MM5 NCAR/Penn State model generated icing forecasts and observations. Preprints, *Fifth Intl. Conf. on Aviation Weather Systems*, 65–69. Vienna, VA: American Meteorological Society.

Reisner, J., R. J. Rasmussen, and R. T. Bruintjes. 1998. Explicit forecasting of supercooled liquid water in winter storms using the MM5 mesoscale model. *Quarterly Journal of the Royal Meteorological Society* 124B: 1071–1107.

Rodean, H. C. 1996. *Stochastic Lagrangian Models of Turbulent Diffusion*. Boston, MA: American Meteorological Society.

Ryall, D. B., and R. H. Maryon. 1997. Validation of the UK Met Office's NAME model against the ETEX dataset. In *ETEX Symposium on Long-Range Atmospheric Transport, Model Verification and Emergency Response*, ed. K. Nodop, 151–154. European Commission EUR 17346.

Saltbones, J., A. Foss, and J. Bartnicki. 1996. A real time dispersion model for severe nuclear accidents, tested in the European tracer experiment. *Systems Analysis and Modelling Simulation* 25: 263–279.

Sienfeld, J. H., and S. Pandis. 1998. *Atmospheric Chemistry and Physics*. New York, NY: Wiley, 1326.

Seland, Ø., A. van Pul, A. Sorteberg, and J.-P. Tuovinen. 1995. Implementation of resistance dry deposition module and a variable local correction factor in the Lagrangian EMEP model. EMEP/MSC-W Report 3/95.

Smolarkiewicz, P. K., and G. A. Grell. 1992. A class of monotone interpolation schemes. *Journal of Computational Physics* 101: 431–440.

Stauffer, D. R., and N. L. Seaman. 1990. Use of four-dimensional data assimilation in a limited area mesoscale model: Part I. Experiments with synoptic-scale data. *Monthly Weather Review* 118: 1250–1277.

Stauffer, D. R., N. L. Seaman, and F. S. Binkowski. 1991. Use of four-dimensional data assimilation in a limited-area mesoscale model: Part II. Effects of data assimilation within the planetary boundary layer. *Monthly Weather Review* 119: 734–754.

Stauffer, D. R., and N. L. Seaman. 1994. Multiscale four-dimensional data assimilation. *Journal of Applied Meteorology* 33: 416–434.

Stockwell, W. R. 1986. A homogeneous gas phase mechanism for use in a regional acid deposition model. *Atmospheric Environment* 20: 1,615–1,632.

Stockwell, W. R., P. Middleton, and J. S. Chang. 1990. The second generation regional acid deposition model chemical mechanism for regional air quality modeling. *Journal of Geophysical Research* 95 (D10): 16,343–16,367.

Stohl, A., and D. J. Thomson. 1999. A density correction for Lagrangian particle dispersion models. *Boundary Layer Meteorology* 90: 155–167.

Sutton, O. G. 1953. *Micrometeorology*, 343 pp. New York, NY: McGraw-Hill.

Tao, W.-K., and J. Simpson. 1993. Goddard cumulus ensemble model: Part I: Model description. *Terrestrial, Atmospheric and Oceanic Sciences* 4: 35–72.

Thomson, D. J. 1987. Criteria for the selection of stochastic models of particle trajectories in turbulent flows. *Journal of Fluid Mechanics* 180: 529–556.

Venckatram, A., and J. Pleim. 1999. The electrical analogy does not apply to modeling dry deposition of particles. *Atmospheric Environment* 33: 3075–3076.

Visschedijk, A. J. H., P. Y. J. Zandveld, and H. A. C. Denier van der Gon. 2007. A High Resolution Gridded European Emission Database for the EU Integrate Project GEMS, TNO-report 2007-A-R0233/B, Apeldoorn, the Netherlands.

Wesley, M. L. 1989. Parameterization of surface resistance to gaseous dry deposition in regional-scale numerical models. *Atmospheric Environment* 23 (6): 1293–1304.

Whitby, K. T. 1978. The physical characteristics of sulfur aerosols. *Atmospheric Environment* 12: 135–159.

Whitten, G. Z., H. Hogo, and J. P. Killus. 1980. The carbon bond mechanism: A condensed kinetic mechanism for photochemical smog. *Environmental Science & Technology* 14: 690–701.

WHO. 2000. Fact Sheet Number 187, World Health Organization.

WHO. 2004. *Health Aspects of Air Pollution*. Results from the WHO Project Systematic Review of Health Aspects of Air Pollution in Europe, World Health Organization.

Zannetti, P. 1992. Particle modelling and its application for simulating air pollution phenomena. In *Environmental Modelling*, eds. P. Melli and P. Zannetti. Southampton, UK: Computational Mechanisms Publications.

Zhang, D.-L., and R. A. Anthes. 1982. A high-resolution model of the planetary boundary layer—Sensitivity tests and comparisons with SESAME-79 data. *Journal of Applied Meteorology* 21: 1594–1609.

Zhang, D.-L., H.-R. Chang, N. L. Seaman, T. T. Warner, and J. M. Fritsch. 1986. A two-way interactive nesting procedure with variable terrain resolution. *Monthly Weather Review* 114: 1330–1339.

20

Acidification

Tomáš Navrátil

CONTENTS

20.1 Introduction 558
20.2 Emissions and Deposition 559
    20.2.1 Acid Deposition 560
    20.2.2 Effects of Acid Deposition 561
        20.2.2.1 Depletion of Base Cations and Mobilization of Al
            in Soils 561
        20.2.2.2 Accumulation of Sulfur in Soils 561
        20.2.2.3 Accumulation of Nitrogen in Soils 562
        20.2.2.4 Effects of Acidic Deposition on Forests 563
        20.2.2.5 Effects of Acidic Deposition on Surface Water 563
    20.2.3 Importance of Land Use 564
20.3 Ecosystem Recovery 564
20.4 Acidification Models 565
    20.4.1 Steady-State Models 565
    20.4.2 Dynamic Models—Basic Concepts and Equations 567
        20.4.2.1 Basic Equations 568
        20.4.2.2 From Steady-State to Dynamic Simulations 570
        20.4.2.3 Buffer Processes 571
        20.4.2.4 From Soils to Surface Waters 574
    20.4.3 Selected Dynamic Models 575
        20.4.3.1 VSD Model 575
        20.4.3.2 SMART Model 577
        20.4.3.3 SAFE Model 578
        20.4.3.4 MAGIC Model 580
20.5 Modeling and Results 582
    20.5.1 Calibration 584
    20.5.2 Applications 585
    20.5.3 Uncertainties 590
20.6 Summary 591
References 592

20.1 Introduction

In environmental science, the term acidification describes the natural process occurring in the environment that has been enormously accelerated by human activity.

The term acid rain has been used to describe the deposition of acidic components, but the more appropriate term is acid precipitation. The composition of acidic deposition includes ions, gases, and particles derived from gaseous emissions of sulfur dioxide (SO_2), nitrogen oxides (NO_x), and ammonia (NH_3), and from particulate emissions of acidifying and neutralizing compounds. In the second half of the twentieth century, acidic deposition has emerged as one of the most significant environmental stresses affecting forested landscapes and aquatic ecosystems in North America, Europe, and Asia.

Acidification is a complex problem because it can originate from air pollution, and it affects large geographic areas. Because of its connection with human activity and climate, it significantly varies in time and space. Acidification alters the biogeochemical cycles of many elements in the environment and contributes directly or indirectly to biological stress, ecosystem degradation, and loss of biodiversity.

Scientists have provided the concepts and tools for calculating the effects of acid deposition on ecosystems. These were later used for negotiation of ceilings for national emissions of pollutants by a series of protocols under the United Nations Economic Commission for Europe Convention on Long-Range Transboundary Air Pollution. The critical loads calculated by steady-state models defined the deposition load that an ecosystem can tolerate without being damaged, and has provided an effect-based concept for determining the required reductions. The information that the steady-state concept did not provide is the time needed for the recovery of damaged ecosystems. In further evolution of the concept, critical loads calculated using steady-state models will be added with target loads, which require the use of dynamic models. Application of dynamic soil chemistry models will yield answers to issues such as:

- How much time is needed for a specific site to recover with a certain level of deposition?
- How much should the emissions be reduced today to achieve recovery by a certain year?

The objectives of dynamic modeling have evolved from studying the acidification effects and recovery processes in soils and natural waters to calculating the target loads. The results of future dynamic modeling will impact the future pollution reductions.

20.2 Emissions and Deposition

Air pollution by acidic substances has been recognized as an environmental problem since the nineteenth century (Smith 1872). In the middle of twentieth century, the gathered evidence signaled that acidification will represent a problem not only in the vicinity of industrialized areas, but also in very remote sites. Acidic deposition emerged as an ecological issue in the late 1960s and early 1970s with reports from Scandinavia and the United States (Odèn 1968; Likens et al. 1972). But progress has been made since then. The main environmental acidifiers of anthropogenic origin are sulfur (S) and nitrogen (N) compounds. The links between emissions of SO_2, NO_x, and NH_y and acidification of soils and waters are well known (Grennfelt et al. 1995; Satake et al. 2001).

The dominant anthropogenic source of sulfur emissions has been the burning of fossil fuels (Grubler 1998). Sulfur emissions are dominated by coal burning with small contributions from burning of oil products. The dominant form of anthropogenic sulfur emissions are airborne emissions of SO_2. The natural sources of sulfur emissions with usually trivial contribution to the total emissions include the weathering of rocks and soil, and volatile biogenic sulfur emissions from land and the oceans as well as volcanoes (Benkovitz et al. 1996).

Sulfur emission inventories are developed on a regular basis in a number of regions, including within the European Monitoring and Evaluation Programme (EMEP) and CORe INventory AIR emissions (CORINAIR) programs in Europe and the National Acid Precipitation Assessment Program (NAPAP) in North America. Emission inventories are also available for Asia, where emission growth rates have been particularly high (Streets et al. 2000).

As a result of sulfur abatement efforts, emissions in the OECD countries, particularly in Europe and Japan have declined. Emissions in Central and Eastern Europe have also declined significantly because of replacement of coal by other fuels, and the economic recession since the early 1990s (Grubler 2002; Kopáček and Veselý 2005). Accelerated economic development in many parts of Asia resulted in recent fast growth of sulfur emissions (Streets et al. 2000). Estimates for Asia indicate a decline of emission growth rates as a result of the introduction of sulfur control legislation (Grubler 1998; Streets et al. 2000). Historical sulfur emission inventories were developed for Europe (Mylona 1993, 1996) as well as for the United States (Husar 1994; Gschwandtner et al. 1985; EPA 1995).

Emissions of nitrogen compounds include many forms, and originate from a wide selection of natural and anthropogenic sources. The main molecular forms of nitrogen emissions are nitrogen oxides (NO and NO_2 usually summed as NO_x) and ammonia (NH_3).

For emissions of nitrogen oxides (NO_x), the dominant anthropogenic source category is the burning of fossil fuels, most notably of automobile fuel and in the generation of electricity. The dominating automobile ownership and electricity consumption in industrialized countries make them major emitters (Grubler 2002).

The dominant source of anthropogenic emissions of ammonia (NH_3) is agriculture, including emissions from animal manure and fertilizers. NH_3 emissions are attributed to dominance from agricultural sources concentrated in developing countries (Parashar et al. 1998). Rising animal populations and increasing fertilizer use are the consequence of increasing agricultural production for steeply growing populations.

The most detailed projections of nitrogen emissions have been developed under the auspices of the Intergovernmental Panel on Climate Change Special Report on Emissions Scenarios (Nakicenovic et al. 1996). Nitrogen emissions may increase over time with the increasing demand for power.

20.2.1 Acid Deposition

Increased emissions of main acidifiers especially S and N connected to the industrial development occurring in different times and different parts of the world result in increased deposition acid compounds.

The acid deposition or deposition of the main acidifiers occurs before all as "wet deposition" and "dry deposition." Wet deposition of a particular element includes concentration of elements in rain, fog, dew, and snow, although dry deposition includes concentrations of elements deposited via particles and gasses (Driscoll et al. 2001). At selected sites, the wet and dry precipitation has been evaluated by collecting various forms of precipitation with a given time step. The purpose of such measurements is to quantify the pollutant loadings to open sites, lakes, and forests.

Dry deposition is especially important in forested areas. The process of dry deposition in the case of sulfur consists in adsorption of gaseous SO_2 on respiratory organs of the woody plants. Adsorbed SO_2 oxidates onto sulfuric acid (H_2SO_4) and the acid is washed with the next rain onto the soil. The process of dry deposition has been especially important in areas with high concentrations of SO_2 in the ambient air. The rainwater sampled below the canopy is referred to as throughfall.

Both wet and dry deposition of main acidifiers S and N, together with other parameters (such as concentrations of calcium, magnesium, potassium, sodium, chloride, fluoride, etc.), are widely monitored at individual sites or in networks such as National Deposition Monitoring Program (NADP) in United States and EMEP in Europe. Long-term records on deposition fluxes are usually lacking for most sites. But observed relationships between current emissions and current deposition, plus estimates of past emissions, enable us to reconstruct historical patterns of atmospheric deposition of S and N.

20.2.2 Effects of Acid Deposition

It was observed that vegetation canopies amplify acid deposition, so special attention has been paid to acidification of forest soils. Tree canopies scavenge aerosols, particles, and gases containing acidifying substances such as S and N compounds and thus multiply acid deposition on an open field. Because of this and other processes such as uptake of nutrients by vegetation, the chemistry of soil has been significantly changed in some sites. Because of the chemical equilibrium between the soil and soil solutions, changes in soil chemistry were reprinted in changes of soil solution chemical composition, which consequently caused changes in the chemistry of groundwater and runoff. The next few subsections will emphasize some of the effects of acidification with a primary focus on forest ecosystems.

20.2.2.1 Depletion of Base Cations and Mobilization of Al in Soils

Numerous experimental and field studies have shown that acidic deposition has changed the chemical composition of soils (e.g., Fernandez et al. 2003). Acidic deposition has increased the concentrations of protons (H^+) and strong acid anions (SO_4^{2-} and NO_3^-) in soils, which has led to increased rates of leaching of base cations and to the associated acidification of soils. Acidification causes depleting the content of available plant nutrient cations in soil by increasing the mobility of Al and by increasing the S and N content. Plant nutrient cations are usually referred to as base cations (BC), a group that includes Ca^{2+}, Mg^{2+}, K^+, and Na^+. In modeling, Na^+ can be sometimes excluded because it is not a major nutrient (Martinson 2004). Base saturation describes the degree of saturation of soil adsorption complex with exchangeable base cations or cations other than H^+ or Al^{3+}. It is expressed as a percentage of the total cation exchange capacity.

If the supply of base cations in soils is sufficient, the acidity of the soil water will be effectively neutralized. But if soil base saturation decreases below 20%, then neutralization will be incomplete and atmospheric deposition of strong acids will cause the mobilization and leaching of Al and H^+ (Cronan and Schofield 1990).

Mineral weathering is the primary source of base cations in most forest ecosystems, although atmospheric deposition provides important inputs especially to sites with low rates of supply from mineral sources (Driscoll et al. 2001). In acid-sensitive areas, rates of base cation supply through chemical weathering are not sufficient to keep pace with leaching rates accelerated by acidic deposition.

20.2.2.2 Accumulation of Sulfur in Soils

Sulfate adsorption on soil particles is an important process concerning the processes of acidification and recovery (Christophersen and Wright 1981;

Cosby et al. 1986; Karltun 1995). Some soils have considerable capacity to adsorb anions such as sulfate (SO_4^{2-}). The process of adsorption temporarily delays the adverse effects of acidification by removing SO_4^{2-} and H^+ from the solution. Because of the equilibrium between the amount of SO_4^{2-} in the solution and the amount adsorbed, the adsorbed sulfate will be released when the atmospheric inputs of SO_4^{2-} will decrease. This release has been observed in a number of catchments in Europe and North America, which will delay recovery (Driscoll et al. 2002; Stoddard et al. 1999; Hruska et al. 2002; Folster et al. 2003; Forsius et al. 2003).

There are also other processes involving sulfate that may influence recovery. The organic sulfate pool in forest soils is large (Fuller et al. 1986; Mitchell and Fuller 1988), and it has been shown that anthropogenic sulfur can be cycled through the organic sulfur pool before being released to soil solution and stream water (Alewell 2001).

20.2.2.3 Accumulation of Nitrogen in Soils

Nitrogen is considered the growth-limiting nutrient for temperate forest vegetation, and its retention by forest ecosystems is high. Because of this, concentrations of NO_3^- are often very low in surface waters draining the forest landscapes. Research results indicate that atmospherically deposited nitrogen has accumulated in soils, and some forest ecosystems have exhibited diminished retention of N inputs. Increased stream N export (usually in the form of NO_3^-) has been observed by monitoring the chemistry of first-order streams draining the forests but also at experimentally manipulated sites (Norton et al. 1994).

Increased losses of NO_3^- to surface waters may indicate changes in the strength of plant and soil microbial N sinks in forest watersheds. Because microbial processes are highly sensitive to temperature and fluctuations in microbial immobilization and mineralization, response to climate variability affects NO_3^- losses via drainage waters (Driscoll et al. 2001).

Despite the linkage between the atmospheric deposition of NH_4^+ and NO_3^- and the loss of NO_3^- from forest ecosystems (Dise and Wright 1995), future effects of atmospheric N deposition on forest N cycling and surface water acidification are likely to be controlled by climate, forest history, and forest type (Aber et al. 1998; Lovett et al. 2000). The complexity of linkages of NO_3^- loss to climatic variation, to land-use history, and to vegetation type has slowed efforts to predict how acid neutralizing capacity (ANC) in surface waters will respond to changes in atmospheric N deposition. It is apparent that additional NH_4^+ and NO_3^- inputs to forests will increase leaching losses of NO_3^-, whereas reductions in NO_x and NH_3 emissions and subsequent N deposition would result in long-term decreases of catchment acidification (Driscoll et al. 2003).

20.2.2.4 Effects of Acidic Deposition on Forests

Researchers noted extensive dieback or reduced growth in coniferous and some deciduous stands back in the twentieth century (Siccama et al. 1982; Houston 1999). Investigations of the effects of acidic deposition on trees identified both the direct effects of acidic precipitation and cloud water on foliage and the indirect effects from chemical changes in soils that alter nutrient uptake by roots. The physiological functioning of tree roots is especially sensitive to high concentrations of Al in the soil solution, respectively to low ratio of base cations and aluminum (Bc/Al). Aluminum adsorbed on root membranes inhibits the active transport of base cations. Deficiencies of Ca^{2+} and Mg^{2+} were reported to be a result of increased Al^{3+} concentrations in the soil solution.

Aluminum can block the uptake of Ca^{2+} and Mg^{2+}, which can lead to reduced growth and increased susceptibility of trees to stress. Cronan and Grigal (1995) concluded that a Ca/Al ratio of less than 1.0 in soil solution indicates a 50% probability of impaired growth. For Ca/Al less than 0.5, the probability of impaired growth increases to 90%. Hruška et al. (2001) found a statistical relationship between spruce defoliation and crown structure transformation with Ca/Al and (Ca + Mg + K)/Al in soil solutions of organic horizons.

20.2.2.5 Effects of Acidic Deposition on Surface Water

Acidification of surface waters occurs in areas where acidic deposition has been high and the pools of base cations in the catchment soil and bedrock are poor. Acidification of the surface waters has been observed in a number of areas, including high-altitude lakes in mountain regions, lakes, and streams in some forest areas, and surface waters (Warfvinge et al. 2000).

As the water originating from acidic deposition drains through the soils in ecosystem and pools of base cations become depleted, the pH of percolating solutes decreases causing increased leaching of Al. The acidified soil solution emerging into the surface water causes increased concentrations of Al in streams or lakes, which are known to be the primary cause of fish mortality. The controls on aluminum release caused by decreased pH are cation exchange, dissolution of aluminum solid phase, usually referred to as gibbsite = $Al(OH)_3$ with variable solubility constants (Warfvinge et al. 2000).

Changes in the buffering capacity of surface water or soil solutions have often been used as a measure of acidification or recovery. The concept of ANC is a further refinement of the alkalinity. There are two ways of defining the ANC. By the definition it is the difference between the sum of BC concentrations and sum of strong acid anions (SAA) concentrations:

$$ANC = [BC] - [SAA] \qquad (20.1)$$

where BC = $[Ca^{2+}]$ + $[Mg^{2+}]$ + $[K^+]$ + $[Na^+]$ and SAA = $[F^-]$ + $[Cl^-]$ + $[NO_3^-]$ + $[SO_4^{2-}]$.

The second definition of ANC is based on the difference between weak acids and cations of weak bases:

$$ANC = [HCO_3^-] + [A^-] - [H^+] - [Al^{n+}] \qquad (20.2)$$

where $[A^-]$ denotes the concentration of organic acids and $[Al^{n+}]$ concentrations of all forms of positively charged aluminum ions. All other ions can usually be disregarded in acid-sensitive waters. This equation enables us to calculate the pH for any given ANC, and also to analyze causes of low pH in a particular sampled stream or lake (Warfvinge et al. 2000).

20.2.3 Importance of Land Use

Large quantities of nitrogen and base cations accumulate in wood, bark, and respiratory organs of trees in the forests. With timber production and other associated practices in forest management, large quantities of nutrients are being removed from the forest ecosystems. The removal of base cations from the forest ecosystems with harvested trees or their compartments has an acidifying effect. Increased extraction of biomass containing significant amounts of base cations will delay recovery from acidification (Warfvinge et al. 2000).

20.3 Ecosystem Recovery

When the term recovery is used, it is usually understood as a reversal of processes caused by acidification. In brief, chemical recovery of acidified soils may be indicated by increase in the size of base cation pool and in cases where stream or lake recovery occurs when pH values decreases, causing lower Al concentrations. In terms of ecosystems, chemical recovery is usually followed by biological recovery.

There is often a long delay between the deposition of acidifying compounds and their effects on soils and waters, because of reactions that buffer the acidifying input. This buffering capacity consisting of several processes varies between sites or regions. Two important soil processes that buffer acidifying deposition and determine the dynamics of acidification and recovery are cation exchange and sulfate adsorption (Warfvinge et al. 2000). Because of this, some sites are more sensitive to acidification than others. When acid loading decreases, the buffering capacity of the acidified soil must be at least partially restored to achieve significant recovery (Martinson 2004).

In the 1990s, several roof projects in Europe were initiated for studying the reversal of acidification (Wright and Rasmussen 1998; Hultberg and Skeffington 1998). By covering forest sites with roofs and thus removing all anthropogenic deposition, these large-scale field experiments have provided a view into the future and an important understanding of recovery processes (Martinson 2004).

20.4 Acidification Models

Numerous models for simulating the acidification of soils and surface waters have been developed during the past decades (Tiktak and Van Grinsven 1995). These models cover a wide spectrum of applications and objectives. Some of them are research models that have been developed for a certain project or site. These scientific single-site or single-problem models usually require detailed input data. Other models were designed to be easily applicable at many sites by minimizing input requirements and with emphasis on user-friendliness. The term *model* refers in general to a model system, that is, a set of software and databases that consists of preprocessors for input data and calibration, postprocessors for the model output, and usually in the smallest part of the actual model itself (Posch et al. 2003b).

Models of soil and surface water acidification have been built for research and learning purposes. The response of soil, soil solutions, and surface waters to acid deposition has been simulated using steady-state and dynamic models.

Model projections of steady-state chemistry have been used to evaluate whether certain areas receive loads of acidity that they can sustain, assuming constant load of acidity and climate. Thus, the modeled results are impossible to validate.

But dynamic models, including long-term processes, which were based on earlier steady-state versions, may be validated using current field data. Thus, the dynamic perspective of acidification models completes and extends the more commonly used steady-state modeling approaches (Posch et al. 2003b). Dynamic models also offer the opportunity to evaluate hypothetical situations that would occur under various deposition scenarios or ecosystem (forest) management scenarios.

20.4.1 Steady-State Models

Steady-state models calculate deposition levels that will not cause harmful effects to ecosystems in a steady state. Thus, these models do not contain finite time scale effects such as cation exchange or sulfur adsorption, and require much less input data than the dynamic models. Steady state in flow systems describes situation when concentrations no longer change with time, because chemical reactions are counteracted by the in- and outflow of the system. Steady state should not be confused with chemical equilibrium, which is a situation where the overall reaction rate is zero (Alveteg and Sverdrup 2002).

Initial steady-state models used empirical data as input to understand regional lake chemistry (Henriksen 1980) and changes in chemistry of streams (Christophersen and Neal 1990). Empirical equilibrium models use steady-state chemical weathering and ionic electroneutrality to calculate a steady-state alkalinity given an acid deposition rate. Simple mass balance models use data on inputs and outputs from an ecosystem on nutrient and provide critical load of deposition to an ecosystem over the long term.

Two more empirical and one process-oriented model have been used frequently for calculating critical loads of acidifying deposition (S and N) for surface waters (Henriksen and Posch 2001). Both the Steady-State Water Chemistry (SSWC) model and the Empirical Diatom Model allow the calculation of critical loads of acidity and their current exceedances. The First-order Acidity Balance (FAB) model allows the simultaneous calculation of critical loads of acidifying S and N deposition and their exceedances. All three models with their latest modifications were presented in detail by Henrisksen and Posch (2001).

Briefly, the SSWC was derived by Henriksen et al. (1986). The model assumes a steady-state situation where outputs are in balance with inputs. The key to the SSWC model is the calculation of the sustainable supply of ANC, or the inherent buffering capacity of the system. Empirical relationships are used to determine the preindustrial concentration of base cations from weathering.

The Empirical Diatom Model is an alternative approach to the SSWC model and was developed from paleolimnological data (Battarbee et al. 1996; UBA 1996). Diatom communities in cores from acidified lakes usually show that before acidification, the diatom flora changed little over time. The point of acidification is indicated by a shift toward a more acidophilous diatom flora. Diatoms are among the most sensitive indicators of acidification in freshwater ecosystems. The Empirical Diatom model and the SSWC model are similar in the sense that they utilize an estimate of the preacidification base cation/calcium concentration for the calculation of the critical load. Differences between outputs from these models were assessed and discussed in detail by Henrisksen and Posch (2001).

The First-order Acidity Balance (FAB) model for calculating critical loads of sulfur and nitrogen for a lake considers sources and sinks of S and N within the lake and its terrestrial catchment (Posch et al. 1997). The lake and its catchment are assumed small enough to be properly characterized by average soil and lake water properties. Derivation of the FAB model by Henriksen and Posch (2001) goes beyond the derivation presented by Posch et al. (1997) by explicitly distinguishing the direct deposition onto the lake, and by considering the land cover type "bare rock," on which no nitrogen transformations take place.

For soils, there are one-layer steady-state models such as steady-state mass balance model (SSMB) and multilayer models such as more complex PROFILE (Warfvinge and Sverdrup 1992).

The SSMB model assumes a simplified, steady-state input/output description of the most important biogeochemical processes that affect soil acidification. Potential ecosystem inputs include atmospheric deposition and estimated soil base cation weathering rate. Ecosystem outputs and consumption include net removal of nutrients by forest harvesting, nutrient loss through soil leaching, denitrification, and N immobilization. The SSMB model was described in detail by the ICP Mapping Manual (UBA 2004).

PROFILE is a fully integrated process-oriented model. Multilayer models such as PROFILE can calculate the chemical quantities at any depth in the soil profile. The multilayer models thus consider the vertical soil inhomogeneity, but because of this they require a larger number of input data. The basic simplifying assumption of PROFILE is that designated vertical compartments of soil profile that might correspond to the individual soil horizons are homogeneous, and appropriate soil solutions in each horizon are perfectly mixed. The number of chemical reactions is represented by the equilibrium relationship or kinetic equations in each soil compartment. The reaction systems considered in PROFILE are soil solution equilibrium reactions, silicate weathering, uptake of nutrient cations, NO_3^- and NH_4^+, nitrification, and cation exchange reactions. All these processes intersect through soil solution. PROFILE is the only steady-state model up to date, calculating or predicting the weathering rates under field condition from physical and mineralogical properties of soil, whereas other models usually have weathering rate as a calibration parameter. The chemistry of the PROFILE model has been used in the dynamic model SAFE (Soil Acidification in Forest Ecosystems), and a detailed description of the PROFILE model is given by Warfvinge and Sverdrup (1992).

The current version of PROFILE model is 5.0. Graphical user interfaces run under Mac OS and Microsoft Windows. It was programmed in Fortran 95 and later updated using Fortran 2003 extensions. The PROFILE model is available through cooperation from http://www2.chemeng.lth.se/.

20.4.2 Dynamic Models—Basic Concepts and Equations

Dynamic models of acidification are based on the same basic principles as steady-state models. These processes include the charge balance of the ions in the soil solution, mass balances of the various ions, and equilibrium equations. In steady-state models, only infinite sources and sinks are considered such as base cation weathering; the inclusion of the finite sources and sinks of major ions into dynamic models is crucial, because they determine the long-term changes in soil and soil solution chemistry. The most important processes involving finite buffers and time-dependent sources/sinks are cation exchange, nitrogen retention, and sulfate adsorption.

This chapter is based and derived from UBA (2004) and Posch et al. (2003b). More information on the dynamic models VSD, MAGIC, SAFE, and SMART mentioned in this section can be found in Chapter 6.3.

20.4.2.1 Basic Equations

The basic assumption for a modeled ecosystem is a noncalcareous forest soil covered by natural or seminatural vegetation. Most models consider soil as a single homogeneous compartment with a defined depth. This means that internal soil processes such as weathering and uptake are evenly distributed over the soil profile, and all physicochemical constants are assumed uniform in the whole profile.

Another consideration is the simplest possible hydrology. It is assumed that the water leaving the root zone is equal to precipitation minus evapotranspiration; more precisely, percolation is constant through the soil profile and occurs only vertically.

The starting point is the charge balance of the major ions in the soil water, leaching from the root zone:

$$SO_{4,le} + NO_{3,le} - NH_{4,le} - BC_{le} + Cl_{le} = H_{le} + Al_{le} - HCO_{3,le} - RCOO_{le} = -ANC_{le}$$

$$(20.3)$$

where $BC = Ca + Mg + K + Na$ and RCOO stand for the sum of organic anions. Equation 20.31 also defines the ANC. The leaching term is given by:

$$X_{le} = Q[X] \tag{20.4}$$

where $[X]$ is the soil solution concentration (eq m^{-3}) of ion X and Q (m yr^{-1}) is the water leaving the root zone (precipitation minus evapotranspiration). All quantities are expressed in equivalents (moles of charge) per unit area and time (e.g., eq m^{-2} yr^{-1}).

The concentration $[X]$ of an ion in the soil compartment, and thus its leaching X_{le} in the charge balance, is related to the sources and sinks of X via a mass balance equation that describes the change over time of the total amount of ion X per unit area in the soil matrix/soil solution system, X_{tot} (eq m^{-2}):

$$\frac{d}{dt} X_{tot} = X_{in} - X_{le} \tag{20.5}$$

where X_{in} (eq m^{-2} yr^{-1}) is the net input of ion X including sources minus sinks, except leaching, which are specified and discussed below.

Using simplifications similar to those used in the derivation of the SMB model, the net input of sulfate and chloride is given by their deposition:

$$SO_{4,in} = S_{dep} \text{ and } Cl_{in} = Cl_{dep} \tag{20.6}$$

For base cations, the net input is given by (Bc = Ca + Mg + K; BC = Bc + Na):

$$BC_{in} = BC_{dep} + BC_w - Bc_u \tag{20.7}$$

where the subscripts dep, w, and u stand for deposition, weathering, and net uptake, respectively. The S adsorption and cation exchange reactions are not included here; they are included in X_{tot} and described by equilibrium equations (see below). For nitrate and ammonium, the net input consists of (at least) deposition, nitrification, denitrification, net uptake, and immobilization:

$$NO_{3,in} = NO_{x,dep} + NH_{4,ni} - NO_{3,i} - NO_{3,u} - NO_{3,de} \tag{20.8}$$

$$NH_{4,in} = NH_{3,dep} - NH_{4,ni} - NH_{4,i} - NH_{4,u} \tag{20.9}$$

where the subscripts ni, i, and de stand for nitrification, net immobilization, and denitrification, respectively. In the case of complete nitrification, $NH_{4,in} = 0$ and the net input of nitrogen is given by:

$$NO_{3,in} = N_{net} = N_{dep} - N_i - N_u - N_{de} \tag{20.10}$$

In addition to the mass balances, equilibrium equations describe the interaction of the soil solution with air and the soil matrix. The dissolution of (free) Al is modeled by the following equation:

$$[Al] = K \times Al_{ox} \times [H]^\alpha \tag{20.11}$$

where $\alpha > 0$ is a site-dependent exponent. For $\alpha = 3$, this equation represents the gibbsite equilibrium ($KAl_{ox} = K_{gibb}$).

Bicarbonate ions (HCO_3^-) were neglected in the derivation of the SMB model since, for low pH values related to the critical limit for forest soils, the resulting error was considered negligible. However, bicarbonate can be computed as:

$$[HCO_3] = \frac{K_1 K_H p_{CO_2}}{[H]} \tag{20.12}$$

where K_1 is the first dissociation constant, K_H is Henry's constant ($K_1 K_H = 10^{-1.7} = 0.02$ eq^2 m^{-6} atm^{-1} at 8°C) and p_{CO_2} (atm) is the partial pressure of CO_2 in the soil solution. Note that the inclusion of bicarbonates into the charge balance is necessary, if the ANC is to attain positive values. Organic anions have been neglected in the critical load calculations with the assumption that all organic anions in complexes with aluminum, that is, free Al is equal to total Al minus organic anions. But its modeling by equilibrium equations with [H] enables their inclusion.

Thus the ANC concentration can be expressed as a function of [H] alone (see previous definition of ANC, Equation 20.3):

$$\text{ANC}([\,H]) = F_{org}([\,H]) + \frac{K_1 K_H p_{CO_2}}{[H]} - [H] - KAl_{ox}[H]^\alpha = [BC] - [SO_4] - [N] - [Cl]$$

(20.13)

where F_{org} is a function expressing organic anion concentration(s) in terms of [H].

20.4.2.2 From Steady-State to Dynamic Simulations

Steady state means there is no change over time in the total amounts of ions involved (see Equation 20.5):

$$\frac{d}{dt} X_{tot} = 0 \Rightarrow X_{le} = X_{in}$$

(20.14)

To obtain time-dependent solutions of the mass balance equations, the term X_{tot} in Equation 20.5 representing the total amount per unit area of ion X in the soil matrix/soil solution system has to be specified. For ions, which do not interact with the soil matrix, X_{tot} is given by the concentration of ion X in solution:

$$X_{tot} = \Theta \times z \times [X]$$

(20.15)

where z (m) is the considered soil depth (usually representing the root zone) and Θ ($m^3\ m^{-3}$) is the annual average volumetric water content of the soil compartment. The above equation holds for chloride. For every base cation Y participating in cation exchange, Y_{tot} is given by:

$$Y_{tot} = \Theta \times z \times [Y] + \rho \times z \times \text{CEC} \times E_Y$$

(20.16)

where ρ is the soil bulk density (g cm^{-3}), CEC is the cation exchange capacity (meq kg^{-1}), and E_Y is the exchangeable fraction of ion Y.

The long-term changes in soil N pool are mostly caused by net immobilization N_{tot} and given by:

$$N_{tot} = \Theta \times z \times [N] + \rho \times z \times N_{pool}$$

(20.17)

If there is no adsorption or desorption of sulfate, $SO_{4,tot}$ is given by Equation 20.15. If sulfate adsorption cannot be neglected, it is given by:

$$SO_{4,tot} = \Theta \times z \times [SO_4] + \rho \times z \times SO_{4,ad} \qquad (20.18)$$

When the rate of Al leaching is greater than the rate of Al mobilization by weathering of primary minerals, the remaining part of Al has to be supplied from readily available Al pools, such as Al hydroxides. This causes depletion of these minerals, which might induce an increase in Fe buffering, which in turn leads to a decrease in the availability of phosphate (De Vries 1994). Furthermore, the decrease of those pools in podzolic sandy soils may cause a loss in the structure of those soils. The amount of Al is in most models assumed to be infinite and thus no mass balance for Al is considered. The Simulation Model for Acidification's Regional Trends (SMART) model includes an Al balance, and the terms in Equation 20.5 are $Al_{in} = Al_w$, and Al_{tot} is given by:

$$Al_{tot} = \Theta \times z \times [Al] + \rho \times z \times CEC \times E_{Al} + \rho \times z \times Al_{ox} \qquad (20.19)$$

where Al_{ox} (meq kg^{-1}) is the amount of oxalate extractable Al, the pool of readily available Al in the soil.

Inserting these expressions into Equation 20.5 and observing that $X_{le} = Q[X]$, one obtains differential equations for the temporal development of the concentration of the different ions. Only in the simplest cases can these equations be solved analytically. In general, the mass balance equations are discretized and solved numerically, with the solution algorithm selected according to the preferences of model builder.

20.4.2.3 Buffer Processes

Finite buffers have not been included in the derivation of critical loads, since they do not influence the steady state. When investigating the chemistry of soils over time as a function of changing deposition patterns, these finite buffers govern the long-term changes in soil and soil solution chemistry. Considered buffer mechanisms include adsorption/desorption processes, mineralization/immobilization processes, and dissolution/precipitation processes.

20.4.2.3.1 Cation Exchange

The solid phase particles of a soil contain an excess of cations at their surface layer. To maintain electroneutrality, these adsorbed cations cannot be removed from the soil, but they can be exchanged against other cations, for example, those in the soil solution. This process is known as cation exchange; and every soil or soil layer is characterized by the total amount of exchangeable cations per unit mass, by the CEC (measured in meq kg^{-1}).

If X and Y are two cations with charges m and n, then the general form of the equations used to describe the exchange between the liquid-phase concentrations $[X]$ and $[Y]$ and the equivalent fractions E_X and E_Y at the exchange complex is

$$\frac{E_X^i}{E_Y^j} = K_{XY} \frac{[X^{m+}]^n}{[Y^{n+}]^m} \tag{20.20}$$

where K_{XY} is the exchange or selectivity constant, a soil-dependent quantity. Depending on the powers i and j, different models of cation exchange can be distinguished: For $i = n$ and $j = m$, the Gaines–Thomas exchange equations are obtained, whereas for $i = j = mn$, after taking the mnth root, the Gapon exchange equations are obtained.

The number of exchangeable cations considered depends on the purpose and complexity of the particular model. For example, Reuss (1983) considered only the exchange between Al and Ca (or divalent base cations). In general, if the exchange between N ions is considered, $N − 1$, exchange equations (and constants) are required, all the other $(N − 1)(N − 2)/2$ relationships and constants can be easily derived from them. The exchange of protons (H^+) is important, if the cation exchange capacity (CEC) is measured at high pH values (pH = 6.5).

In the case of the Bc–Al–H system, the Gaines–Thomas equations read:

$$\frac{E_{Al}^2}{E_{Bc}^3} = K_{AlBc} \frac{[Al^{3+}]^2}{[Bc^{2+}]^3} \quad \text{and} \quad \frac{E_H^2}{E_{Bc}} = K_{HBc} \frac{[H^+]^2}{[Bc^{2+}]} \tag{20.21}$$

where Bc = Ca + Mg + K, with K treated as divalent. The equation for the exchange of protons against Al can be obtained from previous equations by division:

$$\frac{E_H^3}{E_{Al}} = K_{HAl} \frac{[H^+]^3}{[Al^{3+}]} \quad \text{and} \quad K_{HAl} = \sqrt{K_{HBc}^3/K_{AlBc}} \tag{20.22}$$

The corresponding Gapon exchange equations then read:

$$\frac{E_{Al}}{E_{Bc}} = K_{AlBc} \frac{[Al^{3+}]^2}{[Bc^{2+}]^3} \quad \text{and} \quad \frac{E_H^2}{E_{Bc}} = K_{HBc} \frac{[H^+]^2}{[Bc^{2+}]} \tag{20.23}$$

The H–Al exchange can be obtained by division (with $k_{HAl} = k_{HBc}/k_{AlBc}$). Charge balance requires that the exchangeable fractions add up to one:

$$E_{Bc} + E_{Al} + E_H = 1 \tag{20.24}$$

The sum of the fractions of exchangeable base cations (E_{Bc}) represents the base saturation of the soil. It is the time development of the base saturation that is of interest in dynamic modeling. In the above formulations, the exchange of Na, NH_4 (which can be important in high NH_4 deposition areas), and heavy metals is neglected (or subsumed in the proton fraction).

Individual dynamic models contain different sets of exchange equations. In the VSD model, it is possible to choose between the Gaines–Thomas and the Gapon Bc–Al–H exchange model. In the SMART model, Equation 20.19 was used, but with Ca + Mg instead of Bc, K exchange is being ignored. The SAFE model employs the Gapon exchange equations (Equations 20.21), but with exchange constants $k_{X/Y}' = 1/k_{XY}$. In the MAGIC (Model of Acidification of Groundwater In Catchments) model, the exchange of Al with all four base cations is modeled separately with Gaines–Thomas equations, without explicitly considering H exchange. The ranges of values for the exchange constants for the different model formulations can be found in Chapter 5 of Posch et al. (2003b).

20.4.2.3.2 Nitrogen Immobilization

Several models do include descriptions for mineralization and immobilization, following, for example, first-order kinetics or Michaelis–Menten kinetics. In its most simple form, a net immobilization of nitrogen is included, being the difference between mineralization and immobilization. In the calculation of critical loads, the long-term net immobilization of N is assumed to be constant. It is known that the amount of N immobilized is (at present) in many cases larger than this long-term value. Thus, a submodel describing the nitrogen dynamics in the soil is part of most dynamic models. For example, the MAKEDEP model (Alveteg et al. 1998a; Alveteg and Sverdrup 2002), which is part of the SAFE model system, describes the N dynamics in the soil as a function of forest growth and deposition.

According to Dise et al. (1998) and Gundersen et al. (1998), the forest floor C/N ratios may be used to assess risk for nitrate leaching. Gundersen et al. (1998) suggested threshold values of >30, 25–30, and <25 to separate low, moderate, and high nitrate leaching risk, respectively. This information has been used in SMART (De Vries et al. 1994a; Posch and De Vries 1999) and MAGIC (Cosby et al. 2001) to calculate nitrogen immobilization as a fraction of the net N input, linearly depending on the C/N ratio. In these models, the C/N ratio of the mineral topsoil is used.

Between a maximum, CN_{max}, and a minimum C/N ratio, CN_{min}, the net amount of N immobilized is a linear function of the actual C/N ratio, CN_t:

$$N_{i,t} = \begin{cases} \dfrac{CN_t - CN_{min}}{CN_{max} - CN_{min}} \cdot N_{in,t} & \text{for} \\ N_{in,t} & \text{for} \quad CN_{min} \\ 0 & \text{for} \end{cases} \quad \begin{matrix} CN_t & \geq & CN_{max} \\ & < & CN_t & < & CN_{max} \\ CN_t & \leq & CN_{min} \end{matrix}$$

where $N_{in,t}$ is the available N (e.g., $N_{in,t} = N_{dep,t} - N_{u,t} - N_{i,acc}$). At every time step the amount of immobilized N is added to the amount of N in the top soil, which in turn is used to update the C/N ratio. The total amount immobilized at every tine step is then $N_i = N_{i,acc} + N_{i,t}$. The above equation states that when the C/N ratio reaches a preset minimum value, the annual amount of N immobilized equals the acceptable value $N_{i,acc}$. This formulation is compatible with the critical load formulation for $t \to \infty$.

20.4.2.3.3 Sulfate Adsorption

The amount of sulfate adsorbed, $SO_{4,ad}$ (meq kg^{-1}), is often assumed to be in equilibrium with the solution concentration and is typically described by a Langmuir isotherm (e.g., Cosby et al. 1986):

$$SO_{4,ad} = \frac{[SO_4]}{S_{1/2} + [SO_4]} \cdot S_{max}$$

where S_{max} is the maximum adsorption capacity of sulfur in the soil (meq kg^{-1}) and $S_{1/2}$ is the half-saturation concentration (eq m^{-3}). This adsorption definition has been implemented in dynamic model MAGIC (Cosby et al. 1986). Alternatively, sulfate adsorption could have been modeled using the Freundlich isotherm. But neither of these isotherms includes the pH dependence. More sophisticated pH-dependent adsorption definition is part of the SAFE model (for details, see Martinson 2003; Martinson et al. 2003; or Martinson and Alveteg 2004).

20.4.2.4 From Soils to Surface Waters

The processes discussed so far are assumed to occur in the soil solution while it is in contact with the soil matrix. To calculate surface water concentrations, it is assumed that the water leaves the soil matrix and is exposed to the atmosphere (Cosby et al. 1985a, 1985b, 1985c; Reuss and Johnson 1986). When this occurs, the excess CO_2 in the water degasses. This shifts the carbonate–bicarbonate equilibria and changes the pH (see Equation 20.12).

Surface water concentrations are thus calculated by resolving the system of equations presented above at a lower partial pressure of CO_2 (e.g., mean

pCO_2 of 8×10^{-4} atm for 37 lakes; Cole et al. 1994) while ignoring exchange reactions, nitrogen immobilization, and sulfate adsorption. Since exchanges with the soil matrix are precluded, the concentration of the base cations and the strong acid anions (SO_4^{2-}, NO_3^- and Cl^-) will not change as the soil water becomes surface water. ANC is conservative (see Equation 20.4).

20.4.3 Selected Dynamic Models

The basic processes involved in soil and surface water acidification briefly explained and mathematically introduced in the previous chapter have been summarized and expressed in mathematical form, with emphasis on long-term processes. The resulting equations and generalizations, together with appropriate solution algorithms and input–output routines, have been packaged by the model developers into soil acidification models, usually known by their acronyms (Posch et al. 2003b).

An overview of 16 models was provided by Tiktak and Van Grinsven (1995). These models emphasize either soil chemistry or the interaction with the forest. The soil and surface water acidification models were initially used as tools to investigate what processes drive acidification. When the processes were identified and the models validated, they have been used for analyzing and predicting the effects of different future emission scenarios. Because of the inclusion of dynamic results in integrated assessment, the dynamic models are becoming important policy tools (Posch et al. 2003a; Martinson et al. 2003).

Out of the wider selection of developed dynamic models, some became widely used. Because of their survival in use over time, they were also referred to as survivor models: SMART (de Vries et al. 1989), MAGIC (Cosby et al. 2001), and SAFE (Warfvinge et al. 1993). These older models were currently supplemented with relatively new VSD (Posch et al. 2003b). This selection of models has main features in common. It is their simplicity and possible application on a regional scale.

An overview of models that will be described in further text is given in Table 20.1. The models VSD, SMART, and SAFE are soil-oriented models of increasing complexity, whereas the MAGIC model is generally applied at the catchment level.

20.4.3.1 VSD Model

A very simple dynamic soil acidification model (VSD) can be viewed as the simplest extension of steady-state models for critical load calculations with possibility of regional application (Posch and Reinds 2009). The model requires a minimum set of inputs. The VSD model resembles the model SMART (see Section 20.4.3.2), but has been further simplified to make it compatible with the steady-state SMB model by Posch and Reinds (2009):

TABLE 20.1

Overview of Dynamic Models and Included Processes

Model	Essential Process Descriptions	Layers
VSD	ANC charge balance Mass balances for BC and N (complete nitrification assumed)	One
SMART	VSD model SO_4 sorption—Langmuir isotherm Mass balances for $CaCO_3$ and Al Separate mass balances for NH_4 and NO_3 Nitrification Al complexation with DOC	One
SAFE	VSD model pH-dependent SO_4 sorption Separate weathering calculation Element cycling by litterfall Root decay Mineralization and root uptake	Several
MAGIC	VSD model SO_4 sorption—Langmuir isotherm Al speciation and complexation Aquatic chemistry	Two, mostly one

Source: Posch et al., Manual for dynamic modelling of soil response to atmospheric deposition. RIVM Report 259101012/2003, Working Group on Effects of the CLRTAP, Bilthoven, the Netherlands, 2003. With permission.

1. Neglecting buffering by calcium carbonate and aluminum depletion in highly acidified soils
2. Ignoring sulfate adsorption and Al complexation
3. Assuming complete nitrification (no ammonium leaching)

In the VSD model, the various ecosystem processes have been limited to a few key processes. Processes that are not taken into account in VSD model, are:

1. Canopy interactions
2. Nutrient cycling processes
3. N fixation and NH_4 adsorption
4. Interactions (adsorption, uptake, immobilization, and reduction) of SO_4^{2-}
5. Formation and protonation of organic anions ($RCOO^-$)
6. Complexation of Al with OH^-, SO_4^{2-} and $RCOO^-$

The VSD model consists of a set of mass balance equations describing the soil input–output relationships, and a set of equations describing the rate-limited and equilibrium soil processes. The soil solution chemistry in VSD depends solely on the net element input from the atmosphere (deposition minus net uptake minus net immobilization) and the geochemical interactions in the soil (CO_2 equilibrium, weathering of carbonates and silicates, cation exchange). Soil interactions are described by simple rate-limited (zero-order) reactions (e.g., uptake and silicate weathering) or by equilibrium reactions (e.g., cation exchange). It models the exchange of Al, H, and Ca + Mg + K with possibility to choose Gaines–Thomas or Gapon equations. Solute transport is described by assuming complete mixing of the element input within one homogeneous soil compartment with a constant density and a fixed depth. Since VSD is a single-layer soil model neglecting vertical heterogeneity, it predicts the concentration of the soil water leaving this single layer usually equal to rootzone. The annual water flux percolating from this layer is taken equal to the annual precipitation excess. The time step of the model is 1 year, that is, seasonal variations are not considered.

The VSD core model has been packed into a software package with graphical user interface named VSDStudio. It enables simple forward simulations for a single site at a time with given deposition scenarios, but also automatic Bayesian calibration, computation of critical loads, target loads, and damage/recovery delay times. A detailed description of the VSD model including the charge balance equations, equilibrium equations, mass balance equations, and instructions on initialization and calibration, complemented with screenshots and examples of model outputs, can be found in the work of Posch and Reinds (2009). Details on the implementation of Bayesian calibration for VSD can be found in the work of Reinds et al. (2008).

Very Simple Dynamic model, version 3.1, with graphical user interface runs on PC hardware under Microsoft Windows. It was programmed in Fortran 95 and the graphical user interface in C++. It is publicly available for free at www.trentu.ca/ecosystems/i-likeit or www.mnp.nl/cce.

20.4.3.2 SMART Model

The SMART model is similar to the VSD model, but it includes some extensions (De Vries et al. 1989; Posch et al. 1993). The model SMART predicts changes in soil solution chemistry resulting from acidification processes, and it is designed to be used in semiatural areas. As with the VSD model, the SMART model consists of a set of mass balance equations, describing the soil input–output relationships, and a set of equations describing the rate-limited and equilibrium soil processes. The assumptions and simplifications of processes in SMART were similar to those for the VSD model. The checklist and justifications for them are given by De Vries et al. (1989, 1994a).

SMART is a single-layer model that considers the top meter of the soil assumed equivalent to the root zone, and it works with time steps of 1 year. Apart from pH and N availability, SMART predicts changes in aluminum (Al^{3+}), BC, nitrate (NO_3^-), and sulfate (SO_4^{2-}) concentrations in the soil solution and solid-phase characteristics depicting the acidification status such as carbonate content, base saturation, and readily available Al content. SMART models the exchange of Al, H, and divalent base cations using Gaines–Thomas equations. Sulfate adsorption is modeled using a Langmuir equation and organic acids can be described as mono-, di-, or tri-protic.

SMART includes a balance for carbonate and Al, thus allowing the calculation from calcareous soils to completely acidified soils that do not have an Al buffer left. In this respect, the SMART model is based on the concept of buffer ranges outlined by Ulrich (1981). It includes a description of the Al complexation with organic acids. The SMART model has been developed to be applicable at a regional level, and an early example of an application to Europe can be found in the work of De Vries et al. (1994b).

Kros et al. (1995) derived the SMART2 model from an earlier version of SMART (De Vries et al. 1989). The major enhancements in the SMART2 model were the inclusion of a nutrient cycle and an improved modeling of hydrology.

The SMART2 model, version 3.4, runs on PC hardware under Microsoft Windows 98 or higher. It is possible to obtain executable version of the model by emailing contact person Janet Mol-Dijkstra (Mol-Dijkstra et al. 2006).

20.4.3.3 SAFE Model

The SAFE model has been developed by Warfvinge et al. (1993), and was further improved over time (Alveteg 1998; Alveteg and Sverdrup 2002; Martinson et al. 2003).

SAFE is a multilayer soil chemistry model developed for studies of the effects of acid deposition on soils and soil solutions. The model includes process-oriented descriptions of cation exchange reactions, chemical weathering of minerals, leaching and accumulation of dissolved chemical components, and solution equilibrium reactions involving CO_2, organic acids, and Al species.

All processes included in SAFE have been simplified in some respect (Alveteg 1998). The soil is considered to be a series of continuously stirred tank reactors, where each tank reactor represents one soil layer. Each soil horizon is assumed to be homogeneous and perfectly mixed. The water flowpath is assumed to be downward only. The distribution of uptake with depth and the cation exchange capacity are assumed constant over time.

In case of nitrification, SAFE assumes that all NH_4^+ is either taken up by vegetation or nitrified in the top soil layer. In SAFE, calcium (Ca^{2+}), magnesium (Mg^{2+}), and potassium (K^+) are treated as a lumped divalent base cations (Bc^{2+}). SAFE also includes the effects of deposition, uptake of N and Bc^{2+} from soil solution, nutrient cycling of N and Bc^{2+} as litterfall and canopy exchange, net mineralization, and/or immobilization of N, Bc^{2+}, and sulfate. Fluxes of elements in SAFE can be specified as time series either manually or by a separate submodel MAKEDEP (Alveteg et al. 1998b). Soil solution chemistry in each designated layer is influenced both by input from the layer above and by soil processes in the layer itself. The model is constructed for forest ecosystems, and thus the influence of vegetation on soil solution chemistry is important. Furthermore, time series of nutrient uptake or cycling are SAFE model inputs and can be produced by the reconstruction submodel MAKEDEP.

The simulations in SAFE should always start at a time when it can be assumed that the ecosystem was at or close to steady state. To calculate the initial conditions, it includes the InitSAFE submodel, where a steady-state version of SAFE is used. SAFE is calibrated by varying the initial value of base saturation until maximum possible agreement between measurements and model calculations is achieved. The PreSAFE package contains the init-SAFE, the MAKEDEP model together with some additional routines, and is used to create all the necessary input files for running the SAFE model.

Because SAFE is a multilayer model, it requires chemical and physical data on soil properties for each designated individual soil layer. These data include soil moisture, mineralogy, density, and cation exchange capacity, and are assumed to be constant during the whole simulation period. Time series of data are needed concerning precipitation, atmospheric deposition of major ions, nutrient uptake, nutrient cycling, and net mineralization. Chemical weathering is described by a series of kinetically controlled rate equations and it is calculated with included PROFILE submodel.

In each layer, the water flow is divided into evapotranspiration, horizontal flow, and the residual which continues downward to the next layer. The hydrology is currently assumed to be constant and may be calculated using a separate hydrological model or water flux measurements. The change in soil solution chemistry and subsequent changes in the distribution of elements on the cation exchange matrix are calculated by means of mass balance equations.

Early versions of SAFE did not include sulfate adsorption routine but the first implementation of sulfate adsorption in the SAFE model was carried out by Fumoto and Sverdrup (2000) and later simplified by Martinson et al. (2003). The sulfate adsorption implemented in SAFE is dependent on the pH of soil solution and concentration of sulfate. Sulfate adsorption in SAFE is assumed fully reversible, so the adsorbed sulfate can be released (desorbed) from soil with the progressing simulation. To maintain electroneutrality, a coadsorption of hydrogen ions is associated with the adsorption of sulfate in SAFE.

The main differences of SAFE when compared to other complex models are (Posch et al. 2003b):

1. Weathering of base cations is not supplied by user as a model input, but is calculated with the PROFILE submodel from input on soil mineralogy (Warfvinge and Sverdrup 1992).
2. SAFE is oriented onto soil profiles, in which soil solutions are assumed to move vertically through several soil layers.

3. Cation exchange between Al, H, and base cations (Ca^{2+}, Mg^{2+}, and K^+) is modeled with Gapon exchange reactions, and the exchange between soil matrix and the soil solution is diffusion limited.

4. Sulfur adsorption is pH-dependent, but not only simply described by Langmuir or a Freundlich isotherm.

The SAFE model has been applied to many single sites; regional applications of SAFE have also been carried out for Sweden (Alveteg and Sverdrup 2002) and Switzerland (SAEFL 1998; Kurz et al. 1998; Alveteg et al. 1998a).

SAFE model and graphical user interfaces run on MacOS and Microsoft Windows. It was programmed in Fortran 95 and later updated using Fortran 2003 extensions. The SAFE and PROFILE models are available through cooperation from http://www2.chemeng.lth.se/.

The family of SAFE models has been further extended with model ForSAFE (Wallman et al. 2005) and ForSAFE-VEG (Sverdrup et al. 2007). ForSAFE enables users to simulate dynamic changes of forest ecosystems to environmental changes and combines engines of the tree growth model PnET (Aber and Federer 1992), the soil chemistry model SAFE (Alveteg 1998), decomposition model Decomp (Walse et al. 1998), and the hydrology model PULSE (Lindstrom and Gardelin, 1992). The ForSAFE model was supplemented with a vegetation response module (VEG) and is used to simulate the response of ground vegetation plant groups to nitrogen pollution in relation to climate change or forest management.

20.4.3.4 MAGIC Model

MAGIC is a lumped-parameter model of intermediate complexity. It was developed to predict the long-term effects of acidic deposition on soils and surface water chemistry (Cosby et al. 1985a, 1985b, 1985c, 1986). MAGIC has been modified and extended several times from the original version. In particular organic acids (Cosby et al. 1995) and nitrogen processes (Cosby et al. 2001) have been added to the original model. The latest version, MAGIC7, includes utility for simulating short-term episodic responses in lakes and streams and process-based nitrogen dynamics in soils controlled by soil N pools.

The MAGIC model has been extensively applied and tested since its early versions at numerous sites and in many regions around the world. The model simulates soil solution chemistry and surface water chemistry to forecast the monthly and annual average concentrations of the major ions in lakes and streams. MAGIC represents the catchment with aggregated, uniform soil compartments (one or two) and a surface water compartment that can be either a lake or a stream. The soil layers in MAGIC can be arranged vertically or horizontally to represent important vertical or horizontal flowpaths through the soils. If a lake is simulated, seasonal stratification of the lake can be implemented. Time series inputs to the model include annual or monthly estimates of:

1. Deposition of ions from the atmosphere (wet plus dry deposition)
2. Discharge volumes and flow routing within the catchment
3. Biological production, removal, and transformation of ions
4. Internal sources and sinks of ions from weathering or precipitation reactions
5. Climate data

Constant parameters in the model include physical and chemical characteristics of the soils and surface waters, and thermodynamic constants. The model is calibrated using observed values of surface water and soil chemistry for a specified period.

MAGIC consists of two sections. In the first section, the concentrations of major ions are assumed to be governed by simultaneous reactions involving sulfate adsorption, cation exchange, dissolution–precipitation speciation of aluminum, and dissolution–speciation of inorganic and organic carbon. The second is a mass balance section in which the flux of major ions to and from the soil is assumed to be controlled by atmospheric inputs, chemical weathering inputs, net uptake in biomass, and losses to runoff.

At the heart of MAGIC is the size of the pool of exchangeable base cations in the soil. As the fluxes to and from this pool change over time owing to changes in atmospheric deposition, the chemical equilibrium between soil and soil solution shift to give changes in surface water chemistry. The degree and rate of change in surface water acidity thus depend both on flux factors and the inherent characteristics of the affected soils.

Cosby et al. (2001) expressed the description of the main processes as: "Cation exchange reactions between the soil matrix and soil solution are assumed to result in an equilibrium partitioning of Ca, Mg, Na, K and Al between solid and aqueous phases." The equilibrium expressions for cation exchange are constructed using Gaines–Thomas exchange equations. Base saturation of the soil is defined as the sum of the exchangeable fractions of the base cations. The selectivity coefficients must be calibrated for each aggregated soil layer in the model. The calibration procedure relies on observations of the exchangeable fractions of base cations in soils and measured base cation concentrations in stream water (Cosby et al. 1984, 1985a, 1985b). Anion exchange reactions are assumed to occur only for SO_4 ion. The relationship between dissolved and adsorbed SO_4 is assumed to follow a Langmuir isotherm. MAGIC is a catchment-scale model, so the effective values of aggregated parameters intended to represent large-scale function usually cannot be derived by a direct scaling-up of similar parameters measured in a laboratory setting. The SO_4 adsorption parameters used in MAGIC must be calibrated for each site. Cosby et al. (1986) described a method for calibrating SO_4 adsorption parameters in whole catchment simulations based on input/output budgets and deposition histories for the site. Inorganic Al speciation in soils is described by one reaction describing the combined effects of soil cation exchange and dissolution of a solid phase of $Al(OH)_3$ and reactions involving formation of aqueous complexes of Al. The same reactions are assumed to occur in surface waters with the exception of cation exchange. The Al speciation reactions are represented in the model by a series of equilibrium equations. The Al solubility constants for the soils in the model are represented by aggregated values that account for both cation exchange and solution–dissolution of a solid phase (Cosby et al. 2001). These values used in MAGIC are not necessarily associated with a particular crystalline form of $Al(OH)_3$ and must be selected as part of the calibration process. The constants concerning solubility of Al in MAGIC should be based on local observations of Al dynamics in surface waters. MAGIC7 includes a triprotic organic acid analogue model. The trivalent organic anion reacts with trivalent Al to form organic-Al complexes."

The N dynamics included in MAGIC7 are based conceptually on the empirical model described by Gundersen et al. (1998). The mathematical formulation and process representations of the N dynamics derive from a simplification of the structure of the MERLIN model (Cosby et al. 1997; Emmett et al. 1997).

The inclusion of the N dynamics has not been a basic conceptual modification of the original MAGIC model. Two new variables, soil organic C and soil organic N, were included and observations of these variables will be needed to calibrate model simulations. The added N processes directly control NH_4 and NO_3 ions in soil solution, and thus directly affect ANC and soil base saturation. Through this N processes can have significant effects on long-term and short-term simulation of acidification. The first applications of MAGIC7 utilizing the N dynamics (Wright et al. 1998; Jenkins et al. 2001) proved the consistency of model with the behavior of natural systems.

MAGIC7 model runs on Microsoft Windows. Norwegian Institute for Water Research (NIVA) holds the copyright for MAGIC7 and distributes copies of the executable version of the model. [Contact Richard Wright (richard.wright@niva.no) for more information and pricing. Normal price for a license for the MAGIC software is €1250 and discounts are offered for educational purposes.] MAGIC was originally developed as part of a research project and further developments and improvements to the model have largely been supported through the sale of licenses and in part through new research contracts.

20.5 Modeling and Results

The VSD model will be used primarily as a tool for deriving target loads. As already mentioned, MAGIC has been used more on a catchment level to reconstruct or predict water chemistry of lakes or streams, whereas SAFE and SMART are more soil chemistry–oriented models.

Once selection of the model has been done, it is time to start the most time-consuming process before running the model itself, which is gathering of data. For well-studied or intensively monitored sites most of the input data already exists. Some data from laboratories, literature, or databases can be used directly but other data have to be preprocessed and interpreted, often with the help of other simple models (Posch et al. 2003b). This is true especially in regional assessments where interpolations and transfer functions have to often be used to obtain the necessary input data. When applying model runs onto a larger area, for example, a forest instead of a single tree stand, certain variables should be optimized and averaged to be sufficiently representative. Input data has to be continuously checked for consistency and quality. Wide spread use of dynamic models requires quality datasets to produce trustful results. Nonplausible results may emerge from problems in model construction, model implementation errors or inconsistencies in the model input (Martinson 2004). Data required for dynamic modeling can be divided into individual categories such as data for parameterization, data for calibration, and data for validation.

In SAFE, the data for parameterization include in- and output fluxes, and soil and biomass properties. Calibration data are the base saturations of individual soil layers and as validation data will serve the field soil solution chemistry data. In the case of MAGIC, the validation data will be measured values on stream or lake chemistry. The validity of model results can be then checked by comparing the validation data such as soil solution pH, concentrations of base cations, and Al to the model results.

Selection of input data depends on the model considered. For example, weathering has to be specified as a constant input flux in the SMART and MAGIC model, whereas in the SAFE model it is internally computed from soil properties and depends on the chemical state of the soil (Posch et al. 2003b).

In- and output fluxes are also needed in the simplest steady-state SMB model. Since the SMB model describes steady-state conditions, it requires long-term averages for input fluxes. Short-term variations such as episodic, seasonal, interannual, due to harvest, and as a result of short-term natural perturbations are not considered, but are assumed to be included in the calculation of the long-term mean. Long-term averages are defined as about 100 years to represent at least one rotation period of forests (UBA 2004).

The most important soil parameters for dynamic models are the cation exchange capacity and base saturation and the exchange constants describing cation exchange, as well as parameters describing sulfate adsorption, since these parameters determine the long-term chemical behavior of soils.

Detailed descriptions of the input data for individual models are given by Posch and Reinds (2009) for the VSD model, De Vries et al. (1994a) for the SMART model, Cosby et al. (1985) for the MAGIC model, and Alveteg and Sverdrup (2002) for the SAFE model.

Multilayer models such as SAFE also include canopy interactions and organic inputs by litterfall and root decay. These inputs and outputs are either derived from measurements or can be derived by relationships (transfer functions) with basic land/soil and climate characteristics, such as tree species, soil type, elevation, precipitation, temperature, etc., which are often available in geographic information systems. More information on gathering and preprocessing of data for single or regional studies are given, for example, by Sverdrup et al. (1990), De Vries (1991), SAEFL (1998), Bélanger et al. (2002), and in Mapping Manuals (UBA 1996, 2004).

20.5.1 Calibration

If all input parameters, initial conditions, and driving processes are known, the chosen model would describe the future development of the soil chemical status for any given deposition or management scenario. But in the real world several of the input parameters are poorly known, and thus many models have to be calibrated. The method of calibration varies with the model and the type of application. In standard applications of MAGIC and SAFE models, it is assumed that in preacidification times (about 1850) the input of ions is in steady-state equilibrium with the soil and soil solution chemistry (Posch et al. 2003b). Furthermore, it is assumed that the deposition history of anions and cations has been properly reconstructed.

In SAFE, weathering rates and uptake/net removal of N and base cations are computed within the model. Only simulated present base saturation is matched with observations in every soil layer by adjusting the cation exchange selectivity coefficients. Matching simulated and observed soil solution concentrations is not part of the standard calibration procedure but if done by user it serves as method of validation.

The calibration of MAGIC is a sequential process whereby, first the input and output of conservatively acting ions usually Cl to act conservatively in the catchment are balanced. Next, the anion concentrations in surface waters are matched by adjusting catchment net retention of N and soil adsorption of S. Third, the four individual major base cation concentration in the stream and on the soil solid phase expressed, as a percentage of cation exchange capacity, are matched by adjusting the cation exchange selectivity coefficients and the base cation weathering rates. Finally, surface water pH, Al, and organic anion concentrations are matched by adjusting the aluminum solubility coefficient and total organic acid concentration in surface water (Posch et al. 2003b).

Both in MAGIC and SAFE, automatic calibration routines are part of the overall model system. For the SMART and VSD models, no such automatic model calibration routines are available at present. In site-specific applications, calibration has been carried out by trial and error routine. In European applications the initial base saturation in 1960 has been derived from transfer functions. Reinds et al. (2008) added the Bayesian calibration into the latest version of VSD. Bayesian calibration allows users to simultaneously calibrate a list of parameters, including uncertainties in soil chemistry measurements.

20.5.2 Applications

The steady-state models have been widely used for calculation of critical loads in Europe, Canada, and elsewhere. Critical load assessments were done from single sites through up to regional, continental, and global scales. Several studies have determined critical loads for nitrogen and acidity for terrestrial ecosystems at the European (De Vries et al. 1994b; Kuylenstierna et al. 1998), Southeastern Asian (Hettelingh et al. 1995), Northern Asian (Bashkin et al. 1995; Semenov et al. 2001), and global scale (Kuylenstierna et al. 2001; Bouwman et al. 2002).

Most of the European critical loads are computed with the Simple Mass Balance or related models such as PROFILE (Sverdrup and De Vries 1994). The methods of mapping were described in Mapping Manuals (UBA 2004). Critical loads are compared to depositions of sulfur and nitrogen, calculated with the EMEP transport and dispersion model (Tarrasón et al. 2005), to compute the percentage of ecosystem area exceeded or the exceedance amounts in the 50×50 km^2 EMEP grid cells covering Europe. Exceedances are expressed as an average accumulated exceedance (Posch et al. 2001). Hetteling et al. (2007) calculated that critical loads in 8.5% of the ecosystem area in Europe are exceeded, and for eutrophication this area even reaches 28.5%. These exceedances imply that those ecosystems are sooner or later at risk of being damaged.

Critical loads link deposition to ecosystem effects through soil chemical criteria (critical limits). These limits are based on dose–response relationships between chemical characteristics and ecosystem functioning. A critical load equals the deposition that results in a steady state in an ecosystem compartment (e.g., soil, water, plant) that does not exceed the selected critical limit, thus preventing significant harmful effects on specified sensitive elements of the environment (Nilsson and Grennfelt 1988). Consequently, the selection of the chemical criterion and its critical limit is a crucial step in deriving a critical load, and has to be guided by the negative effect(s) one wants to avoid. To date, mostly soil chemical criteria (e.g., nitrate and aluminum concentrations or aluminum to base cation ratios) have been used to derive critical loads with simple steady-state models (Reinds et al. 2008).

In Sweden, the critical loads were calculated and mapped using SMB and PROFILE models for 1804 individual sites of Swedish forests soils (Sverdrup and Warfvinge 1995). Calculations of critical loads were made using both the soil stability and BC/Al criteria. Differences between the maps resulting from SMB and PROFILE were attributed to ignorance of Al complexing by organic acids in SMB and to inclusion of soil multilayer approach in PROFILE especially with respect to weathering. Exceedance of critical loads in terms of acid deposition has been found in >80% of the Swedish forest resources. The exceedances were mainly due to deposition of S and N, but the role of forestry was concluded to be mostly passive but not insignificant. In Switzerland, Kurz et al. (1998) provided regional application of PROFILE and SMB model with a conclusion that frequently (60% of the sites) critical loads for Swiss forest soils were substantially exceeded by present loads, and further emission reductions were proposed to attain a long-term sustainable forest ecosystem.

Dynamic models can provide information on the timescale of changes assuming the effects of long-term processes to determine a target load, that is, the deposition for which a chosen chemical status is respected from a target year on (UBA 2004).

The results of the VSD model have been compared to outputs of the SAFE model. In a detailed study of 176 sites in Switzerland, Kurz and Posch (2002) have shown that the VSD model, when fed with weathering rates calculated with the PROFILE model, yields very similar results with the one-layer runs of the more complex model SAFE. Staelens et al. (2009) applied dynamic model VSD to 83 forest stands located in the Flemish forest area, Belgium. Results of this study indicate that N and S deposition reductions are needed to allow recovery of the Flemish forest soils.

Dynamic models have been applied to single sites or on a regional level. Among the most important single site applications belong the applications of models onto experimentally manipulated catchments such as roof experiments or chemically acidified catchments. Model SAFE has been applied to Gardsjon and Solling roof experiments (Sverdrup et al. 1998; Walse et al. 1998), proving that the processes included are sufficient to describe the most important parts of the soil solution chemistry. Martinson et al. (2005) reapplied the SAFE model to Solling site to validate the improvement of model predictions after implementation of sulfate adsorption processes. Improved model predictions showed no recovery, based on the criteria of Bc/Al ratio above 1 in the rooting zone, before the year 2050, independent of future deposition cuts.

Furthermore, historic soil data from 1883, 1906, 1964, and 1991 from the Geesecroft Wilderness Experiment at Rothamsted Experimental Station, UK, were used to test the dynamic biogeochemical model SAFE. Results indicated that deposition of acidity because of sulfur and nitrogen emissions during the past 110 years was the major cause of soil acidification. Natural afforestation of the site has also contributed with a significant but smaller amount of acidity input to the soil (Sverdrup et al. 1995).

A review of model MAGIC applications to manipulated and other sites, which were consequently used for model improvement, was carried out by Sullivan (1997). For example, applications of MAGIC to Lake Skjervatjern (Cosby et al. 1995) and Bear Brook experimental catchment (Norton et al. 1994; Cosby et al. 1996, Norton et al. 2001) provided examples of possible overestimation of Al responses in surface water and were corrected by implementation of modified Al algorithm reflecting the field values (Sullivan and

Cosby 1995). Refer to Sullivan (1997) for a detailed discussion on the need for experimental manipulation experiments to increase confidence of model outputs.

With the increasing importance of dynamic models, it was necessary to upscale model applications from small catchments (single-site applications) to whole regions or countries. On the basis of data on lake chemistry gathered in 1974–1975 and 1986, regional application of model MAGIC to 180 lakes in Norway indicated that the proposed 30% deposition reduction will be insufficient to reverse the ongoing regional acidification of lakes, whereas reductions of 50% and 70% would result in the recovery of progressively larger numbers of lakes (Wright et al. 1991). Similarly, model MAGIC has been applied to 94 catchments in Wales, and it was concluded that the recovery will be less marked at forested catchments and some sites will undergo further decrease of ANC even with reduced emissions in the future (Collins and Jenkins 1998). In Canada, regional application of MAGIC to 20 headwater catchments in Nova Scotia and 25 lakes in Ontario indicated the need for further reduction in emissions in order to achieve acceptable chemical criteria for soils and lakes, and to reverse some of the damage from anthropogenic acidification (Aherne et al. 2003; Whitfield et al. 2007).

Outputs from both models, MAGIC and SAFE, were combined in a whole countrywide assessment in Sweden (Sverdrup et al. 2005). MAGIC was applied to 133 lakes and SAFE to 645 productive forest sites, indicating that the threat of further damage from acid deposition to forest soils and lakes has been largely prevented due to reduced emissions. But because of extensive depletion of base cations pools in the soils and the slow rate of base cation replacement through weathering and deposition, the recovery process will extend over several decades or centuries. The chemical recovery of lakes will be significantly faster than the soil recovery.

Application of MAGIC to acid-sensitive headwater lakes and streams in 10 regions disseminated throughout the European continent have been a part of project RECOVER 2010, which aimed on assessing the impact of current and future anthropogenic pressures on sensitive European freshwater ecosystems (Jenkins et al. 2003). Significant recovery toward the preacidification chemistry was predicted by MAGIC in all regions except S England, S Alps, S Norway, and S Sweden. In these areas, further decrease in emissions will be necessary, and recovery of the surface waters will take several decades due to the slow recovery of base cations pools in depleted soils. One region (Finland) included in this study has been modeled with model SMART. The SMART model has been used to predict possible recovery of 36 acid-sensitive lakes in Finland (Posch et al. 2003a). The majority of the catchment soils exhibit very little changes in base saturation, but at the same time trends in lake water ANC were positive and trends of sulfate concentrations were negative. According to the model results for the period between 2010 and 2030, all lakes will reach positive ANC values favoring the recovery of fish populations (Posch et al. 2003a).

In the extremely acidified area known as the Black Triangle located in Central Europe, MAGIC has been applied onto two catchments with contrasting soil base saturation status due to different lithology (Hruska and Kram 2003). MAGIC predicted that due to the considerable decline in deposition of sulfate, concentrations of stream water sulfate will significantly decline at poorly buffered Lysina catchment on granite bedrock (Figure 20.1), but stream pH will not increase significantly in the next 30 years (Figure 20.1). At a well-buffered catchment Pluhuv Bor located on a serpentinite bedrock, mitigation of long-term acid deposition was successful due to the relatively high weathering rates. The results of soil base cation pools deterioration at poorly buffered Lysina catchment were further broken down into evolution of soil base saturation in individual soil layers by the SAFE model (Figure 20.2;

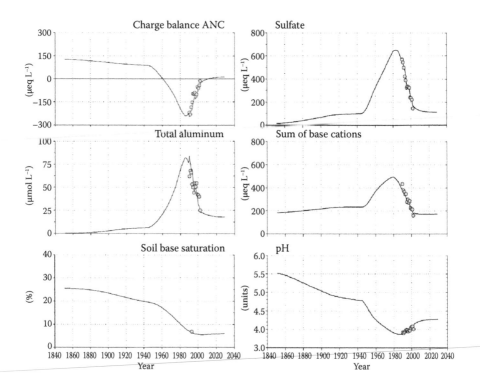

FIGURE 20.1
Example of model MAGIC application on Lysina catchment. (Extracted and modified from Hruska, J., and Kram, P., *Hydrology and Earth System Sciences*, 5, 519–528, 2003.) Individual panels depict the temporal development of ANC, sulfate, total aluminum, base cations, and pH in a stream draining the heavily acidified catchment. Temporal development of soil base saturation indicates significant depletion of base cation pools caused by extreme levels of acid deposition in the period from the 1950s to the 1980s. Gray dots represent discharge-weighted annual means of measured concentrations and pH. Lumped base saturation value representing mass-weighted average of the whole soil profile (0–90 cm).

Navratil et al. 2007). Assuming no changes in deposition since year 2004, significant recovery of soil base saturation has been predicted in top organic soil horizons only, whereas the base saturation of bottom mineral horizons should increase by only 2.3% (Figure 20.2). The Al/Bc ratio in soil solutions of the bottom mineral horizons was predicted to increase over critical value of 1 around year 2060 (Figure 20.2).

In single-site applications of dynamic models, the outputs are usually presented in graphs of the temporal development (Figures 20.1 and 20.2). The most frequently presented data are the most relevant chemical variables in soil solution or stream water (concentrations of Al, SO_4, Bc, etc.), the ratios of solutes such as Al/Bc ratio, and evolution of base saturation. In cases of output from multilayer models, the individual curves of temporal development can be lumped into a single figure (Figure 20.2).

In regional applications, information has to be usually lumped due to a large number of individual modeled sites. This is usually done, for example, by displaying the temporal development of selected percentiles of the cumulative distribution of the variable of interest (see, e.g., Kurz et al. 1998). Sequences of maps depicting aerial situation at certain time intervals can be used for displaying the changes in selected variables such as base saturation, Bc/Al ratio, or pH (see, e.g., Sverdrup et al. 2005).

Some selected examples of steady-state and dynamic model applications out of many were outlined above. The dynamic models need constant

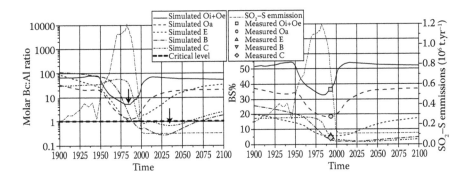

FIGURE 20.2
Example of SAFE model application to Lysina catchment. (Extracted and modified from Navratil et al., *Ecological Modelling*, 205, 410–422, 2007.) Two panels represent temporal development of soil base saturation (right) and Bc/Al soil solution ratio (left) in individual soil layers (five soil horizons). Each panel includes curve of Czech SO_2-S annual emissions. Note the predicted significant recovery of soil base saturation in the top soil horizons (Oi + Oe and Oa), mild recovery in E horizon, and very slow recovery of soil base saturation in the mineral B and C horizons. Mineral horizons represent the greatest soil mass in the soil profile. In the Bc/Al panel, two arrows represent the delay time between the lowest value of Bc/Al ratio in the organic Oi + Oe and mineral C soil horizon. Note that the delay between the peak in S deposition and the lowest Bc/Al ratio value in the mineral C horizon is almost 50 years.

validation and critical review of the modeled results. Well-described single-site applications are useful to evaluate model performance and model alternations. But it is also important to determine the quality lost in regional applications when compared to single-site applications.

20.5.3 Uncertainties

Uncertainty analysis is important to determine the credibility of modeled results. The extent to which results are to be trusted is nevertheless contaminated by many kinds of errors, for example, due to uncertainty of system perception, by insufficient and erroneous data, and in model structure or computer programs (Hettelingh 1990). Uncertainty analysis can be defined as a study with a systematic attempt to explore the range of variation in an output variable generated by quantified uncertainty in data inputs and model parameters.

All models are based on simplifications of natural systems. These simplifications lead to approximation uncertainties (Morgan and Henrion 1990). Systematic errors also arise from bias in measuring instruments used for analysis and can never be completely eliminated. Unknown processes cannot of course be incorporated into models, but their effects may be important. All these types of uncertainty contribute to the overall uncertainty in the results of modeling. An important part of uncertainty analysis is sensitivity analysis. This is an analysis of the extent to which changes in input variables affect changes in outputs. Historically, this was assessed by varying input variables one at a time to obtain single parameter sensitivity analysis.

There are, however, techniques for investigating the sensitivity of all parameters simultaneously as part of an overall uncertainty analysis (Skeffington 2006). Sensitivity analysis can be used to measure which model parameters make the greatest contribution to uncertainty in the final result, and hence where it is best to direct research to reduce these uncertainties. Parameters that turn out to have little influence can be accepted as they are, even though they themselves may be very uncertain.

The most frequently utilized technique for uncertainty analysis is Monte Carlo simulation (Rubinstein 1981). Monte Carlo simulation performs risk analysis by building models of possible results by substituting a range of values of probability distribution for any factor that has inherent uncertainty (Skeffington 2006). It then calculates results over and over, each time using a different set of random values from the probability functions. Depending on the number of uncertainties and the specified ranges, a Monte Carlo simulation could involve thousands or tens of thousands of recalculations. Monte Carlo simulation produces distributions of possible outcome values.

The comprehensive set of uncertainty analyses for critical load models were reported by Barkman (1997) and Barkman and Alveteg (2001), and an overview of the papers dealing with uncertainty in modeling of critical loads was given by Skeffington (2006). Assessment of uncertainty of critical loads derived from dynamic models is at its beginning. Dynamic models are calibrated on current measurements what should lead to narrower uncertainty ranges than unconstrained Monte Carlo analyses of the steady-state models (Skeffington 2006).

20.6 Summary

In past decades of the twentieth century, steady-state acidification models ranging from the simplest Simple Mass Balance model to more complex models such as PROFILE were used for calculation of critical loads and as tools for solving scientific questions. The major criticism of the steady-state models has been the fact that they do not take account of the rate of changes. Results of steady-state models yield information onto the state to which the ecosystem is drifting to, but they give no information on how long it will take or how does the current state compare to steady state (Posch et al. 2003b). For comparing the current state of soil chemistry to the steady state or for evaluation of time when acidification effects will take place, dynamic models must be used.

Out of the numerous developed models, the dynamic models that have survived in wide practical use such as SAFE, SMART, and MAGIC have been complemented with very simple extension of steady-state model VSD. Main focuses of these models differ, but key processes are common. Both MAGIC and SAFE have been applied widely to single sites and in regional studies. MAGIC focuses more on chemistry of surface waters or groundwaters, whereas SMART and SAFE models are oriented onto the soil and soil solution chemistry.

VSD model in its current version will be the primary tool for deriving target loads needed for integrated assessment in Europe. The more complex models such as SAFE will require substantially more input data besides being used for target load calculations. The results of more complex models will be used for validating the modeling concept for calculating target loads by the simpler models.

In particular, regional assessments data for validation may be sparse, and in some cases calibration data could even be lacking. It is, however, important to point out that the trustworthiness of model output will suffer with fewer of the required data available. The quality of the model output is always totally dependent on the quality of the input. Thus, widespread use of dynamic models will require good-quality data sets. It is alarming that as the data requirements have increased over the years, due to the increased use of dynamic models, there has been a trend of decreased funding for monitoring programs.

References

Aber, J. D., and C. A. Federer. 1992. A generalized, lumped-parameter model of photosynthesis, evapotranspiration and net primary production in temperate and boreal forest ecosystems. *Oecologia* 92: 463–474.

Aber, J. D., W. McDowell, K. J. Nadelhoffer, A. Magill, G. Bernston, S. G. Kamakea, W. McNulty, W. Currie, L. Rustad, and I. Fernandez. 1998. Nitrogen saturation in temperature forest ecosystems. *BioScience* 48: 921–934.

Aherne, J., P. J. Dillon, and B. J. Cosby. 2003. Acidification and recovery of aquatic ecosystems in south-central Ontario, Canada: regional application of the MAGIC model. *Hydrology and Earth System Sciences* 7: 561–573.

Alewell, C. 2001. Predicting reversibility of acidification: the European sulfur story. *Water Air and Soil Pollution* 130 (1–4): 1271–1276.

Alveteg, M. 1998. Dynamics of forest soil chemistry. PhD thesis, Reports in Ecology and Environmental Engineering 3, Department of Chemical Engineering II, Lund University, Lund, Sweden, 81.

Alveteg, M., and H. Sverdrup. 2002. Manual for regional assessments using the SAFE model. Lund, Sweden; Department of Chemical Engineering II, Lund University.

Alveteg, M., C. Walse, and H. Sverdrup. 1998a. Evaluating simplifications used in regional applications of the SAFE and MAKEDEP models. *Ecological Modelling* 107: 265–277.

Alveteg, M., C. Walse, and P. Warfvinge. 1998b. Reconstructing historic atmospheric deposition and nutrient uptake from present day values using MAKEDEP. *Water, Air and Soil Pollution* 104 (3–4): 269–283.

Barkman, A. 1997. Applying the critical loads concept: constraints induced by data uncertainty. Reports in Ecology and Environmental Engineering 1, University of Lund, Lund, Sweden.

Barkman, A., and M. Alveteg. 2001. Identifying potentials for reducing uncertainty in critical load calculations using the PROFILE model. *Water, Air and Soil Pollution* 125: 33–54.

Bashkin, V. N., M. Y. Kozlov, I. V. Prepotina, A. Y. Abramychev, and I. S. Dedkova. 1995. Calculation and mapping of critical loads of S, N and acidity on ecosystems of the Northern Asia. *Water, Air and Soil Pollution* 85: 2395–2400.

Battarbee, R. W., T. E. H. Allott, S. Juggins, A. M. Kreiser, C. Curtis, and R. Harriman. 1996. Critical loads of acidity to surface waters – an empirical diatom-based palaeolimnological model. Ambio 25: 366–369.

Bélanger, N., B. Côté, F. Courchesne, J. V. Fyles, P. Warfvinge, and W. H. Hendershot. 2002. Simulation of soil chemistry and nutrient availability in a forested ecosystem of southern Quebec-1. Reconstruction of the time-series files of nutrient cycling using the MAKEDEP model. *Environmental Modelling and Software* 174: 427–445.

Benkovitz, C. M., M. T. Scholtz, J. Pacyna, L. Tarrason, J. Dignon, E. C. Voldner, P. A. Spiro, J. A. Logan, and T. E. Graedel. 1996. Global gridded inventories of anthropogenic emissions of sulfur and nitrogen. *Journal of Geophysical Research* 101 (D22): 29239–29253.

Bouwman, A., D. Van Vuuren, R. Derwent, and M. Posch. 2002. A global analysis of acidification and eutrophication of terrestrial ecosystems. *Water, Air and Soil Pollution* 141 (1): 349–382.

Christophersen, N., and C. Neal. 1990. Linking hydrological, geochemical, and soil chemical processe on the catchment scale: An interplay between field work and modelling. *Water Resources Research* 26: 3077–3086.

Christophersen, N., and R. Wright. 1981. Sulfate budget and a model for sulfate concentrations in streamwater at Birkenes, a small forested catchment in southernmost Norway. *Water Resources Research* 17 (2): 377–389.

Cole, J. J., N. F. Caraco, G. W. Kling, and T. K. Kratz. 1994. Carbon dioxide supersaturation in the surface waters of lakes. *Science* 265: 1568–1570.

Collins, R., and A. Jenkins. 1998. Regional modelling of acidification in Wales; calibration of a spatially distributed model incorporating land use change. *Hydrology and Earth System Sciences* 2: 533–541.

Cosby, B. J., R. C. Ferrier, A. Jenkins, and R. F. Wright. 2001. Modelling the effects of acid deposition: Refinements, adjustments and inclusion of nitrogen dynamics in the MAGIC model. *Hydrology and Earth System Sciences* 5 (3): 499–517.

Cosby, B. J., R. C. Ferrier, A. Jenkins, B. A. Emmett, R. F. Wright, and A. Tietema. 1997. Modelling the ecosystem effects of nitrogen deposition at the catchment scale: Model of ecosystem retention and loss of inorganic nitrogen (MERLIN). *Hydrology and Earth System Sciences* 1: 137–158.

Cosby, B. J., G. M. Hornberger, J. N. Galloway, and R. F. Wright. 1985a. Modelling the effects of acid deposition: Assessment of a lumped parameter model of soil water and streamwater chemistry. *Water Resources Research* 21 (1): 51–63.

Cosby, B. J., G. M. Hornberger, J. N. Galloway, and R. F. Wright. 1985c. Time scales of catchment acidification: A quantitative model for estimating freshwater acidification. *Environmental Science and Technology* 19: 1144–1149.

Cosby, B. J., G. M. Hornberger, R. F. Wright, and J. N. Galloway. 1986. Modelling the effects of acid deposition: Control of long-term sulfate dynamics by soil sulfate adsorption. *Water Resources Research* 22 (8): 1283–1291.

Cosby, B. J., R. F. Wright, and E. Gjessing. 1995. An acidification model (MAGIC) with organic acids evaluated using whole catchment manipulations in Norway. *Journal of Hydrology* 170: 101–122.

Cosby, B. J., S. A. Norton, and J. S. Kahl. 1996. Using paired catchment manipulation experiment to evaluate a catchment-scale biogeochemical model. *The Science of the Total Environment* 183: 49–66.

Cosby, B. J., R. F. Wright, G. M. Hornberger, and J. N. Galloway. 1984. Model of acidification of groundwater in catchments. Project completion report, Proj. E2-14, EPA/NCSU Acid Precipitation Program, Environ. Prot. Agency, N.C. State Univ., October 1984.

Cosby, B. J., R. F. Wright, G. M. Hornberger, and J. N. Galloway. 1985b. Modelling the effects of acid deposition: Estimation of long-term water quality responses in a small forested catchment. *Water Resources Research* 21: 1591–1601.

Cronan, C. S., and D. F. Grigal. 1995. Use of calcium/aluminum ratios as indicators of stress in forest ecosystems. *Journal of Environmental Quality* 24: 209–226.

Cronan, C. S., and C. L. Schofield. 1990. Relationships between aqueous aluminum and acidic deposition in forested watersheds of North America and northern Europe. *Environmental Science and Technology* 24: 1100–1105.

De Vries, W. 1991. Methodologies for the assessment and mapping of critical loads and the impact of abatement strategies on forest soils. 109 pp. DLO Winand Staring Centre for Integrated Land, Soil and Water Research, Report 46, Wageningen, the Netherlands.

De Vries, W. 1994. Soil response to acid deposition at different regional scales. Field and laboratory data, critical loads and model predictions. 487 pp. PhD thesis, Agricultural University, Wageningen, the Netherlands.

De Vries, W., M. Posch, and J. Kämäri. 1989. Simulation of the long-term soil response to acid deposition in various buffer ranges. *Water, Air and Soil Pollution* 48: 349–390.

De Vries, W., G. J. Reinds, and M. Posch. 1994a. Assessment of critical loads and their exceedance on European forests using a one-layer steady-state model. *Water, Air and Soil Pollution* 72 (1–4): 357–394.

De Vries, W., G. J. Reinds, M. Posch, and J. Kämäri. 1994b. Simulation of soil response to acidity deposition scenarios in Europe. *Water, Air and Soil Pollution* 78: 215–246.

Dise, N. B., and R. F. Wright. 1995. Nitrogen leaching from European forests in relation to nitrogen deposition. *Forest Ecology Management* 71: 153–162.

Dise, N. B., E. Matzner, and P. Gundersen. 1998. Synthesis of nitrogen pools and fluxes from European forest ecosystems. *Water, Air and Soil Pollution* 105: 143–154.

Driscoll, C. T., G. B. Lawrence, A. J. Bulger, T. J. Butler, C. S. Cronan, C. Eagar, K. F. Lambert, G. E. Likens, J. L. Stoddard, and K. C. Weathers. 2001. Acidic deposition in the northeastern United States: Sources and inputs, ecosystem effects, and management strategies. *Bioscience* 51: 180–198.

Driscoll, C. T., D. Whitall, J. Aber, E. W. Boyer, M. Astro, C. Cronan, C. L. Goodale, P. Groffman, C. Hopkinson, K. Lambert, G. Lawrence, and S. Ollinger. 2003. Nitrogen pollution in the Northeastern US: Sources, effects, and management options. *Bioscience* 53: 357–374.

Emmett, B. A., B. J. Cosby, R. C. Ferrier, A. Jenkins, A. Tietema, and R. F. Wright. 1997. Modelling the ecosystem effects of nitrogen deposition at the catchment scale: Simulation of nitrogen saturation in a Sitka spruce forest, Aber, Wales, UK. *Biogeochemistry* 38: 129–148.

EPA (Environmental Protection Agency). 1995. *National Air Pollutant Emission Trends 1900–1994*, EPA-454/R-95-011, EPA, Washington, DC.

Fernandez, I. J., L. E. Rustad, S. A. Norton, J. S. Kahl, and B. J. Cosby. 2003. Experimental acidification causes soil base-cation depletion at the Bear Brook Watershed in Maine. *Soil Science Society of America Journal* 67: 1909–1919.

Folster, J., K. Bishop, P. Kram, H. Kvarnas, and A. Wilander. 2003. Time series of long-term annual fluxes in the streamwater of nine forest catchments from the Swedish environmental monitoring program (PMK5). *The Science of the Total Environment* 310: 113–120.

Forsius, M., J. Vourenmaa, J. Mannio, and S. Syri. 2003. Recovery from acidification of Finnish lakes: Regional patterns and relations to emission reduction policy. *Science of the Total Environment* 310: 121–132.

Fuller, R., C. Driscoll, S. Schindler, and M. Mitchell. 1986. A simulation model of sulfur transformations in forested Spodosols. *Biogeochemistry* 2: 313–328.

Fumoto, T., and H. Sverdrup. 2000. Modelling of sulfate adsorption on andisols for implementation in the SAFE Model. *Journal of Environmental Quality* 29: 1248–1290.

Grennfelt, P., H. Rodhe, E. Thornelof, and J. Wisniewski. 1995. Acid Regin '95? *Proceedings from the 5th International Conference in Acidic Deposition, June 26–30, 1995. Water Air and Soil Pollution* 85 (1–4).

Grubler, A. 1998. A review of global and regional sulfur emission scenarios. *Mitigation and Adaptation Strategies for Global Change* 3 (2–4): 383–418.

Grubler, A. 2002. Trends in global emissions: carbon, sulfur, and nitrogen. In: *Encyclopedia of Global Environmental Change*, 35–53. Chichester: John Wiley & Sons Ltd.

Gschwandtner, G., K. C. Gschwandtner, and K. Eldridge. 1985. *Historic Emissions of Sulfur and Nitrogen Oxides in the United States from 1900 to 1980*, Vols. I and II, EPA-600/785-009a and -009b. EPA, Washington, DC.

Gundersen, P., I. Callesen, and W. De Vries. 1998. Nitrate leaching in forest ecosystems is controlled by forest floor C/N ratio. *Environmental Pollution* 102: 403–407.

Gundersen, P., B. A. Emmett, O. J. Kjonaas, C. J. Koopmans, and A. Tietema. 1998. Impact of nitrogen deposition on nitrogen cycling in forests: A synthesis of NITREX data. *Forest Ecology and Management* 101: 37–55.

Henriksen, A.,W. Dickson, and D. F. Brakke. 1986. Estimates of critical loads to surface waters. In *Critical Loads for Sulphur and Nitrogen*, ed. J. Nilsson, 87-120. Copenhagen: Nordic Council of Ministers.

Henriksen, A. 1980. Acidification of freshwaters—a large scale titration. In *Ecological Impacts of Acid Precipitation*, eds. D. Drablos and A. Tollan, 68–74. Oslo: SNSF-project.

Henriksen, A., and M. Posch. 2001. Steady-state models for calculating critical loads of acidity for surface waters. *Water, Air and Soil Pollution Focus* 1: 375–398.

Hettelingh, J. P. 1990. *Uncertainty in Modelling Regional Environmental Systems: The generalization of a watershed acidification model for predicting broad scale effects*. 224 pp. Laxenburg, Austria: IIASA Research Report RR-90-3, International Institute for Applied Systems Analysis.

Hettelingh, J. P., M. Posch, J. Slootweg, G. J. Reinds, T. Spranger, and L. Tarrason. 2007. Critical loads and dynamic modelling to assess European areas at risk of acidification and eutrophication. *Water, Air and Soil Pollution Focus* 7: 379–384.

Hettelingh, J. P., H. Sverdrup, and D. Zhao. 1995. Deriving critical loads for Asia. *Water, Air and Soil Pollution* 85 (4): 2565–2570.

Houston, D. R. 1999. History of sugar maple decline. In *Sugar Maple Ecology and Health: Proceedings of an International Symposium*, eds. S. B. Horsley and R. P. Long, 9–26. Forest Service, Radnor, PA: US Department of Agriculture, General Technical Report NE-261.

Hruska, J., and P. Kram. 2003. Modelling long-term changes in stream water and soil chemistry in catchments with contrasting vulnerability to acidification (Lysina and Pluhuv Bor, Czech Republic). *Hydrology and Earth System Sciences* 5: 519–528.

Hruska, J., P. Cudlin, and P. Kram. 2001. Relationship between Norway spruce status and soil water base cations/aluminium ratios in the Czech Republic. *Water, Air and Soil Pollution* 130: 983–988.

Hruska, J., F. Moldan, and P. Kram. 2002. Recovery from acidification in central Europe—observed and predicted changes of soil and streamwater chemistry in the Lysina catchment, Czech Republic. *Environmental Pollution* 120 (2): 261–274.

Hultberg, H., and R. Skeffington, eds. 1998. *Experimental Reversal of Acid Rain Effects: The Gardsjon Roof Project*. Chichester: John Wiley & Sons Ltd.

Husar, R. 1994. Sulfur and nitrogen emission trends for the U.S. In *Industrial Metabolism: Restructuring for Sustainable Development*, eds. R. U. Ayres and U. E. Simonis, 239–258. Tokyo: United Nations University Press.

Husar, R. B., and J. D. Husar. 1990. Sulfur. In *The Earth as Transformed by Human Action*, eds. B. L. Turner, W. C. Clark, W. R. Kates, J. F. Richards, J. T. Mathews, and W. B. Meyer, 409–421. Cambridge: Cambridge University Press.

Jenkins, A., L. Camarero, B. J. Cosby, R. C. Ferrier, M. Forsius, R. C. Helliwell, J. Kopácek, V. Majer, F. Moldon, M. Posch, M. Rogora, W. Schöpp, and R. F. Wright. 2003. A modelling assessment of acidification and recovery of European surface waters. *Hydrology and Earth System Sciences* 7: 447–455.

Jenkins, A., R. C. Ferrier, and R. C. Helliwell. 2001. Modelling nitrogen dynamics at Lochnagar, N.E. Scotland. *Hydrology and Earth System Sciences* 5: 519–527.

Karltun, E. 1995. Sulfate adsorption on variable-charge minerals in podzolized soils in relation to sulfur deposition and soil acidity. Doctoral thesis, Swedish University of Agricultural Sciences.

Kopáček, J., and J. Veselý. 2005. Sulfur and nitrogen emissions in the Czech Republic and Slovakia from 1850 till 2000. *Atmospheric Environment* 39: 2179–3188.

Kros, J., G. J. Reinds, W. De Vries, J. B. Latou, and M. Bollen. 1995. Modelling the response of terrestrial ecosystems to acidification and desiccation scenarios. *Water, Air and Soil Pollution* 85: 1101–1106.

Kurz, D., and M. Posch. 2002. A comparison of the SAFE and VSD soil dynamic models using Swiss data. Report by EKG Geo-Science and CCE, Bern and Bilthoven, 32 pp.

Kurz, D., M. Alveteg, and H. Sverdrup. 1998. Integrated assessment of soil chemical status: 2. Application of a regionalised model to 622 forested sites in Switzerland. *Water, Air and Soil Pollution* 105: 11–20.

Kuylenstierna, J. C. I., W. K. Hicks, S. Cinderby, and H. Cambridge. 1998. Critical loads for nitrogen deposition and their exceedance at European scale. *Environmental Pollution* 102 (Supp 1): 591–598.

Kuylenstierna, J. C. I., H. Rodhe, S. Cinderby, and K. Hicks. 2001. Acidification in developing countries: Ecosystem sensitivity and the critical load approach on a global scale. *Ambio* 30 (1): 20–28.

Likens, G. E., F. H. Bormann, and N. M. Johnson. 1972. Acid rain. *Environment* 14 (2): 33–40.

Lindstrom, G., and M. Gardelin. 1992. Report from the Swedish integrated groundwater acidification project. In *Modelling Groundwater Response to Acidification*, eds. P. Sanden and P. Warfvinge, 33–36. SMHI, Reports Hydrology.

Lovett, G. M., K. C. Weathers, and W. V. Sobczak. 2000. Nitrogen saturation and retention in forested watersheds of the Catskill Mountains, New York. *Ecological Applications* 10: 73–84.

Martinson, L. 2004. Recovery from acidification—policy oriented dynamic modelling. Doctoral thesis, Department of Chemical Engineering, Lund University, Sweden.

Martinson, L., and M. Alveteg. 2004. The importance of including the pH dependence of sulfate adsorption in a dynamic soil chemistry model. *Water, Air and Soil Pollution* 154: 349–356.

Martinson, L., M. Alveteg, and P. Warfvinge. 2003. Parameterization and evaluation of sulfate adsorption in a dynamic soil chemistry model. *Environmental Pollution* 124: 119–125.

Martinson, L., N. Lamersdorf, and P. Warfiinge. 2005. The Solling Roof revisited— slow recovery from acidification observed and modeled despite a decade of "cleanrain" treatment. *Environmental Pollution* 135: 293–302.

Mitchell, M., and R. Fuller. 1988. Models of sulfur dynamics in forest and grassland ecosystems with emphasis on soil processes. *Biogeochemistry* 5 (1): 133–163.

Mol-Dijkstra, J. P., J. Kros, G. J. Reinds, H. J. J. Wieggers, and M. Posch. 2006. *Model Description and Users' Guide SMART2 version 3.4*. Alterra-raport 1425, Wageningen, Netherlands.

Morgan, M. G., and M. Henrion. 1990. *Uncertainty. A Guide to Dealing with Uncertainty in Quantitative Risk and Policy Analysis*. Cambridge: Cambridge University Press.

Mylona, S. 1993. *Trends in Sulfur Dioxide Emissions, Air Concentrations and Deposition of Sulfur in Europe Since 1880, EMEP/MSC-W Report 2/1993*. Oslo: Norwegian Meteorological Institute.

Mylona, S. 1996. Sulfur dioxide emissions in Europe 1880–1991 and their effect on sulfur concentrations and depositions. *Tellus* 48 (B): 662–689.

Nakicenovic, N., A. Grubler, H. Ishitani, T. Johansson, G. Marland, J. R. Moreira, and H. H. Rogner. 1996. Energy primer. In *Climate Change 1995, Impacts, Adaptations and Mitigation of Climate Change: Scientific–Technical Analyses*, eds. R. T. Watson, M. C. Zinyowera, and R. H. Moss, 75–92. Cambridge: Cambridge University Press.

Navratil, T., D. Kurz, P. Kram, J. Hofmeister, and J. Hruska. 2007. Acidification and recovery of soil at a heavily impacted forest catchment (Lysina, Czech Republic)—SAFE modelling and field results. *Ecological Modelling* 205: 410–422.

Nilsson, J., and P. Grennfelt. 1988. Critical loads for sulfur and nitrogen, 418 pp. Miljø rapport 1988. 15, Copenhagen Denmark Nordic Council of Ministers.

Norton, S. A., J. S. Kahl, I. J. Fernandez, L. E. Rustad, J. P. Schofield, and T. A. Haines. 1994. Response of the West Bear Brook Watershed, Maine, USA, to the addition of $(NH_4)_2SO_4$: 3-year results. *Forest and Ecology Management* 68: 61–73.

Norton, S. A., B. J. Cosby, I. J. Fernandez et al. 2001. Long-term and seasonal variations in CO2: Linkages to catchment alkalinity generation. *Hydrology and Earth System Sciences* 5: 83–91.

Odèn, S. 1968. The acidification of air and precipitation and its consequences on the natural environment, 68 pp. Bulletin 1, Swedish National Science Research Council, Ecology Committee.

Parashar, D. C., U. C. Kulshrestra, and C. Sharma. 1998. Anthropogenic emissions of NO_x, NH_3 and N_2O in India. *Nutrient Cycling in Agroecosystems* 52: 255–259.

Posch, M., and W. De Vries. 1999. Derivation of critical loads by steady-state and dynamic soil models. In *The Impact of Nitrogen Deposition on Natural and Semi-Natural Ecosystems*, ed. S. J. Langan, 213–234. Dordrecht, the Netherlands: Kluwer Academic Publishers.

Posch, M., and G. J. Reinds. 2009. A very simple dynamic soil acidification model for scenario analyses and target load calculations. *Environmental Modelling and Software* 24: 329–340.

Posch, M., M. Forsius, J. Johansson, J. Vuorenmaa, and J. Kamari. 2003a. Modelling the recovery of acid-sensitive Finnish headwater lakes under present emission reduction agreements. *Hydrology and Earth System Sciences* 7: 484–493.

Posch, M., J.-P. Hettelingh, and J. Slootweg. 2003b. Manual for dynamic modelling of soil response to atmospheric deposition, 71 pp. RIVM Report 259101012/2003, Working Group on Effects of the CLRTAP, Bilthoven, the Netherlands.

Posch, M., J. P. Hettelingh, and P. A. M. De Smet. 2001. Characterization of critical load exceedances in Europe. *Water, Air and Soil Pollution* 130: 1139–1144.

Posch, M., J. Kämäri, M. Forsius, A. Henriksen, and A. Wilander. 1997. Exceedance of critical loads for lakes in Finland, Norway and Sweden: Reduction requirements for acidifying nitrogen and sulfur deposition. *Environmental Management* 21 (2): 291–304.

Posch, M., G. J. Reinds, and W. De Vries. 1993. *SMART—A Simulation Model for Acidification's Regional Trends: model description and user manual*, 477, 43 pp. Helsinki, Finland: Mimeograph Series of the National Board of Waters and the Environment.

Reinds, G. J., M. Posch, W. De Vries, J. Slootweg, and J. P. Hettelingh. 2008. Critical loads of sulfur and nitrogen for terrestrial ecosystems in Europe and Northern Asia influenced by different soil chemical criteria. *Water, Air and Soil Pollution* 193: 269–287.

Reinds, G. J., M. Van Oijen, G. B. M. Heuvelink, and H. Kros. 2008. Bayesian calibration of the VSD soil acidification model using European forest monitoring data. *Geoderma* 146: 475–488.

Reuss, J. O. 1983. Implications of the calcium-aluminum exchange system for the effect of acid precipitation on soils. *Journal of Environmental Quality* 12 (4): 591-595.

Reuss, J. O., and D. W. Johnson. 1986. *Acid Deposition and the Acidification of Soils and Waters*. New York: Springer-Verlag.

Rubinstein, R. Y. 1981. *Simulation and the Monte Carlo Method*. New York, NY: John Wiley & Sons.

SAEFL. 1998. Acidification of Swiss forest soils—Development of a regional dynamic assessment, 115 pp. Environmental Documentation No. 89, SAEFL, Berne, Switzerland.

Satake, K., J. Shindo, T. Takamatsu, T. Nakano, S. Aoki, T. Fukuyama, S. Hatakeyama et al. 2001. Acid rain 2000. *Proceedings from the 6th International Conference on Acidic Deposition: Looking Back to the Past and Thinking of the Future. Water Air and Soil Pollution*, 130.

Semenov, M., V. Bashkin, and H. Sverdrup. 2001. Critical loads of acidity for forest ecosystems of North Asia. *Water, Air and Soil Pollution* 130: 1193–1198.

Siccama, T. G., M. Bliss, and H. W. Vogelmann. 1982. Decline of red spruce in the Green Mountains of Vermont. *Bulletin of the Torrey Botanical Club* 109: 162–168.

Skeffington, R. A. 2006. Quantifying uncertainty in critical loads: (A) literature review. *Water, Air and Soil Pollution* 169: 3–24.

Smith, R. A. 1872. *Air and Rain*. London: Longmans Green and Co.

Staelens, J., J. Neirynck, G. Genouw, and P. Roskams. 2009. Dynamic modelling of target loads of acidifying deposition for forest ecosystems in Flanders (Belgium). *iForest* 2: 30–33.

Stoddard, J. L., D. S. Jefferies, A. Lukewille, T. A. Clair, P. J. Dillon, C. T. Driscoll, M. Forsius et al. 1999. Regional trends in aquatic recovery from acidification in North America and Europe. *Nature* 401: 575–578.

Streets, D. G., N. Y. Tsai, H. Akimoto, and K. Oka. 2000. Sulfur dioxide emissions in Asia in the period 1985–1997. *Atmospheric Environment* 34 (26): 4413–4424.

Sullivan, T. J., and B. J. Cosby. 1995. MAGIC model applications for surface and soil waters as input to the Tracking and Analysis Framework (TAF). Model Documentation. Report for U.S. Department of Energy, DOE/ER/30196-7. E&S Environmental Chemistry, Inc., Corvallis, Oregon.

Sullivan, T. J. 1997. Ecosystem manipulation experimentation as a means of testing a biogeochemical model. *Environmental Management* 21: 15–21.

Sverdrup, H., and P. Warfvinge. 1995. Critical loads of acidity for Swedish forest eco-systems. *Ecological Bulletins* 44: 75–89.

Sverdrup, H., S. Belyazid, B. Nihlgård, and L. Ericson. 2007. Modelling change in ground vegetation response to acid and nitrogen pollution, climate change and forest management in Sweden 1500–2100 A.D. *Water Air Soil Pollution Focus* 7: 163–179.

Sverdrup, H., W. De Vries, and A. Henriksen. 1990. *Mapping Critical Loads*, 124 pp. Copenhagen, Denmark: NORD 1990:98, Nordic Council of Ministers.

Sverdrup, H., L. Martinsson, M. Alveteg, F. Moldan, V. Kronnäs, and J. Munthe. 2005. Modelling recovery of Swedish ecosystems from acidification. *Ambio* 34: 25–31.

Sverdrup, H., P. Warfvinge, L. Blake, and K. Goulding. 1995. Modelling recent and historic soil data from the Rothamsted Experimental Station, England, using SAFE. *Agriculture Ecosystems and Environment* 53: 161–177.

Sverdrup, H., P. Warfvinge, H. Hultberg, I. Andersson, and F. Moldan. 1998. Modelling soil acidification in a roofed catchment: Application of the SAFE model. In *Experimental Reversal of Acid Rain Effects: The Gårdsjön Roof Project*, eds. H. Hultberg and K. Skeffington, 363–382. Chichester: John Wiley & Sons Ltd.

Tarrasón, L., A. Benedictow, H. Fagerli, J. E. Jonson, H. Klein, and M. Van Loon. 2005. *Transboundary acidification, eutrophication and ground level ozone in Europe in 2003*. Norway: Norwegian: EMEP Status report 1/2005, Oslo, Meteorological Institute. http://www.emep.int.

Tiktak, A., and H. J. M. Van Grinsven. 1995. Review of sixteen forest–soil–atmosphere models. *Ecological Modelling* 83: 35–53.

UBA. 1996. *Manual on Methodologies and Criteria for Mapping Critical Levels/Loads and Geographical Areas Where They Are Exceeded*, Texte 71/96, 144 pp. Berlin, Germany: Umweltbundesamt.

UBA. 2004. *Manual on methodologies and criteria for modelling and mapping of critical loads and levels and air pollution effects, risks and trends*. Umweltbundesamt, Dessau, Germany. http://www. icpmapping.org.

Ulrich, B. 1981. Ökologische Gruppierung von Böden nach ihrem chemischen. Bodenzustand. *Zeitschrift Pflanzenernährung und Bodenkdunde* 144: 289–305.

Wallman, P., M. G. E. Svensson, H. Sverdrup, and S. Belyazid. 2005. ForSAFE—an integrated process-oriented forest model for long-term sustainability assessment. *Forest Ecology & Management* 207: 19–36.

Walse, C., B. Berg, and H. Sverdrup. 1998. Review and synthesis of experimental data on organic matter decomposition with respect to the effects of temperature, moisture, and acidity. *Environmental Review* 6: 25–40.

Walse, C., K. Blanck, M. Bredemeier, N. Lamersdorf, P. Warfvinge, and X. P. Xu. 1998. Application of the SAFE model to the Solling clean rain roof experiment. *Forest Ecology and Management* 101: 307–317.

Warfvinge, P., and H. Sverdrup. 1992. Calculating critical loads of acid deposition with PROFILE—a steady-state soil chemistry model. *Water, Air and Soil Pollution* 63: 119–143.

Warfvinge, P., M. Morth, and F. Moldon. 2000. What processes govern recovery? In *Recovery from Acidification in the Natural Environment*, eds. P. Warfvinge and U. Bertills, 23–36. Stockholm: Report 5034, Swedish Environmental Protection Agency.

Warfvinge, P., U. Falkengren-Grerup, H. Sverdrup, and B. Andersen. 1993. Modelling long-term cation supply in acidified forest stands. *Environmental Pollution* 80: 209–221.

Whitfield, C. J., J. Aherne, P. J. Dillon, and S. A. Watmough. 2007. Modelling acidification, recovery, and target loads for headwater catchments in Nova Scotia, Canada. *Hydrology and Earth System Sciences* 11: 951–963.

Wright, R. F., and Rasmussen, L. 1998. Introduction to the NITREX and EXMAN projects. *Forest Ecology and Management* 101: 1–7.

Wright, R. F., B. J. Cosby, and G. M. Hornberger. 1991. A regional model of lake acidification in Southernmost Norway. *Ambio* 20 (6): 222–225.

Wright, R. F., B. A. Emmett, and A. Jenkins. 1998. Acid deposition, land-use change and global change: MAGIC7 model applied to Risdalsheia, Norway (RAIN and CLIMEX projects) and Aber, UK (NITREX project). *Hydrology and Earth System Sciences* 2: 385–397.

Index

A

Acetaldehyde, 515
Aceticlastic methanogenesis, 304
    ammonia inhibition of, 306
Acetogenesis, 304
Acidic deposition, 501, 514
    composition of, 558
    effects
        of accumulation of nitrogen in
            soils, 562
        of accumulation of sulfur in soils,
            561–562
        of depletion of base cations and
            mobilization of Al in soils, 561
        on forest, 563
        on land use, 564
        on surface water, 563–564
Acidic substances, emissions and
        deposition of, 559–564
Acidification
    of aquatic ecosystem, 12
    effect on biogeochemical cycles, 558
    meaning of, 558
    of soils, 561, 575, 586
    of surface waters, 563–564, 575
Acidification models
    applications, 585–590
    calibration methods, 584–585
    dynamic models, 567–568
        basic equations, 568–570
        buffer processes, 571–574
        from soils to surface waters,
            574–575
        steady-state to dynamic
            simulations, 570–571
    selected dynamic models
        MAGIC model, 580–582
        SAFE model, 578–580
        SMART model, 577–578
        VSD model, 575–577
    steady-state, 565–567
    uncertainty analysis of, 590–591

Acidified soils, chemical recovery of,
        564
Acidifying pollutants, 501
Acid neutralizing capacity (ANC), 562,
        563, 564
Acidogenesis, 303–305
Acid rain, 32, 558, 560
Acropora species, 156, 157
Activated sludge, 278, 279, 286
    features of mathematical structure of
        growth rate in, 294
    model extensions focused on
        nitrogen removal
        for nitrite accumulation, 291–295
        for nitrous gas emissions from
            plants, 295–296
    model extensions focused on
        phosphorous removal, 296–298
    see also Activated Sludge Model No. 1
        (ASM1); Activated Sludge
        Model No. 3 (ASM3)
Activated Sludge Model No. 1 (ASM1),
        279
    COD fractionation in, 285–286
    death–regeneration concept of, 288
    definition of, 280
    difference with ASM3, 288
    importance of, 280
    limitations of, 289–291, 308
    matrix notation of, 281–285
    nitrogen components in, 288
    nitrogen fractionation in, 286–287
    substrate flows in, 284
Activated Sludge Model No. 3 (ASM3),
        287–289
Activated sludge reactors, 285, 299
Activation energy, 478–480
    Kissinger–Akahira–Sunose
        (KAS) method for obtaining,
        480–481
Active root biomass, 177, 179
Actual oxygen concentration, 143
Added organic matter (AOM), 216

ADMS 4 Air Pollution Model, 543–545
  airport pollution model, 547–548
  industrial version, 545–546
  output of, 545
  plume–building interactions in, 544
  roads pollution model, 547
  treatment of sea–land internal
        boundary layer in, 545
  urban version, 546–547
Adsorption equilibrium, 394
Adsorption isotherms, 394, 431
Advection–diffusion equation, for
        calculating phytoplankton
        biomass and bloom
        development, 91
Aeration, 143
Aerosols, 500, 514, 517–518
Agricultural management, and crop
        rotation, 230
Agricultural systems, 10
  analysis of simulation results, 231
  basic steps in modeling, 205–206
  overview of DAISY-model code,
        212–213
    carbon balance, 214–217
    heat balance, 214
    nitrogen balance, 217–220
    water balance, 213–214
  variables/parameters, sensitivity,
        and modeling practice
    agricultural management and
        crop rotation, 230
    crop parameterization, 228–230
    soil and soil system, 223–228
    weather and nitrogen deposition,
        221–223
Air dispersion modeling, in coastal
        areas, 545
Air pollution, 454, 500
  by acidic substances, 559
  evaluation of simulation results for
        measurement of, 523
  process for determination of, 513–514
    aerosol dynamics, 517–518
    atmospheric processes, 519–520
    cloud dynamics and chemistry,
        519
    emission, 520–523
    gas-phase chemistry, 514–516

Air pollution modeling
  basic approaches to, 502–503
    boundary conditions, 508–509
    comparison of Eulerian and
        Lagrangian approaches, 507–508
    Eulerian approach, 503–505
    Lagrangian approach, 505–507
    mass conservation relation, 503
    trajectory "puff" and "plume"
        models, 509–513
  classification of, 526–527
  examples of, 523–527
    ADMS 4 air pollution model,
        543–548
    community multiscale air quality
        (CMAQ) system, 528–532
    fifth-generation PSU/NCAR
        mesoscale model (MM5),
        532–540
    SILAM air pollution model,
        548–550
    SMOKE modeling system, 540–542
  turbulent diffusion in atmosphere,
        importance of, 505
Air quality (AQ), 500, 502, 514, 517
Air–water interface, 354, 357
Algae, 51
  excretion, 83
  nitrogen-fixing, 327
Alternating direction implicit (ADI)
        technique, 354
Ammonia, 118, 287, 306, 350, 355
  anthropogenic emissions of, 560
  nitrification of, 358, 366
  primary source of, 364
  uptake by plants, 359
Ammonium adsorption, 218, 227
Ammonium fertilizers, 219
Ammonium nitrogen, 219
Ammonium oxidizing bacteria (AOB),
        291
Anaerobic degradation, 303
Anaerobic digester models
  ADM1 limitations, 308
  principles of, 303–305
Anaerobic digestion (AD), 279
  models for, 305–308
Anaerobic Digestion Model No. 1
        (ADM1), 307

ANAFORE (Analysis of Forest
    Ecosystems) model, 162, 167
  flowchart, 168
ANIMO model, 204
Anoxia, 349
Anthes–Kuo scheme, 534–535
Antibiotics, 426
  contamination of agricultural field
    by, 431
APSIM model, 204, 213
  key functionalities of, 207–211
Aquaculture systems, 10
  ecopath modeling approach for,
    243–244
  effect on marine environment, 242
  mass-balance model for predicting
    impact on coastal ecosystem
    application of, 246–248
    Aranci Bay, 245–246
    introduction of finfish culture,
      248–251
AQUAMOD model, 27
Aquatic ecosystem
  acidification of, 12
  carrying capacity of, 51, 337
  conceptual diagram of nitrogen cycle
    in, 330
  eutrophication of, 335
Aquifers, 69
Arakawa–Schubert scheme, 535
Aranci Bay, 245–246
Arctic tundra wetlands, 258
ArcView software, 543
Arrhenius equation, 305–306, 330, 479, 480
Arrhenius relationship, 83, 142, 359, 365
Arrhenius's law, 479
Arrhenius temperature, 357
Arrhenius type expressions, 290
Artificial neural networks (ANNs), 7,
    140, 426, 481
ASRAM model, for analysis of
    acidification of river and
    streams, 140
ASTER model, 28
Asymmetric Convective Model (ACM), 529
ATEAM project, 461
Atmospheric carbon dioxide, 439
Atmospheric dispersion model, 543
Atmospheric emissions, 520–523

Atmospheric nitrogen, 90
Atmospheric pollutants, 514, 528
Atmospheric processes, for
    determination of air pollution,
    519–520
Atmospheric radiation cooling schemes,
    types of, 534
Atmospheric turbulence, theory of, 505,
    509
Autolysis, 366, 368
Autotrophic biomass, 286
  decay rates, 298
Autotrophic denitrification, 219
Autotrophic respiration, 357
Azores anticyclone, 352

B

Bacterial respiration, 350
BALATON model, for management of
    shallow lakes, 29
Basal metabolism, 83
Base cations (BC), 561
Bayesian belief network (BBN), 155, 156
Bayesian models, applications of, 167
Bayesian statistics, for parameter
    estimation, 169
Bed-shear coefficient, 97
BEEKWAM dynamic reservoir model,
    29
Beer–Lambert equation, 94, 101
Benchmark Simulation Model No. 1
    (BSM1), 310
Benthic invertebrates, 50
Benthic microalgae, 106
Benthic vegetation
  area-based biomass of, 361
  oxygen production by, 367
Betts–Miller scheme, 535
Biochemical oxygen demand (BOD),
    350, 355, 358
Biodegradable matter, 285, 286
Biodegradable organic materials, 348
Bioenergy production, 458
Biofilm reactors, 308
Biogenic emissions, 454
Biogeochemical models, 8, 424, 426
  mass-balance scheme, 20
  use of exergy calculations for, 45

Biological concentration factor (BCF), 396

Biological nitrogen, 280

Biological oxygen demand (BOD), 140, 147

Biological phosphorus (bio-P), 278

Biological wastewater treatment, 278

Biomanipulation, 34
  function of, 41
  of nutrient concentrations, 39
  recommendations for using models
    for managing, 40

Biomass, 38
  autotrophic, 286
  estimation of, 249
  exergy for maintenance of, 43
  of fishes, 39, 41
  heterotrophic, 282, 285, 286, 290
  ichthyic, 246
  of macroalgae, 100, 106
  of phytoplankton, 39, 41, 71, 74
  of seagrass, 106

BIOMATH protocol, 313
  SWOT analysis, 314–315

Blackadar scheme, 536

Blue mussels (*Mytilus edulis*), 402

Bomb calorimetry, 45

Bosmina, 46

Bottom mixed layer (BML), 92, 95

Boundary admixture fluxes, 508

Boundary Conditions (BCON)
    processors, 529

Boundary conditions, to air pollution
    modeling, 508–509

Boussinesq pressure, 354

Bream state of ecosystem, characteristics
    of, 35

Broa Reservoir, energy system language
    model of, 52

Brock model, 101

Bulk Formula scheme, 536

Burdine–Campbell parameters, 225

Burk–Thompson scheme, 536

C

Cadmium, contamination of plants by,
    404–408

Cambridge Environmental Research
    Consultants (CERC), 543
  Emissions Inventory, 546

Canopy structure, 168

Carbonaceous biochemical oxygen
    demand (CBOD), 381

Carbon balance, 214
  contribution of wetlands to, 262–263
  for crop growth, 215–216
  turnover of SOM, 216–217

Carbon cycling, 77, 99, 166, 186, 444, 446

Carbon-dependent mortality rate, 94

Carbon dioxide, 440
  diffusion in sea, 453

Carbon fixation, photosynthetic, 356

Carbon storage, 186, 192

Cardyn forest ecosystem models, 165

Carnivorous fish, 36, 41

Carolina Environmental Program
    (CEP), 540

Catchment acidification, 562

Catchment management, 465–467

Cation exchange, 564, 571–573

Cation exchange capacity (CEC), 572

CB4 mechanism, for determination of
    air pollution, 515–516

CCM2 radiation scheme, 534

Cellular automata approaches, for
    calculation of fire spread, 481

Cellular automata (CA) models, 177, 181

Cellular nitrogen/carbon ratio, 94

CENTURY model, for simulation
    of carbon, nitrogen, and
    phosphorus cycling, 188, 205

CE-QUALICM eutrophication model, 350

CE-QUAL-RIV1 model, for analysis of
    reservoir eutrophication, 28

CERES model, 204

CFD clarifier model, 301

Chapman–Richards model, 170

Chemical kinetics, for thermal
    degradation of fuels, 479

Chemical oxygen demand (COD), 140,
    280
  flux for particulate composite, 304
  fractionation in ASM1, 285–286

Chemical risk assessment, steps for,
    419–424

Chihuahuan Desert grasslands, 186

Chinese oyster (*Ostrea plicatula*), 70, 71

Chlorophyll-*a* (CHL), 360

Chlorophyll-*a*/carbon ratio, 360

Climate change, 444
  influence on wetlands, 263–264
Cloud dynamics and chemistry, for
      determination of air pollution,
      519
Coal burning, 559
Coastal ecosystems
  effect of finfish culture on, 243,
      248–251
  impact of aquaculture as predicted
      by mass-balance model
      application of, 246–248
      Aranci Bay, 245–246
      effects of introduction of finfish
          culture, 248–251
  sustainable management of, 66
Coastal lagoons, 10
Coastal zones, 10
  resources, sustainable use of, 67
COIRON model, 179
Coliform bacteria, 138, 148
COMMAS model, 25
Community Modeling and Analysis
      System (CMAS), 540
Community Multiscale Air Quality
      (CMAQ) System
  chemical transport model
      advection and diffusion, 529
      cloud processes, 531
      gas-phase chemistry, 530
      particle modeling and visibility,
          530–531
      photolysis rates, 531
      plume-in-grid modeling, 530
  output of, 531–532
  structure of, 528–529
Complex dynamic models, for quality
      management of lakes, 23
Constructed wetlands, 264–265
Continuous Emissions Monitoring
      (CEM) data, 542
Continuously stirred tank reactor
      (CSTR), 284
Conversion degree, of fuels, 479
  simplified curve of, 485
Copper contamination, impact of, 12, 398
Coral reef models, 10
  for analysis of response to bleaching
      and mass mortality, 159

  intrinsic properties of, 158
  overview of, 154–156
Core Inventory Air emissions
      (CORINAIR) programs, 522,
      559
CO_2 sequestration, 262
COST Action 728 project, 523
COUPmodel, 204, 213
  key functionalities of, 207–211
Crop growth model, 215–216
  for adjusting uptake of nitrogen,
      229–230
  for analyzing reduction of
      leaf area index (LAI), 229
      photosynthesis, 228–229
Crop rotation, and agricultural
      management, 230
Crown extinction coefficient, 168
Cyanobacteria
  life stages of, 51
  modeling of problem caused by, 54
Cyclophosphamide, 428

D

Daily net leaf photosynthesis, 167
DAISY model code
  overview of, 212–213
      carbon balance, 214–217
      heat balance, 214
      nitrogen balance, 217–220
      validation process, 220–221
      water balance, 213–214
  for simulating nitrate leaching from
      agricultural systems, 204, 206
  tutorial and reference manual, 230
Danida project, Tanzania, 258
Danish Operational Street Pollution
      Model (OSPM), 547
DANUIBA object-oriented generic
      model, 455
Daphnia, 46
Darcy's law, 213
Darwin's theory of natural selection, 43
DATAGRID program, 533, 538
Deforestration, ecological consequences
      of, 167
Degree heating weeks (DHW), 154
Degree of freedom, 299

DELWAG-BLOOM-SWITCH
         model, for management of
         eutrophication control of
         shallow lakes, 25
Denitrification, 147, 231, 330
Detritus
    mass balance for, 361
    sedimentation of, 369
Diazepam, 428
Dirac function, 509
Dissolved inorganic carbon (DIN), 70,
         74, 94
Dissolved oxygen (DO), 140, 357
    in fresh, brackish, and sea water at
         different temperatures and
         chlorinities, 144–145
    suppression grazing, 368
Double Sweep (DS) algorithm, 354
Doyle's approximation, 480
Drainage water, 261
Droop-type function, for calculation of
         cellular increase in carbon, 94
Dry deposition, 543, 560
Dry grasslands, 176
Dry scheme, 537
Dudhia longwave scheme, 534
Dynamic global vegetation models
         (DGVMs), 188–191
DYRESM—WQ dynamic reservoir
         model, 28

E

Eco-exergy, 42, 341, 411
    due to "fuel" value of organic matter,
         45
    Kullback measure of, 44
Ecolecon forest ecosystem models, 163
Ecological–economic models, for
         assessing economic advantages
         of increased transparency in
         water body, 18
Ecological law of thermodynamics
         (ELT), 169, 171
Ecological Modelling (2009), 21, 156, 263
ECOPATH model, for analysis of flows
         of energy and matter along
         estuarine food webs, 67,
         243–244

ECOSIM model, for analysis of flows
         of energy and matter along
         estuarine food webs, 67
Ecosystem eutrophication indexes, 85
Ecosystem mass-balance models, 243
Ecosystem models, 9–11
    application of exergy or eco-exergy
         in, 42
    groups of, 5–6
    self-shading effect of, 46
Ecosystem recovery, process for, 564–565
Ecotone IBM, for examining shifts in
         species dominance between
         grasses and shrubs, 179–180
Ecotoxicological lake models, for
         environmental risk assessment
         of chemicals, 7–8, 29–31
Ecotoxicological models
    characteristics and structure of,
         424–427
    classes of, 426
    development of, 426
    difference from other ecological
         models, 425
    of distribution and effects of heavy
         metals, 393
Ecotrophic efficiency (EE), 246
Eddy diffusion, 93, 529
Eelgrass (*Zostera marina*), 118, 123, 344
Effluent suspended solids (ESS), 299
Empirical Diatom Model, 566
Energy consumption, 445
Energy sources, for streams and rivers,
         137
Energy system language model, of Broa
         Reservoir, 52
ENSEMBLE model, for grasslands, 193
Entropy, 42
Environmental concentration (EEC), 417
Environmental impact assessment
         (EIA), 417
Environmental management
    structural dynamic models (SDMs)
         for, 34
    of toxic substances, 29
    use of models in, 21
Environmental modeling
    for analysis of resource use options,
         462–464

concept of applications in numerical optimum control procedures, 463
conceptual, physical, and mathematical models, 459–460
integrated models for ecosystem service quantification and, 461
model definition, 460
for scenario-based analysis of environmental impacts, 461–462
space–time illustration for, 460
tools and applications for, 464–465
Environmental problems, models of, 11–12
Environmental risk assessments (ERA), 416
application of, 417
ecotoxicological models for, 30
EPIC model, 204
Equilibrium criterion models (EQC), 426
Erie Lake, 2D eutrophication model of, 50
Erken Lake, nitrogen fixation in, 52
Estuaries, 10
environmental problems, 66
models developed between 2000 and 2010, 78–82
models of primary producers in macroalgae, 95–108
microphytobenthos, 108–113
overview of, 77–84
phytoplankton, 84–95
seagrasses, 113–121
sustainable management of, 65
Estuarine-ecosystem models
developed between 2000 and 2010, 78–82
review of, 77
Estuarine phytoplankton, 84
Estuarine–salt marsh ecosystem, 381
Estuarine-system coastal model, Xiangshan Gang (East China sea) case study, 67–76
Ethene, 515, 516
Eulerian approach, to air pollution modeling, 503–505
comparison with Lagrangian approach, 507–508

Eulerian fluid velocities, 507
Eulerian grid, 94
European Inventory of Existing Commercial Chemical Substances, 417
European Monitoring and Evaluation Programme (EMEP), 559, 560
Eutrophication models, 329–341
application of SDMs to improve results of, 341–344
of aquatic ecosystem, 335
conceptual diagram for, 360
future development of, 344–345
main parameters for, 362–363
for management of multispecies fishery, 51
overview of, 329
for oxygen consumption, 367–369
for oxygen processes in water column, 366–367
for oxygen production, 367
for oxygen reaeration, 369–370
Eutrophic ecosystems, 348
Evapotranspiration, 69, 213, 222, 231, 537, 568, 579
Exergy
advantages of, 42
for biogeochemical modeling, 45
for maintenance of biomass, 43
parameter combination giving highest, 48
and stability by different combinations of parameters and conditions, 49
Exposure model, for oxytetracycline on grassland, 429
Extracellular solubilization, 307
EXWAT model, for analysis of chlorophenoles and tetraethyl lead, 140

F

FAO–Penman–Monteith method, 222
FATE model, for simulating groups of individuals in similar life stages, 180
Faviids species, 156, 157
Fermentation, 295, 296, 297, 303–305

Fick's law of diffusion, 103
Fifth-generation PSU/NCAR Mesoscale
        Model (MM5)
    data required to run, 540
    four-dimensional data assimilation
        (FDDA) scheme, 539–540
    meteorology model, 532–533
        atmospheric radiation, 534
        convective parameterization,
            534–535
        lateral boundary conditions, 533
        model physics, 534
        PBL processes, 536
        prognostic equations, 533
        resolvable-scale microphysics
            schemes, 537
        surface layer processes, 536–537
    meteorology model preprocessing,
        537–539
    nesting, 539
Finfish aquaculture, 242
    effect on coastal ecosystems, 243,
        248–251
Finnish three dimensional water
        quality-transport model, 28
Finn's cycling index (FCI), 251
Fire brands, 489–491
Fire spread *see* Flame propagation
Fire spread modeling
    simulation model and mathematical
        analogs for, 481–482
        fire brands, 489–491
        geometry of fire spread, 482–483
        heat influence, 485–489
        stochastic wind effects, 483–484
    simulation results for, 491–492
First-order Acidity Balance (FAB)
        model, 566
First-order kinetic law, 357
Fish farms, 243, 250
Fish population, effect of pH of water
        on, 33
Flame propagation
    algorithm for computation of, 489
    temperature changes in, 491
Flocculation zone, 298
FLOWSTAR advanced airflow model, 547
Fluid hydrodynamics, Navier–Stokes
        equations for, 300

Foliage distribution, 169
Food and Agriculture Organization
        (FAO), 221
Food chain, effect of pH of water on, 33
Forest BGC forest ecosystem models, 166
Forest ecosystem models, 10
    ANAFORE (Analysis of Forest
            Ecosystems), 162
    Cardyn, 165
    Ecolecon, 165
    Forest BGC, 166
    Formix, 163
    GLOBTREE, 166
    GROMAP, 163
    issues covered by, 161–162
    list of models representing, 164–165
    overview of, 163–167
    PnEt, 163
    and remote sensing, 162
    TEMFES, 166
    Treedyn 3, 163
    TRIPLEX, 166
Forests
    effects of acidic deposition on, 563
    watersheds, 562
Formaldehyde, 515, 516, 541
Formix forest ecosystem models, 163
Fossil fuels, 445, 452, 559
    combustion of, 263, 517
Four-dimensional data assimilation
        (FDDA) scheme, 539–540
Free energy, 42
    of ecosystem, 43
Freshwater lake ice, model for analyzing
        climate change scenarios for,
        52
Freshwater runoff, 84
Freundlich's adsorption isotherm, 394
Fritsch–Chappel scheme, 535
Fucus vesiculosus, 98
Fuels
    hazards, 478–479
    simplified temperature curve of, 490
    thermal degradation of, 478–481
Fugacity models, 426, 428
    of distribution of hormones in
            environment, 429
Fuzzy logic, 186
Fuzzy models, 7

G

Gaines–Thomas equations, 572, 573, 577, 581
Gamma ray dose, 545
Gapon exchange equations, 572, 577
Gas-phase chemistry, for determination of air pollution, 514–515, 530
  CB4 mechanism, 515–516
  RADM2 mechanism, 516
  SAPRC-97 mechanism, 516
Gaussian diffusion, 549
Gaussian distribution, 509, 510
Gaussian turbulence, 512–513
Gauss theorem, 504
Generic Ecological Model (GEM), 77
Geographical information system (GIS), 7, 96, 162
  for quality management of lakes, 23–24
GFMOLP (fuzzy multi-objective program), for optimal planning of reservoir, 25
Gilthead sea bream (*Sparus auratus*), 246
GIRL OLGA model, for cost minimization of eutrophication abatement, 25
Global Average Models (GAM), 439, 440, 444
  STELLA software for presentation of, 447–454
Global Climate Model (GCM) grid cell, 181
Global warming, 12
  consequences for nature, 454–455
  five consequences of, 446
  overview of models for, 440–446
Global warming potential (GWP), 262
GLOBTREE forest ecosystem models, 166
Glumsø eutrophication model, 331, 336
  levels of development for, 340
  revision by applications for other case studies, 340
Glumsø Lake, 27, 331, 338
  SDM approach to modeling of, 343
  validation of prognosis for, 339
Glycogen, 297

Goal function
  for development of SDMs, 39, 41
  in ecological models, 39
  for minimizing or maximizing critical water quality variables, 24
Gracilaria verrucosa (seaweed species), 98
Grassland simulation models, 10
  future challenges, 192–194
  overview of, 176–177
    biogeochemical models, 186–188
    demographic models, 177–182
    dynamic global vegetation models (DGVMs), 188–191
    physical models, 184–186
    physiological models, 182–184
  recommendations for selection of, 191
  specific models and their key characteristics, 178
Gravitational settling method, for treatment of wastewater, 298–299
Grazing mortality, 84
Greenhouse effect, 114, 262, 440, 444, 454
Greenhouse gas (GHG), 12, 114, 295, 441, 442, 453
Grell scheme, 535
GROMAP forest ecosystem models, 163
Gross productivity (GP), definition of, 77
Groundwater leaching, of pesticides, 430
GUEST model, 204
Gujer matrix, 280
GUMBO model, 461
Gustafsson's model, for exchange between soil water and soil for metals, 404

H

Habitat suitability values, 471
HarmoniQua project, 232
Heat balance, 214
Heat influence
  as function of distance, 487
  rate, 488

Heat influence zone, of burning sites, 483
Heat of combustion, 486
Heavy metals, 261, 502
   air pollution caused by, 501
   biological concentration factor, 396
   conceptual diagram depicting influence of copper on ecosystem, 409
   contamination of plants by cadmium, 404–408
   ecotoxicological models of distribution and effects of, 393–397
   overview of models for, 397–404
   STELLA diagram for model of, 405
   structurally dynamic model of, 408–411
   see also Toxic pollutants
Henry's law, 143
Heterotrophic biomass, 282, 285, 286, 290
Heterotrophic respiration, 357–359, 367
High-performance computing (HPC) sparse-matrix algorithms, 540
Hooghoudt's equation, 214
HSG protocol, 313
   SWOT analysis, 314–315
Human–environment relationships, 66
Human health, impact of chemicals on, 419
Huygens' wavelet principle, in fire spread modeling, 481
HYBRID model, for simulating physiological and biogeochemical relationships, 181
Hydraulic retention time (HRT), 285
Hydrobia ulvae, 112, 113
Hydrodynamic-ecological models, for water quality management of lakes and reservoirs, 26
Hydrogenotrophic methanogenesis, 304
Hydrogen-utilizing methanogenic organisms, 307
Hydrolysis, 303
Hydrostatic approximation theory, 519
Hydrostatic pressure, 354
Hydroxyl radical (HO), 516, 517

Hydynamic–ecological models, 28–29
Hypoxia, 122, 348, 349, 370, 382–383

I

Ichthyic biomass, 246
ICP Mapping Manual, 567
Individual-based models (IBMs), 7, 177, 427
Informal priority setting (IPS), 417
Initial Conditions (ICON) processors, 529
InitSAFE model, 579
Inorganic nitrogen, 94, 118, 121, 307, 361, 369
Inorganic phosphorus (IP), 336, 361, 364, 379, 382
Integrated Generic Bay Ecosystem Model (IGBEM), 77
Integrated Process Rates (IPRs), 532
Integrated Reaction Rates (IRRs), 532
Intergovernmental Panel on Climate Change (IPCC), 442
   Special Report on Emissions Scenarios, 560
International Lake Environment Committee (ILEC), 33
International Society for Ecological Modelling (ISEM), 21
International Water Association (IWA), 309
INTERP program, 533, 538–539
Isoprene, 515, 516, 530
Ivermectin, 428
IWA Task Group
   on Benchmarking of Control Strategies for WWTPs, 311
   for Good Modelling Practice, 310, 316

K

Kain–Fritsch scheme, 535
Kissinger–Akahira–Sunose (KAS) method, for obtaining activation energy, 480–481
Krone's equation, for calculating rates of deposition of sediments, 97
Kullback measure of eco-exergy, 44

L

Lagrangian approach, to air pollution
modeling, 505–507
comparison with Eulerian approach,
507–508
Lagrangian particle tracking model, 92,
505
Lake acidification
biological models of, 33
causes of, 32
models for analyzing effects of acid
rain on soil chemistry, 32–33
Lake ecosystem, 40, 41, 51
Lake fisheries models, 33–34
Lake morphometry, 26
LAKE ONTARIO model, 28
Lake quality management models, 10
communication between decision-
maker and modeler, 21
construction of, 19
data collection program, 20
ecotoxicological, 29–31
factors influencing use of, 23
integration of hydrodynamics and
ecology, 26–29
latest development in, 50–55
as management tools, 17–26
mass-balance scheme, 20
multigoal optimization, 25
multivariable optimization, 25
Lakes and reservoirs
environmental management
problems of, 15–16
hydrodynamic–ecological models
for water quality management
of, 26
latest development in modeling of,
50–55
management models used in water
quality management of, 22
Lambert–Beer equation, 94, 101
LAMOS model, for simulating effects of
grazing frequency, 180
Land resource allocations, 458
Land use distribution patterns,
comparison of, 466
Land use management, 458
catchment, 465–467

ecosystem services and trade-off
analysis for, 467–469
optimal configuration of species
habitats, 469–471
Pareto-frontiers for optimum
land use configurations,
471–473
Land use processor (LUPROC), 529
Langevin equation, 549
Langmuir's adsorption isotherm, 394,
574, 581
Leaching, 69, 165, 204, 206, 219–220, 226,
561, 571
Leaf area distribution, 215, 216
Leaf area index (LAI), 165, 166, 213, 215,
223
Leaf carbon, 165
Light extinction coefficient, 103
Lignocellulosic fuels, thermal
degradation of, 478–479
Littoral zones, for fishery, 51
Livestock grazing, effects on grassland
production, 184
LUPOlib (Land Use Pattern
Optimisation library) software
package, 464
Lysina catchment
MAGIC model application to, 588
SAFE model application to, 589

M

Macroalgae, 350
biomass dynamics and productivity,
96, 106
conceptual diagram of spore and
adult macroalgae productivity
model, 100
models for
life stages and spatial
discrimination, 99–103
mechanistic approach, 103–108
turbidity at intertidal area, 96–99
types of, 95–96
Macroalgal mats, 112
MACRO model, 204, 213
key functionalities of, 207–211
Macroscale atmospheric processes,
519

MAGIC (Model of Acidification of Groundwater In Catchments) model, 573, 574, 580–582, 584, 588
MAGIC7 model, 582
Major Elevated Point Source Emitters (MEPSEs), 530
MAKEDEP model, 573, 578
Makkink equation, 222
Manila clam (*Tapes philippinarum*), 70
Manning roughness coefficient, 98
MapInfo Geographical Information System (GIS) software, 543
Marine aquaculture, 241
Marine ecosystems, 10, 242
Marine phytoplankton, 84
Mass-balance model, for predicting impact of aquaculture on coastal ecosystem
  application of
    first mass-balance scenario, 246–247
    second mass-balance scenario, 247–248
    Aranci Bay, 245–246
  effects of introduction of finfish culture, 248–251
Mass conservation relation, for air pollution modeling, 503
Mass transfer limit, 105
Maximum permissible level (MPL), 421
Mean habitat suitability indices, of focal species, 470, 471
Medium Range Forecast (MRF) model scheme, 536
Mellor–Yamada formulas, 536
MELODIA model, 28
Mercury model of Mex Bay, Egypt, 403
MERLIN model, 582
Mesoscale atmospheric processes, 520
Mesotidal system, 96
Metal sorption in soils, model of, 403
Meteorological Models of the Green House Effect (MGHE), 439, 440, 442
  distribution of global temperature changes predicted by, 445
  schematic structure of, 444
Meteorology–Chemistry Interface Processor (MCIP), 529

Methane
  atmospheric oxidation of, 262
  hydrogenotrophic methanogenesis, 304
  production in wetlands, 262
Methanogenesis, 304–307
Methyl mercury, 398
Michaelis–Menten equation, 332, 337, 338, 394
Michaelis–Menten functions, 71, 104, 142, 147
Michaelis–Menten kinetics, 219, 573
Microbial maintenance respiration, 216
Microphytobenthos
  activity respiration, 83
  models for
    conceptual diagram of, 110
    ERSEM, 110, 111
    Hydrobia ulvae, microphytobenthos, and mudflat's erosion, 112–113
    productivity, 109–111
  nutrient limitation factor, 111
  production rate, 108
Microscale atmospheric processes, 519–520
Mike3 flow model, hydrostatic version of, 354
MIKE model, 28
MINLAKE model, for estimation of effect of climate changes on U.S. lakes, 28
Mixed-Phase scheme, 537
Mixed-Phase with Graupel scheme, 537
Molecular diffusion, 105, 505
Mondego estuary, 102
Monin–Obukhov length, 543
Monod growth equation, 106
Monod kinetics, 305
Monte Carlo simulations, 465
Mortality, types of, 84
Mualem–van Genuchten parameters, 225
Muddy clam (*T. granosa*), 70, 71
Multielement geochemical model, 382
Multi species fishery models, for management of fishery, 54
Municipal wastewater treatment systems, 291

Mussels, 54
Mytilus edulis (blue mussels), 402

N

NASA/Goddard Microphysics scheme, 537
Nash–Sutcliffe efficiency index, 69–70
National Acid Precipitation Assessment Program (NAPAP), North America, 559
National Centers for Environmental Prediction's (NCEP) MRF model, 536
National Deposition Monitoring Program (NADP), United States, 560
Natural capital, 64
Natural systems, sustainable management of, 64
Natural wetlands, 265–266
Navier–Stokes equations, 354
    for fluid hydrodynamics, 300
NCSOIL model, 204
Net primary production, definition of, 77, 85
Newtonian relaxation, 539
Nitrate nitrogen, 219–220
Nitrification, 219, 350
    of ammonium, 142
Nitrite oxidizing bacteria (NOB), 291
Nitrogen balance, 217
    for ammonium nitrogen, 219
    for nitrate nitrogen, 219–220
    for organic nitrogen, 218
    for plant organic nitrogen, 218
    for soil organic nitrogen, 219
Nitrogen compounds, emissions of, 559
Nitrogen cycling, 166
    conceptual diagram of, 330
    in marine systems, 86
    prognostic variables for, 89
Nitrogen fixation, 330
    first-order reaction, 54
    in Lake Erken, 52
Nitrogen-fixing algae, 327
Nitrogen fractionation, in ASM1, 286–287
Nitrogen immobilization, 573–574

Nitrogen in soils, accumulation of, 562
Nitrogen isotopes, phytoplankton model, 86–90
Nitrogen oxides, 32
Nitrous gas emissions from plants, 295–296
NLEAP model, 204
Non-adverse effect level (NAEL), 424
Nonbiodegradable organic matter, 285
Non-effect concentration (NEC), 419
Non-observed adverse effect level (NOAEL), 424
Nonpredatory mortality, 84
Nonreactive carbon atoms, 515
Normalized Difference Vegetation Index (NDVI), 184
Nuclear energy, 445
Nudging *see* Newtonian relaxation
Nutrient cycling, 182, 186, 192, 327, 349, 360, 579
Nutrient-dependent growth limitation, 365, 367
Nutrient levels, effect on lake or reservoir eutrophication, 35
Nutrient stress excretion, 84
Nutrient stress lysis, 84
Nutrient uptake, 103, 105, 111, 465, 563, 579

O

Ohio River Oil Spill Model, for analyzing effect of oil spills, 140
Oil-fired power plants, 521
Olaquindox, 430
Optimization models, for quality management of lakes, 24
Ordinary differential equations (ODEs), 281, 284, 285
Organic acids, 307, 564, 578, 580, 582, 584, 585
Organic carbon, 191, 216, 218, 219, 226, 278, 280, 517, 581
Organic fertilizers, 214, 216, 218, 230
Organic matter, 251
    biochemical degradation of, 357
    inert soluble, 286
    loadings, 349

Organic matter (*continued*)
model applied to assess the amount
of BOD$_5$ removed by wetland,
262
pools and subpools of, 217
removal efficiencies of, 264
Organic nitrogen (ORN), 86, 218, 219,
259, 264, 266, 274, 286, 356, 382
Organic pollutants, 12, 501
Organic wastes, pollution caused by, 348
Osnabruck biosphere model, 170
Oxygen balance, 139, 349, 366, 367, 369
Oxygen depletion, in streams and
rivers, 139, 143
Oxygen reaeration, 369–370
Oxytetracycline, 429, 431
Ozone photochemistry, 530

P

PAMOLARE model, 343
Pamolare 2 model, for restoration and
management of wetlands, 260
Pamolare software package, 49, 53
Pareto curve, 474
Pareto-frontiers, for optimum land use
configurations, 471–473
Partial differential equation (PDE), 300
Particulate matter (PM), 501
Particulate organic matter (POM), 72,
139, 146, 265
Particulate organic nitrogen (PON),
265–266
Particulate organic phosphorus (POP), 266
Patuxent catchment, 474
PCLAKE model, for management of
shallow lakes, 29
Péclet number, 93, 113
Pelagic ecosystem, 111
Pesticide model, for assessing removal
of pesticides by constructed or
recovered wetland, 263
Pesticides, 263
groundwater leaching of, 430
PESTLA model, 204
Petersen matrix, 280
Pharmaceutical models
conceptual diagram of, 433
of environment, 431

illustrative example of, 431–432
equations, 432–437
overview of, 427–431
Pharmaceuticals, decomposition of, 427
pH-effect model, for analyzing effects of
pH of water on food chain, 33
Phosphorus cycle, 188, 331, 381, 444, 447,
452
Phosphorus sediment, profile for Lake
Esrom in Denmark, 337
Photolysis processor (JPROC), 529
Photon flux density (PFD), 101
Photo-oxidants, 501
Photosynthesis, 101, 139, 215, 330
daily net leaf, 167
Photosynthetically available radiation
(PAR), 103
Photosynthetic carbon fixation, 356
Phytoplankton, 36, 40
biomass, 39, 41, 71, 74
carbon loading, 348
composition of, 326
denitrification rate, 87
DIN (ammonium + nitrate)
consumption by, 90
functional groups of, 51
growth plotted *vs.* light intensity, 334
growth rate, 47
isotopic fractioning coefficient, 88
marine and estuarine, 84–85
models for
conceptual diagram of, 87
deep channels and broad shallow,
90–91
growth, 93–95
motility, 91–93
nitrogen isotopes, 86–90
natural mortality of, 366
nitrification rates, 87
$^{15}N/^{14}N$ signal of, 86
photosynthesis and respiration, 350,
356
P–I (production–irradiance) curve for
determining production of, 93
as primary producers in estuaries,
84–95
schematic layout of computational grid
of hydrodynamic model of, 88
two-step model for growth of, 333

Phytoplankton carbon (PC), 361
Phytoplankton nitrogen (PN), 361
Phytoplankton phosphorus (PP), 361
Phytoremediation, 404
P–I (production–irradiance) curve,
    for determination of
    phytoplankton production, 93
Piecewise parabolic method (PPM), 529
Pike state of ecosystem, characteristics
    of, 35
Piscivorous fishes, 50
Planetary boundary layer (PBL), 529
Planktivorous fish, 35–37, 41
Planktothrix rubescens, 51
Plant functional types (PFTs), 182
Plant nutrient cations, 561
Plant organic nitrogen, 218
Plant–soil water dynamics, 182
Plant-wide model, for wastewater
    treatment, 279, 309–310
Pleim–Xiu land–surface scheme, 536, 537
Plug-flow reactors, 308
Plume-in-Grid (PinG) modeling, 530
Plume models, for air pollution
    modeling, 512–513
PnEt forest ecosystem models, 163
Point source model, for flame
    propagation, 485
  geometry of radiation in, 486
Point sources of pollution, control of, 20
Pollination, 474
Pollution problems, models of, 6
Polyhydroxy alkanoates, 297
Polyphosphate accumulating organisms
    (PAOs)
  in bio-P models, 288
  substrate flows for storage and
    growth in
    ASM2 model, 296
    TUDP model, 297
Population dynamic models, 3, 7, 8, 270,
    393, 398, 404, 426
Porites species, 156, 157
Predicted environmental concentration
    (PEC), 421
Predicted non-effect level (PNEL), 421
PreSAFE package, 579
Prescriptive models, for calculating lake
    water quality, 24

Princeton Ocean Model (POM), 111
Probability density function, 506
PROFILE model, 567, 580, 585, 586
Progesterone, 428
PROTECH-C model, 29
PRZM model, 204
Puff models, for air pollution modeling,
    510–512

Q

Q^2 learning (Qualitatively faithful
    Quantitative prediction),
    for interpretation of lake
    eutrophication data, 52
Quantitative structure activity research
    (QSAR), 30
Quasi-Steady State Approximation
    (QSSA), 530

R

Radioactive species, decay of, 545
RADM2 mechanism, for determination
    of air pollution, 516
Rainwater, 264
Random walk equation, 93
RAWINS program, 538
Razor clam (*Sinonovacula constricta*), 70,
    71
Redox reactions, 291
  in nitrification and denitrification
    process, 292
Regional Oxidant Model, 530
Remote sensing, 98, 114, 157, 162
Renewable energy, 445, 452, 454
Resolvable-Scale Microphysics schemes,
    537
Respiration, 216
RESQUAL model, 28
Ria de Aveiro lagoon
  anthropogenic nutrients in, 353
  CHL-*a* distributions in, 378
  description of model for, 354–355
    eutrophication model, 360–370
    water quality model, 355–360
  dissolved oxygen (DO) distribution,
    350, 370, 373, 376
  eutrophication processes in, 351

Ria de Aveiro lagoon (*continued*)
  hydrological characterization of, 352
  maps of currents field in, 371
  as mesotrophic shallow estuarine
      system, 353
  organic contamination, 351
  oxygen conditions in, 350
  salinity values of, 352–353
  study area, 351–354
  time series of water level for, 372
Richards' equation, 213, 214
RISK-N model, 204
River and stream models, 10
  equations of, 150–151
  of medium complexity, 140–151
  overview of, 138–140
  STELLA software, use of, 140
    conceptual diagram of river
        model, 147, 148
Riverine nutrient loading, 349
Root density distribution, 216
Root to shoot ratios, 179
ROSETTA computer program, for
      estimating soil hydraulic
      parameters, 225
Runge–Kutta method, 361, 543
RZWQM model, 204, 213, 230
  key functionalities of, 207–211

S

SAFE model system, 573, 578–580, 584,
      586, 589
SALMO model, 28
SALMOSED model, 28
Salt marshes, 113
  effects of sea-level rise in, 114–117
  intertidal, 381
  roles in stabilizing sediments, 123
SAPRC-97 mechanism, for determination
      of air pollution, 516
SAVANNA model, for studying effects
      of climate change on grazing
      system, 182–184
Scenedesmus (green algae), 339
SDM-PAMOLARE model, 343
Sea bass (*Dicentrarchus labrax*), 246
Seagrasses
  biomass of, 106

characteristics of, 113–114
  effect of light on productivity of, 120
  models for
    conceptual diagram of, 119
    effects of sea-level rise in salt-
        marsh areas, 114–117
    Z. noltii and nitrogen cycle,
        117–121
  mortality processes of, 121
  photosynthesis of, 119–120
  role in
    carbon fixation, 117
    nutrient budget of coastal marine
        ecosystems, 114
Sea lettuce (*Ulva lactuca*), 344
Sea-surface temperature, 538
Secchi disk depth, 361
Sediment oxygen demand (SOD), 138,
      350, 361
  Arrhenius dependence function for,
      369
  and carbon mineralization, 368
  dissolved oxygen (DO) distributions
      for, 379
  for various sediment types, 146
Selected Nomenclature for Air Pollution
      (SNAP), 522, 523
Semiarid grasslands, 179
Sequencing batch reactor (SBR), 297
Settler models, for wastewater
      treatment, 299–303
Sewage treatment plant, 428
Shallow convection scheme, 535
Shellfish
  aquaculture, 70, 242
  factors influencing growth of, 71
  model for simulating population
      growth of, 71, 74
  seeding and harvesting strategies, 71
  standard simulation outputs and
      data for harvesting of, 76
Shellfish individual growth model
      (ShellSIM), 70
Shortwave radiation scheme, 534
SILAM Air Pollution Model, 548–549
  advection–diffusion scheme of, 549
  Lagrangian advection algorithm
      with Monte Carlo random
      walk, 549–550

Silver carp (*Hypophthalmichthys molitrix*), 344
Simple Ice scheme, 537
Simple static calculation models, for quality management of lakes, 23
Simulation Model for Acidification's Regional Trends (SMART) model, 571
Single fluid particle, meaning of, 505–506
Sludge blanket height (SBH), 299
Sludge production, 281, 299
Sludge retention time (SRT), 291
SMART model, for predicting solution chemistry resulting from acidification processes, 577–578
SMOKE Modeling System, 540–541
  features of SMOKE version 2.0, 541–542
Søbygaard Lake
  mode equations for, 47
  phosphorus nutrient load of, 46
Soil Acidification in Forest Ecosystems (SAFE) model, 567
Soil adsorption, 561, 584
Soil and Water Assessment Tool (SWAT) model, 68
Soil carbon, 217
Soil hydrology, 227
Soil Map of the World, 32
Soil microbial biomass (SMB), 216
Soil–moisture retention curves
  closed-form hydraulic conductivity functions based on, 225
  mathematical forms describing, 225
SOIL-N model, 204
Soil nutrient, 69
Soil organic matter (SOM), 188, 216–217, 223, 226
Soil–plant–atmosphere continuum (SPAC), 205, 206
  schematic representation of, 212
Soils
  acidification of, 561, 575, 586
  classification of, 32
  hydraulic conductivity of, 223
  organic nitrogen, 219
  pH and buffer capacity, 32

Soil system, definition of, 223–228
Soil water, 32, 69, 163, 177, 179, 213
  characteristic of, 223
  Gustafsson's model, 404
  heat capacity and thermal conductivity, 214
  heavy metals and, 405
  model of metal sorption and, 403
  SOILWAT model of, 184
  toxic substances dissolved in, 404, 432, 436
  and water balance, 213–214
SOILWAT model, of soil water dynamics, 179, 184
Solid phase digesters, 308
Solid-state reactions, 479
Solid waste pollution, 11
Sparse Matrix Vectorized Gear (SMVGEAR) algorithm, 530
Spatial models, for understanding ecosystem reactions, 7
Spring-neap tidal cycles, 349
Stable precipitation scheme, 537
Steady-state mass balance model (SSMB), 567
Steady-state models, 426, 565–567
Steady-State Water Chemistry (SSWC) model, 566
Steele's equation, for productivity with photoinhibition, 71
Stefan–Boltzmann equation, 441
STELLA diagram
  of cadmium model, 405
  of model developed for environmental risk assessment of growth promoters, 430
  model equations (STELLA format), 406
STELLA software
  application for modeling rivers and streams, 10, 140
  components of conceptual diagram used in, 141, 147, 148
  Global Average Model presented by use of, 447–453
STEP model, 184
STEPPE model, 179
Stochastic models, 7
Stoichiometric nutrient recycling theory, 54

Stoke's law, 97
STOWA protocol, 313
   SWOT analysis, 314–315
Street canyons, 546
Streeter–Phelps model, 138–139, 141
Strong acid anions (SAA), 563
Structurally dynamic models (SDMs), 7,
   17, 397, 427, 454
   for analysis of ecosystem properties
      of lakes, 34–49
   application to improve results of
      eutrophication models, 341–344
   for development of better
      eutrophication models, 327
   goal function for development of,
      39, 41
   of heavy metals, 408–411
   procedure applied for development
      of, 342
   for restoration of lake by
      biomanipulation, 53
Structure activity research (SAR)
      methods, for estimating
      pollution caused by toxic
      substances, 30
Substrate-based Monod-type kinetics, 307
Subsurface chlorophyll, 93
Subsurface water runoff, 69
Subsurface wetlands, 258, 264, 269
Sulfate adsorption, 564, 574
Sulfur dioxide, 517, 521, 545, 560
Sulfur emissions, 559
Sulfuric acid (H_2SO_4), 560
Sulfur in soils, accumulation of, 561–562
Sulfur oxides, 32
SUNDIAL model, 204
Supernatant–sludge interface, 301
Surface radiation schemes, 534
Surface water
   acidification, 563–564, 575
   concentrations, 574
   effects of acidic deposition on,
      563–564
   runoff, 69
Surface wetlands, 258
Surfer contour-plotting package, 543
Suspended particulate matter (SPM),
   71, 72
Symbiodinium spp., 153

T

Taihu Lake, 52
Tanzanian case study, for design of
      wetland, 274–275
Task Group for Good Modelling
      Practice (GMP), 309
Task Group on Benchmarking of
      Control Strategies for WWTPs,
      309
Task Group on Design and Operations
      Uncertainty (DOUTGroup), 309
Technical University Delft Phosphorus
      (TUDP) model, 297
TEMFES forest ecosystem models, 166
TERRAIN program, 538, 539, 540
Testosterone, 428
Thau lagoon system, 117
Thermal degradation, of fuels, 478–481
Thermal infrared radiation, 440
Thermal radiation flux, 485
Thermodynamic equilibrium, of
      ecosystem, 42
   Darwin's theory of natural selection
      and, 43
Thermodynamic forest growth model,
      169–171
Tidal cycle, 96, 98
Tidal flushing, 382
Tolerable daily intake (TDI), 424
Toluene, 515
TOPMODEL model, for simulation of
      soil moisture dynamics, 186
Total heat released rate, 485
Total phosphorus (TP), 266
   concentration, 36–37
Toxic organic compounds, 404
Toxic pollutants, 502
Toxic substance models
   characteristic properties of, 30
   difference from other ecological
      models, 424–425
Toxic substances
   models for environmental
      management of, 29–31
   organic, 12
   pollution caused by, 11
Traffic-induced turbulence, 546
Transition probability density, 507

Stoke's law, 97
STOWA protocol, 313
    SWOT analysis, 314–315
Street canyons, 546
Streeter–Phelps model, 138–139, 141
Strong acid anions (SAA), 563
Structurally dynamic models (SDMs), 7,
        17, 397, 427, 454
    for analysis of ecosystem properties
        of lakes, 34–49
    application to improve results of
        eutrophication models, 341–344
    for development of better
        eutrophication models, 327
    goal function for development of,
        39, 41
    of heavy metals, 408–411
    procedure applied for development
        of, 342
    for restoration of lake by
        biomanipulation, 53
Structure activity research (SAR)
        methods, for estimating
        pollution caused by toxic
        substances, 30
Substrate-based Monod-type kinetics, 307
Subsurface chlorophyll, 93
Subsurface water runoff, 69
Subsurface wetlands, 258, 264, 269
Sulfate adsorption, 564, 574
Sulfur dioxide, 517, 521, 545, 560
Sulfur emissions, 559
Sulfuric acid (H_2SO_4), 560
Sulfur in soils, accumulation of, 561–562
Sulfur oxides, 32
SUNDIAL model, 204
Supernatant–sludge interface, 301
Surface radiation schemes, 534
Surface water
    acidification, 563–564, 575
    concentrations, 574
    effects of acidic deposition on,
        563–564
    runoff, 69
Surface wetlands, 258
Surfer contour-plotting package, 543
Suspended particulate matter (SPM),
        71, 72
Symbiodinium spp., 153

T

Taihu Lake, 52
Tanzanian case study, for design of
        wetland, 274–275
Task Group for Good Modelling
        Practice (GMP), 309
Task Group on Benchmarking of
        Control Strategies for WWTPs,
        309
Task Group on Design and Operations
        Uncertainty (DOUTGroup), 309
Technical University Delft Phosphorus
        (TUDP) model, 297
TEMFES forest ecosystem models, 166
TERRAIN program, 538, 539, 540
Testosterone, 428
Thau lagoon system, 117
Thermal degradation, of fuels, 478–481
Thermal infrared radiation, 440
Thermal radiation flux, 485
Thermodynamic equilibrium, of
        ecosystem, 42
    Darwin's theory of natural selection
        and, 43
Thermodynamic forest growth model,
        169–171
Tidal cycle, 96, 98
Tidal flushing, 382
Tolerable daily intake (TDI), 424
Toluene, 515
TOPMODEL model, for simulation of
        soil moisture dynamics, 186
Total heat released rate, 485
Total phosphorus (TP), 266
    concentration, 36–37
Toxic organic compounds, 404
Toxic pollutants, 502
Toxic substance models
    characteristic properties of, 30
    difference from other ecological
        models, 424–425
Toxic substances
    models for environmental
        management of, 29–31
    organic, 12
    pollution caused by, 11
Traffic-induced turbulence, 546
Transition probability density, 507

Silver carp (*Hypophthalmichthys molitrix*), 344
Simple Ice scheme, 537
Simple static calculation models, for quality management of lakes, 23
Simulation Model for Acidification's Regional Trends (SMART) model, 571
Single fluid particle, meaning of, 505–506
Sludge blanket height (SBH), 299
Sludge production, 281, 299
Sludge retention time (SRT), 291
SMART model, for predicting solution chemistry resulting from acidification processes, 577–578
SMOKE Modeling System, 540–541
features of SMOKE version 2.0, 541–542
Søbygaard Lake
mode equations for, 47
phosphorus nutrient load of, 46
Soil Acidification in Forest Ecosystems (SAFE) model, 567
Soil adsorption, 561, 584
Soil and Water Assessment Tool (SWAT) model, 68
Soil carbon, 217
Soil hydrology, 227
Soil Map of the World, 32
Soil microbial biomass (SMB), 216
Soil–moisture retention curves
closed-form hydraulic conductivity functions based on, 225
mathematical forms describing, 225
SOIL-N model, 204
Soil nutrient, 69
Soil organic matter (SOM), 188, 216–217, 223, 226
Soil–plant–atmosphere continuum (SPAC), 205, 206
schematic representation of, 212
Soils
acidification of, 561, 575, 586
classification of, 32
hydraulic conductivity of, 223
organic nitrogen, 219
pH and buffer capacity, 32

Soil system, definition of, 223–228
Soil water, 32, 69, 163, 177, 179, 213
characteristic of, 223
Gustafsson's model, 404
heat capacity and thermal conductivity, 214
heavy metals and, 405
model of metal sorption and, 403
SOILWAT model of, 184
toxic substances dissolved in, 404, 432, 436
and water balance, 213–214
SOILWAT model, of soil water dynamics, 179, 184
Solid phase digesters, 308
Solid-state reactions, 479
Solid waste pollution, 11
Sparse Matrix Vectorized Gear (SMVGEAR) algorithm, 530
Spatial models, for understanding ecosystem reactions, 7
Spring-neap tidal cycles, 349
Stable precipitation scheme, 537
Steady-state mass balance model (SSMB), 567
Steady-state models, 426, 565–567
Steady-State Water Chemistry (SSWC) model, 566
Steele's equation, for productivity with photoinhibition, 71
Stefan–Boltzmann equation, 441
STELLA diagram
of cadmium model, 405
of model developed for environmental risk assessment of growth promoters, 430
model equations (STELLA format), 406
STELLA software
application for modeling rivers and streams, 10, 140
components of conceptual diagram used in, 141, 147, 148
Global Average Model presented by use of, 447–453
STEP model, 184
STEPPE model, 179
Stochastic models, 7
Stoichiometric nutrient recycling theory, 54

Treedyn 3 forest ecosystem models, 163
TRIPLEX forest ecosystem models, 166
Trophic aggregations, calculation of, 249
Turbulent diffusion flux, 504
Turbulent fluids, 507
Turbulent kinetic energy, 354
Tylosine, 430, 431, 434, 435, 437

U

UK National Air Quality Standard,
    544–545
ULTIMATE-QUICKEST scheme, for
    integration of advection–
    diffusion equation in space–
    time domain, 354
Ulva lactuca (sea lettuce), 98, 344
Ulva rigida (macroalgae), 123
United Nations Economic Commission
    for Europe Convention on
    Long-Range Transboundary
    Air Pollution, 558
United Nations Environment
    Programme (UNEP), 258
    International Environment
        Technology Centre (IETC), 33
University Corporation for Atmospheric
    Research (UCAR), 532

V

van't Hoff equation, 306
Vegetation–herbivore interactions, 184
Vegetation response module (VEG), 580
Vesilind equation, 302
Veterinary medicine, 424, 427
    pathways into environment, 428
Volatile organic compounds (VOC), 501,
    502
VSD model, for calculation of soil
    acidification, 575–577, 586

W

Warm Rain scheme, 537
WASP4 model, 28
Wastewater management, 41
Wastewater systems, 11
Wastewater treatment, 258, 278

Wastewater treatment plants (WWTPs),
    278, 292
    Activated Sludge Model No. 1
        (ASM1), 279
    guidelines for application of models
        for, 312–316
    model development issues, 308–309
        good modeling practice, 310–311
        plant-wide models, 309–310
        uncertainty and sensitivity
            analysis, 311–312
    modeling sedimentation
        processes for gravitational
        settling, 298–299 settler
        models, 299–303
    plant-wide models, 279, 309–310
    schematic layout of, 279
    Task Group on Benchmarking of
        Control Strategies for, 309
Water anoxia, 348
Water balance, 68, 163, 188, 213–214, 231
Water column, 348, 349, 354, 356, 360,
    366, 379, 383
    oxygen processes in, 366–367
Water cycle, 166, 186, 454
Water pollutants, sources of, 20
Water quality management models, 18,
    349, 355–360
    for analyzing water quality
        influenced by human activities,
        355
    integration of hydrodynamics and
        ecology, 26–29
    main parameters for, 356
    optimization of, 25
Watershed, estimation of pollution
    sources in, 22
Water temperature stratification, 27
Water vapor, 214, 440, 445, 517, 534, 537
WEPP models, 204
WERF protocol, 313
    SWOT analysis, 314–315
Wet deposition, 223, 531, 543, 548, 560
Wetland Modelling (1988), 258
Wetland models, 10
    basic design of, 264–266
    differential equations in, 272–273
    for environmental management, 259
    forcing functions of, 268–269

Wetland models (*continued*)
 groups of processes included in,
  259–260
 overview of, 258–260
 parameters of, 271–272
 process equations, 269–271
 recent development in
  construction of wetlands for
   control of nonpoint sources,
   260–261
  contribution of wetlands to
   carbon balance, 262–263
  increase in modeling experience,
   260
  influence of climate changes
   on wetlands and wetland
   processes, 263–264
 state variables of, 266–268
 Tanzanian case study, 274–275
Wetlands
 contribution to carbon balance,
  262–263
 growth of plants and trees in, 263
 influence of climate changes on,
  263–264
 methane production in, 262
 nitrogen model, 262
 Pamolare 2 model for restoration and
  management of, 260
 recovery or construction of, 260–261
 for treatment of agricultural
  drainage water, 261
 types of, 258
Wildland fire modeling, 478, 479
Wind-based ellipse model, for
  determination of heat influence
  zone of burning sites, 483
Wood quality and forest ecosystem
  functioning model, 167–169
World Health Organization (WHO), 500

X

Xiangshan Gang bay (East China Sea)
 aquatic resource submodel, 70–71
 biogeochemical submodel, 71–72
 catchment submodel, 68–70

development scenarios for
  sustainable management of, 74
 ecosystem simulation of, 72–74
 estuarine-system coastal model
  aim of, 67
  conceptual diagram, model
   structure, and main equations,
   68–72
  parameters for standard
   simulation, 73
  problems and development
   scenerios, 67
  results of study, 72–76
  study system, 68
 hydrodynamic submodel, 70
 land use and aquaculture practices,
  impact of, 74
 location and physical characteristics
  of, 68
 multilayered ecosystem model, 66, 68
  conceptual diagram of, 69
Xylene, 515

Y

Yamartino–Blackman cubic algorithm,
  529
Yulin offshore industrial park, Taiwan,
  382

Z

Zooplanktons, 12, 36, 40, 46
 carbon, 365
 grazing of phytoplankton by, 337,
  368
 oxygen consumption by respiration
  of, 367
 predation from planktivorous fish, 41
Zostera marina (eelgrass), 118, 123, 344
Zostera noltii (salt marsh plant species),
  115, 117–118
 effect of temperature on physiology
  of, 120
 nutrient redistribution and
  reclamation capacity of, 120
 sexual reproduction of, 121

9 780367 865832